ANNUAL REVIEW OF
ECOLOGY AND SYSTEMATICS

ANNUAL REVIEW OF ECOLOGY AND SYSTEMATICS

RICHARD F. JOHNSTON, *Editor*
University of Kansas

PETER W. FRANK, *Associate Editor*
University of Oregon

CHARLES D. MICHENER, *Associate Editor*
University of Kansas

VOLUME 6

1975

ANNUAL REVIEWS INC.   4139 EL CAMINO WAY   PALO ALTO, CALIFORNIA 94306

ANNUAL REVIEWS INC.
Palo Alto, California USA

International Standard Book Number: 08243-1406-9
Library of Congress Catalog Card Number: 71-135616

Annual Reviews Inc. and the Editors of its publications assume no
responsibility for the statements expressed by the contributors to this Review.

REPRINTS

The conspicuous number aligned in the margin with the title of each article in this
volume is a key for use in ordering reprints. Available reprints are priced at the
uniform rate of $1 each postpaid. Effective January 1, 1975, the minimum acceptable
reprint order is 10 reprints and/or $10.00, prepaid. A quantity discount is available.

PRINTED AND BOUND IN THE UNITED STATES OF AMERICA

PREFACE

Good review articles are always contributions to the history of their fields of knowledge. Reviews are deliberately restricted topically, most cover a limited time period, and they should be ideal sources for science historians of the future. Few review articles, certainly in *ARES,* are historical beyond their assigned time limits, and most concern the most recent developments in the field. This is fine for our purposes, but it is not otherwise fortunate, because the history of ecology and systematics is not yet written. Such a lack guarantees a shortsightedness that ranges from the trivial to the profound. And some ecologists still think of their work as an academic exercise that must not be encroached upon by persons doing anthropology, sociology, economics, animal behavior, or, for that matter, systematics. But however much we need a formal history, *ARES* is not the vehicle for it. Even so, we think there is space for biography, and we believe that *ARES* will be a source for the future historians.

The first of what we hope may be a sequence of short biographies appears in this volume with Stephen D. Fretwell's perceptive and idiosyncratic view of Robert MacArthur. MacArthur did charismatic science and changed the outlook of virtually every ecologist and systematist practicing today. The integration of the man and the ecologist is something that MacArthur achieved wonderfully, and to know that he did so is helpful; it means that it is possible to do good science, to be transcendently innovative, to train good students well, and at the same time remain a dedicated humanist. This message is in retrospect clear enough in MacArthur's preface to *Geographical Ecology.* But humanistic messages in science can be confused with subjectivity and are mostly ignored.

The makeup of the remainder of this volume is varied, reflecting current topical interests and availability of authors. Assessments of ecosystem studies are presented by R. G. Wiegert who deals with simulation models, by T. Platt and K. L. Denman on treatment of periodicities stemming from nonlinear characteristics of ecosystem components, by R. E. Hungate concerning microbial ecosystems of ruminant mammals, and by E. G. Leigh, Jr., on tropical rain forests. Reviews at the level of communities and populations are marshalled by G. L. Bush on animal speciation, L. E. Gilbert and M. C. Singer on ecology of butterflies, M. Dickeman on the demography of infanticide in man, P. R. Ehrlich on population biology of reef fishes, L. R. Fox on cannibalism, which is to say intraspecific predation, B. Heinrich on evolution of pollination systems, and R. K. Colwell and E. R. Fuentes on experimental studies of the niche. New behavioro-ecologic syntheses are presented by K. Immelmann on the ecology of imprinting and early learning and R. Jander on orientation ecology. The volume is rounded off by D. Livingstone's review of work on ecology of the African Pleistocene.

We thank the authors for their contributions and cooperation. H. A. Mooney and P. R. Ehrlich assisted us in discussions of possible topics and authors at the time the volume was planned. Kathleen Gardner and Susan Futterman provided timely and sturdy assistance at Palo Alto. We appreciate the time and effort associated with all such cooperation.

We hope that our readers will continue to send us suggestions, including prospective topics and authors.

THE EDITORS AND THE EDITORIAL COMMITTEE

ANNUAL REVIEWS INC. is a nonprofit corporation established to promote the advancement of the sciences. Beginning in 1932 with the *Annual Review of Biochemistry*, the Company has pursued as its principal function the publication of high quality, reasonably priced Annual Review volumes. The volumes are organized by Editors and Editorial Committees who invite qualified authors to contribute critical articles reviewing significant developments within each major discipline.

Annual Reviews Inc. is administered by a Board of Directors whose members serve without compensation.

Annual Reviews are published in the following sciences: Anthropology, Astronomy and Astrophysics, Biochemistry, Biophysics and Bioengineering, Earth and Planetary Sciences, Ecology and Systematics, Entomology, Fluid Mechanics, Genetics, Materials Science, Medicine, Microbiology, Nuclear Science, Pharmacology, Physical Chemistry, Physiology, Phytopathology, Plant Physiology, Psychology, and Sociology. The *Annual Review of Energy* will begin publication in 1976. In addition, two special volumes have been published by Annual Reviews Inc.: *History of Entomology* (1973) and *The Excitement and Fascination of Science* (1965).

CONTENTS

THE IMPACT OF ROBERT MACARTHUR ON ECOLOGY, *Stephen D. Fretwell* 1

ECOLOGICAL SIGNIFICANCE OF IMPRINTING AND EARLY LEARNING,
Klaus Immelmann 15

THE RUMEN MICROBIAL ECOSYSTEM, *R. E. Hungate* 39

STRUCTURE AND CLIMATE IN TROPICAL RAIN FOREST, *Egbert Giles
Leigh, Jr.* 67

CANNIBALISM IN NATURAL POPULATIONS, *Laurel R. Fox* 87

DEMOGRAPHIC CONSEQUENCES OF INFANTICIDE IN MAN, *Mildred
Dickeman* 107

ENERGETICS OF POLLINATION, *Bernd Heinrich* 139

ECOLOGICAL ASPECTS OF SPATIAL ORIENTATION, *Rudolf Jander* 171

SPECTRAL ANALYSIS IN ECOLOGY, *Trevor Platt and Kenneth L. Denman* 189

THE POPULATION BIOLOGY OF CORAL REEF FISHES, *Paul R. Ehrlich* 211

LATE QUATERNARY CLIMATIC CHANGE IN AFRICA, *D. A. Livingstone* 249

EXPERIMENTAL STUDIES OF THE NICHE, *Robert K. Colwell and Eduardo
R. Fuentes* 281

SIMULATION MODELS OF ECOSYSTEMS, *Richard G. Wiegert* 311

MODES OF ANIMAL SPECIATION, *Guy L. Bush* 339

BUTTERFLY ECOLOGY, *Lawrence E. Gilbert and Michael C. Singer* 365

INDEXES

AUTHOR INDEX 399

SUBJECT INDEX 411

CUMULATIVE INDEX OF CONTRIBUTING AUTHORS, VOLUMES 2–6 419

CUMULATIVE INDEX OF CHAPTER TITLES, VOLUMES 2–6 420

THE IMPACT OF ROBERT ❖4084
MACARTHUR ON ECOLOGY

Stephen D. Fretwell
Division of Biology, Kansas State University, Manhattan, Kansas 66506

Every word of God is tested.

Proverbs 30:5

PROLOGUE

Scientists are responsible for truth, knowledge, wisdom, and understanding.

Truth is what is—it is the underlying reality of all existence. Knowledge is what we think we know about truth. Knowledge, however, is always an imperfect assessment, and is always subject to revision and improvement. The realization that there are discrepancies and weaknesses in knowledge is wisdom. Wisdom leads to a process, called the philosophy of science, through which knowledge is modified to better fit the truth. Philosophy means the love of wisdom, and doctors of philosophy are supposed, before all else, to be experts in wisdom. Understanding, as defined in Job (28:28), is the effort to avoid evil. We may think of understanding as what we use in order to adequately apply our wisdom and our knowledge in guiding our actions. While applied scientists seek understanding, basic scientists seek knowledge.

Dr. Robert MacArthur has made a dramatic impact on ecology because, to him, all of this was second nature.

INTRODUCTION

There are several ways of looking at what Robert MacArthur's life and work mean to ecology. First, he affected the way many other ecologists and students think and work, and so has had a broad, but indirect influence via their work. Second, his own work has directly affected the dogma and frontiers of ecology, so that discussions of many subjects are now incomplete without considering ideas that he advanced.

Third, MacArthur's lifestyle and attitude provided unique examples to guide ecologists in those dimensions of their lives that affect their work, but are usually not codified in print.

I look at these areas in order, mostly trying to recount my personal view of the history of his work and influence. My central thesis is that Robert MacArthur had a major impact on the way we do things, both as people and as ecologists, rather surpassing that which he taught us directly about the natural world. I attempt to express what I learned about his style, and his spirit of doing things, not only in what I say, but also in the way that I say it. There is a certain unfairness about this. I spent far less time with him than some, and I am less steeped in the details of his work than many. However, my personal perspectives are both mystical and broad, sufficiently so, I hope, to allow me to organize appropriately the experience I do have.

HIS EFFECT ON ECOLOGISTS

The Personal Touch

MacArthur's effect on other ecologists was both personal and indirect. My own conversations with him were (to me) significant and life-shaping. However, even when I had not met him, his published work reflected an attitude that was and is changing the field. I treat these two aspects separately, the first briefly, for fewer people have enjoyed his personal presence.

There is a mysterious connection between love and truth that almost borders on equivalence. Somehow, love is enlightening or mind brightening. MacArthur, in most of my personal encounters with him, was a very loving person. He was patient, kind, tactful, joyous, responsive, openly human. This had the effect, as J. T. Bonner put it, of making a conversation with him seem especially clear and significant, and of making one feel very bright (introductory remarks at the MacArthur Symposium, November 1973). I rarely thought so clearly as when I talked with MacArthur.

Robert clearly loved most of the things he encountered in life. He loved his family, nature, and any exchange of ideas that sought to resolve some honest and tangible confusion. He loved elegant mathematical structures, and patterns in nature. He wore his genius lightly, and shared it easily. He welcomed insights and abandoned inappropriate ideas. He honored the search for truth so deeply that digressions (small talk) were unwelcome; however, he accepted both honest confusion or a tangible discovery equally. This attracted me to visit with him, but only after I had developed some "good" questions, or could present a developed idea. The effect of all this on the careers of the scientists who knew Robert is difficult to estimate or discuss. Two things seem certain: each of us responded uniquely, becoming more ourselves. And, each of us became better at what we wanted to contribute. I believe this indirect influence on the careers of his colleagues and students was profound.

Hypothesis Testing and Fame

On a broader scale, MacArthur has helped transform American ecology by his effect on scientific attitude and methodology. I look at this transformation from the perspectives provided in part by H. R. van der Vaart and H. L. Lucas at North Carolina State University in Raleigh, who developed in me a great interest in the methods of science, especially the philosophy of science that is formally called the hypothetico-deductive (H-D) method. This method has had great success in many fields of science (physics, chemistry). But, prior to MacArthur's 1957 paper (13) on relative abundance, it had been little used in the study of natural history (about 5% of papers in *Ecology* from 1950–1956 tested predictions, compared to almost 50% nowadays). MacArthur accomplished two things with respect to furthering the use of this method: he provided acceptable examples of how one should do H-D science and he provided leadership and protection to those who wanted to work this way. As a result, many now proceed using this philosophical approach. The reader is invited to compare recent issues of *Ecology* with issues from, say, 1955, before MacArthur began publishing. Nowadays, almost half the papers contain tests of explicit or indirect predictions from theory.

I focus, in this section, on the way that MacArthur's influence was developed, rather than look too closely at his methods themselves. I take this emphasis because the methods are well known and established elsewhere (e.g. see Tricker, 37). MacArthur's main contribution here was his leadership in getting this approach accepted.

The issues were (and still, in part, are) these: Should scientists be allowed to make mistakes in print? How extensive is one's responsibility to previous literature and scholarship? The first issue is usually precipitated by so called "weak" tests of theories. These are insubstantial data contributions in response to a theoretical prediction, confirming the prediction but only raising the plausibility of the theory a modest amount. Critics claim that such "tests" mislead the naive into believing that the theory is proved. One hears that science reaches "conclusions," i.e. ideas are proved correct, and that then (and only then) should work be published. The second issue pertains to rejection of clever or interesting statements because they are not interpreted in light of most of the previously published reports that seem related.

MacArthur never really discussed these issues; he just took a position. The position he took was intellectually sound, but socially risky. He simply went ahead making and publishing predictions and weak tests of predictions. And so, like any good H-D scientist, he crawled out on a limb in print. His mathematics was sound and useful, but not so to the extent it might have been. And his data, although edifying and encouraging, were usually limited and open to a variety of interpretations. Nor did he overwhelm possible critics, or play citation politics, by including "all the right" references.

The rest of us did not have to wait until an endless debate on scholarly principles was settled before we dared to seriously consider using weak data to test loosely formulated models or to discuss a new idea without a lengthy library research. MacArthur was famous enough to silence most criticism and provided an outstanding example of success, doing just what he sensed needed doing to excite himself and his many reasonable colleagues. From a position of authority, he ignored, and so took responsibility for our ignoring, the reviewer of research who looks for and at mistakes instead of assessing progress. Thus his contribution to the people who let themselves be positively affected by him was both to point the way for a radically different perspective on how ecologists should proceed and to get famous enough so that we could follow this direction without undue harassment.

In order to understand this aspect of MacArthur's impact, we must first look at the process of publication, the fame-making machinery in the scientific community. There is a difference between the current system, by which scholarly papers are rejected and accepted, and the ordinary process of public censorship, where the expression of painful ideas is openly suppressed. But the difference varies from discipline to discipline, and from time to time, and is never as large as we would like to think. Too few reviewers of papers (for ecological journals, anyway) will advocate the acceptance of a paper they think is "wrong"; not many realize that, almost always, any innovative idea will be thought of as wrong by most other scientists. Not all wrong ideas are innovative, of course, but all truly innovative contributions must, on first reading, appear wrong. Scientific papers should never be rejected because someone is found who disagrees with the ideas presented. Quite the contrary, a truly scientific journal might reject as unnecessary any paper that failed to elicit such criticism. Too rarely, nowadays, are papers rejected for the right reasons (internal inconsistencies, unclear writing, or poor scholarship leading to redundancies); most are merely disagreed with.

MacArthur bypassed this problem, using some hints and some help from G. E. "Concluding remarks" Hutchinson. The National Academy of Sciences (NAS) once had the enlightened view that any idea that impressed one intelligent scientist (i.e. a NAS member) was worthy of publication. Many of MacArthur's significant ideas were expressed in the journal of that society (the Proceedings of the NAS) with Hutchinson's blessing. Having bypassed the normal review process, he was well on his way to becoming famous.

In appreciating MacArthur's rise to fame, we must never overlook the differences between his style of working and the style of most established ecologists. His papers began in speculation and ended in data, instead of the other way around. The math was often fuzzy or incomplete, and the data oversimplified and limited. It was easy for anyone with a lifetime commitment to the heavily descriptive or elegantly formal ecology of the times to be upset by the style. And, rejecting (I would say, censoring) one of his papers must have been easy, for there were plenty of flaws to catch. Without the Proceedings of the NAS, I wonder how far MacArthur would have gotten.

That he suffered at the hands of reviewers was obvious from several of his actions. Specifically, the formation of the Princeton Monograph Series was partially an effort to provide an outlet for the sort of research that he stimulated and encouraged. The *Journal of Theoretical Population Biology,* formed in 1970, was another effort to save us from, as he put it, "all that gas." He once related to me the history of his 1967 paper with Levins (24) on limiting similarities, published in the *American Naturalist.* When the work was first submitted, it contained a mathematical mistake that completely reversed the conclusion. While the paper was being reviewed, MacArthur discovered the error, rewrote the paper and resubmitted. The editor meanwhile had received the review of the erroneous draft. The error had not been caught, but the reviewer still recommended that the paper not be published. His grounds were that the conclusions (now known to be reversed) were intuitively obvious! The editor had no choice but to accept the corrected version, since its findings had been proved to be clearly counter-intuitive.

So, leaning on the Proceedings of the NAS and on "special" publications, MacArthur got to the top. Some called him a charlatan, others called attention to some of his mistakes. But many simply delighted in each new writing, encouraged to find such fresh air blowing in the top ranks of American ecology. With this sort of person leading, I, for one, felt more free to carry on the same way; we no longer needed to be so worried about criticism and rejection.

Some special comment should be made with reference to MacArthur's example in using the literature. We ask: how are we to use the literature? There is a great expanse between the minimal and the maximal number of references that are possible in a paper. Ecology as a whole has tended toward the maximal, but MacArthur leaned the other way, and reasonably so. The process of prediction making requires such strict logic that peripheral references merely distract. There is also the danger of high scholarship, which tries to settle issues by argument instead of by empirical testing. Few scientists realize how antithetic are scholarship and science. A Ph.D. who has been taught in a rich tradition of scholarship is tempted to use its techniques inappropriately, as an end in themselves, instead of using the literature just to clarify the process of making real-world tests. Consistent with his use of H-D science, MacArthur rarely used a broad literature base in his work.

In brief, MacArthur was a lover, in the highest sense of the term. Thus he attracted, enlightened, and stimulated many ecologists in personal exchanges. He gave his genius up, and it spread. He also accepted the intellectually sound but risky idea that science should proceed by the testing of hypotheses in print, even if it means that published mistakes may be made. He both provided examples of this idea and worked to silence or defeat the power of criticism that would prevent its wider adoption and acceptance. He became famous largely on his own authority, often bypassing the normal publication process, and thus provided leadership and support for those who agreed that H-D science was acceptable. He also freed us to use the literature in ways appropriate to predictive science.

IMPACT ON ECOLOGICAL THOUGHT

Although MacArthur's main contributions were (in my opinion) to the methodology (the wisdom, the philosophy) of science in ecology, the fact remains that what we call the body of knowledge of ecology has been significantly affected by his particular discoveries. I now look closely at these contributions. In order to do so, I first describe briefly the general classes of dogmatic ecological knowledge, and some of what may be called the frontiers of ecological knowledge. Then, armed with this overview of the field as a whole, we can look at MacArthur's explicit contributions.

The Dogma of Ecology

Some central concepts of modern ecology are:

1. *Succession, i.e. ecosystem ontogeny*   So far, we think that succession proceeds because dominant life forms pollute their environment, making it unsuitable for their own continued existence, and inviting invasion from other species.

2. *Energy flow and nutrient cycling in ecosystems (ecosystem physiology)*   The methods of systems analysis and the theories of multiple simultaneous causation allow us to discuss the flow of energy and nutrients through an ecosystem. This flow drives an ecosystem's dynamics and connects all of its elements. Some of the counter-intuitive consequences of environmental actions are discovered here.

3. *Community diversity*   Information, opportunity, and glory are all to be found in species-rich communities. Why are some communities richer than others? They live in more structured environments and have different balances of immigration, speciation, and extinction rates. Intraspecific morphological variation plays a role, perhaps, and predation pressure.

4. *Population regulation*   The population size at a particular time is the result of a sequence of previous increases and decreases in numbers. These changes in size are dependent, in turn, on changes in the growth rates. Growth rates are usually correlated with population size (density-dependence) through the effects of predation, disease, societal pressure, and contest or scramble competition. These factors may stabilize a population or drive it to extinction.

5. *Evolutionary ecology*   Everything has evolved strategically to optimize fitness, and there is great beauty and wonder in coming to know "why" nature is shaped as it is.

6. *Population genetics*   What do evolution and population regulation do to gene frequencies? They drive some to zero, others to "fixation," and leave still others at various intermediate levels, either by heterosis or by frequency-dependent selection.

7. *Ecological management*   We want to understand ecology, to bring more desirable results from resource exploitation, and to minimize environmental impact. Also, we need to conceptualize (in the silence of God) what *is* desirable.

Now we can ask: how is the content of ecology different, having experienced Robert MacArthur's input? The dogma that appears to have been most heavily influenced by his writings is the area of community diversity or richness. The question of numbers of species, though earlier asked by Hutchinson (9), was given some substantial answers by MacArthur (17).

Second in importance, in my opinion, are MacArthur's contributions to the area of evolutionary ecology. Lack (11), with his clutch size research, had a great influence in establishing this field. But MacArthur also was an early leader in "strategic" analysis. One of his most important contributions in this area is the distinction between r versus k selection, which he was early (1961) in noticing (15). Also, his work on the evolution of generalist versus specialist feeders was seminal, although others were also contributing here.

In the field of population regulation, MacArthur mostly clarified the idea of scramble competition, a basic assumption in most of his serious diversity modeling (e.g. with Levins, 23). In the process of explaining diversity, he developed some clear statements of how competitors might reduce each other's resources. Andrewartha & Birch (1) had earlier raised questions that demanded the sorts of precise statements found in MacArthur & Levins (23).

Rosenzweig's predator-prey work, done with MacArthur (34), also made an important advance in population regulation theory by providing a needed analytical tool. It is interesting to note that Kolmogoroff (reviewed in Rescigno & Richardson, 32) had early accomplished a similar development. Kolmogoroff's work is rather more elegant, but was obscurely published. Still, there is a simplicity about the Rosenzweig-MacArthur formulation that makes it one of the most cited textbook graphs, one of the most explicitly tested predator-prey theories (Maley, 28), and a most useful theory for extending our grasp of new problems (e.g. see Rosenzweig, 33).

In ecosystem theory, only one idea of MacArthur's is much noticed: his 1955 paper (12) showing that diversity in an ecosystem enhances its intrinsic stability. This, it seems, was what everyone wanted to hear in the then-budding ecology movement, so MacArthur's "proof" was grabbed up. May (29) recently found some potential exceptions to this dogma. Still, this report by MacArthur is one of the most cited of all his papers. It appears, incidentally, that MacArthur's early interest in diversity hinged around the relationship between diversity and stability (Hutchinson, 9; MacArthur & MacArthur, 25).

MacArthur only touched on community succession and his work may not significantly influence conventional treatment of that subject. He explored, with Wilson (27), the properties of good island colonizers, mentioning that the theories that were developing should apply to successional communities. He noticed that bird species of secondary mainland habitats, being the best colonizers, were most likely to appear on islands. He and Horn (8) showed that if there were a trade-off between competitive ability and colonizing ability, a so-called harlequin environment could be stably subdivided by two similar species, one sedentary and one mobile. In time, the sedentary species replaces the mobile, in a given habitat.

MacArthur said very little about energy flow and nutrient cycling. In his book with Connell (22, p. 179) there is a review of the subject with a clever and concise formulation relating relative biomasses in the links of a food chain to body sizes and turnover rates. This little model is an excellent one for dealing with inverted food chains, but is both little noticed and seldom used.

In population genetics, MacArthur made a couple of rather esoteric contributions (16; and with Wilson, 27) that are generally overlooked nowadays, although they foreshadowed much current research having to do with ecotypic selection and frequency-dependent selection. It is hard to pinpoint the specific impact on the field that these models made; yet one mention of them was in MacArthur & Wilson's widely cited book. The failure of many workers in this field (myself included) to cite this work may have been due to our inability to understand it fully, at least until after we had "rediscovered" similar ideas.

Generally, those of MacArthur's contributions that have appeared in textbooks concern diversity, stability, community dynamics, and r versus k selection. His work in nearly all other fields of ecology has been or should be helpful (or even seminal) in more complete treatments; yet his impact in these areas will probably be less recognized. However, MacArthur typically is one of the most cited authors in current ecology texts; usually only the author of a given text has more listings in the bibliography than MacArthur.

Frontiers of Ecology

It remains, then, to discuss MacArthur's "trend-setting" activities: his effect on "new" ideas not yet part of the dogma of ecology.

Some frontiers of ecological research today are:

(a) density-dependent versus density-independent factors in the evolution and regulation of populations;

(b) the nature of competition: social intolerance versus resource exploitation versus altruistic management as competitive mechanisms;

(c) species diversity: invadable versus "packed" communities; the shapes of extinction, immigration, and speciation curves;

(d) generalist versus specialist feeding strategies—who does which and why?;

(e) the response of ecosystems and populations to variable environments, including those that are seasonally (or predictably) variable and those that vary stochastically;

(f) genetic and morphologic variability *within* populations: ecotypic versus archetypic selection;

(g) models for population prediction and management;

(h) the structure of ecosystems—complex systems models for the prediction of counter-intuitive effects;

(i) material flux through ecosystems, limiting energy transfers;

(j) the evolution of reproductive rates and other life history parameters.

(May I be forgiven for those I have overlooked.)

MacArthur has had three sorts of effects on developing ideas in ecology: 1. He has provided didactic development and tests of certain ideas that were generally accepted by those who understood them, but were not widely understood. 2. He has asked questions that had not been asked before, and stimulated other ecologists to think about them. 3. He has discovered new patterns and relationships in nature that probably will become part of the dogma.

First, let us consider density-dependent versus density-independent sorts of population dynamics. This area received a good deal of attention from the nearly simultaneous publication of books in 1954 by Lack and Andrewartha & Birch, and is still painfully polarized. MacArthur first entered it by "showing" that warblers were regulated in a density-dependent fashion (14). He used a clever statistical analysis of runs of population increases and decreases to show that an increase was usually followed by a decrease, instead of another increase. This analysis foreshadowed Tanner's (36) extensive correlational analysis and all the statistical debate that has followed. For all its weaknesses, this sort of analysis of population fluctuations promises to unravel the way density-dependent and density-independent effects cooperate in regulating numbers, as is evidenced by the excellent analysis by Varley, Gradwell & Hassell (38).

MacArthur had very little to say about mechanisms of competition (21, pp. 25–28). As noted above for competition dogma, MacArthur clarified resource exploitation (scramble) competition theory, but he really said very little to contrast this theory with the alternatives. The major contributors here were Nicholson (30) and Wynne-Edwards (39). The work of the latter author is vastly underappreciated and misunderstood, due, I think, to some inappropriate attacks by evolutionary ecologists. MacArthur [with Connell (22, p. 140); see first quote in Appendix] was early in encouraging a temperate attitude towards Wynne-Edward's ideas.

Most of the people now working on species diversity are working on concepts that were first stated by MacArthur and his colleagues. Perhaps ten percent of the papers in recent issues of the Journal of *Ecology* deal with this subject. MacArthur's impact here is so conspicuous that it requires little analysis.

The problem of specialist and generalist feeding strategies was first raised by MacArthur & MacArthur (25). MacArthur & Pianka (26) then provided an early answer to that question, but others (e.g. Emlen, 5) reached similar conclusions at about the same time. Tests have been slow in coming, and few patterns have been discovered.

The problem of variable (both seasonal and stochastic) environments, which is another question of population regulation, was raised by Lack (11), who suggested that post-breeding survival (i.e. periods of negative population growth rates) regulated populations. Andrewartha & Birch (1) made a similar point, a bit less explicitly. However, MacArthur's widely imitated early study on warblers (14) emphasized coexistence and competition (i.e. regulation) in the breeding season (positive population growth rates). This emphasis was not conceptual, however, for MacArthur expended considerable energy studying wintering warblers as well as breeding ones. The winter studies were simply less productive and so received less attention. However, many of those following this study have worked only in the breeding season, assuming that populations are regulated or limited then. Hespenheide (7) has shown that these studies need not assume strict breeding limitation, but do require some breeding regulation. Few studies to date support even that assumption, except perhaps for nest sites (Fretwell, 3; Krebs, 10), which, however, are not usually emphasized in competition studies (but see Hespenheide, 6). Thus MacArthur somewhat inadvertently has led workers into making an a priori unlikely assumption, which has made an issue of seasonal regulation.

I have already touched on MacArthur's contributions to the archetypic versus ecotypic selection problem (reviewed by Rothstein, 35). I repeat it here, since population genetics arrives as theoretical dogma before it becomes an empirical frontier. MacArthur personally contributed to my report on this latter area (2, as cited), and even predicted some of the sorts of tests since confirmed by Rothstein (Fretwell, 4). (He successfully predicted that pygmy nuthatches would be less variable than brown-headed nuthatches, since the former have more competitors.)

MacArthur contributed almost nothing to the area of ecologic management. I believe he rather disapproved of using ecology theory for management, impact statements, and the like. He seemed to feel that nature enriched the naturalist, and hence the world, in spiritual ways, so that there was a greater harvest of peace and truth than of lumber in a forest.

I daresay, also, that few of the current systems modelers, including those working on nutrient cycling, are using MacArthurian theory or philosophy. In this area of research, descriptive thoroughness is optimized at the sacrifice, perhaps, of elegance, simplicity, and interest.

Research on life history evolution, although not begun by MacArthur, certainly was boosted substantially by his 1961 paper, "Population effects of natural selection," (15) and by his monograph with Wilson (27). This work developed the concepts of r and k selection. This distinction (actually it is a continuous dimension, see Pianka, 31) promises to play a significant role in most areas of ecology, since it is tied simultaneously to environmental stability, predator-prey interactions, and succession.

MISCELLANEOUS IMPACTS

`My intent in this section is to present some things that I learned (or am learning) from the way MacArthur did things. I suspect that his spirit and/or example is affecting others similarly. Discussing these issues, all a bit subtle or personal, should accomplish one objective: by openly describing the example provided, it will be easier for others to follow. As many ecologists as possible should have access to these details, lest MacArthur's disciples unknowingly leave out something essential in developing a predictive ecology.

Leadership and Reform

MacArthur's position towards "the establishment," i.e. authority, and leadership in science was enlightening. He clearly trusted and respected the system and was hopeful that, with proper leadership, it could be made more effective in accomplishing science. Although he did things differently, he did try to move so that the system was not offended. So, he succeeded in having an impact. Insofar as optimism can be cultivated or passed on, his effect on what might be called "revolution" in ecological method was therefore a temperate one. I do not believe he thought much about this; yet I wonder at his brilliance in balancing innovation with conservatism.

Being Wrong or Awkward in Print

Conventionally, mistakes in print are regarded as disasters. Yet MacArthur made several published mistakes (18). He offended in many ways the sanctity of print, and was caught at it. Some of his best works are flawed in terms of style and symbolic or typographical error. We are inclined to forgive this, in deference to his great genius. I elevate it to the place of a lesson about misplaced emphasis. In all research reporting, over-attention to detail can obscure the spirit of curiosity and wonder upon which good basic research depends. MacArthur's work survived all these errors, which does not justify them exactly, but makes inexcusable the report that is deadly dull but otherwise correct. Better to spend (as MacArthur did) one's time and energy being interesting; if something has to be sacrificed, let it be the exactness. Clearly, one can contribute as much with such an approach.

How to Recognize a "Good" Ecologist

MacArthur thought highly of some of those ecologists that he knew, and he was concerned about the problem of identifying good ecologists. In a commentary (20), he emphasized the importance of field work. He later told G. Lark (personal communication) that a good test for an ecologist was to walk with the person through a field and see how many questions he asked. This, presumably, measures curiosity; love of and attention to nature. I doubt that this test should stand alone; but I do not doubt that the things it measures are generally neglected.

Open-Mindedness before Skepticism

The characteristics of a scientist are open-mindedness, empiricism, and skepticism. MacArthur advocated that these be in balance, re-emphasizing the importance of open-mindedness during a time when skepticism was dominant. However, he also put them in the above order, putting skepticism last, and only as a response after data had been gathered to test an idea (empiricism). I believe that his attitude towards these traits has helped restore, even in these competitive days, a very sweet mood of tolerance, widespread "wait and see" attitude. There was a bristling arrogance that once stalked meetings of ecologists with cries of "unwarranted speculation!" and "but what about. . .?" Such questions are now confined to private conversations among those coming in from the greatest distance.

Love Your Wife and Family First

Maybe I'm the only one affected by this example that MacArthur set, but it was the first and hardest thing he taught me. When I knew him (in 1968–1969) he was above all a family man; he spent much time at home, and he surely loved his wife and placed her first. This is, admittedly, a personal question, yet, it is one that must pervade every scientist's life: to marry or not, and, if married, how much of one's time that could be spent in research should be spent instead with one's family. And where does one's mind rest? I had the distinct impression that MacArthur worked on ecology mostly when his family got tired of his hanging around. I suspect that he had the freedom to do what he did for ecology, going against all convention, because it was all secondary to him anyway; or maybe he just spent so much time with his family that he never really had time to learn the conventions! Yet, even so, look at what he accomplished!

One can argue (I often do) that it was his genius that allowed him to so lightly toss off pieces of research and so to be home more. However, I am dissatisfied with this interpretation, which, in any case, gets me nowhere. Just as plausibly and much more usefully, I now believe that MacArthur's personal priorities (family before profession) were part of what made him so successful. For, in inspecting his work, the freedom is more conspicuous and more unique than the brilliance; the freedom, then, is the major contribution. And such freedom very cogently follows from lowering research and publication into a second (or lower) place.

The human and philosophical contributions of an eminent scientist's life are usually reserved for biographies, where their influence on other scientists is, at best, tardy. MacArthur's contributions here may be critical to a successful continuation of his style of ecology, and the science cannot move too quickly into a more effective and engaging method. So I discuss the attitudes and personal factors that I think were a part of his accomplishing what he did. I emphasize his respect for the "system," his emphasis on curiosity and open-mindedness, and his love of family, with the hope that other scientists seeking a worthy example will have a better view of the man.

SUMMARY

I have suggested that MacArthur's major impact was in methodology and spirit; he taught us how to be basic scientists, what to value, what to do. But he also contributed grandly to our science in increasing significantly our grasp of diversity and stability and community structure, with lesser impacts in most other areas of ecology.

Before he died in 1972, his work had drifted into some rather sophisticated mathematical structures. But his last book (21) written, in part, to summarize his life's work, proved that his heart was still in nature, and that both theory and data are merely tools to improve understanding of that on which the eye fell of its own accord.

APPENDIX
Some quotes of interest:

In a section titled *Group Selection:* "Perhaps the biggest unsolved problem of natural selection—the problem that more than any other makes evolutionists get angry and say something irrational—is concerned with whether . . . (individual selection) . . . is the only one possible" (in MacArthur & Connell, 22). "Much of modern ecology has to be done in the field, and should be taught there" (20). ". . . by no means all ecologists will favor such field stations and, as long as they aren't made to feel inferior by staying behind, will keep ecology one of the components of the community of scholars" (20).

"It is the purpose of this note to raise a problem, and to show by means of an example how interesting that problem is" (19).

Literature Cited

1. Andrewartha, H. G., Birch, L. C. 1954. *The Distribution and Abundance of Animals.* Chicago: Univ. Chicago Press. 782 pp.
2. Fretwell, S. D. 1969. Ecotypic variation in the non-breeding season in migratory populations: a study of tarsal length in some fringillidae. *Evolution* 23:406–20
3. Fretwell, S. D. 1972. *Populations in a Seasonal Environment.* Princeton, NJ: Princeton Univ. Press. 217 pp.
4. Fretwell, S. D. 1973. *Ecotypic versus archetypic selection, and the niche breadth of species.* Presented at the MacArthur Memorial Symposium, November 1973

5. Emlen, J. M. 1966. The role of time and energy in food preference. *Am. Nat.* 100:611–17
6. Hespenheide, H. A. 1964. Competition and the genus *Tyrannus. Wilson Bull.* 76:265–81
7. Hespenheide, H. A. 1973. Ecological inferences from morphological data. *Ann. Rev. Ecol. Syst.* 4:213–29
8. Horn, H. S., MacArthur, R. H. 1972. Competition among fugitive species in a harlequin environment. *Ecology* 53:749–52
9. Hutchinson, G. E. 1959. Homage to Santa Rosalia or Why are there so many

kinds of animals? *Am. Nat.* 93:145–59
10. Krebs, J. R. 1971. Territory and breeding density in the Great Tit, *Parus major* L. *Ecology* 52:1–22
11. Lack, D. 1954. *The Natural Regulation of Animal Numbers.* Oxford: Clarendon. 343 pp.
12. MacArthur, R. H. 1955. Fluctuations of animal populations and a measure of community stability. *Ecology* 36:533–36
13. MacArthur, R. H. 1957. On the relative abundance of bird species. *Proc. Nat. Acad. Sci. USA* 43:293–95
14. MacArthur, R. H. 1958. Population ecology of some warblers of northeastern coniferous forests. *Ecology* 39:599–619
15. MacArthur, R. H. 1961. Population effects of natural selection. *Am. Nat.* 95:195–99
16. MacArthur, R. H. 1962. Some generalized theorems of natural selection. *Proc. Nat. Acad. Sci. USA* 48:1893–97
17. MacArthur, R. H. 1965. Patterns of species diversity. *Biol. Rev.* 40:510–33
18. MacArthur, R. H. 1966. Note on Mrs. Pielou's comments. *Ecology* 47:1074
19. MacArthur, R. H. 1968. Selection for life tables in periodic environments. *Am. Nat.* 102:381–83
20. MacArthur, R. H. 1969. Commentary: The ecologists' telescope. *Ecology* 50(3):352
21. MacArthur, R. H. 1972. *Geographical Ecology.* New York: Harper and Row. 269 pp.
22. MacArthur, R. H., Connell, J. 1966. *The Biology of Populations.* New York: Wiley. 200 pp.
23. MacArthur, R. H., Levins, R. 1964. Competition, habitat selection, and character displacement in a patchy environment. *Proc. Nat. Acad. Sci. USA* 51:1207–10
24. MacArthur, R. H., Levins, R. 1967. The limiting similarity, convergence, and divergence of coexisting species. *Am. Nat.* 101:377–85

25. MacArthur, R. H., MacArthur, J. W. 1961. On bird species diversity. *Ecology* 42:594–98
26. MacArthur, R. H., Pianka, E. R. 1966. On optimal use of a patchy environment. *Am. Nat.* 100:603–9
27. MacArthur, R. H., Wilson, E. O. 1967. *The Theory of Island Biogeography.* Princeton, NJ: Princeton Univ. Press. 203 pp.
28. Maley, E. 1967. A laboratory study of the interaction between the predatory rotifer *Asplanchna* and *Paramecium.* *Ecology* 50:59–73
29. May, R. M. 1971. Stability in multispecies community models. *Math. Biosci.* 12:59–79
30. Nicholson, A. J. 1954. An outline of the dynamics of animal populations. *Aust. J. Zool.* 2:9–65
31. Pianka, E. R. 1974. *Evolutionary Ecology.* New York: Harper & Row. 350 pp.
32. Rescigno, A., Richardson, I. W. 1967. The struggle for life: I. Two species. *Bull. Math. Biophys.* 29:377–88
33. Rosenzweig, M. L. 1971. The paradox of enrichment: destabilization of exploitation ecosystems in ecological time. *Science* 171:385–87
34. Rosenzweig, M. L., MacArthur, R. H. 1963. Graphical representation and stability conditions of predator prey interactions. *Am. Nat.* 97:209–23
35. Rothstein, S. I. 1973. The niche variation model—is it valid? *Am. Nat.* 107:598–620
36. Tanner, J. T. 1966. Effects of population density on growth rates of animal populations. *Ecology* 47:733–45
37. Tricker, R. A. R. 1965. *The Assessment of Scientific Speculation.* New York: American Elsevier. 200 pp.
38. Varley, G. C., Gradwell, G. R., Hassell, M. P. 1973. *Insect Population Ecology.* Oxford: Blackwell. 212 pp.
39. Wynne-Edwards, V. C. 1962. *Animal Dispersion in Relation to Social Behavior.* New York: Hafner. 653 pp.

ECOLOGICAL SIGNIFICANCE OF IMPRINTING AND EARLY LEARNING[1]

Klaus Immelmann

Department of Ethology, University of Bielefeld, Bielefeld, West Germany

INTRODUCTION

Recent years have seen a considerable expansion in the study of imprinting, both in the groups of animals studied and the motivational systems and the kinds of problems involved. Whereas early work on imprinting was restricted almost entirely to birds, some recent investigations have also been carried out with insects, fish, and mammals. And whereas the early findings referred only to the acquisition of object preferences in the social sphere, studies in the meantime have been extended to a wide variety of other behavioral characters (5, 40). Moreover, apart from collecting further data about the various aspects of the phenomenon of imprinting itself, research now also covers the physiological background of the imprinting process, e.g. its possible relationships with the maturation of sensory organs and pathways (8, 74), with brain maturation (49, 86), with biochemical changes in the brain (7, 37, 83), and with the maturation of hormonal systems (50). Finally, it is being realized more and more that imprinting may also be of considerable ecological and evolutionary significance. It is this aspect that has found increasing attention during recent years and that seems to justify a review in a series mainly devoted to the field of ecology.

WHAT HAS BEEN CALLED IMPRINTING

There are two "classical" cases of imprinting: imprinting of the following reaction of precocial birds and sexual imprinting. For many years, studies have concentrated mainly on the following reaction and on the stimuli eliciting this response. As a consequence, general reviews of the context and characteristics of imprinting also refer mainly to filial imprinting, and this in turn has frequently led to the impression that the latter represents the most typical example of the phenomenon as a whole.

---

[1]Dedicated to Professor Bernhard Rensch on the occasion of his seventy-fifth birthday.

15

It has been found, however, that besides its importance for infant-mother relations, the influence of early experience may be equally important with respect to certain aspects of adult behavior, especially with regard to the determination of sexual preferences, but also to several other aspects of social and other behavior. It even appears that some of these processes represent better examples of imprinting than the following reaction (see review in 40).

As a consequence, the term *imprinting* has been applied to quite a variety of phenomena, including the acquisition of a number of other, nonsocial object preferences, mainly in the ecological sphere (see Ecological Imprinting). Early experience has likewise proved to exert a crucial and permanent influence in a variety of other phenomena, such as the development of normal copulatory behavior, the development of contact behavior, the organization of maternal behavior, the degree of socialization and aggressive behavior, and the establishment of diurnal rhythms. As even these phenomena show some similarity to the main characters of imprinting (see The Characteristics of Imprinting), the terms *imprinting, imprinting-like process,* or *learning process akin to imprinting* have frequently been used in the literature (40). Even the very early and very rapid song learning occurring in some species of passerine birds (57, 64, 89, 90), as well as the "clicking into place" of certain sequences of movements after only one or very few successful attempts, have repeatedly been compared to imprinting. For the latter example, Leyhausen (51) has introduced the term *motoric imprinting.*

Finally, similar importance of early experience has also been found in human infants, and the word *imprinting* is now being employed quite commonly by psychologists and psychiatrists to characterize the origin of such phenomena as hospitalism, homosexuality, and certain particularities of the ontogenetic development of language (9, 49).

"ECOLOGICAL" IMPRINTING

For the present review, those imprinting or imprinting-like processes that refer to the ecology of a species are certainly of special interest and hence deserve a more detailed treatment. The most important examples of ecological imprinting are concerned with food preferences, selection of a home area, and habitat preferences, as well as with host selection in parasitic animals.

Food Preferences

In several species of animals it has been shown that raising an individual from birth on a restricted diet of only one kind of food can lead to strong food preferences or even to imprinting-like fixations. Burghardt & Hess (16) raised three groups of naive snapping turtles (*Chelydra serpentina*) on either fish, earth worms, or horse meat. When tested after 12 daily feedings, the animals preferred the diet to which they were accustomed. After another period of 12 days during which they were fed a different food they still preferred their original diet when again given a choice. These results were supplemented by Burghardt (15), who demonstrated that an imprinting-like effect can be found after only one meal. On the other hand, he also revealed natural food preferences in inexperienced young that make some kinds of food more attractive than others. Another species of turtle, the diamondback terrapin (*Malaclemys centrata*) has likewise been found to develop firmly established food preferences very early in life (1).

Food preferences of birds have been studied by Rabinowitch (75, 76). From his experiments with ring-billed gulls, herring gulls, and zebra finches, he concluded that foods to which individual birds had become familiar during early life tended to be selected preferentially when subsequently tested in a choice situation. In domestic chicks, Hess (33) found a period of maximum effectiveness for food reward with a peak around the 4th day of life, whereas at the age of 7 or 9 days the animals showed no effect of the food reward experience. [For review of further data on the establishment of food preferences in reptiles and birds, see Hess (34).]

In mammals, data on the European polecat (*Putorius putorius*) suggest that the smell of the prey serving as a sign stimulus for prey recognition has to be learned during a sensitive period occurring at 2–3 months of age. After this period, the animals display only slight interest in new odors and learn to recognize the smell of a new prey only if no familiar prey is available. As soon as they are given a choice, they again prefer the odor or prey encountered before the end of the sensitive period (2). The occurrence of food preferences based on early experience of course does not exclude the possibility that in other species such experience does not exert a permanent influence, as has been described, for example, in the rat (14).

Selection of Home Area

As in food choice, preferences for a certain area or locality often seem to depend on early experience. One of the most intensely studied examples is provided by the North Pacific salmons (*Oncorhynchus* ssp.), which, after spending 2–7 intervening years at sea, have been found to return to the streams or riverlets in which they were raised, irrespective of the rivers from which their parents came. Hasler (31) showed that fishes are able to discriminate successfully between chemical differences of various streams and that they may retain a particular stream odor in their memory. In extinction tests he also found that retention of learning is longer in fishes trained when young than in those trained in senility. Hasler assumed, therefore, that a young salmon becomes imprinted to the odor of its parent stream, and that such "olfactory imprinting" is an essential mechanism employed by the sexually mature fish during the return river migration, when it swims against the current and rejects all odors until it finally arrives at the home riverlet (31). An experimental study with *Rana cascadae* tadpoles indicated that early experience may also exert an influence on habitat selection in amphibians: laboratory-hatched animals established a clear preference for a certain substrate pattern during two weeks of experience (92).

In birds, Löhrl (53) found that the collared flycatcher (*Ficedula albicollis*) returns during spring migration to the area to which it was exposed after fledging during the preceding year. He also reported that it is the period immediately before the onset of fall migration that is essential, and that during that time a period of about two weeks is sufficient for this type of locality imprinting. In the reed warbler (*Acrocephalus scirpaceus*), banded individuals have also been found to return for breeding to the area where the post-fledging period was spent, whether or not it was the natal area (17). Egg transfer experiments have been carried out with the short-tailed shearwater (*Puffinus tenuirostris*). This species is known to return to its natal island after its extensive migratory circuit in the Pacific Ocean. The experiments showed that such return tendency has no hereditary basis, but is due to locality imprinting that appears to take place during a restricted period when the young begin their nocturnal emergences from the nesting burrows (81).

Similar findings have been reported for a great number of species, suggesting that site tenacity may be based, at least in part, on early experience, i.e. that the birds tend to return to that particular area with which they have become acquainted after being fledged or, in transfer experiments, released (for review, see Drost, 25; Hildén, 35). Even recurrences within the same winter quarters year after year that have been found in a number of Palaearctic migrants in Africa and South Asia (for review, see Moreau, 63) have been suggested to be due to locality imprinting during the first period of sojourn within the particular area (see, for example, Lord Medway, 62). Tenacity to the one locality, although of course on a spatially smaller scale, is also known in insects. Circumstantial evidence indicates that many insects, above all a great many aculeate Hymenoptera, become attached to the immediate locality first perceived by the newly emerged adult (see review in Thorpe, 90).

Habitat Preferences

For many reasons, especially in view of the possible evolutionary consequences (p. 8), mechanisms of habitat selection in animals are of general interest. Experimental investigations carried out in this field show that hereditary as well as experiential factors may be involved and that in the latter case the early experience is of greatest importance. A rather strong innate preference has been found by Wecker (91) in the deer mouse (*Peromyscus maniculatus bairdi*). In his frequently cited paper, he reported that the choice of the field environment is predetermined genetically, but can still be reinforced through early experience with the normal field habitat. Experience with another type of environment, on the other hand, has been found to be insufficient to reverse the normal preference for the field habitat. A similar case of interaction between heredity and experience has been reported for the chipping sparrow (*Spizella passerina*). In young kaspar hauser birds raised in the laboratory, Klopfer (43) found the same preference for pine over oak leaves as exists in wild-trapped adults. This preference for a particular type of foliage, however, can be altered through exposure to a different type during the first 2 or 3 months of life.

An obviously much stronger influence of early experience has been described by Drost (26), who transported large numbers of herring gulls (*Larus argentatus*) from the coast of the North Sea into the interior of western Germany where they were raised in zoological gardens. During their first autumn these animals disappeared, but some of them returned at an age of 3–4 years and settled down for breeding. This, without doubt, resembles the cases of site tenacity described above. However, as several cases of nesting were observed far from the rearing places, but in a similar environment, the birds seem to have been attracted by a particular type of inland habitat rather than by the actual locality.

Hess (34) suggested the possibility of imprinting to the nest site type of habitat in mallard ducks. He found that when individuals hatched in an incubator were given a choice between nesting sites on the ground and elevated nest boxes, they chose the boxes, whereas birds hatched in natural nests preferred to nest on the ground. The results of another extensive study of a possible correlation between nesting habitats and early experience in mallards, however, do not support this hypothesis (Bjärvall, 11). Obviously, the problem needs further investigation.

In addition to the experimental investigations mentioned, there are a large number of field observations that also suggest a strong influence of early experience upon the subsequent selection of a particular type of habitat or nest site. An impressive example is provided by the European mistle thrush (*Turdus viscivorus*). Originally a forest inhabitant, the species has recently invaded open parkland areas in large parts of northwestern Europe. Peitzmeier (72) collected good evidence that the parkland form did not originate from local stock, but from a parkland population in northern France that spread in a northeasterly direction with a speed of about 5–9.5 km per year until it reached the North Sea shore in northern Germany. As a consequence, two separate populations with a mosaic pattern of distribution are currently to be found. In a detailed discussion of the relevant data, Peitzmeier arrived at the conclusion that the respective preferences of the populations have to be regarded as a result of habitat imprinting.

Many other cases of a possible influence of early experience upon habitat selection have been summarized and discussed by Thorpe (90), Hildén (35), Klopfer & Hailman (45), and Braestrup (12, 13). However, there is also some contrary evidence. In the pied flycatcher (*Fidedula hypoleuca*), Berndt & Winkel (10) found that 70–80% of the birds chose deciduous woodlands for breeding irrespective of whether they had been raised in deciduous or in coniferous woodland. Coal tits (*Parus ater*) and blue tits (*P. caeruleus*) can likewise select their usual habitat without any previous experience of it. (70).

In some cases, especially in insects, "locality imprinting" and "habitat imprinting" probably cannot be differentiated completely, as the return to one locality most likely is not based exclusively on the specific characters of that particular place, but may also involve a general preference for a certain type of habitat (90). As a consequence, true habitat imprinting should only be assumed if a permanent preference for a certain type of environment is found at localities not identical with the place of birth.

Host Imprinting

Another nonsocial sphere in which several cases of imprinting-like processes have been described is the establishment of host preferences in parasitic animals. In the European cuckoo (*Cuculus canorus*) it has long been known that each female keeps to a particular species of host, and it is widely believed that this species is the same as the fosterer which originally reared the female. In this case, it would seem likely that such specificity is based on early experience, and the term *host imprinting* or *host fixation* has, therefore, been used by many authors (reviews in 30, 47, 48). It must be stressed, however, that evidence is merely observational so far and any definite proof of host imprinting by means of long-term studies on individually marked birds is still lacking. If early experience is in fact essential, another open question is whether the female really becomes imprinted on the host species itself, or whether imprinting on the habitat, nest site, or kind of nest of the host may also be involved.

Individual host specificity or host preferences have also been described in several other cuckoos, like the jacobine cuckoo (*Clamator jacobinus*) (52) and some of the species of glossy cuckoos (genus *Chrysococcyx*) (27), and it has likewise been assumed that these preferences may partially depend upon the female's own foster parents. Again, no definite proof is available. In contrast to the cuckoos, the evidence for the occurrence of host imprinting is fairly definite in another group of parasitic birds, the viduine finches (subfamily Viduinae of the Ploceidae). Host specificity in these birds is much stronger than in the cuckoos. Their parasitism is restricted to members of the family Estrildidae (waxbills), and each species or even subspecies has been found to parasitize only one species or subspecies of host. Such specialization is based on early experience: during their period of parental care, the young viduines form a specific host search image, and during their own period of reproductive activity they are attracted by members of their foster parents' species. The females even seem to come into breeding condition and ovulate only when stimulated by observing the reproductive activities of a pair of the host species. The song of the male viduine finch is composed of two sets of elements. One of them comprises the entire vocal repertoire of the host species and is indistinguishable from the original. This probably also contributes to attracting the female that prefers to mate with males whose song resemble most perfectly the vocalizations she was exposed to during parental care by the foster parents. In the viduine finches, rigorous imprinting on the host species has even led to parallel subspecies and species formation in parasites and hosts (p. 34) and to the subsequent evolution of morphological adaptations (mouth markings, plumage color) in the parasite nestlings (for details, see Nicolai, 65, 66; Payne, most recently, 71).

Some kind of imprinting may also be involved in host selection by parasitic insects. A frequently cited example has been provided by the experimental studies with *Nemeritis canescens* carried out by Thorpe and his co-workers. This ichneumonid is parasitic on the larva of the moth *Ephestia kühniella*. Using an olfactometer, Thorpe was able to show that for oviposition, the female finds its host entirely by the sense of smell and that there is a strong innate response to the odor of the natural host. However, if the female has been reared artificially, and shortly after emergence as an adult has been exposed to the larval odor of a strange species of Lepidoptera, the attractiveness of the *Ephestia* odor is significantly reduced when the strange odor is offered as a counter-attraction. Obviously, as in the selection of habitat (p. 18), there seems to be a hereditary basis for host selection, but there is also an early period in adult life when some characteristics of host odor may be learned (see Thorpe, 90, for review and further examples).

THE CHARACTERISTICS OF IMPRINTING

The phenomena that have been called imprinting considerably extend the original Lorenzian concept that referred exclusively to the acquisition of particular social object preferences.[2] The question arises, therefore, as to whether, even with the wider usage of the term, some common distinguishing characters of these processes can be found. A brief discussion of this question seems justified even in a review paper concentrating on the ecological significance of imprinting because, as I hope to point out, it is just the typical attributes of imprinting that are responsible for its great importance in the ecological sphere.

In his Kumpan paper, Lorenz (54) mentioned four main criteria characteristic of imprinting:
1. It can take place only during a restricted time period of the individual's life, the sensitive period.
2. It is irreversible—that is, it cannot be forgotten.
3. It involves learning of supra-individual, species-specific characters.
4. It may be completed at a time when the appropriate reaction itself is not yet performed.

Sensitive Periods

The occurrence of sensitive periods has always been regarded as one of the most important characteristics of imprinting. In the details of such periods, differences between species as well as between different motivational systems within a species occur. For example, sensitive periods may be very brief and last only for several hours or days, as in filial imprinting of ducklings (32, 79), or they may be rather extended and last for several months, as in sexual imprinting in grayleg geese (78). Furthermore, their onset and termination may be rather rigid or more gradual.

Such quantitative differences seem to be unimportant, however, if one considers the essential feature to which the concept of sensitive periods refers. This is merely the fact that certain kinds of experience are most effective during specific developmental stages so that the same amount of experience leads to different results at different ages. And in this very broad sense, this statement can be applied to all learning processes that have been called imprinting.

Irreversibility

A second important characteristic of imprinting is its great persistence, usually referred to as irreversibility (although this term has been frequently misunderstood). The reasons for such controversies are due mainly to methical reasons and to misinterpretations of the original statements made by Lorenz (for review see Immelmann, 40). For example, irreversibility in the original sense had always only been understood with regard to certain preferences but never in an absolute way. In this

[2]The original definition given by Lorenz reads: "The process of acquisition of the object of instinctive behaviour patterns oriented towards conspecifics which are initially incorporated without the object" (translation published in Lorenz, 55, p. 246).

sense, lifelong irreversibility has really been proved in some cases of sexual imprinting (39). In many other contexts (e.g. host imprinting or social deprivation), it has likewise been found that the influence of early experience resists extinction to a high degree. In filial imprinting, on the other hand, the question of irreversibility is irrelevant, as the reaction itself—the following response—lasts only for a rather short time.

The degree of persistence, however, again depends on the specific biological demands and may be different in different species or motivational systems. In sexual imprinting, for which the characteristics of the "object," i.e. of the conspecific mate(s), remain the same for the whole life span of the imprinted animal, great stability is strongly selected for in order to promote sexual isolation. In host imprinting of parasitic animals, a high degree of persistence may be advantageous, too. For so-called "food imprinting," on the other hand, because of possible changes in the food situation, rigid invariability would even be disadvantageous, except perhaps in a few strongly monophagous species, as, for example, some parasitic insects (cf Mayr, 60).

Unfortunately, in most studies on food imprinting the relevant preferences have been followed up only for days or weeks. Few data—like those published by Rabinowitch (76) on food preferences in the zebra finch—are available that cover a period of several months. The durability of early food preferences, therefore, in most cases still remains obscure, although Hess (34) concluded that in domestic chicks modification of innate pecking preferences during the sensitive period is "apparently permanent."

In locality and habitat imprinting, for which several long-term studies have been carried out, the situation seems to be somewhat intermediate. Especially during the reproductive period, strong selection mechanisms, but perhaps not as rigid as in sexual imprinting, could be advantageous, whereas outside this period a higher degree of variability may be selected for, although even then many cases of habitat and site tenacity have been described (p. 17).

Again one must ask for the essential feature to which the concept of irreversibility refers, and this obviously has been frequently misunderstood in the literature (for discussion, see Immelmann, 40). This concept, if it can be claimed to have any general validity at all, does not relate to the absolute length of time the relevant preference or other modifications of behavior are maintained. Rather, it refers to the fact that a particular experience during the sensitive period will have, even in the face of considerable experience with other objects, a more permanent effect than the same experience encountered at a later time. In this broad sense, it applies to all cases of imprinting described so far, even to the sometimes rather short-term preferences for certain kinds of food for which definite periods of maximum effectiveness have been reported (cf Hess, 34).

The somewhat misleading term *irreversibility,* however, should for general use perhaps be replaced by the term *persistence.* It ought to be retained only for those cases in which an alteration of particular preferences through subsequent experiences has definitely been proved to be no longer possible, as, for example, in sexual imprinting in the zebra finch (39).

Generalization

With regard to the amount of generalization, which has been mentioned as another attribute of imprinting, the variation within the different types of imprinting is even greater than for the first two characteristics mentioned. In sexual imprinting, for example, it has been found that in all cross-fostering experiments, sexual preferences for the foster parents' species are not restricted to particular individuals, but, in spite of sometimes rather pronounced individual differences, apply to all members of that particular species. The same applies to host imprinting. A certain degree of generalization is also found in food or habitat imprinting where the young animal learns to recognize a certain class of objects rather than only one specific single object. In filial imprinting, on the other hand, preferences as a rule really do seem to be restricted to one particular individual.

It follows that the degree of generalization again depends on the specific biological demands on the particular imprinting process. Similar differences in stimulus-generalization are known to occur after discrimination learning in many other situations (cf Bateson, 6). Therefore, it can be concluded that the amount of generalization, in contrast to the first two characteristics mentioned, cannot be regarded as a specific attribute of imprinting.

Temporal Relation Between Imprinting and Performance

The fourth characteristic of imprinting—its completion at a time when the appropriate reaction itself is not yet performed—that has some general importance with regard to the difficult problem of reward, was thought by Lorenz to be applicable only to sexual imprinting. It may, in a certain way, also apply to host imprinting, as well as to some cases of song learning. In filial imprinting, on the other hand, the reaction itself—the following response—is more or less fully developed during the sensitive period. The same probably holds true for food imprinting and for many other early learning processes (for review, see Immelmann, 40).

It follows that the temporal relation between imprinting and performance, and hence the kind of reinforcement needed, again cannot be regarded as a specific characteristic of imprinting.

Other Characteristics

Several further possible features of imprinting have been mentioned and discussed during the years following the publication of Lorenz's Kumpan paper. They refer to the "effort" expended by an animal during imprinting, the effect of punishment, the influence of drugs, or the consequences of socialization (for review see, Hess, 33, 34).

These criteria are more specific, however, and are valid or have been tested only for particular kinds of imprinting, mainly for the following reaction of nidifugous birds. As a consequence, in all probability these claims again cannot be used for a general characterization of the phenomenon of imprinting as a whole.

To summarize, it can be concluded that, although all four characteristics given by Lorenz are fully realized in some cases of sexual imprinting and perhaps also in some other imprinting situations (cf Immelmann, 39), there are only two criteria that really hold true for the phenomenon as a whole: the existence of a sensitive period and the subsequent stability of the result of experience gained during that period.

The Possible Background of Imprinting

The existence of these two common criteria of which the second is only a consequence of the first, has been interpreted in many different ways, and quite a number of explanations for the occurrence of sensitive periods have been suggested. Almost all of them agree that such periods must be related in one way or another to maturational processes within the organism itself.[3] However, the degree to which maturation is involved is judged differently, and some authors have quite rightly warned of a purely maturational view.

Sensitive periods are certainly due to a combination of internal processes coupled with the role of the environment. However, it seems that the relative importance of internal and external control not only varies between different kinds of imprinting, but also between the onset and the end of the sensitive period. There seems to be general agreement that the onset of the period is largely dependent on the development of the animal as a whole, mainly on sensory and neural development, and, especially in filial imprinting, on an increase in locomotor ability (for discussion, see Bateson, 6; Immelmann, 40).

The determinants of the end of sensitive periods, on the other hand, seem to be manifold and to include a greater overall importance of environmental factors. Important external influences known to contribute to determining the duration of the sensitive period are the amount of previous experience with the same or other objects and any kind of social stimulation, as well as various kinds of other sensory stimulation prior to or during the sensitive period. Despite these external influences, however, maturational processes may also play an important role. This can be concluded from the fact that under natural conditions the duration of sensitive periods is adapted very accurately to the biology and the specific ecological demands of a species (cf p. 24). During evolution, such adaptation probably can be achieved most properly through a connection of optimal periods for learning with maturational events. It is to be expected, therefore, that selection will favor the development of genetically based maturational processes that, within certain limits, determine the age at which a sensitive period closes.

Definite proof that internal processes are involved can only be gained experimentally. If, under otherwise identical conditions, a certain length of exposure at a certain stage of ontogenetic development leads to a more permanent result than exactly the same (or greater) length of exposure to the same object at another stage, this can be due only to internal changes in learning ability. Such internally determined alterations with age have been proved experimentally for sexual imprinting

---

[3]In his Kumpan paper, Lorenz (54) already stated that imprinting "depends on a quite definite physiological developmental condition in the young bird" (translation published in Lorenz, 55, p. 127).

in zebra finches (Immelmann, in preparation). But, although it has not been formulated in this particular way, the same correlation becomes evident from the results of many investigations on filial, sexual, and food imprinting, as well as on song learning in birds.

As far as the kind of maturational processes is concerned, no definite conclusions are possible yet. As the perceptual and locomotor development of the organism in almost all cases is completed long before the end of sensitive periods, a correlation with brain development seems to be most likely. For many animals, the number of neurons and synapses is not complete at birth, but the division of neurons and the formation of new synapses continue for several weeks, months, or (in the case of larger mammals and man) years. It is possible, therefore, that the great sensitivity early in life and the great stability of early experience could be correlated with the formation of new synapses, and that the end of the sensitive period could be associated with the end of mitotic activity or with other kinds of morphological, physiological, or biochemical neural changes. But, again, this is still speculative and further research, especially histological and biochemical investigations of brain development, is urgently needed (37, 40, 86).

The fact that, in all probability, there is a correlation between brain maturation and sensitivity to certain environmental influences does not allow any conclusions as to the nature of the age-dependent parameters of learning. Experimental investigations with zebra finches suggest that it is retention rather than acquisition of information that changes with age: zebra finches raised by another species of estrildid finches, the so-called Bengalese or society finch (*Lonchura striata* var. dom.), always prove to be sexually imprinted on the foster parents' species (38). Subsequent experience with their own species does not lead to any permanent alteration of this preference, even if it lasts for several years and even if no contact with the foster parents' species is possible during that time. If, after separation from its conspecifics, the bird is tested again in the same double-choice situation as prior to its intraspecific contact, it will again prefer members of the foster parents' species (39). This preference for the alien species, however, may not be visible from the very beginning of a new series of tests, but may reappear only gradually. During the first experiments, the bird very often prefers the conspecific female, and it is only in the course of several days or sometimes weeks that the original preference for the foster parents' species is reestablished (Immelmann, in preparation).

Such temporary reversal of preferences during intraspecific contact clearly indicates that the bird is perfectly able to establish new sexual preferences even during adult life, i.e. a considerable amount of time after the end of the sensitive period. The only difference, therefore, refers to the stability of such preferences. In the absence of the preferred object, i.e. without constant reinforcement, a sexual preference established during the sensitive period is maintained indefinitely, whereas a preference established by adult birds is lost again within a couple of days or weeks.

What seems to be age-dependent, therefore, is not, in the first place, the learning ability of the individual, but rather its ability to establish long-term memory, i.e. its capacity for permanent preservation of acquired information in the nervous system. If one looks at the two qualities found to be the only ones that really characterize imprinting in general, it is quite obvious that they may refer in greater degree to the retention than to the acquisition of information. All that the various phenomena called imprinting really have in common seems to be the stability of information storage that expresses itself in the combination of the characters' "sensitive period" and "persistence." Obviously, this is ultimately a consequence of the relation between the input of new information and the amount of information already stored in the nervous system. And this relation necessarily changes with age.

Although lifelong irreversibility or even long-lasting stability has not been proved experimentally in many of the instances that have been called imprinting, especially in the ecological sphere, age-dependent differences in the retention of information have always been found. Thus the term *imprinting* in the above-mentioned sense seems to be justified. And it is just the temporal differences in information storage that are responsible for the great ecological significance of imprinting in general.

BIOLOGICAL SIGNIFICANCE OF IMPRINTING

In order to arrive at a better understanding of the biological significance of imprinting, the two common characteristics discussed in the previous chapter should be used as a starting point for the discussion. The following problems seem to be of particular interest:
1. Why do so many behavioral characters in the broadest sense have to be learned, i.e. why does the environment exert such an essential influence?
2. Why is imprinting involved in these learning processes, i.e. what selection pressures have led to the evolution of sensitive periods and subsequent persistence?
3. What are the ecological consequences of imprinting?

Biological Advantage of Learned Characters

The most significant advantage of learned characters over characters based on genetic factors is their greater adaptability. This becomes apparent, for example, if one looks at object preferences: if a species possesses an "open program" of information, i.e. if recognition of essential "objects" (conspecifics, food, details of habitat) is based on learning, any changes in the environment of the particular species, including the appearance of the species itself, are followed automatically by a corresponding change in the relevant object preferences. In the case of a "closed program," on the other hand, where the relevant preferences are coded in the gene pool of the species, such alterations would depend upon genetic change.

The very conservative nature of obviously innate preferences has been demonstrated quite distinctly in certain populations of the three-spined stickleback (*Gasterosteus aculeatus*) in western North America. In these populations, the male nuptial color is a deep black rather than the normal bright red of the underside. This difference is probably the result of strong selection pressure provided by an endemic predatory fish that feeds on young sticklebacks concentrated near the male's nest. McPhail (61), who tested female mate preference in a choice situation between red or black males, was able to show that females from the black populations demonstrate a distinct preference for red males; the preference is almost as strong as that of females from red populations. The age of the black populations has been estimated to be 6000–8000 years. This means that this time span was insufficient to alter substantially the original mate preferences of the females.

If, on the contrary, the choice of a mate were determined through sexual imprinting, preferences could be altered from one generation to the next. Such quick adaptability is of special importance in any rapidly evolving group of animals in which the appearance of a species may change in a comparatively short time. In this case, learning of the species-specific characters may be an effective method of ensuring "correct" species recognition and pairing, the necessity of which, especially under conditions of sympatry, is evident. Probably it is not accidental, therefore, that those groups that provide the most typical examples of sexual imprinting, i.e. for which learning of species-specific characters seems to be a widespread phenomenon (e.g. ducks and geese, gallinaceous birds, pigeons and doves, estrildid finches, and cichlid fishes), are also known for their extensive adaptive radiation with several closely related and often very similar species occurring in the same area. It seems to be possible that an open ontogeny with reference to species recognition may be one of the preconditions for rapid evolutionary changes.

Quick adaptability does not exclude the possiblity that, provided there is no need for any alteration, acquired information may be preserved over many generations as persistently as any innate kind of information. The males of four species of introduced European birds in New Zealand, for example, still possess songs that are not different from those of males in Europe, although the introduced species were brought to the islands about 100 years ago. The songs of at least two of these species (blackbird, *Turdus merula,* and chaffinch, *Frinjilla coelebs*) have to be learned (88).

A second advantage of an open program may be connected with the fact that, in general, the amount of information that can be stored in the genome tends to be smaller than the possible amount of information able to be stored in the memory. Consequently, object preferences based on learning tend to include more details and will thus be more precise than those based on genetic factors (cf Mayr, 60). This question is discussed in more detail in Immelmann (41).

Biological Advantage of Sensitive Periods

From the facts discussed above, the specific advantages of an open ontogeny, during which many behavioral attributes depend on personal experience, are beyond doubt. On the other hand, this system also contains a certain risk, as there is always the danger that for some reason or the other a "wrong" kind of information will be acquired. It is to be expected, therefore, that selection will favor the evolution of mechanisms that contribute to minimize this particular danger, i.e. mechanisms that help to secure the acquisition of the appropriate sort of information.

Among the possible mechanisms, there are two that can be expected to be of particular effectiveness: the restriction of maximum learning capacity to those periods during ontogeny when the opportunity to acquire biologically relevant information is greatest, and a stable storage of information acquired during that particular period in order to prevent a detrimental influence of subsequent and possibly less appropriate information. The two mechanisms are of course identical with the attributes really characterizing the phenomenon of imprinting on the whole. This may explain why imprinting is involved in so many of the early learning processes and what selection pressures have led to the evolution of sensitive periods and subsequent persistence.

The most favorable learning period, in general, is early in the individual's life, while the young animal is still a member of the family group. At this time, its opportunities to learn species-specific characteristics, to acquire species-specific calls or songs, or to learn certain features of the species-specific environment are greater than at a later time when it has to live on its own or in less tight social units. Furthermore, the certainty of adequately learning the correct features is also greatest during this very early period.

These theoretical considerations are in accord with the wealth of information about the early occurrence and generally short duration of sensitive periods for all processes that have been called imprinting or have been compared to imprinting. Furthermore, the available evidence suggests that the time of occurrence and the duration of sensitive periods are closely correlated with the specific conditions and biological demands on the particular learning process. For sexual imprinting, for example, Schutz (78), comparing ducks and geese, found that sensitive periods closely correspond to the duration of parental care and consequently are much later and much more extended in geese than in ducks. An adaptive component of sensitive periods has also been found in song learning (p. 30).

The examination of the great importance of an "open program" in general and of early learning periods in particular also throws some light on the biological significance of parental care in animals: in addition to supplying the offspring with food, shelter, body heat, etc, parental care may also be essential in providing a young animal with the opportunity to acquire relevant species-specific information (cf Mayr, 60).

In reply to the second question on p. 20, imprinting offers a number of very distinct advantages. Information transfer through imprinting may contain a great number of details. It is quickly adaptable to any changes in the social or nonsocial environment of the organism. As a consequence of the early sensitive periods it very often (e.g. in host imprinting or sexual imprinting) ensures the availability of relevant information well before its first application. Finally, due to its great persistence, it has very stable and long-lasting effects for any one individual. Altogether, therefore, imprinting presents a combination of attributes that otherwise are distributed to different mechanisms of information transfer, as the first two of the advantages mentioned are typical for learned, but the third and fourth attributes for genetically coded, information.

Ecological Consequences of Imprinting

Due to its specific advantages, imprinting may be of considerable ecological significance. Interestingly enough, this applies not only to "ecological" imprinting (cf that section), but also to a number of "nonecological" imprinting phenomena, such as sexual imprinting or imprinting-like song learning.

SIGNIFICANCE OF "ECOLOGICAL" IMPRINTING   With regard to ecological adaptations, imprinting results in a situation somehow intermediate between the two "classical" types of stenophagic or stenoecious and euryphagic or euryecious species or subspecies that are adapted to a narrow or a wide range of ecological conditions, respectively. If certain ecological preferences, e.g. for a certain kind of habitat or a particular species of host, are established through imprinting, genetic plasticity of the species is combined with individual phenotypic stability. In other words, the gene pool permits a large range of ecological adaptations, whereas the single individual, according to its early experience, may be highly specialized.

A situation like this secures an optimal adaptation to the relevant ecological conditions. This refers to the temporal as well as to the spatial utilization of a particular environment.

Temporal   If the environment of a population or species remains stable, the behavioral features acquired through early experience may be transferred unchanged from one generation to the next. An example from the "nonecological" sphere has already been mentioned (p. 27). But if, as a consequence of a change in the environmental conditions, an alteration of relevant preferences or other adaptive features becomes necessary, this change is achieved almost automatically from one generation to the other when the offspring are exposed to the new conditions.

This may be relevant in two situations: (*a*) in the case of an alteration in the original environment of the species (e.g. occurrence of a competitive species, change in prevailing food plants or prey animals, change in the vegetation of an area, disappearance of a particular host) or (*b*) in the case of the colonization of new habitats, the utilization of hitherto unexploited food sources, the extension of parasitism to a new species of host, or similar invasions into new ecological niches. The fairly rapid colonization of urban and other man-made habitats that has occurred in several species of birds (12, 87) has frequently been supposed to be based on some kind of habitat imprinting.

A very impressive example is provided by the collared dove (*Strertopelia decaocto*). This species, which has a wide range in southern Asia, seems to have developed a very early and strong preference for human settlements and has been introduced by man into many hitherto uninhabited areas. In southeastern Europe, the spread of the species was correlated with Turkish rule over wide parts of the Balkan peninsula because the Turks tolerated the doves even when they were nesting on buildings in villages and towns. When, during the nineteenth century, the Turks left most of southeastern Europe, the doves lost their protectors and in many areas they quickly disappeared. Early during the 20th century, however, the species began to increase in numbers again and in an unparalleled expansion spread over most parts of southern, eastern, and central Europe within only few decades. Circumstantial evidence indicates that this expansion started with a shift from nesting on buildings to nesting in trees close to buildings, sites much safer from nest destruction by humans. Such imprinting to the relevant environmental features has enabled the species to rapidly spread over densely populated areas in large parts of Europe (36, 85).

In connection with habitat imprinting, there are even some practical implications with regard to wildlife conservation. If the decrease of a particular species within a given area is due to the destruction of its specific habitat, some offspring of that particular population could be raised in captivity and have imprinted on them the features of another "safer" type of habitat; subsequently, they could be released into this new type of environment. This is at present done in Austria with the great European bustard (*Otis tarda*) (56).

Spatial   In the case of different specializations based on early experience, whole populations of euryecious species may be composed of individuals with different ecological adaptations. This situation may express itself either in a mosaic distribution of individuals with different requirements, as, for example, in female cuckoos that are host-specific in their parasitism as individuals only (p. 19), or in the existence of entire populations with more or less uniform ecological requirements, as, for example, in the European mistle thrush (p. 19). These populations, however, may again be distributed over the area of the species in a mosaic fashion.

For a subdivision of a species or population based on behavioral characters, for example on different food or habitat preferences, the term *polyethism* or *behavioral polymorphism* has been introduced.

ECOLOGICAL SIGNIFICANCE OF "NONECOLOGICAL" IMPRINTING   In a population or species of animals consisting of different "ecological types" of individuals that, due to differences in early experience, are adapted to different ecological conditions, it is to be expected that selection will favor any genetical alterations that reinforce the advantageous behavioral adaptations so that genotypic differences will develop eventually. The evidence discussed in the previous section clearly supports the view expressed by Mayr (60)—that in most cases changes in behavior, such as preferences for a new habitat, food, or kind of host, represent the initial step in the subdivision of a species, and that morphological changes are only a consequence of the new selection pressure caused by the preceding behavioral shifts.

As soon as the population is divided into such groups of individuals with different ecological adaptations, selection will also favor the evolution of any characters that enable the animals to distinguish between members of their own and another group. Such ability has at least three advantages: it contributes to restrict matings to members of the same population and thus helps to maintain the identity of gene pools of groups of individuals adapted to local ecological conditions; it may have a guidance function to the "correct" type of environment; and, finally, it may lead to a reduction of mutual aggression between members of different populations. The latter could be advantageous in those cases in which competition between the populations is reduced and a certain degree of sympatry is possible or—in the light of a better overall utilization of a given area by a particular species—even favorable.

In the mechanisms that enable the animals to recognize members of their own population early experience may again be involved; here, "nonecological" imprinting processes, like song learning and sexual imprinting, come into play. Vocalizations and general appearance are the main features known to serve for mutual recognition, and knowledge of these characters is frequently based on early learning.

Song learning   Most information on song learning comes from birds in which a system of dialects characterizes different populations. Here, early experience may be doubly involved: in the ontogenetic development of song itself, and in the recognition of a particular type of song by the female or by another male. The great ecological importance of dialects becomes apparent from the following examples: In the European ortolan bunting (*Emberiza hortulana*), several "ecological types" with rather different environmental claims have been found (18). In all probability, the respective ecological preferences are based on early experience. At least in some cases they have been observed to be paralleled by a corresponding system of clearly differentiated regional song dialects, and from circumstantial evidence it seems to be certain that the details of the local dialects are learned during the same period that imprinting on the relevant ecological factors occurs (19). One of the most important functions of these dialects lies in the orientation they offer birds returning from their winter quarters. This is of special importance to younger and less experienced animals; if they prove to be attracted by the type of song they heard and (in the case of the males) learned in their youth, they may be guided to settle for breeding in those habitats to which they are, due to their own early experience, most properly adapted. If, as seems probable, the female's choice of mate is likewise affected by song types heard in youth, such song dialects together with habitat imprinting also promote reproductive isolation between different populations.

It is to be expected that other systems of song dialects serve a similar biological function. This has been suggested, e.g. for the white-crowned sparrow (*Zonotrichia leucophrys*) (59). In the chingolo or rufous-collared sparrow (*Z. capensis*), King (42) demonstrated statistically significant correlations between the frequency of specific song themes and certain types of habitat. This again may be the result of a certain habitat preference being associated with specific themes, although King stated that his data do not yet permit definite conclusions.

As a necessary precondition for the ecological effectiveness of a system of dialects, song has to be learned while the young birds are still within their natal area; this may be a reason why, at least in one of the species known to have local dialects, the white-crowned sparrow, song learning is restricted in an imprinting-like manner to a brief period from about 10–100 days of age (58). The rather strong selection pressure towards an early sensitive period for song learning in this species has been demonstrated by Baptista (4). In the San Francisco Bay area, California, he found a number of males of the sedentary population (*Z. leucophrys nuttalli*) singing songs typical of the migratory subspecies *Z. leucophrys pugetensis,* which winters in that particular area. Probably these males were born very early or very late in the season and thus were exposed to the songs of the wintering birds. This exposure, supposedly together with the death by disease or predator of the natural tutor, the father, has led to the acquisition of the "wrong" type of dialect. (For further discussion of the functional significance of song dialects and of the function and nature of sensitive periods for vocal learning, see 42, 46, 58, 59, 67–69.) An adaptive restriction of song learning in time has also been found with regard to vocal mimicry of viduine finches (p. 20) (71).

Sexual imprinting   In the same sense as early experience with a certain type of song, any other mechanism that attracts the young to the parental characters can also help to preserve the traditional ecological adaptations of a population or subpopulation. In this respect, even sexual imprinting may be of ecological significance.

In birds, the splitting of a species based on behavioral characters has often been found to be correlated with differences in plumage color. Such polymorphism combined with a recognition mechanism based on early experience may contribute to achieving a subdivision of gene pools within a population by means of nonrandom mating. A well known example is provided by the North American snow goose (*Chen caerulescens*), which in its smaller subspecies is polymorphic and has two color phases, a blue form and a white form. In mixed populations, birds of the same morph mate with one another more frequently than with birds of the opposite morph (21, 22). This system of positive assortative mating is based on sexual imprinting, as the males select mates that have plumage patterns similar to those of one of their parents (23, 24). Between the two morphs there are slight differences in breeding biology, the white one being better adapted to a rigorous environment. Under the present amelioration in climate, the blue phase is favored and is rapidly increasing in numbers. Altogether the situation presents a system of balanced polymorphism, which, according to the climate, in some years or decades favors the blue morph and, in others, the white (20).

Apart from the lesser snow goose, assortative mating has also been described in the dark and light color forms of the arctic skua (*Stercorarius parasiticus*), and, although the data about the possible origin and extent of nonrandom mating are still very meager, it seems possible that a similar situation might also exist in other polymorphic species. (For a detailed discussion of polymorphism in birds, see Selander, 80).

POSSIBLE SIGNIFICANCE OF INDIVIDUAL DIFFERENCES IN IMPRINTABILITY
For many species of birds, experimental investigations have revealed a certain amount of individual variation in imprintability. Schutz (77) found that, despite identical rearing conditions, some cross-fostered male mallards paired with the foster species while others formed pairs with members of their own species. The same has been reported by Goodwin for female blueheaded waxbills (*Uraeginthus cyanocephalus*) (28). In zebra finches, individual differences occur in the duration of the sensitive period and in the amount of social experience needed for sexual imprinting, as well as in the exclusiveness with which sexual responses are restricted to only one object (Immelmann, 38, and in preparation). Similar results are available for filial imprinting (29, 82).

Baptista (4) assumed genetically determined individual differences with regard to song learning dispositions of white-crowned sparrows. Such differences might be the reason for the fact that, of the naive fledglings exposed to the song from the resident as well as from a wintering migratory subspecies (p. 29), some learn the song of the former and some of the latter. A similar behavioral polymorphism with regard to song learning abilities seems to occur in the house finch (*Carpodacus mexicanus*) (3).

In domestic ducks, Klopfer & Gottlieb (44) found individual differences in responses to different kinds of signals, some animals being more susceptible to visual, and others to auditory, stimuli. Provided the number of possible learned responses is limited for any one individual, a "division of labor" with different individuals learning different signals would lead to an increase in the overall response repertoire of the group of ducklings as a whole.

In ecological imprinting processes, individual differences have not yet been proved experimentally. There are numerous indications, however, that a similar polyethism may also occur. In six species of glossy cuckoos, for example, some females are host specific and others are not. Both types are found together in the same populations (27). As host specificity is supposed to be based on early experience (p. 19), such differences may be due to individual differences in female imprintability.

A system like this might be favorable for the species as a whole. Those individuals that imprint strongly, e.g. on a certain type of habitat, will very strictly adhere to that environment. They will be able to adapt themselves very neatly to its specific conditions and will thus fill their particular ecological niche very properly. Those animals, on the other hand, that do not imprint very strongly represent some kind of a "pool," the members of which are able to colonize new niches, e.g. in a changing environment or in marginal zones of the distributional area, although as individuals these "nonspecialists" might be less successful in their niches than the "specialists." Experimental investigations along this line are highly desirable.

EVOLUTIONARY CONSIDERATIONS

The restriction of gene flow within a population or between various populations necessarily must have evolutionary consequences (73). If reproductive isolation becomes more and more complete, the ecological preferences that initially were based on early experience, together with assortative mating that may be dependent on early learning, will be able to initiate speciation. An interesting example is provided by the parasitic African indigo birds (genus *Hypochera*). The taxonomy of this viduine genus has always been a puzzle to systematists because there are several forms which exist sympatrically without interbreeding but which, nevertheless, have been shown to intergrade through intermediate "races." The sympatric forms have been found to be separated by ethological barriers on the basis of fixations to different hosts, which originate from host imprinting and vocal mimicry (p. 20). Obviously, therefore, early experience leads to an initial separation of different "host populations," which, in some cases, have reached more or less complete reproductive isolation. This in turn has promoted the development of slight morphological differences. The fact that there is no general agreement on the number of recognizable species clearly indicates that several of these "forms" are just at the point of reaching species level. Race, species, and genus formation among the viduines has occurred parallel to that of their hosts (reviews in 65, 66, 71). For a population that acquires reproductive isolation through the process of learning, Payne (71) has introduced the term *cultural species.*

To summarize, it can be concluded that early experience has two possible evolutionary functions: initially it leads to the formation of habitat and other ecological preferences in any one individual through "ecological imprinting"; as soon as natural selection has led to the evolution of slightly different gene pools, adapting groups of individuals to local conditions, it also serves to preserve such gene pools by means of sexual imprinting or imprinting-like song learning. This leads to the splitting of a species into a continuous or mosaic system of populations with habitat-linked differences in various characters and may finally be a first step in speciation. In this respect, even "nonecological" imprinting is of great relevance in maintaining adaptations to particular environmental conditions and thus may be of great ecological significance. A more detailed discussion of the possible evolutionary consequences of imprinting is given by Payne (71) and Immelmann (41).

Literature Cited

1. Allen, J. F., Littleford, R. A. 1955. Observations on the feeding habits and growth of immature Diamondback Terrapins. *Herpetologia* 11:77–80
2. Apfelbach, R. 1973. Olfactory sign stimulus for prey selection in polecats (*Putorius putorius* L.). *Z. Tierpsychol.* 33:270–73
3. Baptista, L. F. 1972. Wild House Finch sings White-crowned Sparrow song. *Z. Tierpsychol.* 30:266–70
4. Baptista, L. F. 1974. The effects of songs of wintering White-crowned Sparrows on song development in sedentary populations of the species. *Z. Tierpsychol.* 34:147–71
5. Bateson, P. P. G. 1966. The characteristics and context of imprinting. *Biol. Rev.* 41:177–220
6. Bateson, P. P. G. 1972. The formation of social attachments in young birds. *Proc. XV Int. Ornithol. Congr., Den Haag, 1970,* 303–15
7. Bateson, P. P. G., Horn, G., Rose, S. P. R. 1969. Effects of an imprinting procedure on regional incorporation of tritiated lysine into protein of chick brain. *Nature* 223:534–35
8. Bateson, P. P. G., Wainwright, A. A. P. 1972. The effects of prior exposure to light on the imprinting process in domestic chicks. *Behaviour* 42:279–90
9. Beadle, M. 1970. *A Child's Mind.* New York: Doubleday
10. Berndt, R., Winkel, W. 1974. Gibt es beim Trauerschnaepper *Ficedula hypoleuca* eine Praegung auf den Geburtsbiotop? *J. Ornithol.* 115: In press
11. Bjärvall, A. 1973. Nest-site selection by year-old female mallards *Anas platyrhynchos* in relation to the locality of their hatching. *Int. Zoo Yearb.* 13: 23–27
12. Braestrup, F. W. 1968. Evolution der Wirbeltiere. Oekologische und ethologische Gesichtspunkte. *Zool. Anz.* 181: 1–22
13. Braestrup, F. W. 1971. The evolutionary significance of learning. *Vidensk. Medd. Dan. Naturhist. Foren. Khobenhavn.* 134:89–102
14. Bronson, G. 1966. Evidence of the lack of influence of early diet on adult food preferences in rats. *J. Comp. Physiol. Psychol.* 62:162–64
15. Burghardt, G. M. 1967. The primacy effect of the first feeding experience in the snapping turtle. *Psychonomic Sci.* 7:383–84
16. Burghardt, G. M., Hess, E. H. 1966. Food imprinting in the Snapping Turtle, *Chelydra serpentina. Science* 151: 108–9
17. Catchpole, C. K. 1972. A comparative study of territory in the Reed warbler (*Acrocephalus scirpaceus*) and Sedge warbler (*A. schoenobaenus*). *J. Zool.* 166:213–31

18. Conrads, K. 1968. Zur Oekologie des Ortolans (*Emberiza hortulana*) am Rande der Westfaelischen Bucht. *Vogelwelt* 2:7–21

19. Conrads, K., Conrads, W. 1971. Regionaldialekte des Ortolans (*Emberiza hortulana*) in Deutschland. *Vogelwelt* 92:81–100

20. Cooch, F. G. 1961. Ecological aspects of the Blue Snow Goose complex. *Auk* 78:72–89

21. Cooch, F. G., Beardmore, J. A. 1959. Assortative mating and reciprocal difference in the Blue-Snow Goose complex. *Nature* 183:1833–34

22. Cooke, F., Cooch, F. G. 1968. The genetics of polymorphism in the goose *Anser caerulescens*. *Evolution* 22:289–300

23. Cooke, F., McNally, C. M. 1975. Mate selection and colour preferences in Lesser Snow Geese. *Behaviour*. In press

24. Cooke, F., Mirsky, P. J., Seiger, M. B. 1972. Colour preferences in the lesser snow goose and their possible role in mate selection. *Can. J. Zool.* 50:529–36

25. Drost, R. 1951. Study of bird migration 1938–1950. *Proc. Xth Int. Ornithol. Congr. Uppsala, 1950,* 216–40

26. Drost, R. 1958. Ueber die Ansiedlung von jung ins Binnenland verfrachteten Silbermoewen (*Larus argentatus*). *Vogelwarte* 19:169–73

27. Friedmann, H. 1968. The evolutionary history of the avian genus *Chrysococcyx*. *US Nat. Mus. Bull.* No. 265

28. Goodwin, D. 1971. Imprinting, or otherwise, in some cross-fostered Red-cheeked and Blue-headed Cordonbleus. *Avic. Mag.* 77:26–31

29. Goodwin, E. B., Hess, E. H. 1969. Innate visual form preferences in the imprinting behavior of hatching chicks. *Behaviour* 34:238–54

30. Harrison, C. J. D. 1968. Egg mimicry in British cuckoos. *Bird Study* 15:22–28

31. Hasler, A. D. 1966. *Underwater Guideposts—Homing of Salmon,* Madison, Wis.: Univ. Wis. Press

32. Hess, E. H. 1959. Imprinting. *Science* 130:133–41

33. Hess, E. H. 1964. Imprinting in birds. *Science* 146:1128–39

34. Hess, E. H. 1973. *Imprinting. Early Experience and the Developmental Psychobiology of Attachment.* New York: Van Nostrand. 472 pp.

35. Hildén, O. 1965. Habitat selection in birds. *Ann. Zool. Fenn.* 2:53–75

36. Hofstetter, F. B. 1960. Moegliche Faktoren bei der Ausbreitung von *Strepto-*

pelia d. decaocto Friv. *Proc. XII Int. Ornithol. Congr., Helsinki, 1958,* 298–309

37. Horn, G., Rose, S. P. R., Bateson, P. P. G. 1973. Experience and plasticity in the central nervous system. *Science* 181:506–14

38. Immelmann, K. 1969. Ueber den Einfluss fruehkindlicher Erfahrungen auf die geschlechtliche Objektfixierung bei Estrildiden. *Z. Tierpsychol.* 26:677–91

39. Immelmann, K. 1972. The influence of early experience upon the development of social behaviour in estrildine finches. *Proc. XV Int. Ornithol. Congr., Den Haag, 1970,* 291–313

40. Immelmann, K. 1972. Sexual and other long-term aspects of imprinting in birds and other species. *Adv. Study Behav.* 4:147–74

41. Immelmann, K. 1974. The evolutionary significance of early experience. *Tinbergen-Festschrift.* ed. A. Manning. Oxford: Clarendon

42. King, J. R. 1972. Variation in the song of the Rufous-collared Sparrow, *Zonotrichia capensis,* in Northwestern Argentina. *Z. Tierpsychol.* 30:344–73

43. Klopfer, P. 1963. Behavioral aspects of habitat selection: the role of early experience. *Wilson Bull.* 75:15–22

44. Klopfer, P., Gottlieb, G. 1962. Imprinting and behavioral polymorphism. *J. Comp. Psychol.* 55:126–30

45. Klopfer, P. H., Hailman, J. P. 1965. Habitat selection in birds. *Adv. Study Behav.* 1:279–303

46. Konishi, M. 1965. The role of auditory feedback in the control of vocalization in the White-crowned Sparrow. *Z. Tierpsychol.* 22: 770–83

47. Lack, D. 1963. Cuckoo hosts in England. *Bird Study* 10:185–201

48. Lack, D. 1968. *Ecological Adaptations for Breeding in Birds.* London: Methuen. 409 pp.

49. Lenneberg, E. H. 1967. *Biological Foundations of Language.* New York: Wiley

50. Levine, S., Lewis, G. W. 1959. Critical period for the effects of infantile experience on the maturation of a stress response. *Science* 129:42–43

51. Leyhausen, P. 1965. Ueber die Funktion der relativen Stimmungshierarchie. *Z. Tierpsychol.* 22:412–94

52. Liversidge, R. 1970. The biology of the Jacobine Cuckoo *Clamator jacobinus*. *Ostrich Suppl.* 8:117–37

53. Löhrl, H. 1959. Zur Frage des Zeitpunktes einer Praegung auf die Heimatregion beim Halsbandschnaep-

per (*Ficedula albicollis*). *J. Ornithol.*
100:132–40

54. Lorenz, K. 1935. Der Kumpan in der
Umwelt des Vogels. *J. Ornithol.*
83:137–213, 289–413

55. Lorenz, K. 1970. *Studies in Animal and
Human Behavior,* Vol. 1. Cambridge:
Harvard Univ. Press

56. Lukschanderl, L. 1971. Zur Verbreitung und Oekologie der Gross-
trappe (*Otis tarda*) in Oesterreich. *J.
Ornithol.* 112:70–93

57. Marler, P. 1963. Inheritance and learning in the development of animal vocalizations. In *Acoustic Behaviour of Animals,* ed. R.-G. Busnel, 228–243. Amsterdam: Elsevier

58. Marler, P. 1967. Comparative study of
song development in sparrows. *Proc.
XIV Int. Ornithol. Congr., Oxford,
1966,* 231–44

59. Marler, P., Tamura, M. 1962. Song
"dialects" in three populations of
White-crowned Sparrows. *Condor* 64:
368–77

60. Mayr, E. 1970. Evolution und Verhalten. *Verh. Dtsch. Zool. Ges. Koeln,*
322–36

61. McPhail, J. D. 1969. Predation and
the evolution of a stickleback (*Gasterosteus*) *J. Fish. Res. Bd. Can.* 26:
3183–3208

62. Medway, Lord 1970. A ringing study of
the migratory Brown Shrike in West
Malaysia. *Ibis* 112:184–98

63. Moreau, R. E. 1972. *The Palaearctic-
African Bird Migration Systems.* London: Academic

64. Mulligan, J. A. 1966. Singing behavior
and its development in the Song Sparrow, *Melospiza melodia. Univ. Calif.
Publ. Zool.* 81:1–76

65. Nicolai, J. 1964. Der Brutparasitismus
der Viduinae als ethologisches Problem.
Z. Tierpsychol. 21:129–204

66. Nicolai, J. 1967. Rassen- und Artbildung in der Viduinengattung *Hypochera. J. Ornithol.* 108:309–19

67. Nottebohm, F. 1969. The song of the
Chingolo, *Zonotrichia capensis,* in Argentina: description and evaluation of a
system of dialects. *Condor* 71:299–315

68. Nottebohm, F. 1970. Ontogeny of bird
song. *Science* 167:950–56

69. Nottebohm, F., Selander, R. K. 1972.
Vocal dialects and gene frequencies in
the Chingolo Sparrow (*Zonotrichia capensis*). *Condor* 74:137–43

70. Partridge, L. 1974. Habitat selection in
titmice. *Nature* 247:573–74

71. Payne, R. B. 1973. Behavior, mimetic
songs and song dialects, and relationships of the parasitic Indigobirds
(*Vidua*) of Africa. *Ornithol. Monogr.*
11:1–333

72. Peitzmeier, J. 1951. Zum oekologischen
Verhalten der Misteldrossel (*Turdus v.
viscivorus* L.) in Nordwesteuropa.
Bonn. Zool. Beitr. 2:217–24

73. Pimentel, D., Smith, G. J. C., Soans, J.
1967. A population model of sympatric
speciation. *Am. Nat.* 101:493–504

74. Poulson, G. W. 1965. Maturation of
evoked responses in the duckling. *Exp.
Neurol.* 11:324–33

75. Rabinowitch, V. 1968. The role of experience in the development of food preferences in gull chicks. *Anim. Behav.*
16:425–28

76. Rabinowitch, V. 1969. The role of experience in the development and retention
of seed preferences in Zebra Finches.
Behaviour 33:222–36

77. Schutz, F. 1965. Sexuelle Praegung bei
Anatiden. *Z. Tierpsychol.* 22:50–103

78. Schutz, F. 1970. Zur sexuellen Praeg-
barkeit und sensiblen Phase von Gaen-
sen und der Bedeutung der Farbe des
Praegungsobjektes. *Verh. Dtsch. Zool.
Ges. Wuerzburg 1969,* 301–6

79. Schutz, F. 1972. Frueherfahrung und
Verhalten. *Verh. Schweiz. Naturforsch.
Ges.* 71–84

80. Selander, R. K. 1971. Systematics and
speciation in birds. In *Avian Biology,*
ed. D. S. Farner, J. R. King, Vol. I,
57–147. New York: Academic

81. Serventy, D. L. 1967. Aspects of the
population ecology of the Short-tailed
Shearwater *Puffinus tenuirostris. Proc.
XIV Int. Ornithol. Congr. Oxford, 1966,*
165–190

82. Sluckin, W. 1973. *Imprinting and Early
Learning.* Chicago: Aldine 2nd ed.

83. Smith, F. V., Nott, K. H., Yarwood, A.
1970. Brain protein synthesis and the
approach response of chicks to a visual
stimulus. *Brain Res.* 21:79–90

84. Sossinka, R. 1972. Besonderheiten in
der sexuellen Entwicklung des Ze-
brafinken *Taeniopygia guttata castanotis*
(Gould). *J. Ornithol.* 113:29–36

85. Stresemann, E., Nowak, E. 1958. Die
Ausbreitung der Tuerkentaube in
Asien und Europa. *J. Ornithol.*
99:243–96

86. Strobel, M. G., Baker, D. G., Macdonald, G. E. 1967. The effect of embry-
onic X-irridation on the approach and
following response in newly hatched
chicks. *Can. J. Psychol.* 21:322–28

87. Tast, J. 1968. Changes in the distribution, habitat requirements and nest-sites of the Linnet, *Carduelis cannabina* (L.), in Finland. *Ann. Zool. Fenn.* 5:159–78

88. Thielcke, G. 1974. Stabilitaet erlernter Singvogel-Gesaenge trotz vollstaendiger geographischer Isolation. *Vogelwarte* 27:209–15

89. Thorpe, W. H. 1958. The learning of song patterns by birds, with special reference to the song of the Chaffinch *Fringilla coelebs*. *Ibis* 100:535–70

90. Thorpe, W. H. 1963. *Learning and Instinct in Animals.* London: Methuen. 2nd ed.

91. Wecker, S. C. 1963. The role of early experience in habitat selection by the Prairie Deer Mouse, *Peromyscus maniculatus bairdi. Ecol. Monogr.* 33:307–25

92. Wiens, J. A. 1972. Anuran habitat selection: early experience and substrate selection in *Rana cascadae* tadpoles. *Anim. Behav.* 20:218–20

THE RUMEN MICROBIAL ECOSYSTEM[1]

R. E. Hungate
Department of Bacteriology, University of California, Davis, California 95616

Ruminants are earth's dominant herbivores, due in part to the evolution within this group of a mechanism utilizing microorganisms to digest plant components not susceptible to attack by ruminant enzymes.

THE RUMEN RETICULUM

Ingested food is retained in the capacious rumen reticulum and there undergoes extensive microbial digestion and fermentation. Mastication during feeding comminutes the food into particles varying in volume from one to 10^3 mm^3 and as long as 10 cm, the degree of comminution being greater for young, lush, green feeds than for older and drier vegetation. The particle size is further reduced by rumination, especially with dry feeds; digesta in the reticulum (see Figure 1) are regurgitated—the liquid swallowed, the solids rechewed—and returned to the rumen. Only particles with a volume of about 5 mm^3 or less can pass from the reticulum into the next compartment, the omasum. Large particles are usually more abundant in the dorsal rumen contents, particularly when the feed is unchopped hay.

  The contents of the rumen comprise 8–15% (w/w) of the total animal and contain 10–18% (w/v) dry matter. Contractions of the rumen wall at 1–2 min intervals mix the contents; after digestion has proceeded for some time, or when the feed has been ground and pelleted, the rumen contents are fairly homogeneous.

  Saliva containing 100–140 millimolar (mM) bicarbonate, chiefly sodium bicarbonate, and 10–50 mM phosphate is secreted more or less continuously by the salivary glands, most rapidly during feed consumption and rumination, but also at other times. The total daily quantity is approximately 1–3 times the volume of the rumen contents. The relatively large quantity of alkaline saliva partially neutralizes

[1]This article describes the rumen ecosystem as a basis for comparison with other ecosystems. The description is based primarily on the author's experience. The references are a small fraction of the total literature, most of which is cited in more extensive reviews (42, 56, 96, 111, 133).

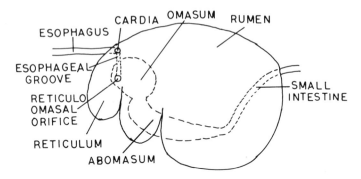

Figure 1   Diagram of the ruminant stomach.

the fermentation acids and maintains favorable conditions for their continued production. After feed is consumed, the pH of rumen contents decreases, the extent depending on the nature of the feed and the interval between feedings, then gradually rises to the prefeeding pH of about 6.5. The undissociated acids are inhibitory to many rumen bacteria and probably to many exotic microbes (24). The saliva contains about 10 mM nitrogen as urea, the latter decomposing to CO_2 and NH_3 in the rumen.

At intervals depending on the rate of fermentation, gas is belched (eructated) through the esophagus and mouth, some of it momentarily entering the lungs (57).

In grazing ruminants, forage is ingested during approximately one third of the 24-hr day, chiefly in morning and evening, with another third spent in rumination if the feed is coarse. The rumen never empties, even in starvation. Under experimental and some production conditions, the feed can be provided at intervals as short as 1 hr (140), supplying substrate almost continuously. The temperature and the absence of O_2 are other constant features of the rumen environment.

At periodic intervals the orifice between the reticulum and the omasum opens, and liquid and small particles of digesta flow into the omasum. Larger particles are retained by conical tooth-like cornified epithelial projections around the orifice; they cannot proceed further along the alimentary tract until rumination has comminuted them to a passable size. In the omasum, the fermentation acids and bicarbonate are absorbed by leaf-like layers of tissue (81) between which the digesta pass. Their absorption reduces the buffering capacity of the digesta (von Engelhardt & Moir, in 111), permitting greater acidity in the next compartment of the stomach, the abomasum, in which typical acid gastric digestion occurs. Posterior to the omasum the ruminant alimentary tract is comparable to that in most other animals except that the pancreatic juice contains an exceptionally high concentration of ribonuclease (16), an adaptation to the abundant ribosomes in the bacteria to be digested.

The reticulum has an esophageal groove connecting the entrance from the esophagus with the exit to the omasum. The act of suckling causes a reflex contraction of ridges along the groove, twisting them into a closed tube connecting esophagus and omasum. Milk is shunted past the rumen into the omasum, and through it into the abomasum, thus escaping fermentation. The reflex can be retained in the adult stage by continued suckling (127).

The net effect of the various supply and removal systems of the ruminant stomach is to provide ideal conditions for microbial growth. A teeming microcosm of protozoa and bacteria develops in the rumen, and the products of their activities constitute the principal food of the host. These products are 1. the fermentation acids, absorbed and oxidized by the host, and 2. the microbial cells passing from the rumen and digested in the abomasum and intestine.

THE PROTOZOA

Between 10^{10} and 10^{11} bacteria and 10^5 to 10^6 protozoa inhabit each milliliter of rumen contents. The protozoa include a few small flagellates, but consist chiefly of two ciliate groups, holotrichs and entodiniomorphs (55, 124). The morphological distinctiveness of the entodiniomorphs and their restriction to the rumen attest to a long evolution isolated in the rumen habitat. The entodiniomorphs include some of the most complex protozoa, a seeming anomaly to the usual reduced somatic differentiation in parasites. But the rumen protozoa are not parasites living off host tissues; they are free-living, sequestered in a highly favorable environment external to the host tissues, subjected to intense competition from myriads of accompanying microbes. Their morphological diversity and complexity can be compared to that of tropical ecosystems, where favorable environments support great productivity and species diversity.

The holotrichs use soluble sugars and convert them rapidly into starch (129), chemically indistinguishable from plant starch (67), then use the stored polysaccharide during periods when sugars are not available. The hypothesis that sugar concentration in the rumen limits the rate of the rumen fermentation is supported by the observation that when the holotrichs are provided with excess sugar they store starch until they burst (146). They have not evolved a mechanism to control the storage process.

The holotrichs are normally abundant in the rumen of animals on forage rations. They may be prominent also in animals fed a predominantly starch diet, presumably because the concentration of soluble carbohydrates is kept fairly high by rapid digestion of starch by exocellular bacterial enzymes, or possibly by leakage of sugars from starch-filled entodiniomorphs. An attraction of highly motile rumen bacteria to the vicinity of the mouths of entodiniomorphs has been observed by the author and interpreted as evidence for a loss of sugar to the environment, and Coleman (46a) has shown that much of the digested starch appears in the medium as sugar.

Early results on clone cultures of several species of entodiniomorphs (91) indicated that cellulose is digested and some of the sugar formed is stored as starch (90). In *Entodinium,* the results were negative, although a weak cellulase in mixed species of *Entodinium* separated from rumen contents was found subsequently (22). The amount of cellulose digested by the rumen protozoa is probably small compared to that digested by the bacteria, since the actively cellulolytic protozoa constitute only a small part of the protozoan population.

Many of the entodiniomorphs ingest starch, often so rapidly that, within a few seconds after it is supplied, several grains can be seen in the endoplasmic sac of most individuals. As with cellulose, much of the sugar produced is stored as small starch grains in the ectoplasm and used as a reserve carbohydrate.

Early microscopic observations (73) indicated that bacteria are ingested by the protozoa. Removal of the protozoa (defaunation) increased the numbers of bacteria as compared to faunated controls (61). Recent in vitro results of Coleman (46, 47, 111), from feeding isotopically labeled bacteria, substantiate the earlier findings, though *Epidinium,* ingesting relatively few bacteria, may derive its protein from ingested plant materials. The latter lose their green color within a few minutes of ingestion. Some of the entodiniomorphs prey on others, and only certain combinations of species occur in a particular animal (60, 61, 120).

Clone cultures of most of the entodiniomorph species have been obtained where attempted, but none have been gnotobiotic. The holotrichs have been more difficult to culture, though *Dasytricha* has been grown with as few as three accompanying bacteria (44).

THE BACTERIA

The rumen bacteria are almost all obligately anaerobic; the culture count for aerobic and euryoxic bacteria is about 1000 times less than that for obligate anaerobes. In cattle on forage rations, the culture count is usually only 10–30% of the direct count, but on concentrate rations most of the bacteria seen can be cultivated.

The oxidation-reduction potential of rumen contents as measured with a platinum electrode is as low as -0.35 V, a potential at which the concentration of O_2, if in equilibrium with the system, is less than 10^{-22} M. In practice, such low potentials are obtained by rigorous elimination and exclusion of air during preparation of culture media (98), coupled with use of reducing agents such as cysteine, sodium sulfide, or tetrathionate (115). Techniques for excluding air are essential; many rumen bacteria are instantly killed by O_2. The originally successful anaerobic methods (93) involved exclusion of air from the roll tube equivalent of a pour plate. Subsequently, syringe injection (99), plastic anaerobic glove boxes (9), and techniques and instruments for anaerobic streaking have been developed (83).

As would be expected from the composition of plants, carbohydrate is the chief substrate used by rumen bacteria, and, as in the protozoa, it can serve also as a storage reserve (41, 86, 110). Other nutritional requirements (26) are peculiar to the rumen. Success in cultivation depended originally on provision of rumen fluid as part of the medium; subsequently, equally effective artificial media have been developed (32) as required factors in rumen fluid have been identified (31).

A few rumen bacteria can grow on an inorganic medium plus carbohydrate (80), but most strains need also acetic, isobutyric, isovaleric, and 2-methylbutyric acids. The latter three are formed in the rumen by bacterial fermenters of valine, leucine, and isoleucine, respectively, and in turn are used by other bacteria for resynthesis of these same amino acids (4, 5). The pathway of resynthesis is usually the reverse of breakdown. The acids are needed also for synthesis of certain plasmalogens composing the bacterial structure (6, 153). Acetic acid is assimilated into cell material (64). Another interesting nutrient in the rumen is coenzyme M, required by *Methanobacterium ruminantium,* and recently identified as 2-mercaptoethanesulfonic acid (150).

Amino acids and soluble proteins are rapidly fermented to form various acidic fermentation products, NH_3, H_2, and CO_2 (35, 39, 117). The rumen concentration of amino acids is very low (135, 160). Extensive studies by Bryant (103) have shown that the predominant nitrogen source is ammonia. Pure cultures of some bacteria for which NH_3 is essential can assimilate small amounts of amino acids (80, 89), and for a few bacteria, amino acids are essential (28). Oligopeptides are more readily assimilated than are amino acids (134, 159).

Carbon dioxide is required by the bacterial species, forming succinate as an important fermentation product (34). Pyruvate is carboxylated to form oxaloacetate, converted further to malate, fumarate, and succinate. The moles of succinate formed (and CO_2 fixed) amount to as much as 80% of the moles of glucose fermented (93). Carbon dioxide is an essential nutrient for a number of other rumen bacteria not producing succinate (50), but the amount required is much less. Sodium and some other inorganic ions are essential for many rumen bacteria (31).

Of the cellulolytic bacteria, *Bacteroides succinogenes* adheres most closely to the fibrous substrate (118); no extracellular cellulase has been found in most of the strains studied. Glucose, cellobiose, cellulose, and pectin are fermented. Other di- and polysaccharides are used by some strains, but this species is predominantly a stenotrophic digester of fiber. *Ruminococcus albus,* fermenting only cellobiose, cellulose, and xylose, must also be a digester of fiber in the rumen. Cells of this species appear to require some sort of preferred position near the cellulose to be digested, since they are unable to develop to any extent in cellulose agar inoculated with a mixed culture, but can grow in liquid under the same circumstances. Pure cultures grow well in cellulose agar, completely digesting the cellulose. Some strains of *R. albus* produce a constitutive cellulase with a molecular weight of about 2×10^6, so large that it sediments at earth's gravity, producing an elliptical rather than a circular zone of clearing of the cellulose in the thin layer of cellulose agar that lines the wall of vertically incubated roll tubes. It is difficult to conceive how such a large exocellular enzyme could be ecologically advantageous. Other cellulases are also formed by these strains, with the particular combination of enzymes varying from one strain to another and with time in the same strain.

Some cellulolytic bacteria are less stenotrophic. *Clostridium lochheadii* can use starch and the derivatives from it, and is also very proteolytic. Its cellulase is so active that the organism can grow in cellulose agar in the presence of many other bacteria. *Eubacterium cellulosolvens* is cellulolytic, but can use also a number of sugars not derived from cellulose. *Butyrivibrio fibrisolvens* is distinctly eurytrophic. Many strains are cellulolytic, but some are very weakly so, and all strains can use a wide variety of sugars, as well as inulin, xylan, pectin, and starch.

Noncellulolytic stenotrophic bacteria include *Veillonella alkalescens* and *Megasphaera elsdenii,* using lactate as an energy source (15), and *Methanobacterium ruminantium* and *M. mobilis,* capable of using only CO_2 plus H_2 or formate.

Bacteroides amylophilus and *Succinimonas amylolytica* are extreme stenotrophs, using only starch and its derived saccharides. *Anaerovibrio lipolytica* digests fats and ferments the glycerol (77). A recently discovered new species of *Desulfovibrio* from the rumen (88) is unique in being able to metabolize sugars, but, aside from sulfate, its rumen substrate is unknown. An anaerobic rumen mycoplasma appears to grow at the expense of nonviable bacteria in the rumen (142).

Besides *Butyrivibrio,* other important eurytrophic rumen bacteria are *Selenomonas ruminantium, Bacteroides ruminicola, Lachnospira multiparus,* and *Streptococcus bovis. Selenomonas* is the only abundant organism forming propionate in the bovine rumen, and is presumably responsible for decarboxylation of the rumen succinate to propionate and CO_2; in the sheep this function may be served by *Veillonella alkalescens.* In mixed cultures, *Selenomonas* forms propionate from the succinate produced by *Bacteroides succinogenes* from cellulose. Even though *Bacteroides* does not produce an extracellular cellulase, some product is formed from cellulose and released in sufficient amounts to support the development of the noncellulolytic *Selenomonas* to the extent of about 25% of the mixed culture (Wolin et al, in 111).

Bacteroides ruminicola, one of the most abundant eurytrophs of the rumen (29), includes the subspecies *brevis,* requiring heme (33) for synthesis of a cytochrome used anaerobically for phosphorylation during the reduction of fumarate to succinate (C. A. Reddy, Rumen Function Conference, Chicago, 1973, unpublished). The subspecies *ruminicola* synthesizes the heme. Strains of *Lachnospira multiparus* show much variability in their carbohydrate substrates, but all ferment pectin, esculin, and salicin. In agar, this organism grows as filamentous chains of rods that seem well adapted to penetrate the pectic middle lamella of plant cell walls. It is often quite abundant in cattle fed on fresh alfalfa. *Streptococcus bovis* ferments starch and a number of disaccharides but is not abundant in the normal rumen.

Some large prokaryotes can be microscopically identified. A large selenomonad retains its size in pure culture (138). Quin's oval and Eadie's oval, two large prokaryotic rumen organisms that have not been named, have been grown in vitro for an extended time (125, 126), but in the presence of other bacteria. Many of their characteristics have been studied through microscopic examination and biochemical tests of organisms obtained directly from the rumen (23, 65).

THE FERMENTATION

The carbohydrate substrates in the rumen are fermented to form microbial cells and microbial waste products, chiefly acetic, propionic and butyric acids, CO_2, and methane. These volatile fatty acids (VFA), though microbial waste products, are essential to the host. They are absorbed and oxidized. Propionate is a key product because it is the only fermentation acid convertible into carbohydrate by the ruminant. Though ruminant needs for carbohydrate are relatively less than those of nonruminants, some carbohydrate is essential, especially during periods of lactation.

When the rumen microbes are grown in pure culture, the waste products accumulating are more diverse than those mentioned above. Various species form lactate, ethanol, succinate, formate, or H_2, as well as one or more of the end products found in the rumen. Are these other fermentation products not formed in the rumen, or are they formed, yet fail to accumulate because they are used by accompanying organisms?

This question has been answered for lactate, ethanol, and succinate by measuring their concentration (pool size) in the rumen liquid and the turnover rate constant of the pool, calculated from the rate of disappearance of added isotope. The product of the two values is the turnover or flux, i. e. the amount of the intermediate produced and used. Results for these possible intermediate materials are shown in Table 1, together with the product formed and its percentage of the rumen formation of that product.

Lactate is converted chiefly to acetate and is relatively unimportant in relation to the total amount of acetate formed (107), except on concentrate rations, where it is more important as an intermediate and may accumulate to some extent immediately after feeding (121).

Ethanol is also converted chiefly to acetic acid, but the rate is very slow; ethanol seldom accumulates unless the rumen is quite acid (7). In contrast, succinate as an intermediate accounts for a significant part of the propionate formed in the rumen (21). As much as one fourth of it is assimilated into microbial cell material, possibly in the synthesis of glutamic acid (63).

The turnover rate constants of H_2 and formate could not be measured because H_2 exchanges rapidly with water, and formate with CO_2 and numerous other cell constituents. The values in Table 1 have been calculated indirectly as follows: The rate of methanogenesis by rumen liquid was shown to be a nearly linear function of the concentration of dissolved H_2 (97) when formate, H_2, or glucose was fed at a low constant rate. The concentrations of dissolved H_2 and formate were determined from their concentration in sterile balanced salt solution in dialysis sacs equilibrated with the incubating rumen liquid. From the values obtained plus those from additional measurements at higher concentrations, a K_s (substrate half-saturation constant) of 1 μM was calculated for H_2 and 30 μM for formate (104). Their average concentrations in dialysate equilibrated with rumen contents in vivo were 1 μM and 12 μM, respectively. Thus H_2 was metabolized in the rumen at half-maximal rate, and formate at 12/30 of half-maximum. In earlier studies (39), the maximal rates were 1.42 $\mu mol \cdot g^{-1} \cdot min^{-1}$ for H_2 and 0.63 for formate. From these values the rumen rate of conversion is 710 nmol $H_2 \cdot g^{-1} \cdot min^{-1}$ and 126 nmol formate $\cdot g^{-1} \cdot min^{-1}$, giving the pool turnover rate constants of 710/min and 10/min, respec-

Table 1   Parameters for possible intermediates in the rumen fermentation

Intermediate	Concentration nmol ml^{-1}	Turnover rate constant min^{-1}	Flux nmol ml^{-1} min^{-1}	Product formed	Rumen production accounted for %
lactate	<12	0.03	0.36	acetate	<1
ethanol	not measurable	0.003		acetate	
succinate	4	10	40	propionate	33
H_2	1	710	710	methane	100
formate	12	10	126	H_2	18

tively, included in Table 1. Formate forms 18% of the H_2 produced in the rumen. The rest comes from the split of pyruvate into acetyl-CoA, H_2, and CO_2, and from NADH. The metabolism of the methanogenic bacteria keeps the rumen concentration of H_2 so low that production of H_2 from NADH is exergonic (Wolin, in 111), and H_2 is drained into the methanogenic sink.

The K_s of a pure culture of *Methanobacterium ruminantium* is 1 μM for H_2 and >200 μM for formate. This high K_s for formate, as compared to the 30 μM for rumen contents, indicates that formate is not a substrate for the methanogenic bacteria in the rumen population even though they can use it axenically. Hydrogen gas appears to be the sole substrate for methanogenesis in the natural ecosystem.

The unimportance of ethanol and lactate in the normal rumen fermentation can be accounted for by the biochemistry of anaerobic production of ATP. During the conversion of hexose to lactate, pyruvate is formed via glycolysis, with net production of one molecule of ATP (2 ATP/glucose) and NADH + H^+, the latter then reducing pyruvate to lactate. In the ethanolic fermentation, the pyruvate is split to acetyl-CoA, CO_2, and H_2, and this hydrogen, plus that from glycolysis, reduces the acetyl-CoA to ethanol. In both reactions, pyruvate or its product uses all the available hydrogen, and the hexose is stoichiometrically recovered as lactate or as ethanol plus CO_2. With methanogenesis as a repository for hydrogen in the rumen, pyruvate is not needed for this purpose; through conversion to acetyl phosphate it can yield an additional ATP (4ATP/hexose). The cell yield is doubled, even without taking into account the quantity of methanogenic cells formed.

Experimental support for this increased cell yield through methanogenesis has been demonstrated in two-membered cultures of *Methanobacterium ruminantium* grown with *Ruminococcus flavefaciens* on cellulose (Wolin, in 111). Similarly, *Vibrio succinogenes* uses the H_2 from *Ruminococcus albus* (106) to reduce fumarate to succinate, significantly increasing the cell yield from hexose.

Other microbial interactions demonstrated experimentally with two-membered cultures include the production by *Bacteroides ruminicola* of sufficient methionine to support *Ruminococcus flavefaciens* as 20–40% of the total population (85). *Bacteroides succinogenes,* growing on cellulose, gratuitously produces enough sugar to support growth of *Selenomonas* as 20% of the total population, with the leaked sugar probably a greater percentage since *Selenomonas* is much the larger cell (144). This leakage must occur also in the rumen since much of the cellulose is digested by unattached cells (1).

Because methane is not metabolizable as a source of energy by the ruminant (57), its production, though increasing the microbial cell crop, decreases the oxidizable substrate available to the ruminant. The overall effect of rumen methanogenesis depends on whether metabolism and growth of the host are limited by the availability of ATP derived through oxidation of VFA or by the amount of microbial protein available.

Supplementation of the diet with readily digestible protein cannot be used to test whether ruminant growth is limited by the amount of available microbial protein. Fed protein is rapidly digested (20) and fermented in the rumen. To feed additional protein, it must be protected in some way or introduced posterior to the rumen. The latter has been accomplished by rearing a sheep to maturity on milk suckled from a bottle (127), thereby preserving the reflex closure of the esophageal groove and shunting the protein directly to the abomasum. At several months of age the sheep was fed also the usual forage rations. Additional suckled protein increased nitrogen assimilation (128). Also, venous infusion of amino acids can increase the protein in milk (66). Protein treated with formaldehyde (74) or tannin (59) is less digestible in the rumen, but is digestible in the abomasum and small intestine and can increase growth and wool production. Maximal microbial cell production from available carbohydrate usually favors ruminant growth, though in some cases the propionate needed for carbohydrate production may be limiting.

Formation of succinate instead of propionate as a final product of most pure cultures of rumen bacteria (except *Selenomonas* and *Veillonella*) contrasts with the formation of propionic acid as the chief product of *Propionibacterium* growing in cheese, the best-known microbial ecosystem in which propionic acid is a final product. In both fermentations, pyruvate must be converted to a 4C-dicarboxylic acid (108, 131). In *Propionibacterium,* a 1C unit is transferred from methylmalonyl-CoA to pyruvate to form oxaloacetate without release of CO_2 as an intermediate. Because CO_2 is so abundant in the rumen, such conservation has no selective value and the CO_2 remains in the dicarboxylic acid as succinate. Succinate may arise by a pathway not requiring ATP, possibly by reductive carboxylation of pyruvate to malate (108) or by carboxylation of phosphoenol pyruvate, with later transfer of $\sim$P to guanosine diphosphate (87), preserving its energy.

THE CONTINUOUS FERMENTATION MODEL

The rumen is not as precisely regulated as a continuous laboratory fermentor (78), but a model for continuous fermentation can be advantageously applied and its discrepancies noted.

With animals fed once daily at a specific time, variations from the model occur within the day, but the rumen is in approximately the same state at a particular time on any day. The fluctuations in rates during the day can be estimated from numerous measurements and the average rate compared with integrated parameters such as feed-feces difference or amount of methane produced in the intact animal.

By feeding equal aliquots of the daily feed at frequent, evenly spaced intervals, a close approach to a continuous and constant rumen fermentation is attained. To reduce sampling error, Sutherland et al (147) mixed the rumen contents within a fistulated sheep with a pump before sampling and showed that the VFA and ammonia concentrations remained constant within narrow limits. Though such heroic methods for homogenization prior to sampling are not usually feasible, neither are they essential for most purposes; the normal mixing by rumen contractions maintains enough uniformity that a fairly representative composite can be obtained by sampling from different parts of the rumen (25, 62, 101, 148).

In the continuous fermentation model (95), the rate at which liquid enters the fermentor (rumen), i.e. the feed or dilution rate, is the same as the rate at which it leaves, the volume (and presumably everything else) remaining constant. Instant mixing of any entering material is assumed. The fraction entering or leaving per unit time is the specific turnover rate constant, μ. Its reciprocal, $1/\mu$, is the average retention time, the time required to feed a volume equal to that in the fermentor. For the rumen, the total volume can be expressed as 1; μ, the fraction/unit time, can refer to the microbiota or liquid leaving the rumen, $-\mu$, or to the fraction of feed entering, $+\mu$. With dry feeds, the rate of secretion of saliva determines the specific turnover rate constant for rumen fluid, since rumen absorption of ingested water tends to keep the osmotic pressure of rumen contents about the same as that of blood (157, 158).

For any fed material not produced or used in the rumen, μ represents the rate of entrance or exit, but the residence time of any particular batch of material varies from zero for that leaving immediately after entering, to almost infinity for that leaving last. The rate of exit for a soluble marker such as polyethylene glycol (PEG) shows this variation in residence time and discloses the turnover rate constant, μ. PEG, added to rumen contents and instantly mixed, leaves the rumen at a rate depending on its concentration and on μ, i.e. $dx/dt = \mu x$. If the initial concentration is taken as unity, the fraction x remaining in the rumen at any time t is $x = e^{-\mu t}$. If x_0 is taken as an initial absolute amount, the amount remaining at any time is $x_0 e^{-\mu t}$.

The curve $x = e^{-\mu t}$ shows the fraction of soluble pulse marker such as PEG remaining in the rumen at any time after its addition. It is plotted as curve A in Figure 2 for a rumen containing 30 liters of fluid in an animal secreting 30 liters of saliva per day, including any consumed water passing through the rumen, i.e. $\mu = 1/\text{day}$ or $1/24$ hr, the average residence time being one day. The rate of exit determines the residence time, represented by the curve, $dx/dt = \mu x$, or, since $x = e^{-\mu t}$, $dx/dt = \mu e^{-\mu t}$. This is plotted as curve B in Figure 2, and shows the distribution of residence time for the fluid in the rumen during the steady state. The shape of the curve is the same as that of A, but the absolute values are less by the factor μ and refer to a rate rather than to an amount.

The area under curve B, $\quad \int (dx/dt)dt = \int_0^\infty \mu e^{-\mu t} = 1,$ \hfill 1.

represents the total initial material. The fraction of material with a residence time of t is this same integral between the values of $t = 0$ and $t = t$, or $1 - e^{-\mu t}$.

PEG is generally accepted as a good (14) but not infallible (43) marker for determining μ for the liquid in the rumen. It also has been assumed (96) to measure the turnover rate constant for the rumen microbes and small plant particles; this includes the assumption that the liquid and the exiting small microbes and particles are representative of these materials in the total rumen.

Passage of the large particles has been more difficult to assess. Dyed hay has been used (38) for the more resistant fibrous material, as has its lignin content (13). Chromic oxide gives smaller μ values than PEG, and presumably marks the solids, but it is uncertain exactly which components are represented.

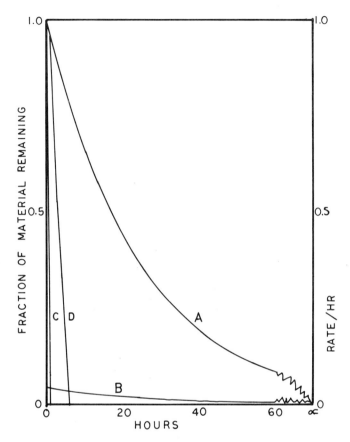

Figure 2 Curves showing the fraction of material remaining in the rumen at various times (left-hand ordinate) and the residence time distribution (right-hand ordinate) for a rumen with a specific turnover rate constant of 1/24 hr. Curve *A* represents the fraction remaining when the material disappears only by passage, and curve *B*, the residence time distribution curve when disappearance is only by passage. Curve *C* illustrates the fraction remaining when the soluble materials leave by fermentation and by passage, and curve *D*, the fraction remaining when the intermediate materials leave by fermentation and by passage.

From the amount of solid digesta in the rumen and the amount of feed consumed per day, an average turnover rate constant can be calculated by dividing the daily weight of food solids by the weight of residual food solids in the rumen, provided that the extent to which the solids have undergone digestion is known, i.e. that the weight of the rumen solids can be corrected back to the weight of the food initially containing them. The following model has been developed for estimating the extent of digestion of forage in a continuous fermentation system.

EXTENT OF DIGESTION: MODEL

The time course of digestion of alfalfa hay has been followed in vitro by subjecting small amounts to attack by an excess of rumen microorganisms under conditions of limiting substrate, as in the normal rumen (96). By difference from a control without added forage, the time course of the microbial utilization can be estimated from the disappearance of forage with time or, more easily, by the fermentation gas produced. This has been done for alfalfa hay (curve A in Figure 3). It is assumed that this curve shows for any particular time, t, the amount of hay that has disappeared due to microbial utilization (measured by fermentation products formed). If the initial amount of the fermentable food is taken as unity, the total amount of fermentation products formed can also be considered as unity, and the fraction of food unfermented at any time is 1–(the fraction fermented). This is shown as curve B in Figure 3. The fraction of fermentable food in the rumen at any time is the product of the fraction not yet passed multiplied by the fraction not yet fermented, i.e. $x = fx_p \cdot fx_d$, in which $fx_p = e^{-\mu t}$. The subscripts are used to distinguish disappearance by passage from disappearance by fermentation.

The term fx_d is the equation represented by curve B in Figure 3. This equation is not known, but it has been shown (2, 96) that the microbes attack the soluble fermentable materials in the hay during the first hour after ingestion, and that the fermentation products appear at an almost linear rate. A negligible amount of fiber disappears during this time.

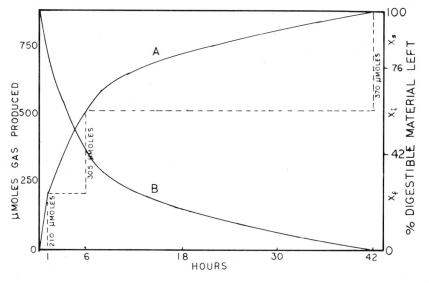

Figure 3   Curves showing the amount of gas produced during a 42-hr fermentation of 137.5 mg of ground alfalfa hay (curve *A*) and the fraction of digestible material remaining (curve *B*).

The rate of fermentation is almost linear also in the period 1–6 hr after ingestion. The explanation for this near linear rate is not clear, but is believed to result in part from fermentation of easily digested insoluble components of the hay, and in part from a decreasing rate of metabolism of starch stored during the first hour, coupled with a gradually increasing rate of attack on the fiber.

The final part of the curve represents the gradually decreasing rate at which fiber is digested between 6 and 42 hr, after which further fermentation is assumed to be negligible.

Because the equation for the curve fx_d is not known, and because two parts of the curve correlate with identifiable components of the hay, curve B is assumed to result from utilization of three different categories of materials in the feed—soluble, intermediate, and fiber; the fraction of each that escapes digestion due to passage from the rumen is calculated separately. The initial amount of each is assumed to be unity. The fraction of the soluble materials is designated x_s, the intermediate material x_i, and the fiber x_f.

The equation showing the fraction of soluble material unfermented is $x_s = 1 - t$, for values of $t \leq 1$, and applies to the first hour after the hay is ingested. The equation, $x_i = (6 - t)/5$, for values $1 < t < 6$, covers the second near linear portion of curve B and shows the unfermented fraction of the intermediate material. The final portion of the curve, in which the fiber is fermented, agrees well with a harmonic series in which the decrease in the rate of fermentation in each successive hour after 6 hr is proportional to $1/1, 1/2, 1/3, \ldots 1/36$, respectively, the material being regarded as completely digested after 42 hr (96). To plot this model curve, the sum of the rate-decreases during the 36 hr was regarded as the initial rate. The rate during each hour was then calculated, the rates summed, and the fraction of the sum constituted by each rate taken to represent the fraction of the total fiber digested during that hour. The theoretical curve closely matched the 6–42 hr portion of curve B in Figure 3 (96).

By considering simultaneously the disappearance from the rumen due to both turnover and fermentation, the fraction of a given ingested amount of soluble material remaining in the rumen at any time during the first hour after its ingestion is $x_s = e^{-\mu t}(1 - t)$, plotted as curve C in Figure 2. The rumen residence-time distribution curve for this material is $\mu x_s = \mu e^{-\mu t}(1 - t)$, and the area under this curve,

$$\int_0^1 \mu e^{-\mu t}(1 - t) \mathrm{d}t = \underbrace{[e^{-\mu t}(t + 1/\mu - 1)]}_{t=1} - \underbrace{[e^{-\mu t}(t + 1/\mu - 1)]}_{t=0} \qquad 2.$$

is the fraction of the soluble material passing from the rumen without being fermented. The fraction fermented, $1 - \int_0^1 \mu e^{-\mu t}(1 - t)\mathrm{d}t$, is shown in Table 2 for μ values of $1/12$, $1/24$, $0.93/24$, and $1/48$ hr.

In similar fashion, the amount of the intermediate material, x_i, in the rumen between 1 and 6 hr is $e^{-\mu t}(6 - t)/5$, curve D in Figure 2, and the residence-time distribution curve is $\mu e^{-\mu t}(6 - t)/5$. The integral,

$$\int_1^6 \mu e^{-\mu t}(6 - t)/5 dt =$$

$$\underbrace{[e^{-\mu t}(t/5 + 1/5\mu -6/5)]}_{t = 6} - \underbrace{[e^{-\mu t}(t/5 + 1/5\mu -6/5)]}_{t = 1},\qquad\qquad 3.$$

gives the fraction of the intermediate material left at 1 hr that escapes digestion by passage between 1 and 6 hr. During the first hour, the fraction, $1 - e^{-\mu}$, of this material will leave the rumen before any of its digestion begins, making the total fraction escaping utilization

$$1 - e^{-\mu} + \int_1^6 \mu e^{-\mu t} (6 - t)/5 dt \qquad\qquad 4.$$

The fractions of the intermediate material utilized at the four average retention times are shown in Table 2.

Table 2 Utilization of alfalfa hay at several average retention times

Retention time, $1/\mu$	Utilization of different components			Utilization of total digestible material
	soluble	intermediate	fiber	
hr	%	%	%	%
12	96	75	34	54
24	98	87	56	76.6
24/0.93	98	87.5	58	77.6
48	99	93	73	86

Finally, the fiber fraction leaving the rumen without being digested equals the areas under the curves A–C of Figure 4, divided by the areas, $\int_0^\infty e^{-\mu t}dt$, under the curves labeled 1/12, 1/24, and 1/48, respectively. These latter areas represent the fraction of fiber in the rumen if no utilization occurs, and they are equal to the areas of the rectangles 0, 1.0, L, 12; 0, 1.0, M, 24; and 0, 1.0, N, 48, respectively. The fraction of the fiber digested and fermented at different turnover rates, i.e. one minus the fraction leaving the rumen without being digested, is also recorded in Table 2.

RELATIVE UTILITY OF DIFFERENT FORAGE COMPONENTS: MODEL

Inspection of the values in Table 2 brings out the differences in the utility to the host of the various feed components at different average residence times. The soluble components are almost completely utilized at all turnover rates, with only a 3% difference in the availability of this fraction between average retention times of 12 and 48 hr. In contrast the difference for the fiber is 39%.

On the assumption that the amount of fermentation gas is proportional to the amount of substrate utilized, the gas production during 0–1, 1–6, and 6–42 hr, respectively, indicates the relative amount of the fermentation supported by particular fractions. From Figure 3, the relative gas productions from 137.5 mg of alfalfa

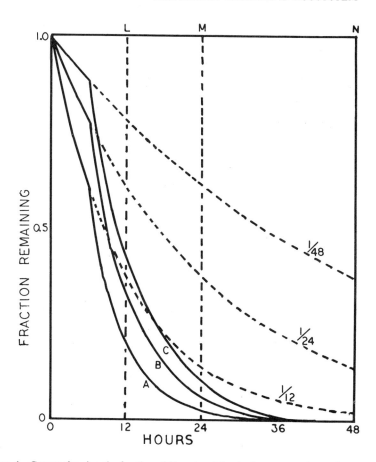

Figure 4 Curves showing the fraction of fiber remaining in the rumen when only passage is considered, at rates of 1/12, 1/24, and 1/48 hr, and when both passage and digestion are considered, curves *A, B,* and *C,* respectively.

meal were 210 μmol (0.24 of the total) for the solubles, 305 μmol (0.34 of the total) for the intermediate material, and 370 μmol (0.42 of the total) for the fiber. The fiber supports the most microbial fermentation, and the soluble materials the least, but the rate is 210 μmol/hr for the solubles, 61 for the intermediate material, and only 10 for the fiber. The easily digestible and fermentable components of plant material are most rapidly and completely utilized.

Maximum productivity is obtained with forages of easy and high digestibility, not only because of the ready availability of the carbohydrate, but also because such forages usually contain a good balance of minerals, proteins, and other essential nutrients. The turnover rate is faster, with a greater production of microbial cells and fermentation acids and a decreased fraction of substrate expended for "maintenance" (H. R. Isaacson, Rumen Function Conf., Chicago, 1973, unpublished). Also, easily digestible components escaping the rumen fermentation can be salvaged by digestion and absorption during further passage through the host.

Another factor favoring easily digested foods for high production is the fact that rumen "fill" influences appetite (11, 48). With less digestible forages, the undigested residue is a greater fraction of the feed consumed; disappearance by fermentation is less, passage is slowed because of the greater proportion of large particles, and new food cannot enter as rapidly. The fill slows ingestion. The better feeds not only give a greater yield of valuable products per unit of feed consumed, they also can be consumed in greater amounts.

Lush spring forage supports optimal growth of ruminants in both the feral and domesticated state. But this growth is seasonal and usually brief, with intervening periods of drought when only less digestible forages, often not nutritionally well balanced, are available. Under these conditions, plant fibers play a relatively greater role, nourishing the animal until conditions improve. With greater fill and longer retention time more fiber is digested, but the value to the animal is not proportionately increased, due to rumination and other costs of handling the greater bulk for a longer time. Grinding and pelleting of feed increase consumption, rate of passage, and host assimilation (69, 70).

The surface area of fiber exposed to enzymatic attack limits the rate of digestion. Rumination increases this area, and extremely fine grinding (pebble milling) even breaks the bonds between lignin and other fiber components (cellulose and hemicellulose) (19), rendering them almost completely digestible (51). Digestion and rumen-mixing movements can alone comminute digesta enough to let it pass through the alimentary canal, though at a much reduced rate. When a sheep's mouth was tied shut except during periods of feeding (132), rumination was prevented, yet slow passage of digesta continued.

Inability of fiber-digesting bacteria to absorb all of the soluble sugars produced by their carbohydrases seems a disadvantage for their speed of growth; however, a more varied population is supported, with greater potential for provision of needed growth factors that may be advantageous in the total ecosystem. Under poor nutritional conditions, the increased maintenance costs reduce the fraction of substrate converted into microbial cells, but increase the VFA. The supply of oxidizable acids to the host is relatively better than the supply of protein; the availability of the acids diminishes the use of proteins for energy.

Realization that many natural and domesticated fibrous feeds are not nutritionally optimal focuses attention on the desirability of increasing the digestibility of fiber. Chemical treatment with alkali (82) or other delignifying agents can be effective (52), but increases the cost. Crops might be selected for higher digestibility of fiber. Since the long-term value of ruminants in our economy lies in their utilization of fiber, the possibilities for selection of improved heritable digestibility need exploration.

INFORMATION DERIVED THROUGH APPLICATION OF RUMEN MODELS

The assumption for a continuous fermentation model that each component of the digesta has the same average rate of passage is not applicable to the rumen (72). Mathematical modeling for different rates of passage of different components is more realistic and can be accomplished with computer simulation techniques being developed (14, 71). But even with the simple model, some insight into actual rumen function can be obtained.

From the continuous fermentation equation for rate of entrance or exit, $dx/dt = \mu x$, it is evident that since $\mu = (dx/dt)/x$, μ is the daily amount of material entering or leaving divided by the amount of material in the rumen. This calculation is particularly valuable in the case of the rumen solids, for which there is no reliable marker. The feed consumed, commonly recorded in terms of weight per day, gives a measure of rate of entrance. The fecal solids are a parameter of the rate of exit. If it is assumed that the amount of digestion posterior to the omasum is negligible, the composition of the feces is an index to the composition of the rumen digesta, provided that the material leaving the rumen is representative of the total solids. Actually, digestion posterior to the omasum amounts to 5–15% of the total, depending on the nature of the ration (152); greater errors occur with more rapid turnover rates.

The turnover rate of the liquid and solids in a continuously fed ruminant have been estimated without taking any postruminal digestion into account. A sheep weighing 55 kg was fed a daily ration of 960 g alfalfa pellets in 12 equal portions administered at 2-hr intervals (105). The daily fecal output was 369 g, of which 50 g was estimated to be bacteria, based on the 4.69 g fecal nitrogen, leaving 319 g as the food residues in the feces. The PEG turnover rate was 2.27/day, and the PEG liquid volume was 3.77 liters. By microscopic measuring and counting of the bacteria and protozoa, the volume of microorganisms in the rumen was calculated to be 0.93 liters, giving 4.7 liters as the total rumen volume.

The rumen dry matter, exclusive of 57 g VFA, 128 g microbial cells, and 43 g minerals, was 343 g. This 343 g must leave at a rate of 0.93/day, an average rate for all solids components, to give the 319 g of food residues in the feces. Fibrous particles too large to pass the reticulo-omasal orifice have a turnover rate even slower than the average, increasing the extent to which fiber is digested. But most of the rumen solids are finely divided. Their slower passage as compared to liquid may be due to their higher specific gravity.

The 319 g of undigested food residues in the feces constitute 33% of the food consumed, for a feed-feces difference of 67% (digestibility as usually calculated for the animal, but less than the 42-hr digestibility by the microbes). From Table 2, the turnover rate constant of 0.93/24 hr should allow utilization of 0.776 of the food digestible within 42 hr, giving 0.67/0.776 = 0.86 as the fraction digestible within 42 hr. The undigestible food in the feces constitutes 0.14 of the feed and 0.42 of the feces; 0.58 of the fecal dry weight is digestible. A report (M. P. Bryant, personal communication) that about 50% of the material in bovine feces is broken down during a long-term methanogenic fermentation of feces is in fair agreement with this estimate.

The turnover rate constant of the microbial cells is of the same order of magnitude as that for the rumen solids, in contradiction to the earlier assumed equivalence with the liquid turnover rate (96). The dry weight and nitrogen content of the microbes in the sheep rumen were estimated to be 128 and 12 g, respectively, as compared to 14 g N found by analysis of the total rumen contents (105). Most of this nitrogen was in the solids fraction, and presumably chiefly in the microbes (123). If this pool turned over with PEG, the supply of microbial nitrogen to the host would be 2.27×13 g $= 29.6$ g N, as compared to 23.5 g N in the feed. Without significant quantities of urea supplied to the rumen through the saliva and rumen wall, it is apparent that the rumen microbial population could not grow at the PEG turnover rate.

The actual yield of rumen microbes to the sheep was estimated from an equation to show the materials consumed and formed in the rumen fermentation. The material fermented was the feed-feces difference in ash, C, H, and N, with the material unaccounted for assumed to be O. The products formed were estimated by measuring the rates of CO_2 and CH_4 production during short in vitro incubations of weighed representative samples of rumen contents. The rates of NH_3 (8) and VFA production were estimated by the zero time rate method (36). This method, verified experimentally (102), assumes that when a sample is rapidly removed from the rumen and incubated anaerobically at rumen temperature, its composition does not instantly change, but changes gradually as conditions deteriorate from those of the rumen. By measuring the concentration of a component of the sample at the instant of removal and after several short sequential periods of incubation, the rate of use or production at zero time can be obtained graphically from the zero time slope of the curve for concentration against time.

The following empirical formula represents the materials used and produced daily in the sheep:

$$C_{20.03}H_{36.99}O_{17.405}N_{1.345} + 5.65 \; H_2O \longrightarrow C_{12}H_{24}O_{10.1} + 0.83 \; CH_4 + 2.76 \; CO_2$$

$$\text{feed-feces difference} \qquad\qquad\qquad \text{VFA}$$

5.

$$+ \; 0.50 \; NH_3 + C_{4.44}H_{8.88}O_{2.35}N_{0.785}$$
$$\text{microbial cells (11 g N)}$$

The feed-feces C, not accounted for in other products, was assumed to be in the microbial cells. The elemental composition of the microbes was estimated from reports of the C, H, and ash of bacteria [(116); see also (54) for a different estimate]. The N content of the bacteria was assumed to be 10.5%, and of the protozoa 8%, for an average of 9.39% N (11 g total N in the rumen microbes), based on the relative volumes of the protozoa and bacteria. On the assumption that the rest of the ash-free content was oxygen, errors in the estimated composition of the microbial cells or in other missed fermentation products show up as discrepancies in the balance for H. This discrepancy was 1.4%. A nitrogen balance is not too meaningful because of the urea entering the rumen from the saliva and the blood; however, the 7 g N as NH_3, the 11 g N in the microbial cells, and the 4.69 g N in the feces represent 96.5% of the 23.52 g N in the daily feed. The N in the feces is chiefly in the bodies of the microorganisms developing in the small and large intestine at the expense of food residues and host cell contents released during digestion. The fairly high microbial content of bovine feces has led to the practice of using them as part of the ration, usually after ensiling.

The 12 g of rumen microbial nitrogen, estimated from the size and volume measurements of the bacteria and protozoa, would turn over 0.92/day to supply the 11 g of microbial N of Equation 5, or the 14 g N in the rumen (by analysis) would turn over 0.79/day. Both estimates indicate that the rate of passage of the rumen microbes resembles the average rate for the solids rather than that for the PEG liquid; their concentration in the liquid is less than in the remaining contents. This is confirmed by higher counts of protozoa in the total contents than in samples containing only liquid and fine particles (105), and by the significantly higher fermentation rate of in vitro incubated total contents as compared to rumen liquid (96).

Retention in the rumen of a larger microbial population than that expected from the PEG turnover rate is comparable to the recycling of the cells produced in a continuous fermentor; instead of being recycled, their exit is retarded. The larger population is less productive per unit cell, but has a greater capacity for converting fresh ingesta. From the standpoint of the animal's maximum productivity, longer retention of microbes is disadvantageous because of increased expenditure of microbial cell material for maintenance.

MICROBIAL CELL YIELD IN RELATION TO SUBSTRATE FERMENTED

If the carbon in the daily feed-feces difference of 20.03 mol is assumed to be in hexose, the yield of microbial cells per mol of hexose is 50 g. From the value of 10.5 g cells/mol ATP (18), 4.5 ATP/hexose were available in the rumen fermentation of the experimental sheep. This is consistent with 4 ATP/hexose fermented to acetate, CO_2, and methane, or to acetate, CO_2, and propionate (79). Additional ATP may be conserved by a phosphorolysis of cellobiose or higher glucose polymers (10), and some ATP is obtained in methanogenesis. At a net average continuous growth rate of about 1 μ per day, the rumen population uses most of the available ATP for growth. Growth yields have been estimated also in experiments based on rumen microbial assimilation of phospholipids (30), diaminopimelic acid (62), and sulfur-35 (119).

When the experimental sheep was fed the same daily quantity of feed in two equal portions 12 hr apart, the rumen volume was greater, with a slower PEG turnover rate constant, and the net productivity of microbial cells was only a little more than half that with the 2-hr feeding of the same ration (105). Also, in practice, frequent feeding gives greater weight gains (68, 130).

DYNAMICS OF THE RUMEN ECOSYSTEM

Sudden administration of grain or glucose to a ruminant animal previously on a dried forage ration often leads to death of the animal within as short a time as 18 hr (99). The cause is an explosive growth (from 10^7 to 10^{10} cells/ml) of *Streptococcus bovis*, dividing at intervals as brief as 20 min, and forming abundant lactic acid, which accumulates because the rumen is relatively impermeable to it and because the numbers of lactacidiclastic microbes are insufficient to convert the increased lactate to VFA. The pH of the rumen contents quickly (within 6 hr) drops to 4.5, and most rumen bacteria and protozoa are killed. Also the rumen tissues may be irreparably damaged.

The phenomenon of lactic toxicity discloses that maximum ATP production per unit time is actually the key element in microbial competition for substrate, rather than maximum ATP/hexose. With excess substrate, *S. bovis*, though producing only 2 ATP/hexose, can process hexose faster than its competitors, producing more ATP per unit time, and outgrowing them. The extreme comparison is with the protozoa; some of them may divide within as short a time as 6 hr, but within that period, *S. bovis* can form 262,144 cells. When hexose is limiting, ability to form more ATP per hexose gives more ATP per unit time. On the basis that metabolic reactions, such as ATP production, represent biochemical work (96), ATP production per unit time has the dimensions (work per unit time) of power. Microbial competitive success depends on the power for growth in the given environment. This same feature characterizes the survival success of all living forms.

In animals well adapted to a high grain ration, the numbers of *S. bovis* are only 10^7/ml, about the same as in the hay-fed animal. During gradual adaptation to the grain, amylolytic components of the microbiota (including the protozoa) and the lactacidiclastic bacteria increase sufficiently to keep hexose and lactate at limiting concentrations (121, 122). Efficiency in ATP production/hexose determines the ATP per unit time, and the population is stable. In the new balance, lactate is more important as an intermediate than it is in hay-fed animals, and the proportion of lactacidiclastic bacteria is greater. Hexose concentration comes closer to saturating the enzymes of the microbiota; the system is less stable, but productivity is high.

Another deviation from the normal rumen ecosystem occurs in bloat (45). Rumen contents become a viscous mass in which small bubbles of fermentation gas fail to coalesce to form the large gas volume normally eructated from the top of the rumen. The digesta expand like bread dough and in 30 min to 1 hr the resulting pressure may cut off the main circulation and respiration, killing the animal. Ingested fresh legumes are a common cause of this frothy bloat, but it is encountered also in feedlot animals on a high grain ration. Factors concerned in bloat are the kind and amount of feed, the kind and amount of saliva, the particle size of the digesta (40), and genetic characteristics of the animal. A role of microorganisms in bloat has not been generally accepted, but mucinolytic and capsule-producing bacteria and lysing holotrich protozoa have been postulated to be important.

INHIBITION OF METHANOGENESIS

The possibility of increasing ruminant productivity by eliminating rumen methanogenesis has led to tests in vitro and in vivo (49, 53, 75) on inhibition by methane analogs (17) and long-chain polyunsaturated fatty acids (113, 155). Inhibition of methanogenesis is almost immediate upon addition of chloroform or chloral hydrate (112). The latter is converted in part to chloroform in the rumen (136, 137). Hydrogen gas accumulates immediately, and the ratio of propionic to acetic acid increases, due to the increased H_2 (136), in part used for the reduction of 4-carbon dicarboxylic acids to succinate.

With continued administration of methane inhibitors, the H_2 production diminishes, but the ratio of propionate to acetate remains high (143). Many rumen organisms producing H_2 are inhibited by the increased H_2 concentration. Those producing no H_2, but gaining a high ATP yield through the propionate-succinate fermentation are selected, and the proportion of propionate in the fermentation products increases significantly. The marked changes following inhibition of methanogenesis indicate the important role it plays in the normal fermentation.

Since propionate is the only rumen fermentation acid convertible to hexose in the ruminant, feeding trials have been conducted to see if ruminant performance is increased by inhibiting methane production. Some results show improved performance (151), others no effect (156).

CONSTANCY OF THE MICROBIOTA

Ruminants in different parts of the world, under different conditions of climate, feed, and breed, are all supported by a fermentation in which VFA, carbon dioxide, methane, and microbial cells are produced at comparable rates and in roughly the same proportions. The ecosystem in each animal includes a great variety of niches, each theoretically occupied by an organism performing a particular activity, their sum characterizing the whole. Are there identical niches in the various rumens? The question is in part fruitless because of inability to identify all factors characterizing the niche. Individual differences in animals, in feed under various climatic conditions, and in climate itself, make academic the question whether two niches are identical, even in the same animal at different times. If each kind of organism occupies a unique niche, the variations in the composition of the microbial population are an index to the variation in the niches. Knowledge of the rumen cellulolytic bacteria provides an example of this variability (Kistner, in 56).

Bacteroides succinogenes and *Ruminococcus albus* have been reported as the most important rumen digesters of cellulose in Texas (92), the Netherlands (145), Maryland (27), South Africa (109, 154), Japan (76), and Kenya (100). Kistner reported (in 56) that they occur in all rumens but in widely varying numbers. At various times and at various localities, other species of cellulolytic bacteria have predominated. In the Okanogan valley of Washington, *Clostridium lochheadii* was most common (94). In the Netherlands, *Eubacterium cellulosolvens* was usually isolated (139).

Cellulomonas is found in Japan (118), and a cellulolytic *Micromonospora* in Poland (114). In Scotland, a very large cellulolytic coccus was identified (12) as an important cellulose digester on the basis of its location in pits digested out of the cell walls of ingested grasses. *Clostridium longisporum* occurs in small numbers in cattle in California, and still another sporeformer has been noted in certain sheep in South Africa (56). These observations suggest that in the rumen there is both similarity and a great deal of variation in the kind and extent of niches occupied by cellulolytic bacteria.

In spite of this variability in biotic composition, the proportions of VFA, CO_2, and CH_4 remain the same. Substrate is limiting. The maximum possible biological work (ATP production per unit time) is accomplished in the redistribution of the chemical elements of the substrate into that combination of products, including microbial cells, characterizing the rumen fermentation.

Maximal production of ATP per unit time is a principal factor selecting the components of the rumen population, but within its restrictions, innumerable secondary factors such as kind and amounts of noncarbohydrate nutrients, the particular protozoa present, and the numbers and kinds of lytic bacteria and viruses (84, 149) affect the growth rate of individual species and strains. These factors constantly select for new strains and the composition of the microbial population fluctuates with time, even on a constant feeding regime; however, the total microbial cell crop and other fermentation products are remarkably constant.

Literature Cited

1. Akin, D. E., Burdick, D., Michaels, G. E. 1974. Rumen bacterial interrelationships with plant tissue during degradation revealed by transmission electron microscopy. *Appl. Microbiol.* 27:1149–56
2. Alexander, C. L. et al 1963. Rumen VFA production from C^{14} labeled alfalfa *J. Anim. Sci.* 22:831 (Abstr.)
3. Allison, M. J. 1969. Biosynthesis of amino acids by ruminal microorganisms. *J. Anim. Sci.* 29:797–807
4. Allison, M. J., Peel, J. L. 1971. The biosynthesis of valine from isobutyrate by *Peptostreptococcus elsdenii* and *Bacteroides ruminicola. Biochem. J.* 121: 431–37
5. Allison, M. J., Robinson, I. M. 1970. Biosynthesis of α-ketoglutarate by the reductive carboxylation of succinate in *Bacteroides ruminicola. J. Bacteriol.* 104:50–56
6. Allison, M. J., Bryant, M. P., Katz, I., Keeney, M. 1962. Metabolic function of branched-chain volatile fatty acids, growth factors for ruminococci. II. Biosynthesis of higher branched-chain fatty acids and aldehydes. *J. Bacteriol.* 83:1084–93
7. Allison, M. J., Dougherty, R. W., Bucklin, J. A., Snyder, E. E. 1964. Ethanol accumulation in the rumen after overfeeding with readily fermentable carbohydrate. *Science* 144:54–55
8. Al-Rabbat, M. F., Baldwin, R. L., Weir, W. C. 1971. *In vitro* [15]Nitrogen-tracer technique for some kinetic measures of ruminal ammonia. *J. Dairy Sci.* 54:1150–61
9. Aranki, A., Syed, S. A., Kenney, E. B., Freter, R. 1969. Isolation of anaerobic bacteria from human gingiva and mouse cecum by means of a simplified glove box procedure. *Appl. Microbiol.* 17:568–76
10. Ayers, W. A. 1959. Phosphorolysis and synthesis of cellobiose by cell extracts from *Ruminococcus flavefaciens. J. Biol. Chem.* 234:2819–22
11. Baile, C. A., Mayer, J. 1966. Hyperphagia in ruminants induced by a depressant. *Science* 151:458–59
12. Baker, F., Nasr, H. 1947. Microscopy in the investigation of starch and cellulose breakdown in the digestive tract. *J. R. Microsc. Soc.* 67:27–42
13. Balch, C. C. 1957. Use of the lignin-ratio technique for determining the ex-

tent of digestion in the reticulo-rumen of the cow. *Brit. J. Nutr.* 11:213–27
14. Baldwin, R. L., Smith, N. E. 1971. Application of a simulation modeling technique in analyses of dynamic aspects of animal energetics. *Fed. Proc.* 30:1459–65
15. Baldwin, R. L., Wood, W. A., Emery, R. S. 1965. Lactate metabolism by *Peptostreptococus elsdenii:* Evidence for lactyl coenzyme A dehydrase. *Biochim. Biophys. Acta* 97:202–13
16. Barnard, E. A. 1969. Biological function of pancreatic ribonuclease. *Nature* 221:340–44
17. Bauchop, T. 1967. Inhibition of rumen methanogenesis by methane analogs. *J. Bacteriol.* 94:171–75
18. Bauchop, T., Elsden, S. R. 1960. The growth of microorganisms in relation to their energy supply. *J. Gen. Microbiol.* 23:457–70
19. Beveridge, R. J., Richards, G. N. 1973. Digestion of polysaccharide constituents of tropical pasture herbage in the bovine rumen. IV. The hydrolysis of hemicelluloses from spear grass by cellfree enzyme systems from rumen fluid. *Carbohyd. Res.* 29:79–87
20. Blackburn, T. H., Hullah, W. A. 1974. The cell-bound protease of *Bacteroides amylophilus* H 18. *Can. J. Microbiol.* 20:435–41
21. Blackburn, T. H., Hungate, R. E. 1963. Succinic acid turnover and propionic production in the bovine rumen. *Appl. Microbiol.* 11:132–35
22. Bonhomme, A. 1973. *Contribution a l'étude de la physiologie des ciliés entodiniomorphes endocommensaux des ruminants et des équidés.* Thèse, Docteur des Sciences Naturelles. Université de Reims, France
23. Brough, B. E., Howard, B. H. 1971. The biochemistry of the rumen bacterium "Quin's Oval": II. The carbohydrase enzymes. *N. Z. J. Sci.* 13:576–83
24. Brownlie, L. E., Grau, F. H. 1967. Effects of food intake on growth and survival of salmonellas and *Escherichia coli* in the bovine rumen. *J. Gen. Microbiol.* 46:125–34
25. Bryant, M. A. M. 1964. Variations in the pH and volatile fatty acid concentration within the bovine reticulo-rumen. *N. Z. J. Agric. Res.* 7:694–706
26. Bryant, M. P. 1973. Nutritional requirements of the predominant rumen cellulolytic bacteria. *Fed. Proc.* 32:1809–13

27. Bryant, M. P., Burkey, L. A. 1953. Numbers and some predominant groups of bacteria in the rumen of cows fed different rations. *J. Dairy Sci.* 36:218–24
28. Bryant, M. P., Robinson, I. M. 1962. Some nutritional characteristics of predominant culturable ruminal bacteria. *J. Bacteriol.* 84:605–14
29. Bryant, M. P., Small, N., Bouma, C., Chu, H. 1958. *Bacteroides ruminicola* N. Sp., and *Succinimonas amylolytica* the new genus and species. Species of succinic acid-producing anaerobic bacteria of the bovine rumen. *J. Bacteriol.* 76:15–23
30. Bucholtz, H. F., Bergen, W. G. 1973. Microbial phospholipid synthesis as a marker for microbial protein synthesis in the rumen. *Appl. Microbiol.* 25:504–13
31. Caldwell, D. R., Arcand, C. 1974. Inorganic and metal-organic growth requirements of the genus *Bacteroides*. *J. Bacteriol.* 120:322–33
32. Caldwell, D. R., Bryant, M. P. 1966. Medium without rumen fluid for nonselective enumeration and isolation of rumen bacteria. *Appl. Microbiol.* 14:794–801
33. Caldwell, D. R., White, D. C., Bryant, M. P., Doetsch, R. N. 1965. Specificity of the heme requirement for growth of *Bacteroides ruminicola*. *J. Bacteriol.* 90:1645–54
34. Caldwell, D. R., Keeney, M., van Soest, P. J. 1969. Effects of carbon dioxide on growth and maltose fermentation in *Bacteroides amylophilus*. *J. Bacteriol.* 98:668–76
35. Candlish, E., Devlin, T. J., LaCroix, L. J. 1970. Tryptophan utilization by rumen microorganisms *in vitro*. *Can. J. Anim. Sci.* 50:331–35
36. Carroll, E. J., Hungate, R. E. 1954. The magnitude of the microbial fermentation in the bovine rumen. *Appl. Microbiol.* 2:205–14
37. Carroll, E. J., Hungate, R. E. 1955. Formate dissimilation and methane production in bovine rumen contents. *Arch. Biochem. Biophys.* 56:525–36
38. Castle, E. J. 1956. The rate of passage of foodstuffs through the alimentary tract of the goat. 1. Studies on adult animals fed on hay and concentrates. *Brit. J. Nutr.* 10:15–23
39. Chalupa, W. 1974. Amino acid degradation by rumen microbes. *Fed. Proc.* 33:708 (Abstr.)

40. Cheng, K. J., Hironaka R. 1973. Influence of feed particle size on pH, carbohydrate content, and viscosity of rumen fluid. *Can. J. Anim. Sci.* 53:417–22

41. Cheng, K. J., Hironaka, R., Roberts, D. W. A., Costerton, J. W. 1973. Cytoplasmic glycogen inclusions in cells of anaerobic gram-negative rumen bacteria. *Can. J. Microbiol.* 19:1501–6

42. Church, D. C. 1972. *Digestion, Physiology and Nutrition of Ruminants.* Corvallis, Oregon: Oregon State University Book Stores. 3 vols.

43. Clark, J. L., Hembry, F. G., Thompson, G. B., Preston, R. L. 1972. Ration effect on polyethylene glycol as a rumen marker. *J. Dairy Sci.* 55:1160–64

44. Clarke, R. T. J., Hungate, R. E. 1966. Culture of the rumen holotrich ciliate *Dasytricha ruminantium* Schuberg. *Appl. Microbiol.* 14:340–45

45. Clarke, R. T. J., Reid, C. S. W. 1974. Foamy bloat of cattle. A review. *J. Dairy Sci.* 57:753–85

46. Coleman, G. S. 1968. The metabolism of bacterial nucleic acid and of free components of nucleic acid by the rumen ciliate *Entodinium caudatum*. *J. Gen. Microbiol.* 54:83–96

46a. Coleman, G. S. 1969. The metabolism of starch, glucose, and some other sugars by the rumen ciliate *Entodinium caudatum*. *J. Gen. Microbiol.* 57: 303–32

47. Coleman, G. S. 1973. The metabolism of bacteria and amino acids by three species of Epidinia isolated from the rumen and cultivated *in vitro*. *J. Gen. Microbiol.* 77:viii (Abstr.)

48. Cowsert, R. L., Montgomery, M. J. 1969. Effect of varying forage-to-concentrate ratio of isonitrogenous rations on feed intake by ruminants. *J. Dairy Sci.* 52:64–67

49. Czerkawski, J. W. 1973. Effect of linseed oil fatty acids and linseed oil on rumen fermentation in sheep. *J. Agric. Sci.* 81:517–31

50. Dehority, B. A. 1971. Carbon dioxide requirement of various species of rumen bacteria. *J. Bacteriol.* 105:70–76

51. Dehority, B. A., Johnson, R. R. 1961. Effect of particle size upon the in vitro cellulose digestibility of forages by rumen bacteria. *J. Dairy Sci.* 44:2242–49

52. Dekker, R. F. H. 1973. Effect of delignification on the in vitro rumen digestion of polysaccharides of bagasse. *J. Sci. Food Agric.* 24:375–79

53. Demeyer, D. I., Henderickx, H. K. 1967. The effect of C_{18} unsaturated fatty acids on methane production *in vitro* by mixed rumen bacteria. *Biochim. Biophys. Acta* 137:484–89

54. Demeyer, D. I., Henderickx, H. K., van Nevel, C. J. 1972. The elementary composition of mixed rumen microbes and its use in fermentation balances. *Proc. World Congr. Anim. Feed., 2nd, Madrid.* 5:21–25

55. Dogiel, V. A. 1927. Monographie der Familie Ophryoscolecidae. Teil I. *Arch. Protistenk.* 59:1–288

56. Dougherty, R. W., Ed. 1965. Physiology of Digestion in the Ruminant. *Proc. Int. Symp. Ruminant Physiol., 2nd.* Washington D.C.: Butterworth

57. Dougherty, R. W., Allison, M. J., Mullenax, C. H. 1964. Physiological disposition of C^{14}-labeled rumen gases in sheep and goats. *Am. J. Physiol.* 207:1181–88

58. Dougherty, R. W., O'Toole, J. J., Allison, M. J. 1967. Oxidation of intraarterially administered carbon^{14}-labeled methane in sheep. *Proc. Soc. Exp. Biol. Med.* 124:1155–57

59. Driedger, A., Hatfield, E. E. 1972. Influence of tannins on the nutritive value of soybean meal for ruminants. *J. Anim. Sci.* 34:465–68

60. Eadie, J. M. 1967. Studies on the ecology of certain rumen ciliate protozoa. *J. Gen. Microbiol.* 47:175–94

61. Eadie, J. M., Hobson, P. N. 1962. Effect of the presence or absence of rumen ciliate protozoa on the total rumen bacterial count in lambs. *Nature* 193:503–5

62. el-Din, M. Z., el-Shazly, K. 1969. Evaluation of a method of measuring fermentation rates and net growth of rumen microorganisms. *Appl. Microbiol.* 17:801–4

63. Emmanuel, B., Milligan, L. P. 1972. Enzymes of the conversion of succinate to glutamate in extracts of rumen microorganisms. *Can. J. Biochem.* 50:1–8

64. Emmanuel, B., Milligan, L. P., Turner, B. V. 1974. The metabolism of acetate by rumen microorganisms. *Can. J. Microbiol.* 20:183–85

65. Every, D. D., Howard, B. H. 1971. The biochemistry of the rumen bacterium "Quin's Oval": III. The storage polysaccharide. *N. Z. J. Sci.* 13:584–90

66. Fisher, L. J. 1972. Responses of lactating cows to the intravenous infusion of amino acids. *Can. J. Anim. Sci.* 52: 377–84

67. Forsyth, G., Hirst, E. L. 1953. Protozool polysaccharides. Structure of the polysaccharide produced by the holo-

trich ciliates present in sheep's rumen. *J. Chem. Soc.* 2132–35

68. Gordon, J. G., Tribe, D. E. 1952. The importance to sheep of frequent feeding. *Brit. J. Nutr.* 6:89–93

69. Greenhalgh, J. F. D., Reid, G. W. 1973. The effects of pelleting various diets on intake and digestibility in sheep and cattle. *Anim. Prod.* 16:223–33

70. Grieve, C. M., Robblee, A. R., Berg, R. T., McElroy, L. W. 1963. Native lowland hay in pelleted and non-pelleted rations for sheep. 1. Effects on feed consumption, rate of gain, efficiency of feed utilization and digestibility. *Can. J. Anim. Sci.* 43:189–95

71. Grovum, W. L., Phillips, G. D. 1973. Rate of passage of digesta in sheep. 5. Theoretical considerations based on a physical model and computer simulation. *Brit. J. Nutr.* 30:377–90

72. Grovum, W. L., Williams, V. F. 1973. Rate of passage of digesta in sheep. 4. Passage of marker through the alimentary tract and the biological relevance of rate-constants derived from the changes in concentration of the marker in faeces. *Brit. J. Nutr.* 30:313–29

73. Gutierrez, J. 1958. Observations on bacterial feeding by the rumen ciliate *Isotricha prostoma. J. Protozool.* 5: 122–26

74. Hemsley, J. A., Rees, P. J., Downes, A. M. 1973. Influence of various formaldehyde treatments on the nutritional value of casein for wool growth. *Austr. J. Biol. Sci.* 26:961–72

75. Henderson, C. 1973. The effects of fatty acids on pure cultures of rumen bacteria. *J. Agric. Sci.* 81:107–12

76. Higuchi, M., Uemura, T. 1965. Studies on the cellulose fermentation in the rumen. III. Isolation of anaerobic cellulolytic bacteria from the rumen of sheep. *J. Agric. Chem. Soc. J.* 39:95–101

77. Hobson, P. N., Mann, S. O. 1961. The isolation of glycerol-fermenting and lipolytic bacteria from the rumen of the sheep. *J. Gen. Microbiol.* 25:227–38

78. Hobson, P. N., Summers, R. 1967. The continuous culture of anaerobic bacteria. *J. Gen. Microbiol.* 47:53–65

79. Hobson, P. N., Summers, R. 1972. ATP pool and growth yield in *Selenomonas ruminantium. J. Gen. Microbiol.* 70: 351–60

80. Hobson, P. N., McDougall, I. J., Summers, R. 1967. The nitrogen sources of *Bacteroides amylophilus. J. Gen. Microbiol.* 50:i (Abstr.)

81. Hofmann, R. 1969. Zur Topographie und Morphologie des Wiederkauermagens im Hinblick auf seine Funktion. Nach vergleichenden Untersuchungen an Material ostafrikanischer Wildarten. *Zentralbl. Veterinaermed. Beih.* 10: 180 pp.

82. Hogan, J. P., Weston, R. H. 1971. The utilization of alkali treated straw by sheep. *Austr. J. Agric. Res.* 22:951–62

83. Holdeman, L. V., Moore, W. E. C. 1972. *Anaerobe Laboratory Manual.* Blacksburg, VA: Virginia Polytechnic Institute Anaerobe Laboratory. 130 pp.

84. Hoogenraad, N. J., Hird, F. J. R., Holmes, I., Millis, N. F. 1967. Bacteriophages in rumen contents of sheep. *J. Gen. Virol.* 1:575–76

85. Hoover, W. H., Lipari, J. J. 1971. Pure and mixed continuous culture of two rumen anaerobes. *J. Dairy Sci.* 54: 1662–68

86. Hopgood, M. F., Walker, D. J. 1967. Succinic acid production by rumen bacteria. I. Isolation and metabolism of *Ruminococcus flavefaciens. Austr. J. Biol. Sci.* 20:165–82

87. Hopgood, M. F., Walker, D. J. 1969. Succinic acid production by rumen bacteria. III. Enzymic studies on the formation of succinate by *Ruminococcus flavefaciens. Austr. J. Biol. Sci.* 22: 1413–24

88. Huisingh, J., McNeill, J. J., Matrone, G. 1974. Sulfate reduction by a *Desulfovibrio* species isolated from sheep rumen. *Appl. Microbiol.* 28:489–97

89. Hullah, W. A., Blackburn, T. H. 1971. Uptake and incorporation of amino acids and peptides by *Bacteroides amylophilus. Appl. Microbiol.* 21: 187–91

90. Hungate, R. E. 1942. The culture of *Eudiplodinium neglectum,* with experiments on the digestion of cellulose. *Biol. Bull.* 83:303–19

91. Hungate, R. E. 1943. Further experiments on cellulose digestion by the protozoa in the rumen of cattle. *Biol. Bull.* 84:157–63

92. Hungate, R. E. 1947. Studies in cellulose fermentation. III. The culture and isolation of cellulose-decomposing bacteria from the rumen of cattle. *J. Bacteriol.* 53:631–45

93. Hungate, R. E. 1950. The anaerobic mesophilic cellulolytic bacteria. *Bacteriol. Rev.* 14:1–49

94. Hungate, R. E. 1957. Micro-organisms in the rumen of cattle fed a constant ration. *Can. J. Microbiol.* 3:289–311

95. Hungate, R. E. 1963. Polysaccharide storage and growth efficiency in *Ruminococcus albus. J. Bacteriol.* 86: 848–54

96. Hungate, R. E. 1966. *The Rumen and Its Microbes.* N.Y.: Academic. 533 pp.

97. Hungate, R. E. 1967. Hydrogen as an intermediate in the rumen fermentation. *Arch. Mikrobiol.* 59:158–64

98. Hungate, R. E. 1968. A roll tube method for cultivation of strict anaerobes. *Methods in Microbiology* 3B ed. R. Norris, D. W. Ribbons, Chap. IV. London: Academic

99. Hungate, R. E., Dougherty, R. W., Bryant, M. P., Cello, R. M. 1952. Microbiological and physiological changes associated with acute indigestion in sheep. *Cornell Vet.* 42:423–49

100. Hungate, R. E., Phillips, G. D., MacGregor, A., Hungate, D. P., Buechner, H. K. 1959. Microbial fermentation in certain mammals. *Science* 130:1192–94

101. Hungate, R. E., Phillips, G. D., Hungate, D. P., Macgregor, A. 1960. A comparison of the rumen fermentation in European and zebu cattle. *J. Agric. Sci.* 54:196–201

102. Hungate, R. E., Mah, R. A., Simesen, M. 1961. Rates of production of individual volatile fatty acids in the rumen of lactating cows. *Appl. Microbiol.* 9:554–61

103. Hungate, R. E., Bryant, M. P., Mah, R. A. 1964. The rumen bacteria and protozoa. *Ann. Rev. Microbiol.* 18: 131–66

104. Hungate, R. E., Smith, W., Bauchop, T., Yu, I., Rabinowitz, J. C. 1970. Formate as an intermediate in the rumen fermentation. *J. Bacteriol.* 102:384–97

105. Hungate, R. E., Reichl, J., Prins, R. A. 1971. Parameters of rumen fermentation in a continuously fed sheep: evidence of a microbial rumination pool. *Appl. Microbiol.* 22:1104–13

106. Ianotti, E. L., Kafkewitz, D., Wolin, M. J., Bryant, M. P. 1973. Glucose fermentation products of *Ruminococcus albus* grown in continuous culture with *Vibrio succinogenes:* Changes caused by interspecies transfer. *J. Bacteriol.* 114:1231–40

107. Jayasuriya, G. C. N., Hungate, R. E. 1959. Lactate conversions in the bovine rumen. *Arch. Biochem. Biophys.* 82: 274–87

108. Joyner, A. E., Baldwin, R. L. 1966. Enzymatic studies of pure cultures of rumen microorganisms. *J. Bacteriol.* 92:1321–30

109. Kistner, A., Gouws, L. 1964. Cellulolytic cocci occurring in the rumen of sheep conditioned to lucerne hay. *J. Gen. Microbiol.* 34:447–58

110. McAllan, A. B., Smith, R. H. 1974. Carbohydrate metabolism in the ruminant: Bacterial carbohydrates formed in the rumen and their contribution to digesta entering the duodenum. *Brit. J. Nutr.* 31:77–88

111. McDonald, I. W. 1975. *Digestion and Metabolism in the Ruminant. Proc. Int. Symp. Ruminant Physiol.* Armidale, Australia: Univ. New England. 4th ed.

112. Marty, R. J., Demeyer, D. I. 1973. The effect of inhibitors of methane production on fermentation pattern and stoichiometry in vitro using rumen contents from sheep given molasses. *Brit. J. Nutr.* 30:369–76

113. Marwaha, S. R., Kochara, A. S., Bhatia, I. S. 1973. An "in vivo" study on the effect of different types of dietary lipids on the microbial population in the rumen. *Indian J. Nutr. Diet.* 10:27–30

114. Matuszynska, G. M., Janota-Bassalak, L. 1974. A cellulolytic rumen bacterium, *Micromonospora ruminantium* sp. nov. *J. Gen. Microbiol.* 82:57–65

115. Matuszynska, G. M., Pietraszek, A. 1972. The use of dithiothreitol in in vitro studies on rumen microflora. *Acta Microbiol. Polon. Ser. B* 3:171–73

116. Mayberry, W. R., Prochazka, G. J., Payne, W. J. 1968. Factors derived from studies of aerobic growth in minimal media. *J. Bacteriol.* 96:1424–26

117. Menahan, L. A., Schultz, L. H. 1963. Effect of leucine and valine administration on VFA. *J. Dairy Sci.* 46:640 (Abstr.)

118. Minato, H., Endo, A., Higuchi, M., Ootomo, Y., Uemura, T. 1966. Ecological treatise on the rumen fermentation. I. The fractionation of bacteria attached to the rumen digesta solids. *J. Gen. Appl. Microbiol.* 12:39–52

119. Nader, C. J., Walker, D. J. 1970. Metabolic fate of cysteine and methionine in rumen digesta. *Appl. Microbiol.* 20: 677–81

120. Nakamura, K., Kanegasaki, S. 1969. Densities of ruminal protozoa of sheep established under different dietary conditions. *J. Dairy Sci.* 52:250–55

121. Nakamura, K., Takahashi, H. 1971. Role of lactate as an intermediate of fatty acid fermentation in the sheep rumen. *J. Gen. Appl. Microbiol.* 17:319–28

122. Nakamura, K., Kanegasaki, S., Takahashi, H. 1971. Adaptation of

ruminal bacteria to concentrated feed. *J. Gen. Appl. Microbiol.* 17:13–27

123. Nikolic, J. A., Jovanovic, M. 1973. Preliminary study in the use of different methods for determining the proportion of bacterial nitrogen in the total nitrogen of rumen contents. *J. Agric. Sci.* 81:1–7

124. Noirot-Timothee, C. 1960. *Étude d'une famille de cilliés: les "ophryoscolecidae".* *Structures et ultrastructures.* Ph.D. thesis, Ser. A, No. 3454. Masson et Cie, Paris: Univ. Paris

125. Orpin, C. G. 1972. The culture of the rumen organism Eadie's Oval in vitro. *J. Gen. Microbiol.* 70:321–29

126. Orpin, C. G. 1972. The culture in vitro of the rumen bacterium Quin's Oval. *J. Gen. Microbiol.* 73:523–30

127. Orskov, E. R., Benzie, D., Kay, R. N. B. 1970. The effects of feeding procedure on closure of the esophageal groove in young sheep. *Brit. J. Nutr.* 24:785–95

128. Orskov, E. R., Fraser, C., Pirie, R. 1973. The effect of bypassing the rumen with supplements of protein and energy on intake of concentrates by sheep. *Brit. J. Nutr.* 30:361–67

129. Oxford, A. E. 1955. Rumen ciliate protozoa: Their chemical composition, metabolism, requirements for maintenance and culture, and physiological significance for the host. *Exp. Parasitol.* 4:569–605

130. Pant, H. C., Ray, A. 1972. Effect of frequency of feeding on the rumen microbial activity. II. Concentration of total and particulate nitrogen in the rumen liquor. *Indian J. Anim. Sci.* 41:650–53

131. Paynter, M. J. B., Elsden, S. R. 1970. Mechanism of propionate formation in *Selenomonas ruminantium.* *J. Gen. Microbiol.* 61:1–8

132. Pearce, G. R., Moir, R. J. 1964. Rumination in sheep. I. The influence of rumination and grinding upon the passage and digestion of food. *Austr. J. Agric. Res.* 15:635–44

133. Phillipson, A. T., Ed. 1970. *Physiology of Digestion and Metabolism in the Ruminant. Proc. Int. Symp. Ruminant Physiol, 3rd.* Newcastle upon Tyne, England: Oriel

134. Pittman, K. A., Lakshaman, S., Bryant, M. P. 1967. Oligopeptide uptake by *Bacteroides ruminicola.* *J. Bacteriol.* 93:1499–1508

135. Portugal, A. V. 1966. Some aspects of nitrogen metabolism in the rumen. *Congr. Mund. Aliment. Anim., Madrid,* 25 pp.

136. Prins, R. A. 1965. Action of chloral hydrate on rumen microorganisms in vitro. *J. Dairy Sci.* 48:991–92

137. Prins, R. A. 1970. Methanogenesis and propionate production in the rumen as influenced by therapeutics against ketosis. *Z. Tierphysiol. Tierehrnähr. Futtermittelk.* 26:147–51

138. Prins, R. A. 1971. Isolation, culture and fermentation characteristics of *Selenomonas ruminantium* var. *bryanti* var. n. from the rumen of sheep. *J. Bacteriol.* 105:820–25

139. Prins, R. A., van Vugt, F., Hungate, R. E., van Vorstenbosch, C. J. A. H. V. 1972. A comparison of strains of *Eubacterium cellulosolvens* from the rumen. *Antonie van Leeuwenhoek J. Microbiol. Serol.* 38:153–61

140. Prior, R. L. 1974. Effects of frequency of feeding soy or urea containing diets on nitrogen metabolism and urinary citric acid excretion in lambs. *Fed. Proc.* 33:707 (Abstr.)

141. Reid, C. S. W. 1965. Quantitative studies of digestion in the reticulo-rumen. I Total removal and return of digesta for quantitative sampling in studies of digestion in the reticulo-rumen of cattle. *Proc. N.Z. Soc. Anim. Prod.* 25:65–84

142. Robinson, J. P., Hungate, R. E. 1973. *Acholeplasma bactoplasticum* sp.n., an anaerobic mycoplasma from the bovine rumen. *Int. J. Syst. Bacteriol.* 23:171–81

143. Rufener, W. H., Wolin, M. J. 1968. Effect of CCl_4 on CH_4 and volatile acid production in continuous cultures of rumen organisms and in a sheep rumen. *Appl. Microbiol.* 16:1955–56

144. Scheifinger, C. C., Wolin, M. J. 1973. Propionate formation from cellulose and soluble sugars by combined cultures of *Bacteroides succinogenes* and *Selenomonas ruminantium.* *Appl. Microbiol.* 26:789–95

145. Sijpesteijn, A. K. 1949. Cellulose-decomposing bacteria from the rumen of cattle. *Antonie van Leeuwenhoek J. Microbiol. Serol.* 15:49–52

146. Sugden, B., Oxford, A. E. 1952. Some cultural studies with holotrich ciliate protozoa of the sheep's rumen. *J. Gen. Microbiol.* 7:145–53

147. Sutherland, T. M., Ellis, W. C., Reid, R. S., Murray, M. G. 1962. A method of circulating and sampling the rumen contents of sheep fed on ground, pelleted foods. *Brit. J. Nutr.* 16:603–14

148. Taljaard, T. L. 1972. Representative rumen sampling. *J. S. Afr. Vet. Assoc.* 43:65–69

149. Tarakanov, B. V. 1971. Bacteriophages of cattle rumen. *Mikrobiologiya* 40: 544–50

150. Taylor, C. D., McBride, B. C., Wolfe, R. S., Bryant, M. P. 1974. Coenzyme M, essential for growth of a rumen strain of *Methanobacterium ruminantium. J. Bacteriol.* 120:974–45

151. Trei, J. E., Scott, G. C., Parish, R. C. 1972. Influence of methane inhibition on energetic efficiency of lambs. *J. Anim. Sci.* 34:510–15

152. Ulyatt, M. J., Macrae, J. C. 1974. Quantitative digestion of fresh herbage by sheep. I. The sites of digestion of organic matter, energy, readily fermentable carbohydrate, structural carbohydrate, and lipid. *J. Agric. Sci.* 82:295–307

153. van Golde, L. M. G., Prins, R. A., Franklin-Klein, W., Akkermans-Kruyswijk, J. 1973. Phosphatidylserine and its plasmalogen analogue as major lipid constituents in *Megasphaera elsdenii. Biochim. Biophys. Acta* 326: 314–23

154. van Gylswyk, N. O., Labuschagne, J. P. L. 1971. Relative efficiency of pure cultures of different species of cellulolytic rumen bacteria in solubilizing cellulose in vitro. *J. Gen. Microbiol.* 66:109–13

155. van Nevel, C. J., Demeyer, D. I., Henderickx, H. K. 1971. Effect of fatty acid derivatives on rumen methane and propionate in vitro. *Appl. Microbiol.* 21: 365–66

156. van Nevel, C. J., Demeyer, D. I., Cottyn, B. G., Boucque, C. V. 1973. Incorporation of linseed oil hydrolysate and sodium sulfite in rations for beef production 2. Effect on rumen fermentation in vivo. *Z. Tierphysiol. Tierernähr. Futtermittelk.* 31:66–71

157. Warner, A. C. I., Stacy, B. D. 1968. The fate of water in the rumen. 1. A critical appraisal of the use of soluble markers. *Brit. J. Nutr.* 22:369–87

158. Warner, A. C. I., Stacy, B. D. 1968. The fate of water in the rumen. 2. Water balances throughout the feeding cycle in sheep. *Brit. J. Nutr.* 22:389–410

159. Wright, D. E. 1967. Metabolism of peptides by rumen microorganisms. *Appl. Microbiol.* 15:547–50

160. Wright, D. E., Hungate, R. E. 1967. Amino acid concentrations in rumen fluid. *Appl. Microbiol.* 15:148–51

STRUCTURE AND CLIMATE IN TROPICAL RAIN FOREST

＊4087

Egbert Giles Leigh, Jr.
Smithsonian Tropical Research Institute, Balboa, Canal Zone

INTRODUCTION

The similarity of tropical rain forests of comparable habitat around the world has struck many observers. This review first discusses similarities in lowland tropical rain forests (lowland forests between the tropics, which in most months receive more rain than they can transpire). Primary concern is with "normal" rain forest, not swamps or heath forest. These forests are also of similar productivity, which reflects an extraordinary similarity of climate, particularly with respect to evaporation: productivity of these forests appears to vary seasonally with transpiration rate. Second, this review discusses montane rain forest, in whose frequently saturated climates structure is often very obviously related to transpiration rate.

A marine biologist might well object to treating forest structure without mentioning herbivores; therefore, I estimate the impact of vertebrate and insect on fruit and leaf in the old successional forest of Barro Colorado Island, Panama, then explore the rhythms of abundance and scarcity by which tropical forests defend themselves.

LOWLAND RAIN FOREST

Physiognomy

Lowland tropical rain forests may be distinguished from most temperate formations by the conspicuous presence of thick lianes and buttressed trees, the prevalence of trees with columnar, smooth-barked boles, and the occurrence, however infrequent, of ant-defended plants (27, 58, 62, 71). One may look in vain among lowland rain forests for differences such as exist between some mesic temperate forests, say the redwoods of California or riverbottom hemlock forest of the northeastern United States on the one hand, and the beech or oak forests of Europe on the other. Temperate forests differ partly because temperature as well as rainfall varies with the season; dry winters favor deciduous forest, dry summers evergreen. Yet, in the tropics there is no difference in shape or leaf form between deciduous and evergreen trees to equal that between hardwoods and conifers in the north temperate zone. Thus it is first necessary to document the similarity of lowland rain forests.

Although the forest canopy may occasionally be dominated by a leguminous tree with minute leaflets, published tables and diagrams (25, 58) agree that the median area of a leaf or leaflet (weighted by species) lies between 20.25 and 182.25 cm^2 in nearly all lowland tropical rain forests. In these forests a leaf's area is normally about a third the square of its length, so the median leaf length is between 8 and 24 cm. Data (14) on trees of the Mucambo forest near Belem, Brazil, suggest that the average leaf length for simple-leaved trees capable of growing 30 m high is ~12.5 cm when weighted by species and ~10.6 cm when weighted by the basal area per hectare (ha) of these species. The average length of leaves sampled randomly from the forest floor lies between 7 and 13 cm for the 15 lowland rain forests around the world that I have studied.

As in the temperate zone, it is undoubtedly true that understory leaves, and indeed shade leaves of any sort, except those of seedlings and ground-herbs, are generally larger than the sunlit leaves of the canopy, but quantitative data on the subject do not abound (54).

The prevalence of entire-margined leaves with acuminate tips is one of the most striking features of tropical forest (58, 71). My own data suggest that acuminate tips, especially the pronounced "drip tips," are more prevalent among understory saplings than among fallen leaves on the forest floor. The convergence in leaf form among the trees of unrelated plant families in one forest, as well as the similarity between forests, is most striking when compared with the diversity of leaf form in temperate forests.

In understory saplings of lowland rain forest, leaves tend to be arranged alternately or oppositely along horizontal twigs (Figure 1) to form horizontal sprays of foliage (42). Understory plants, for which this is not true, will frequently spiral their leaves (Figure 2) atop an unbranched stem. Similarly, the understory pawpaws (*Asimina*), spicebush (*Lindera*), hawthorns (*Crataegus*), and ironwood (*Carpinus*)

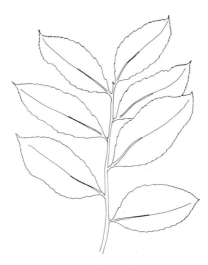

Figure 1   Leaves arranged alternately in a horizontal spray.

of the largely virgin temperate zone riverbottom forest of the Congaree below Columbia, South Carolina, all arrange their leaves alternately in horizontal sprays; the same holds for understory plants in other climax hardwood forests of the eastern United States. By contrast, the sweetgums (*Liquidambar*), oaks (*Quercus*), sourgums (*Nyssa*), and sycamores (*Platanus*) that dominate the canopy of the Congaree all spiral their leaves around the tips of their twigs, forming quasi-rosettes: such arrangements better suit the inclined leaf poses of canopy foliage (26; 71, p. 111). Such spiral, or, more rarely, decussate (Figure 3), leaf arrangements also appear to prevail among canopy trees in the tropics.

Many writers see lowland tropical forest as composed of three stories of trees. Trees fall continually, so the layers are blurred and impossible to distinguish objectively (25). Possibly all one can say is that a random point on the forest floor will

Figure 2   Spirally arranged leaves.

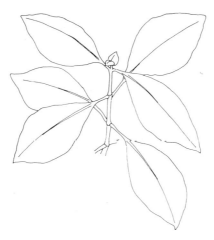

Figure 3   Decussately arranged leaves.

lie, on the average, under the crowns of three trees. Stratification is more obvious in climax temperate forests, for example in the 6 m ironwood layer under the 30 m oaks in the Berling tract near Arlington, Virginia, or the 1–2 m pawpaws or spicebush and the 8–15 m holly (*Ilex*) layer under the 40 m oaks of the virgin portions of the Congaree forest. Stratification is clearest in very old regrowth from cleared areas: in old farm sites in the Congaree bottom a layer of 45 m sweetgums overlays an even, sometimes parklike, layer of 8–10 m ironwoods and an herb layer of *Rhus toxicodendron*. J. W. Terborgh (in preparation) found that the topmost subcanopy layer in such forests is just far enough below the canopy that every point on an imaginary plane at that level will sometime during the day be lit by sunlight entering between the canopy crowns. Perhaps an extension of such arguments will show why there ideally should be three tree layers and one shrub layer in lowland tropical rain forests.

Lowland tropical rain forest is generally between 30 and 60 m high (58); virgin mesic temperate forest is rarely shorter, but in Australia and California (and perhaps in Lebanon in former times) it has attained heights of 90 m or more.

Coverage of 1/40 ha plots by foliage less than 60 cm from the ground lies between 10 and 50% for the lowland rain forests I have studied; the coverage is generally quite patchy and variable within a forest. Such sparse cover suggests either intense herbivory, or that competition among ground herbs is primarily for water or nutrients (so that, as in deserts, roots touch even though leaves do not), or that light is only barely sufficient for growth under the conditions prevailing at the forest floor. If the last is true, then doubling light increases growth far more than doubling available water, phosphorus, etc. Coverage would then be very sensitive to available light, while increase of water would enable plants to grow in only slightly darker places. These questions can be investigated experimentally after the method of the intertidal ecologist (16).

Now, however, all one can say is:

1. the median light intensity at an average point on most rain forest floors is ∿0.5% of that above the canopy, although intensities up to 1% and one Sumatran rain forest with an intensity of 0.1% have been recorded (71, p. 83f);
2. sunflecks (too rare to affect the median) make average light intensity at the forest floor about three times the median (1, p. 33; 18; 63, pp. 40–55; 75);
3. light intensities are equally low for climax hardwood forest in the eastern United States and have been shown to be near the threshold of profitable photosynthesis for the understory plants there (56, 76). Competition for light, as measured by how little light is allowed to reach the forest floor, is apparently comparably intense in lowland tropical rain forests around the world, and in many mesic temperate forests as well.

It is amusing to note that in the climax Douglas fir forest on Mt. Lemmon, Arizona, which is as dense and leafy as eastern hemlock (*Tsuga*) forests, far more light reaches the forest floor (78), as if, in this mountain "island" surrounded by desert, competition is oriented around water and trees do not deliberately adapt to shade competitors. The same is undoubtedly true for climax mesic *Eucalyptus* in Australia, whose leaves hang vertically. Tropical forests are more similarly shady than a comparable series of mesic temperate forests.

Biomass and Production

The volume of timber per hectare is far more similar for different tropical forests than for a comparable series of virgin mesic temperate forests (Table 1). This is due less to difference in rates of wood production, which vary greatly from year to year but are roughly comparable for tropical and climax temperate forest (Table 1), than to the lower tree fall rates prevailing in some temperate forests. Tree fall rates in the lowland wet tropics generally exceed 1% a year for all sizes of tree, as we may judge either from Table 2 or from the ratio of standing crop to production in Table 1. Redwoods, by contrast, live over a thousand years.

Leaves generate the income needed for all other plant functions, so leaf fall changes much less from year to year (Table 3) than wood increment (Table 1) or fruit and flower fall (Table 4). Probably for the same reason, tropical and mesic temperate forests seem to support a similar area of leaves per unit ground area; tropical foliage, however, is heavier. On Barro Colorado Island, Panama, 8 ha of leaves fall per hectare of ground per year, including the holes eaten by insects (E. G. Leigh, in preparation), and a hectare of intact fallen leaves weighs about a ton when dried. At Hubbard Brook, New Hampshire, where a hectare of dried leaves weighs a half ton (77), 2.9 tons (24) or over 6 ha (counting holes) of leaves fall per hectare of ground per year. Assuming leaves are renewed once a year in both places, Barro Colorado and Hubbard Brook support leaf area indices of 8 and 6 respectively. The best direct estimates of leaf area index for a tropical forest come from the submontane rain forest of El Verde, Puerto Rico (51; pp. I-209). These values cluster about 6. Higher leaf areas are quoted for other tropical forests (52),

Table 1 Wood production in cubic meters per hectare per year, and standing crop of wood in cubic meters per hectare, in some tropical and temperate forests

Locality and authority	Production	Standing crop
Puerto Rico 1946–1956 (8)	9.7	300
Puerto Rico 1956–1966 (8)	4.0	332
Ivory Coast 1956–1960 (49)	13.1	525
Thailand 1962–1965 (40)	5.1	290
Cove forest, Tennessee (77)	5.4	851
Redwood forest, California (76)	13.1	6950
Hubbard Brook, New Hampshire (77)	3.8	204

Table 2 Tree fall rates in various tropical forests

Locality and authority	Number of trees watched	Duration of watch (yr)	Number of deaths	Average lifetime
Ulu Gombak, Malaya (46)	59	6	4	89
Ivory Coast (49)	130	5	9	72
El Verde, Puerto Rico (8)	720	20	166	87
BCI, Panama[a]	486	2.033	44	45

[a] G. G. Montgomery, in preparation.

Table 3 Leaf litter fall in dry tons per hectare for two successive years of sample in tropical and mesic temperate forests

Locality and authority	Leaf fall in		Annual rainfall (cm)
	1st year	2nd year	
Manaus, Brazil (41)	6.4	4.8	178
El Verde, Puerto Rico (79)	4.0	5.3	370
Banco, Ivory Coast: Plateau (3)	8.2	9.2	210
Thalweg (3)	7.9	7.3	210
Yapo, Ivory Coast: Plateau (3)	6.6	7.6	180
Thalweg (3)	5.7	6.8	180
BCI, Panama: Plateau[a]	5.8	6.3	250
BCI, Panama: Watershed (66)	7.5	6.9	260
Olokemeji, Nigeria (33)	4.5	4.7	120
Omo, Nigeria (33)	7.2		210
Ipassa, Gabon[b]	6.5		176
Pasoh, Malaya (44)	7.5		210
Glendon Hall, Toronto, Canada (7)	3.2	3.1	63
Great Divide, Victoria, Australia (7)	4.2		
Brookhaven, New York (81)	2.3	2.7	120

[a] R. Foster, personal communication.
[b] A. Hladik, personal communication.

Table 4 Fall in successive years from September, 1971 to August, 1974 of various categories of litter on Barro Colorado Island, in dry tons per hectare (N. Smythe, 66 and unpublished)

Year	Leaves	Twigs	Flowers	Fruit
1971–1972	7.8	1.9	0.5	1.4
1972–1973	7.5	1.6	0.3	0.5
1973–1974	6.3	1.5	0.2	0.6

but they are based on predicting a tree's leaf area from its diameter—a very inaccurate and untrustworthy business.

Vines may play an important role in tropical forest production. In the Ipassa rain forest of Gabon, which is much torn up by elephants and is thus (like disturbed forests elsewhere), inordinately full of vines; 40% of the leaf fall is from vines (28).

Transpiration and the Similarity of Lowland Rain Forest

To demonstrate that the similarity of lowland rain forests is due to the similar evaporative power of their climates, it is necessary to show that similar amounts of water are indeed evaporated or transpired in these climates, and to explain why similar evapotranspiration should matter so much for these forests.

I have defined rain forest as forest that, most months of the year, receives as much or more rain than it can use. On Barro Colorado Island, Panama, once the soil is saturated, rainfall exceeds runoff (measured as negative rainfall) by 10–20 cm a month, no matter how much rain falls (N. Smythe, personal communication); the remaining rainwater must either evaporate or be transpired. When the rains stop but the soil is still wet, west African rain forest transpires at a maximum rate of 15 cm a month (4). Walter (71, p. 212) presented a diagram suggesting that 100 cm of evenly spaced rainfall supports forest similar to that maintained by 6 or 7 quite wet months alternating with 5 or 6 months of drought, while 80 cm of evenly spread rainfall is correspondingly worth about 5 very wet months, as if the forest cannot use more than 16 cm of rain a month.

On the other hand, few rain forests receive less than 10 cm of rain a month for more than four months a year (58, p. 141) and most of the wet months are much wetter than this. Evapotranspiration in lowland rain forests thus lies between 120 and 190 cm a year. Table 5 gives rainfall and evapotranspiration for five watersheds covered by mature or old successional forests, suggesting how different the temperate values are, relative to the tropical. A very general convergence in lowland tropical climates conspires to produce this similarity in evapotranspiration: lowland rain forests nearly all receive about 6 hr of bright sunshine a day (58, p. 148), have an average annual temperature near 25°C and a diurnal temperature range near 10°C (71, p. 76), and in clearings relative humidity is nearly saturate at night, descending to near 60% at noon of a sunny day.

Evaporation of water through the stomata draws up water and soil nutrients from the roots, thanks to the cohesion of water molecules in the thin vessels of the xylem (60, pp. 168 and 125ff). Transpiration is accordingly a major factor in nutrient uptake (10, p. 358f; 51, p. I-241). The discovery of crassulacean metabolism, whereby stomata open at night when evaporation is low to take in CO_2, which is then stored as acid until needed for photosynthesis (71, p. 325), cuts harshly at the notion that daytime transpiration is a necessary evil connected with CO_2 uptake. Excessive transpiration is indeed harmful, but stomata shut to prevent it.

In spite of Rosenzweig's suggestive correlation (59) of production and evapotranspiration, there is little direct evidence that in tropical rain forest production actually does parallel evapotranspiration. Bernhard-Reversat, Huttel & Lemee (4) found a correlation of 0.78 between weekly growth of cross-sectional area of tree trunks in a west African rain forest and the amount of water the trees drew from the soil. Brown & Mathews (11, p. 521) found a seasonal rhythm to the growth of trees in Philippine rain forest, with sharp minima in the dry season when water was short

Table 5 Annual rainfall and annual evapotranspiration (in centimeters) in sample tropical and temperate watersheds

Locality and authority	Rainfall	Evapotranspiration
BCI, Panama (66)	250	140
Ulu Gombak, Malaya (5)	250	150
Sogeri Plateau, New Guinea (69)	270	120
Coweeta, North Carolina (38)	180	90
Hubbard Brook, New Hampshire (43)	120	50

and in the cloudiest portion of the rainy season when evaporation was lowest. Stronger evidence for the importance of transpiration is that lowland rain forests in climates of similar evaporative power are much alike, whereas montane rain forests, even forests of identical altitude and rainfall, vary enormously in structure according to the evaporative powers of their climates. We therefore turn to rain forest in the mountains.

MONTANE RAIN FOREST

The Relation Between Climate and Structure

INTRODUCTION   Tropical rain forest changes in many ways with increase of altitude. Leaves, both of canopy and understory plants, get smaller. If the mountains are high enough or close enough to the sea, they attain a cloudy zone where the forest is thickly coated with mosses and/or liverworts, filmy ferns, and other epiphytes. The trees may become more twisted and gnarled, and the forest may become shorter. Stratification often becomes more obvious and the canopy layer more even. Leaves of understory plants tend more and more to be arranged spirally or decussately around erect twigs rather than alternately or oppositely along horizontal ones (cf Figures 1–3). Finally, there may be signs of delayed or interrupted decomposition: the forest may be growing on a thick layer of peat or old logs. Not every montane forest will exhibit all these traits, but they all exhibit some.

GATES'S THEORY OF THE LEAF   I believe that many structural features of montane rain forest are dictated by the fact that transpiration is blocked when the air is saturated and leaf temperature is equal to air temperature. A leaf in a driving mist that does not permit it to warm thus cannot transpire at all, and a shaded leaf in a saturated atmosphere can transpire only very slightly.

To show this, I first outline Gates's (21, 22) theories of transpiration and gas exchange. Gates assumed that a leaf's temperature balances the heat gained from radiation reaching the leaf and the heat lost by reradiation from the leaf (which is more the warmer the leaf), heat lost by convection (which cools the leaf more, the stronger the wind, the smaller the leaf, and the more leaf temperature exceeds ambient), and heat lost through transpiration (which cools the leaf as sweat cools us).

The amount of water transpired per cm^2 of leaf is

$$[_s d_l(T_l) - \text{RH} _s d_a(T_a)]/(r_l + r_a),$$

where RH is relative humidity, T_l and T_a are temperature of leaf and air respectively, $_s d_l$ and $_s d_a$ are the saturation vapor pressures at leaf and air temperatures respectively, r_l measures the resistance of the stomatal layer to transpiration, and r_a measures the analogous resistance of the sourrounding layer of air (the latter is less, the stronger the wind or the smaller the leaf). A computer could find the leaf temperature-balancing heat gain and loss if it were told the leaf's size and "stomatal resistance," and the incident radiation, wind speed, relative humidity, and air temperature in the leaf's immediate environment. Knowing leaf temperature, one can then calculate transpiration rate.

Notice that wind affects transpiration two ways: it cools the leaf by convection and it lowers the resistance to transpiration. In dry air the wind enhances transpiration, but in very humid air the wind blocks it by cooling the leaf and annihilating the saturation differential on which transpiration depends.

THE GRADIENT IN LEAF SIZE   The most universal gradient in tropical rain forest is the tendency of leaf size to decrease with increase of altitude (25). My data for rain forests around the world suggest that average leaf length decreases ∽4.5% for every degree centigrade drop in the average temperature of the coolest month of the year. Temperature is apparently the best predictor of leaf size; I find that neither altitude nor the height of the canopy trees yields additional information on the average size of canopy leaves.

Parkhurst & Loucks (54) assumed that leaf size is adjusted to maximize water use efficiency: that is, to minimize the amount of water evaporated per unit CO_2 uptake. CO_2 uptake per unit leaf area is presumed to be the difference between the CO_2 concentration in the leaf mesophyll and in the surrounding air, divided by the sum of the resistances to CO_2 exchange by the mesophyll, the stomata, and the "boundary layer" of air. This last resistance varies according to wind speed and leaf size. Parkhurst & Loucks calculated the efficiency of water use for a "typical" leaf of "normal" size and resistance to gas and water vapor exchange in different combinations of temperature, humidity, and incident radiation (light intensity). They asked under what combinations of these environmental variables would a slight increase in leaf size increase water use efficiency. They found that larger size was favored in the shade, and that the shade required to favor a larger leaf was deeper, the lower the temperature. They concluded accordingly that shade leaves should be larger than sun leaves, and that average leaf size should be smaller in cooler places. Their conclusion assumes that the closer a given combination of temperature, light, and humidity is to one for which the "normal" leaf is best, the closer the normal leaf is to the optimum size for that combination of circumstances.

EPIPHYTES   If the mountain has a zone frequently bathed in fog, the forest there will usually be thick with moss or moss-like plants, while the forest just below this cloud zone will be rather thick with vascular epiphytes, generally held to be more prevalent in more humid atmospheres (10; 25, p. 596; 31, p. 48f; 74, p. 100). The cloud zone may start below 1000 m as in Puerto Rico or above 2000 m as in some parts of New Guinea. Presumably, the frequent rains and high humidities prevailing below the cloud belt lessen a plant's need for access to the soil and its reservoir of water, permitting epiphytes to replace vines (74). Epiphytes compete effectively for space on the ground in elfin woodland (10, p. 100; 68, p. 26), the most stunted form of cloud forest. It is as if the soil of elfin woodland were only a substrate and not a source of nutrient. Moreover, just as a warm climate encourages poikilothermic lizards over small homoiothermic mammals, the saturated atmosphere of the cloud belt encourages poikilohydric mosses, filmy ferns, etc over their homoiohydric vascular counterparts.

GNARLING AND CORIACEOUS LEAVES Montane forest trees often have gnarled twisted branches with leathery leaves and are often leaning and twisted (6, p. 268; 35; 70, p. 42; 74, p. 109). This gnarling, like that of dwarf bonsai pines cultured in sand, is presumably due to nutrient starvation. Such gnarling also occurs in the lowlands in some particularly nutrient-poor white-sand forests (82). In the mountains, gnarling may occur on thin soils (70), but it also occurs on apparently decent soil atop such mountains at Mt. Maquiling in the Philippines or El Yunque in Puerto Rico, where humidity is normally nearly saturate thanks to a steady sea breeze and frequent, persistent mist and fog (2, 10). The leaves are leathery, enabling them to resist the intense drying—more intense than in the lowlands—that occurs in the mountains when skies clear (6, p. 268f; 70, p. 37; 71, fig. 121).

INTERRUPTION OF THE CARBON CYCLE Montane forests often accumulate fallen logs or peat (68, 74) that, during exceptionally dry spells, make them unusually vulnerable to fire (36). The coriaceous, "xeromorphic" leaves of these habitats are expensive to make. They are accordingly built to last, and filled with poisons to keep off hungry insects (cf 37). These poisons hinder decomposition, especially in cooler environments. In short, the occasional intense drying spells that do occur in the mountains breed adaptations that interrupt the nutrient cycle and make the plant community less productive. Plant communities can "corrupt" this process to their own advantage. The trees of the upper montane forest capping Gunong Ulu Kali, Malaya, drop leaves that decay incompletely, forming peat (74, p. 114f). Their chemistry "encourages" hardpan to form between the peat and the mineral soil, blocking access to the nutrients below (13, p. 127). Upper montane forest with the same species occurs on isolated ridges 700 m below the main belt (74, p. 110), a striking contrast to the usual case where low-altitude patches of elfin forest contain the species appropriate to their low altitude (70, p. 42). Just as the New Jersey pine barrens "encourage" fires to burn out competitors, so might one say that this forest "makes" hardpan to starve competitors of nutrients, incidentally permitting peat to form.

FOREST HEIGHT Forest height generally decreases with increased altitude. This decrease has been ascribed to nutrient starvation (58, p. 355). At 1600 m on Mt. Nimba in West Africa, the forest is 20–30 m high on thick soil, whether in a ravine or on a plateau, whereas it is less than half as high where hardpan occurs slightly below the soil surface (61, pp. 307ff, 323ff). Van Steenis (70, p. 42) stated that Javanese montane forest is dwarfed where the soil is thin and leached of nutrient by heavy rains. He notes that the Japanese horticulturist dwarfs his bonsai plants through a similar nutrient starvation. He feels nutrient starvation must cause the dwarfing because, where the ridge widens out and the soil deepens, trees are taller, even though they are of the same species.

However, a steady humid sea breeze and persistent fog can dwarf even forests on relatively decent soil, presumably by blocking transpiration and, consequently, nutrient uptake. In these areas, it appears that evaporation is simply not strong enough to pull nutrients up to the tops of tall trees. Such humid conditions apparently have stunted the single-story 3 m forest atop the 1100 m El Yunque (50). Blasco (6, p. 162ff) described a two-story, 7–15 m moss forest in the High Wavys mountains of south India, much dwarfed, in spite of adequate soil, by an extremely foggy, seasonally windy climate. Curiously, this forest has the physiognomy of montane forests several hundred meters higher up, but has the species composition of taller forests at the same altitude. Sunny mountaintops higher up have taller forest (6, p. 173ff). Brown (10) described a single-story 10 m forest on good soil atop the 1000 m Mt. Maquiling, also stunted by humid breeze and fog. He noted that forest at the same altitude on the sunnier (but very wet) Mt. Banahao is 20 m tall, whereas the *Podocarpus* forest atop Mt. Banahao at 2100 m is 12 m tall and has two stories.

The stunting of these cloud forests cannot be ascribed simply to low production. A. Hladik (personal communication; 6, p. 185ff) has studied a moss forest at 2100 m in Sri Lanka that is analogous in every way to the cloud forest of the High Wavys mentioned above. This forest drops nearly six dry tons of litter (including twigs, etc) per hectare per year, about half the fall for lowland rain forests. The litter fall from this cloud forest is higher than the falls for most cool temperate forests (7), even though many of these northern forests must be much taller. The problem, rather than the availability of nutrients, seems to be getting the nutrients into the treetops.

Usually the cloud belt extends to the top of the mountain, or high enough up that the vegetation above it is stunted by overfrequent frost or an interrupted nutrient cycle, but it is perhaps comforting to learn of a mountain in New Guinea where the forest above the cloud belt is taller than the moss forest (58, p. 350).

Both breeze and fog are necessary to stunt a forest. The 30 m *Nothofagus* cloud forest atop the 2300 m Mt. Kaindi in New Guinea is said to be so high because that mountaintop is so windless (J. L. Gressitt, personal communication); the height that cloud forest can attain in sheltered ravines is well known (10, 64).

CROWN SHAPE   In nearly all the mossy or dwarfed montane forests I have visited, the trees have "dense crowns with shallow domed subcrowns" (74, p. 109), with "ramifications ultimes très denses" (6, p. 268). The trees are monolayer (cf 34); their leaves are held stiffly inclined, nearly erect, on the vertical twig ends, forming a cushion as on the top of a well-kept boxbush. Leaves thus arranged shelter each other from the wind as much as possible without shading each other excessively, presumably so that the leaves can warm up as much as possible above ambient temperature, permitting transpiration to occur even in a saturated atmosphere. Under extreme conditions, where breeze and fog are frequent, the canopy will be relatively even, generally lacking emergents, although it may not be entirely closed (35, p. 391; 74).

STORY STRUCTURE   Story structure is often much more obvious in montane rain forest than in the lowlands. Atop Mt. Kaindi in New Guinea, the 30 m *Nothofagus* form a high canopy over a mass of shrubbery 3–8 m high plastered over with a climbing bamboo (personal observation). Partly this reflects the relative stasis of montane forest: when trees fall rarely, layering can develop. Brown (10) noticed that in lowland dipterocarp forest the ground cover was mostly seedlings aspiring to grow into the upper stories, whereas in the higher and less productive forest, the ground was covered mostly by herbs specific to the ground layer.

However, the story structure also reflects the influence of the mosses. Where there is sufficient fog to permit thick moss to develop in the shade of the canopy (it can be more than 10 cm thick), the mosses take up water from the fog; the enormous store of water in this wet moss keeps the atmosphere within the forest perpetually humid. This humidity stunts the undergrowth, creating an obvious story structure. Accordingly, the distinguishing feature of the mossy forest atop Mt. Maquiling is not the evaporation rate above the canopy, but rather that evaporation is practically nil at all points below the canopy (Table 6). There is quite enough evaporation atop Mt. Maquiling to permit second-growth plants in the open to grow at a quarter of the very rapid rate displayed by second-growth plants at the base of the mountain (10, p. 168).

The stunting influence of moss may explain some other puzzling features of cloud forest. Blasco (6, p. 168) observed that in the cloud forest of the High Wavys, second growth in a cleared area grew in 20 yr to within 3 m of the height of the original forest. These second-growth trees are not contorted and they have remarkably large leaves. The soil profile under this second growth is identical to that under the primary forest, but there is as yet little moss. Does the accumulation of moss control the form and stature of the forest? Is accumulating moss yet another way by which plants can poison their competitors?

LEAF ARRANGEMENT   The higher the altitude, the greater the proportion of leaves on understory plants that are arranged spirally or decussately about erect twigs rather than alternately or oppositely along horizontal ones (42). Among the forests I have studied, the proportion of spiral and/or decussate arrangements in the understory correlates negatively and rather closely (−0.74) with the logarithm of the height of the forest canopy—a closer correlation than with either altitude or temperature. Such arrangements are typical of the canopy, and the shorter and more montane the forest, the more nearly will the spectrum of angles from which sunlight strikes understory leaves resemble that for canopy leaves.

Higher in the mountains, internode distances on understory plants are longer. Understory leaves now are scattered along their twigs rather than crowded at their tips, presumably because conditions will not support denser foliage.

Table 6 Evaporation in different levels of the forest at various altitudes on Mt. Maquiling, Philippine Islands (2 year averages, in arbitrary units), from data of Brown (10)

Site/Altitude	1050 m	740 m	450 m	300 m
On pole or atop dominant tree	5.3–5.4	7.2	14.7	16.4
In crown of dominant tree	0.6			
Atop second story tree		4.9	6.9	5.3
Near ground	0.4	2.0	3.4	2.7

OTHER EFFECTS OF CLIMATE   So far this review has considered the effects of average temperature, breeze, and fog on montane vegetation, finding that differences in windiness and in the distribution of cloudbelts may explain much of the diversity in structure among montane forests, whereby, for example, El Yunque at 1000 m supports a 3 m single-story moss forest while Mt. Kaindi at 2300 m supports a 30 m rather mossier forest. Yet to be explained are 1. why the canopies of cloud forests so often appear drought-resistant (xeromorphic), and 2. why vegetation above the cloud forest is so frequently more stunted than the cloud forest itself. I can only suggest answers.

1. Under clear skies, radiation during the day and reradiation from the earth during the night are both stronger in the mountains (71, p. 203). A lowland tropical sky is often silvery or grey-blue with water vapor even when the sun is shining (71, p. 79), whereas mountain skies, lacking such obstacles to radiation, are a clear rich blue. The alternation of sun and shade (55, p. 36) thus imposes a greater variation of temperature on canopy foliage in the mountains. In spite of the frequent fog, the vegetation must be able to resist sunburn or occasional excess transpiration.

The temperature variations affecting mountain plants cannot be adequately monitored by the meteorologist's customary screened thermometer. Shreve (64, p. 12) recorded that on a clear night, grass in a clearing in a montane rain forest at 1700 m in the Blue Mountains of Jamaica was exposed to temperatures 6°C lower than those registered by a screened thermometer.

2. Above the cloud zone, skies are often clear and night frosts frequent. The harm these frosts work is illustrated by the rhythm of growth in Andean paramo. It is bathed in cloud during the rainy season, when diurnal temperature range as measured by screened thermometers may be less than 2°C, but the skies are clear in the dry season, when diurnal temperature range may exceed 17°C, with frost every night (72, p. 56). A. Smith (personal communication) informs me that paramo plants grow much better in the rainy season.

Production

How does production vary with altitude? Wood production for various forests can be estimated from data in refs (9, 10, 12, 47). One can calculate the average diameter increased ΔD for trees of a given diameter D, and estimate volume increment for such a tree as $HD\Delta D\pi/4$, where H is the tree's height. Wood production is the sum of these increments for all sizes of tree on the plot. These crude, but comparable, estimates are given in Table 7. Comparable litter fall data are given in Table 8. Through an odd chance, the data mostly record total litter fall rather than leaf litter only. Notice that all the most productive sites are well below the cloud zone, whereas the midmontane forest on Mt. Maquiling is shadowed by the cloud zone, and the others are in or partly above the clouds. These data are scanty, but they suggest the importance of evaporation to production as well as structure.

Summary: Transpiration and Montane Forest Structure

In short, the effects of humid climate mimic those of poor soil: poor soils lack nutrients, whereas foggy, breezy climates block transpiration and prevent nutrient

Table 7 Estimated annual wood production, in cubic meters per hectare; average annual temperature and climate notes for various Philippine and Javanese forests

Site, altitude, and vegetation	Production	Temperature	Climate type
Mt. Maquiling, Philippines (10)			
300–450 m 37 m dipterocarp forest	12.4	23°C	normal lowland
740 m 22 m midmontane forest	4.8	22°C	just below clouds
Tjibodas, Java (12, 47)			
1500 m 50 m high forest	10.7	18°C	well below clouds
Mt. Banahao, Philippines (9)			
2100 m 12 m podocarp forest	~ 3	15°C	very wet: in or above clouds

Table 8 Total litter fall (dry tons per hectare); average annual temperature and forest type for various localities

Site, altitude, authority	forest height	litter fall	average temperature	forest type
BCI, Panama, 100 m (66)	35 m	11.4	26°C	lowland
Rancho Grande, Venezuela, 1100 m (45)	20 m	7.8	19°C	cloud
San Jose, Costa Rica, 1200 m (20)	—	27	20°C	premontane
Horton Plains, Sri Lanka, 2100 m[a]	15 m	5.9	14°C	cloud

[a] A. Hladik, personal communication.

uptake. Shade plants and plants exposed to wind are disproportionately stunted, as one expects in cases where transpiration is responsible. In cloud forest, moreover, tree crowns are designed to minimize the cooling effects of wind. There is also a great increase of often facultative epiphytes at the expense of vines; the climate provides so little power for drawing nutrients from the ground that being rooted in soil is of little use except for support.

ANIMALS AND THE STRUCTURE OF THE FOREST

A marine ecologist (53) acquainted with the very striking effects of starfish on the zonation of a rocky shore might well ask why I have discussed forest structure without mentioning animals. Why have I so resolutely insisted on climate as the primary "cause" of rain forest structure when insect pressure (23) has so much to do with the diversity of the forest and the rarity of specific kinds of trees?

How Much do Animals Eat?

To answer, I first assess the importance of folivores, frugivores, and some of their predators in the food web of old second-growth forest on Barro Colorado Island, Panama, then pass on to the rhythms by which the forest as a whole defends itself against its pests. The chemical defenses by which each species immunizes itself against as many pests as possible are well documented elsewhere (67) and not considered here.

Express the budget in dry weights, and assume the leaves and fruit that sloths and monkeys eat are 2/3 and 3/4 water respectively (Gaulin, personal communication). When more direct estimates of feeding rate are lacking, equations can be used for predicting the standard metabolic rates of birds and mammals from their weights (17, 39). These rates, multiplied by 3 (cf 32), give feeding rates. To convert to dry weight, assume that 5000 kcal corresponds to one dry kg (cf 51, p. I–214).

Seven dry tons of leaves fall per hectare per year (Table 4). Vertebrates in the trees, mostly sloths and howling monkeys, eat about 150 dry kg of leaves per hectare per year. By contrast, a Ceylonese deciduous monsoon forest drops half as much leaf litter, while its leaf monkeys still crop 150 dry kg of leaves per hectare per year (30).

In the secondary forest of Barro Colorado there are about ten sloths per hectare, which together defecate 30 dry kg a year. They defecate only once a week, the better to digest their food (48). Their very low standard metabolic rate accounts for 40 dry kg a year (cf 17). I presume they eat in all 100 dry kg of leaves per hectare per year. There are two thirds of a howling monkey per hectare (15); these monkeys eat 40 dry kg of leaves and 30 kg of fruit per hectare per year (29). Howlers, and probably sloths, prefer young leaves, and they prefer or require specific kinds of leaves (K. Milton, personal communication; 48). When the rains end and new leaves are rare, they perhaps can eat only as fast as they can digest (73). Their numbers seem to be regulated proximately by the opportunities for subadults to establish themselves in troop or territory, and ultimately by the shortage of suitable food in the lean season.

There are 800 kg worth of holes and gaps in the 7 dry tons of leaves that do fall on Barro Colorado (E. G. Leigh, in preparation). Vertebrates cause few of these, for they eat their leaves whole. However, insects do not necessarily eat all 800 kg. Reichle et al (57) showed that insects prefer young leaves, and that in the *Liriodendron* forest they studied, holes made by insects nearly tripled in area as the leaves grew.

To judge from unpublished, rather rough, censuses of E. O. Willis, and Karr's (39) data on bird weights, the insectivorous birds of Barro Colorado, excluding woodpeckers and feeders on litter fauna, eat about 40 dry kg of insects per hectare [those of Hubbard Brook, New Hampshire eat only 8 (cf ref 32)]. Just as it takes ten pounds of feed to make a pound of beef, so we may presume it takes about 400 dry kg of leaf to make the 40 kg of insects the birds eat. How much more leaf goes into the food of bat or spider one cannot say: they presumably eat less than the birds. In any event, the insects' predators seem to be an integral and effective part of the forest's defense against pest pressure.

Fruit consumption begins in the trees. The Jamaican fruit bat, *Artibeus jamaicensis,* eats about 20 dry kg of fruit, mostly figs, per hectare per year (D. Morrison, personal communication; F. J. Bonnacorso, personal communication); other bats together perhaps eat half as much more. Monkeys eat 40 dry kg per hectare per year (29), and frugivorous birds eat 30. In all, vertebrates eat about 150 dry kg of fruit and seed per hectare per year in the trees. Over a half ton more falls to the ground (Table 4). Judging from the metabolism estimates of Eisenberg & Thorington (17), agoutis, pacas, spiny rats, and coatis (the main terrestrial frugivores, whom I presume eat fruit half the year) eat a seventh of the fallen fruit and seed; much of the rest rots (A. Worthington, personal communication) or is spoiled by insects. Some sprouts.

The Community's Defense: the Rhythm of Tropical Forest

So much fruit goes to waste because the trees of Barro Colorado restrict the numbers of frugivores through alternate seasons of fruit abundance and shortage (65). Fruit falls mainly between April and June and in September and October (R. Foster, personal communication). Between November and March terrestrial mammals are hungry and easy to trap (N. Smythe, personal communication), and one year, when unseasonable rains disrupted the fall fruiting, there was mass starvation in the forest (80).

This rhythm is a fundamental feature of Barro Colorado. A sharp dry season lasts from mid-December into April; as it ends, new leaves begin flushing. Sloths give birth in June (48) when succulent new leaves are presumably most abundant. Insects are also peaking then (66), as is bird breeding. Even frugivorous birds breed then, for they feed insects to their young. Presumably, a dry season shortage of new leaves restricts the number of insects and the survival of young sloths and howling monkeys. By the end of rainy season, bird breeding (E. O. Willis, manuscript) has almost entirely ceased.

Sarawak, by contrast, lacks a true dry season; the average rainfall for the driest month at Kuching is 20 cm (19), more than the forest can transpire. Yet the rhythm of leaf flush, fortunately more carefully documented than on Barro Colorado, is equally marked. Leaf flush peaks from November to February with the onset of the northeast monsoon, and new leaves are least common in September when insects are also rarest (19). Birds breed from December to May. Their numbers and breeding rhythm are apparently governed by food availability (19). The September lean season no doubt is meant to restrict insect numbers, or, to be more precise, it is maintained because in September any conspicuous leaf flush would be eaten by hungry insects. Malaysian fruiting rhythms, however, differ greatly from those of Barro Colorado. There is an overall annual fruiting cycle in Malaysian forest (46), but the dipterocarps that dominate these forests tend to fruit gregariously, many species together, every few years, in response to unusually sharp dry spells. Dipterocarps are wind-dispersed, and animals avoid eating the fruit (46); adapting to eat it may be like adapting to eat a distasteful strain of 17-year cicadas. In short, even when meteorological seasons are not obvious, the forest controls its animals partly by imposing lean seasons on them.

ACKNOWLEDGMENT

I have studied rain forests in South and Central America, the West Indies, West Africa, Madagascar, southern Asia, and New Guinea. I have burdened this review with as little unpublished data as possible: these visits, however, have affected my perspective on the literature and widened my knowledge of it. They were made possible by the generosity of the American Philosophical Society, Princeton University, and the Smithsonian Tropical Research Institute.

It has been my good fortune that when I, as others before me, needed assistance the learned did not refuse to help me. Drs. Benjamin Stone and R. H. Whittaker very kindly took the time to answer a host of questions and to forward information not otherwise available to me. Many have permitted me the use of unpublished data that have helped to set findings in the literature in perspective. Dr. N. Smythe allowed me use of the wealth of unpublished records whose collection he has so ably organized on behalf of the Smithsonian Environmental Sciences Program. Sra Mejia, librarian of the Smithsonian Tropical Research Institute, has kindly procured for me copies of some most obscure publications. I. Downs drew the figures.

Finally, I must thank the many friends and colleagues who have patiently read and criticized drafts of this paper. I am particularly grateful to A. R. Kiester for asking one disagreeable but essential question, and to C. Toft for her help in imposing some sense of form and purpose on this paper.

Literature Cited

1. Baur, G. N. 1968. *The Ecological Basis of Rain Forest Management.* Sydney: Government Printer, New South Wales
2. Baynton, H. W. 1968. The ecology of an elfin forest on Puerto Rico 2. The microclimate of Pico del Oeste. *J. Arnold Arbor. Harv. Univ.* 49:419–30
3. Bernhard, F. 1970. Etude de la litière et de sa contribution au cycle des elements mineraux en forêt ombrophile de Côte-d'Ivoire. *Oecol. Plant.* 5:247–66
4. Bernhard-Reversat, F., Huttel, C., Lemee, G. 1972. Quelques aspects de la periodicité ecologique et de l'activité vegetale saisonnière en forêt ombrophile sempervirente de Côte d'Ivoire. *Papers from a Symposium on Tropical Ecology with an Emphasis on Organic Productivity,* ed. P. M. Golley, F. B. Golley, 219–43. Athens, Georgia. 405 pp.
5. Bishop, J. E. 1973. *Limnology of a Small Malayan River Sungei Gombak.* The Hague: Junk. 485 pp.
6. Blasco, F. 1971. *Montagnes du Sud de l'Inde. Forets, Savanes, Ecologie.* Madras: Inst. Francais de Pondichery. 436 pp.
7. Bray, J. R., Gorham, E. 1964. Litter production in forests of the world. *Adv. Ecol. Res.* 2:101–57
8. Briscoe, C. B., Wadsworth, F. H. 1970. Stand structure and yield in the tabonuco forest of Puerto Rico. *A Tropical Rain Forest,* ed. H. T. Odum,

R. Pigeon, B-79-89. Springfield, Va. US AEC
9. Brown, W. H. 1917. The rate of growth of *Podocarpus imbricatus* at the top of Mt. Banahao, Luzon, Philippine Islands. *Philipp. J. Sci. C* 12:317–28
10. Brown, W. H. 1919. *The Vegetation of Philippine Mountains.* Manila: Bur. Sci. 433 pp.
11. Brown, W. H., Mathews, D. M. 1914. Philippine dipterocarp forests. *Philipp. J. Sci. A* 9:413–561
12. Brown, W. H., Yates, H. S. 1917. The rate of growth of some trees on the Gedeh, Java. *Philipp. J. Sci. C* 12:305–10
13. Burgess, P. F. 1969. Ecological factors in hill and mountain forests of Malaya. *Malay. Natur. J.* 22:119–28
14. Cain, S. A., de Oliveira Castro, G. A., Pires, J. M., da Silva, N. T. 1956. Application of some phytosociological techniques to Brazilian rain forest. *Am. J. Bot.* 43:911–41
15. Chivers, D. J. 1969. On the daily behavior and spacing of howling monkey groups. *Folia Primatol.* 10:48–102
16. Connell, J. H. 1961. The influence of interspecific competition and other factors on the distribution of the barnacle *Chthamalus stellatus. Ecology* 42: 710–23
17. Eisenberg, J. F., Thorington, R. W. Jr. 1973. A preliminary analysis of a neo-

tropical mammal fauna. *Biotropica* 5:150–61

18. Evans, G. C. 1956. An area survey method of investigating the distribution of light intensity in woodlands, with particular reference to sunflecks. *J. Ecol.* 44:391–428

19. Fogden, M. P. L. 1972. The seasonality and population dynamics of equatorial forest birds in Sarawak. *Ibis* 114: 307–42

20. Fournier, L. A., de Castro, L. C. 1973. Produccion y descomposicion del mantillo en un bosque secundario humedo de premontano. *Rev. Biol. Trop.* 21: 59–67

21. Gates, D. M. 1968. Energy exchange and ecology. *Bioscience* 18:90–95

22. Gates, D. M. 1969. The ecology of an elfin forest in Puerto Rico. 4. Transpiration rates and temperatures of leaves in a cool humid environment. *J. Arnold Arbor. Harv. Univ.* 50:93–98

23. Gillett, J. B. 1962. Pest pressure, an underestimated factor in evolution. *Syst. Assoc. Publ.* 4:37–46

24. Gosz, J. R., Likens, G. E., Bormann, F. H. 1972. Nutrient content of litter fall on the Hubbard Brook experimental forest, New Hampshire. *Ecology* 53:769–84

25. Grubb, P. J., Lloyd, J. R., Pennington, T. D., Whitmore, T. C. 1963. A comparison of montane and lowland rain forest in Ecuador. I. The forest structure, physiognomy and floristics. *J. Ecol.* 51:567–601

26. Hadfield, W. 1974. Shade in northeast Indian tea plantations. II. Foliar illumination and canopy characteristics. *J. Appl. Ecol.* 11:179–99

27. Henwood, K. 1973. A structural model of forces in buttressed tropical rain forest trees. *Biotropica* 5:83–93

28. Hladik, A. 1974. Importance des lianes dans la production foliare de la forêt equatoriale du Nord-Est du Gabon. *C. R. Acad. Sci. Paris D* 278:2527–30

29. Hladik, A., Hladik, C. M. 1969. Rapports trophiques entre vegetation et primates dans la forêt de Barro Colorado (Panama). *Terre et Vie* 23:25–117

30. Hladik, C. M., Hladik, A. 1972. Disponibilités alimentaires et domaines vitaux des primates à Ceylan. *Terre et Vie* 26:149–215

31. Holdridge, L. R., Grenke, W. C., Hatheway, W. H., Liang, T., Tosi, J. A. Jr. 1971. *Forest Environments in Tropical Life Zones: A Pilot Study.* Oxford: Pergamon. 747 pp.

32. Holmes, R. T., Sturges, F. W. 1973. Annual energy expenditure by the avifauna of a northern hardwoods ecosystem. *Oikos* 24:24–29

33. Hopkins, B. 1966. Vegetation of the Olokemeji Forest Reserve, Nigeria. IV. The litter and soil, with special reference to their seasonal changes. *J. Ecol.* 54:687–703

34. Horn, H. S. 1971. *The Adaptive Geometry of Trees.* Princeton, NJ: Princeton Univ. Press. 144 pp.

35. Howard, R. A. 1968. The ecology of an elfin forest in Puerto Rico. 1. Introduction and composition studies. *J. Arnold Arbor. Harv. Univ.* 49:381–418

36. Humbert, H., Cours Darne, G. 1965. *Notice de la Carte Madagascar.* Madras: Inst. Français de Pondichery. 164 pp.

37. Janzen, D. H. 1974. Tropical blackwater rivers, animals, and mast fruiting by the Dipterocarpaceae. *Biotropica* 6: 69–103

38. Johnson, P. L., Swank, W. T. 1973. Studies of cation budgets in the southern Appalachians on four experimental watersheds with contrasting vegetation. *Ecology* 54:70–80

39. Karr, J. R. 1971. Structure of avian communities in selected Panama and Illinois habitats. *Ecol. Monogr.* 41: 207–33

40. Kira, T., Ogawa, H., Yoda, K., Ogino, K. 1967. Comparative ecological studies on three main types of forest vegetation in Thailand. IV. Dry matter production, with special reference to the Khao Chong rain forest. *Nature and Life in Southeast Asia* 5:149–74

41. Klinge, H., Rodrigues, W. A. 1968. Litter production in an area of Amazonian terra firme forest. *Amazoniana* 1:287–310

42. Leigh, E. G. Jr. 1972. The golden section and spiral leaf-arrangement. *Growth by Intussusception,* ed. E. S. Deevey Jr., 163–76. Hamden, Conn: Archon Books. 441 pp.

43. Likens, G. E., Bormann, F. H., Johnson, N. M., Pierce, R. S. 1967. The calcium, magnesium, potassium and sodium budgets for a small forested ecosystem. *Ecology* 48:772–85

44. Lim, M. T. 1974. *The study of litter decomposition in Pasoh.* Presented at IBP Synthesis Meeting, Kuala Lumpur, August 12–18, 1974. Mimeographed report

45. Medina, E., Zelwer, M. 1972. Soil respi-

ration in tropical plant communities. See Ref. 4:245–67

46. Medway, Lord 1972. Phenology of a tropical rain forest in Malaya. *Biol. J. Linn. Soc.* 4:117–46

47. Meijer, W. 1959. Plant sociological analysis of the montane rainforest near Tjibodas, West Java. *Acta Bot. Neerl.* 8:277–91

48. Montgomery, G. G., Sunquist, M. E. 1974. Impact of sloths on neotropical energy flow and nutrient cycling. *Trends in Tropical Ecology: Ecological Studies IV*, ed. E. Medina, F. B. Golley. New York: Springer

49. Muller, D., Nielsen, J. 1965. Production brute, pertes par respiration et production nette dans la foret ombrophile tropicale. *Forstl. Forsøgsvaes. Dan.* 29:69–160

50. Odum, H. T. 1970. Rain forest structure and mineral-cycling homeostasis. See Ref. 8:H-3-52

51. Odum, H. T. 1970. Summary: an emerging view of the ecological system at El Verde. See Ref. 8,I-191-289

52. Ogawa, H., Yoda, K., Ogino, K., Kira, T. 1965. Comparative studies on three main types of forest vegetation in Thailand. II. Plant biomass. *Nature and Life in Southeast Asia* 4:49–80

53. Paine, R. T. 1974. Intertidal community structure. Experimental studies on the relationship between a dominant competitor and its principal predator. *Oecologia* 15:93–120

54. Parkhurst, D. F., Loucks, O. L. 1972. Optimal leaf size in relation to environment. *J. Ecol.* 60:505–37

55. Paulian, R. 1961 *La Zoogeographie de Madagascar et des Iles Voisines.* Tananarive: L'Institut de Recherche Scientifique. 481 pp.

56. Perry, T. O., Sellers, H. E., Blanchard, C. O. 1969. Estimation of photosynthetically active radiation under a forest canopy with chlorophyll extracts and from basal area measurements. *Ecology* 50:39–44

57. Reichle, D. E., Goldstein, R. A., Van Hook, R. I. Jr., Dodson, G. J. 1973. Analysis of insect consumption in a forest canopy. *Ecology* 54:1076–84

58. Richards, P. W. 1952. *The Tropical Rain Forest.* London: Cambridge Univ. Press. 450 pp.

59. Rosenzweig, M. L. 1968. Net primary productivity of terrestrial communities: prediction from climatological data. *Am. Natur.* 102:67–74

60. Salisbury, F. B., Ross, C. 1969. *Plant Physiology.* Belmont, Calif: Wadsworth. 747 pp.

61. Schnell, R. 1952. *Vegetation et Flore de la Region Montagneuse du Nimba.* Dakar: Inst. Français d'Afrique Noire. 604 pp.

62. Schnell, R. 1970. *Introduction à la Phytogeographie des Pays Tropicaux. Vol. 1. Les Flores–Les Structures.* Paris: Gauthier-Villars. 499 pp.

63. Schulz, J. P. 1960. *Ecological Studies on Rain Forest in Northern Suriname.* Amsterdam: North-Holland. 267 pp.

64. Shreve, F. 1914. A montane rain forest. A contribution to the physiological plant geography of Jamaica. *Publ. Carnegie Inst. Washington* 119:1–110

65. Smythe, N. 1970. Relationships between fruiting seasons and seed dispersal methods in a neotropical forest. *Am. Natur.* 104:25–35

66. Smythe, N. 1974. Terrestrial studies— Barro Colorado Island. *Smithsonian Institution Environmental Sciences Program 1973: Environmental Monitoring and Baseline Data,* ed R. Rubinoff, 1–127. Washington DC: Smithsonian Inst. 465 pp. Mimeographed report

67. Sondheimer, J. B., Simeone, E., Eds. 1970. *Chemical Ecology.* New York: Academic. 336 pp.

68. Terborgh, J. W. 1971. Distribution on environmental gradients: theory and a preliminary interpretation of distributional patterns in the avifauna of the Cordillera Vilcabamba, Peru. *Ecology* 52:23–40

69. Turvey, N. D. 1974. Water in the nutrient cycle of a Papuan rain forest. *Nature* 251:414–15

70. van Steenis, C. G. G. J. 1972. *The Mountain Flora of Java.* Leiden: E. J. Brill

71. Walter, H. 1971. *Ecology of Tropical and Subtropical Vegetation.* Edinburgh: Oliver & Boyd. 539 pp.

72. Walter, H. 1973. *Vegetation of the Earth in Relation to Climate and the Ecophysiological Conditions.* New York: Springer. 237 pp.

73. Westoby, M. 1974. An analysis of diet selection by large generalist herbivores. *Am. Natur.* 108:290–304

74. Whitmore, T. C., Burnham, C. P. 1969. The altitudinal sequence of forests and soils on granite near Kuala Lumpur. *Malay. Natur. J.* 22:99–118

75. Whitmore, T. C., Wong, Y. K. 1959. Patterns of sunfleck and shade light in

tropical rain forest. *Malay. For.*
22:50–62

76. Whittaker, R. H. 1966. Forest dimensions and production in the Great Smoky Mountains. *Ecology* 47:103–21

77. Whittaker, R. H., Bormann, F. H., Likens, G. E., Siccama, T. G. 1974. The Hubbard Brook ecosystem study: forest biomass and production. *Ecol. Monogr.* 44:233–54

78. Whittaker, R. H., Niering, W. 1975. Vegetation of the Santa Catalina Mountains, Arizona. V. Biomass, production and diversity along the elevation gradient. *Ecology.* In press

79. Wiegert, R. G. 1970. Effects of ionizing radiation on leaf fall, decomposition, and litter microarthropods of a montane rain forest. See Ref. 8, H-89-100

80. Willis, E. O. 1974. Populations and local extinctions of birds on Barro Colorado Island, Panama. *Ecol. Monogr.* 44:153–69

81. Woodwell, G. M., Marples, T. G. 1968. The influence of chronic gamma irradiation on production and decay of litter and humus in an oak-pine forest. *Ecology* 49:456–65

82. Takeuchi, M. 1961. The structure of the Amazonian vegetation. III. *Campina* forest in the Rio Negro region. J. Fac. Sci. Univ. Tokyo 3. 8:27–35

CANNIBALISM IN NATURAL POPULATIONS

❖4088

Laurel R. Fox

Department of Environmental Biology, Research School of Biological Sciences, Australian National University, Canberra, A.C.T. 2601

INTRODUCTION

Cannibalism, defined as intraspecific predation, is a behavioral trait found in a wide variety of animals, although most references to this behavior are anecdotal or based on casual laboratory observations. The role of cannibalism in the dynamics of natural populations has been largely neglected; the most elegant and detailed analyses of the population consequences of cannibalism are still provided by laboratory studies of flour beetles (*Tribolium*) that describe the process, examine its interactions with other population processes, and attempt to derive generalities about its effects (60, 78, 88). Therefore, some authors have suggested that cannibalism is an artifact of laboratory systems (20, p. 324) or that it occurs only in cases of severe stress, especially when alternatives, such as dispersal, are not possible (105).

My purpose in this review is to show that cannibalism is a normal phenomenon in many natural populations, to evaluate its possible roles in influencing demographic structure and population processes, and to suggest conditions for, and constraints on, its occurrence. Cannibalism may be an interaction that reduces population size before acute resource shortage causes severe physiological stress, and in this sense its effects may be analogous to those of spacing behavior or dominance hierarchies in some social animals (104). I also discuss cannibalism as a limiting case of predator-prey interactions among potential competitors.

This review is based mainly on field experiments and direct estimates of rates of cannibalism, but I use papers describing laboratory experiments to supplement the discussion of behavioral triggers and genetic components and to fill gaps in the field evidence. I have excluded numerous references with only casual mention of cannibalistic events.

DISTRIBUTION OF CANNIBALISM

Cannibalism has been reported in many groups of animals, e.g. protozoa (54), planaria (4), rotifers (46), snails (72), copepods (3), centipedes (31), mites (24, 98,

87

106), insects (63, 100), fish (66, 91), anurans (12), and birds and mammals (99, 108). Table 1 summarizes the distribution of species for which there are sufficient data to indicate that cannibalism occurs normally in the field. I have grouped these species by their predominant feeding habit as herbivores (including scavengers and a few omnivores) or predators, and by habitat. Perhaps surprisingly, a large proportion of observations of cannibalism among terrestrial animals is for species that are usually "herbivores," including butterfly larvae (14), leaf-eating beetles (35), and bark beetles (6). Several authors have noted that cannibalism may be widespread among herbivorous insects. Kirkpatrick (63, p. 194) stated that cannibalism is rare among terrestrial predatory insects, but that " . . . some plant-feeding insects are inveterate cannibals, even in the presence of an abundance of food." Strawinski (100) mentioned only one cannibalistic species among the 55 carnivorous hemipterans in Poland, but listed 6 of the 67 species (9%) of bugs usually regarded as phytophagous as eating their own species' eggs. Cannibalism is also common among granivores (88). Most species in the category of terrestrial carnivores in Table 1 are insects such as coccinellid beetles (25 species), syrphids (Diptera, 19), and chrysopids (Neuroptera, 71), but the list also includes other arthropods (e.g. 32, 71) and some vertebrates (65, 95).

The second major point of interest in Table 1 is the large proportion of known cannibalistic species that are predators living in fresh water, primarily fish and insects. This might be an artifact of the relative ease of collecting adequate samples for gut analyses in freshwater situations. In contrast, it is frequently difficult to observe or collect large numbers of individuals in terrestrial systems, and one common method for identifying prey, based on antibody reactions (27, 28), cannot detect cannibalism. Another explanation could be that the large proportion of freshwater predators among cannibalistic species might reflect basic differences in mechanisms determining population and community structure in terrestrial and aquatic systems. Freshwater habitats may be less complex faunistically or less stable than terrestrial ones, and many cannibalistic aquatic species do live in relatively ephemeral or nonstructured habitats such as rock pools or leaf axils (84, 87). However, habitats that are spatially more heterogeneous and physically more stable, such as larger ponds or lakes, also support cannibalistic populations (e.g. 66, 70).

Very few marine species are listed in Table 1. There are many anecdotal mentions of occasional cannibalistic acts by marine animals, but detailed feeding records are available for relatively few species. Apart from studies of intertidal molluscs (72, 86), most well documented examples are commercially important species of fish (85).

Table 1 Number of species for which there are sufficient data to indicate that cannibalism occurs normally in the field

Habitat	Trophic Level	
	Herbivores	Predators
Terrestrial	38	38
Freshwater	10	43
Marine	0	8

FACTORS AFFECTING CANNIBALISM

While cannibalism is frequently a response to food or density, other factors may also be important; in many species, several such factors are known to be involved.

Food

Starvation may increase cannibalistic tendencies, but it is not essential for initiating this behavior. Many animals will cannibalize as soon as all other food items are removed, but they may also respond simply to a reduction in the relative availability of alternatives. Cannibalism has been shown to be inversely related to the density of other food in intertidal opisthobranchs (86), freshwater insects (39, 43, 56), flatworms (4), and fish (16, 17, 40, 91, 97), as well as among terrestrial herbivores such as the cotton leafworm, *Spodoptera littoralis* (1), and the black cutworm, *Agrotis ipsilon* (36).

Rates of cannibalism may vary with the food resources present in different sites. Fahy (38) compared the diets of benthic insects in different parts of an oligotrophic stream in Ireland. The upper section was low in detrital content and consequently had smaller herbivore populations than a section further downstream, with higher detrital inputs. The species present were much the same at both sites but the trophic relationships differed. In the upper, poorer area, three of the predatory stoneflies and mayflies were cannibalistic and there were many predatory interactions among the predators themselves. At the more favorable lower site cannibalism did not occur and there were fewer predatory interactions. Perch, *Perca fluviatilis,* may also be more cannibalistic in lakes that are poor in nutrients than in eutrophic situations (2), and among humans, nonritual cannibalism seems to have been most common in nutritionally marginal areas supporting relatively low population densities, where it may have provided 5–10% of the annual protein requirement (30, 96). It has been suggested that cannibalism was less common in those settlements where populations were dense enough to raise their own food and produce a more predictable and adequate supply (30).

Density

The effects of crowding are often confounded with those of food shortage, but in several examples cannibalism was mainly a response to high density. Under crowded conditions survival of litters of house mice, *Mus musculus,* was low even when given excess food and nesting material (98). This mortality was attributed to improper parental care, including cannibalism, although the magnitudes of specific mortality factors were not measured. Similarly, both crowded damselfly larvae, *Lestes nympha* (39), and pike, *Esox lucius* (62), were very cannibalistic in the presence of other food, although the amount of food offered also affected cannibalism rates.

Behavior of Victims

In some situations cannibalism is initiated by particular behavioral patterns of the susceptible individuals. Laboratory experiments on fish in the *Poeciliopsis mona-cha–P. lucida* species complex, in which the relative densities of different age-classes were varied, have shown that mortality may be related more to the responses of the young to density than to the density of cannibalistic adults (101). These fish are viviparous and the young were vulnerable to cannibalism for about 24 hours after birth. Below a critical density the young dispersed throughout the tank and were eaten by females as encountered, but above densities of about 10 per tank the young formed aggregations that stimulated attacks by the adult females. The adults' behavior was not affected by their degree of starvation, and the females remained unsatiated even after eating large numbers of young. In another example, differences in the rates of cannibalism of two closely related species of Lepidoptera may also be explained by differences in larval behavior patterns. Larvae of the corn ear-worm, *Heliothis armigera,* move actively and induce aggressive responses when contacted by other larvae. In contrast, larvae of *H. punctigera* do not move when encountered, and cannibalism rates are much lower than in *H. armigera* (S. Stanley, personal communication).

Stress

Physiological or psychological stress has been associated with cannibalism in only a few examples in field situations. The success of litters of Norway rats, *Rattus norvegicus,* was related to the social rank of their parents. Low ranking, scarred females that were caught in the field ate more than 60% of their young, while unscarred and presumably higher ranking females weaned all of their offspring (8). In the Australian grasshopper, *Phaulacridium vittatum,* mass emigration of older juveniles and young adults occurred if food was seriously depleted; migrating animals showed obvious signs of physiological stress, eating corpses and other migrating grasshoppers. Those remaining at the original site did not seem to be stressed and did not cannibalize (18).

Availability of Victims

In many examples initiation and control of cannibalism has not been ascribed to any obvious factor, and in these cases cannibalism may be a response primarily to the presence of vulnerable individuals. For some species, rates of cannibalism are consistent with simple encounter models in which the probability of attack is proportional to the probability of encountering a vulnerable individual. For instance, at low population densities with abundant alternative food, the freshwater backswimmer, *Notonecta hoffmanni,* cannibalized whenever vulnerable individuals were present (43). In other insects cannibalism rates on eggs and newly hatched young may be determined by the size of egg batches and the time span over which they hatch. There was no cannibalism on eggs or young larvae of coccinellid beetles if they all hatched before the oldest began to search for prey, which it did when about one hour old (59).

Herbivorous insects may cannibalize at low population densities even when their plant food seems to be abundant. For several hours after hatching, uncrowded Monarch and Queen butterfly larvae (*Danaus plexippus* and *D. gilippus berenice*) destroyed nearby eggs, although they were also eating their normal food plant (14); in both these species, as in cutworms, *Agrotis ipsilon* (36), and milkweed bugs,

Lygaeus sp. (C. A. Istock, personal communication), feeding on vulnerable young could not be prevented by satiating the herbivore with other food. Larvae of the corn ear-worm, *Heliothis armigera,* crawled from the silk to the corn husk one to two days after hatching and the first larva to reach the husk ate all subsequent arrivals, even though there was sufficient food for several animals to mature (63). On artificial diets, however, cannibalism in *H. armigera* increased with population density (102).

Refuges may reduce the chances of contact and so reduce cannibalism even in very aggressive species. Mortality of young notonectid nymphs increased as the spatial separation of the different age-classes was reduced in early summer (42), while in the wandering spider, *Pardosa lugubris,* separation of age-classes in both time and space may reduce cannibalistic interactions (34). Among crows, *Corvus corone,* the availability of eggs and nestlings as prey for intruding crows was influenced by the amount of protection provided by parents. Even within territories, parents protected their young better when food was clumped near the nest than when food was more dispersed and the adults forced to forage over a wider area (108).

CANNIBALISM AS A PREDATOR-PREY INTERACTION

Feeding Rates of the Predator

The number of cannibalized individuals in a species' diet is difficult to determine accurately. Most cannibalistic predators have very generalized feeding habits and may eat a large number of alternative food items, and among both herbivorous and predaceous species cannibalism rates may vary appreciably with time. When the annual diet of all age-classes is considered, cannibalism may account for a small part of the species' total food intake. For instance, for both the leech, *Erpodbella octoculata* (37), and the dragonfly, *Pyrrhosoma nymphula* (67), cannibalism provided less than 1% of the species' annual diet; in the backswimmer, *Notonecta hoffmanni,* it was 5% (42), and in three species of wandering spiders of the genus *Pardosa* it was 16–20% (32, 47). The annual rate, however, may fluctuate greatly: during 4 years of sampling yellow bass, *Roccus mississippiensis,* in Clear Lake, Iowa, 1–5% of bass stomachs contained other fish (64), while cannibalism accounted for 20% of the food eaten during a subsequent annual sampling period in the same lake (16). Similar variations in annual rates of cannibalism occurred in walleye, *Stizostedion vitreum* (40).

Rates of cannibalism may also be very variable over short periods of time. In the example of yellow bass mentioned above, in which 20% of the annual diet of adults was young bass, cannibalism occurred almost entirely during the summer (16). Similarly, cannibalism on young walleye usually increased in late summer (17, 40), but the smallest fish also were eaten over the winter months. Cannibalism among perch fry, *(P. fluviatilis),* usually occurred in midsummer although the intensity could vary from 0–40% of the diet within a week (97). Among rotifers (53) and backswimmers (42) cannibalism rates fluctuated between weekly and even daily sampling periods. In all of these examples changes in cannibalism rates appear to have been influenced by rapidly changing age distributions in the populations, as well as by changes in the availability of alternative food. Rapid fluctuations in rates of cannibalism have also been reported for caddisfly larvae (76), an intertidal mollusc (86), and rock pool corixids (87), and similar changes are implicit in other studies (2, 3, 10, 13, 14, 39, 58).

Mortality of Prey

Field estimates of cannibalistic mortality among freshwater organisms range up to 95% of particular age-classes. For instance, 9% of caddisfly pupae in a stream were cannibalized by caddisfly larvae (G. Gallep, personal communication). 6% of eggs were eaten by adult water boatmen (Corixidae) in a British lake (23), while two species of corixids living in rock pools ate 20–50% of their own eggs in particular months (87). Predation by larger copepodids and adults caused about half the observed mortality of small nauplii in a freshwater cyclopoid copepod (75), and cannibalism has been invoked to explain most of the mortality of backswimmer nymphs (Notonectidae) in a lake (70) and in permanent pools of a stream (42), and also among young walleyes (17).

Data for terrestrial species are fewer and harder to evaluate. Edgar (32, 33) found 85% mortality among *Pardosa lugubris* spiderlings; he also observed that cannibalism on young spiders provided 16% of the diet, but he did not discuss any connection between these two observations. Cannibalism has been observed in some carnivorous and herbivorous coccinellid beetles in the field, and Hawkes (51) estimated 25% mortality for eggs; however, Dixon (29) argued that many of the eggs that were eaten were not viable, so that mortality attributable solely to cannibalism is difficult to estimate. Yom-Tov (108) suggested that studies in England, Germany, and Finland all have shown that 75% of crows' eggs and nestlings are eaten by other crows intruding into the parental territories.

DEMOGRAPHIC CONSEQUENCES OF CANNIBALISM

The population consequences of a behavioral trait such as cannibalism may not be immediately obvious because they may involve small changes in the population's age distribution or increases in the cannibals' survivorship or fitness, and average rates of cannibalistic mortality may bear little relationship to the importance of its role as a regulatory mechanism. It is often difficult to evaluate the importance of any mortality factor in field studies because the data frequently are inadequate for estimating population consequences. Data on cannibalism usually are given only as the numbers or percentages of members of the animals' own species found in the diets of a few, frequently large, individuals. Because most of the predatory species studied have generalized feeding habits, it is to be expected that cannibalism would contribute only a small amount to their annual diet.

To predict the consequences of cannibalism one must know at least the age structure of the population and the feeding rates of various age-classes. It is also necessary to reject alternative explanations that lead to similar predictions. For example, if food is in short supply then both cannibalism and starvation may produce similar survivorship patterns, both in descriptive studies where only population size is monitored and in experimental work where numbers of individuals or their food supply are manipulated. Independent measurements of cannibalism are needed to distinguish between these alternatives. Arguments that conclude that cannibalism is not a major cause of mortality are often couched in terms of the "low" proportion of its own species in the cannibals' food. For example, it was concluded that cannibalism was unimportant among dragonfly (92) and damselfly (67) nymphs because of its rarity and because nymphs cannibalized only when hungry; similar arguments have been made for fish (e.g. 44, 62) even when cannibalism was relatively common (5–15%) for short periods of time (74). On the other hand, egg cannibalism was suggested as the major regulatory mechanism for the freshwater leech *E. octoculata,* even though eggs were about 0.2% of the annual diet (37).

Depending on the age composition of the population, even a very low cannibalism rate can cause significant mortality. To illustrate this point Le Cren (69) used Frost's data (44) to estimate the consequences of cannibalism on two-year-old pike. He assumed that a four-year-old pike ate about 50 fish of all species in a year, and given the age distribution (1.5 two-year-olds: 1.0 four-year-olds) in Lake Windermere, and the fact that the smaller pike were about 1% of the diet of larger fish, Le Cren calculated that cannibalism could account for all mortality among the younger class. In another example, it was suggested that less than 3% cannibalism in the diet of adult walleye was more than sufficient to explain the 88% mortality observed among the young fish (17). The effects of cannibalism may also be underestimated because encounters may be restricted to short periods of time or involve a relatively small part of the total population. Egg cannibalism among terrestrial insects might be particularly hard to detect: Brower (14) demonstrated that cannibalism by butterfly larvae occurred for several days after hatching, but that most larvae cannibalized during the first few hours.

Age Structure

Variations in cannibalistic tendencies among strains of *Tribolium castaneum* and *T. confusum* have been used to explain differences in both population size and age structure, although cannibalism is only one of a complex of processes affecting population growth. Egg cannibalism, especially by adult females and larvae, may cause cyclical fluctuations in the abundance of eggs and hence of larvae and pupae (77), while in turn, the intensity of cannibalism on eggs, and also on pupae and young adults, is influenced by the age composition. An occasional age-class of *Tribolium* will survive in large numbers when a particularly fortuitous age structure reduces the abundance of voracious beetles, and outbreak populations may result if pupae are sufficiently abundant to satiate the cannibalistic adults (77, 109).

Similar patterns of cohort dominance have been observed among other animals, and the hypothesis that cannibalism may control the age structure of some populations has been tested experimentally in the field. Alm (2) found that a single year-class could dominate European perch (*Perca fluviatilis*) populations in low nutrient (dystrophic) lakes in Sweden for many years (in one lake a single year-class was still dominant after 15 years, when the study was terminated). In more eutrophic lakes, year-classes were dominant for shorter periods of time. Alm experimentally reduced the size of the dominant year-classes in some dystrophic lakes and found that the numbers of small perch increased in subsequent years and developed into new dominant classes. Because adult fish normally bred each year in every lake, Alm concluded that missing year-classes were eliminated from the dystrophic lakes by cannibalism and that the dominant year-class persisted until the numbers of cannibals were so reduced by other causes (e.g. senescence) that younger fish could survive. In eutrophic situations, the perch had more alternative prey and more predators, but a shorter life span possibly due to increased predation or increased intraspecific competition from the successfully developing young. Perch populations in two limnologically similar English lakes, Windermere and Ullswater, showed much the same patterns, but age-class dominance was more marked in the less dense population in Ullswater, where there were also fewer alternative fish species available as food and no predators (68, 69, 73). The absence of pike in Ullswater may have permitted a greater proportion of adults to survive and become cannibalistic (65), although intense cannibalism by adult perch occurred in Lake Windermere in some months (74).

In water striders (Gerridae), the success of the several generations of eggs that can be produced in a year depends on the presence of older individuals. In natural populations of *Gerris najas* none of the later broods survived, but when the numbers of older striders were artificially reduced the survivorship of young nymphs improved (13). One age-class dominated populations of predaceous diaptomid copepods in most of 19 Canadian lakes and ponds (3); because other predators were rare, it was suggested that the observed age distribution was caused by cannibalism on the later developing, smaller individuals. A more complicated example of cohort dominance involves the chironomid *Chironomus anthracinus*, which has different life cycles at different depths in the same lake (58). In the warmer, shallower areas where growth was rapid and predation intense, chironomids emerged every year, leaving space for the successful establishment of new egg masses dropped by the females. Both growth and predation rates were lower in the deeper water and larvae took two years to develop. The resident population of year-old larvae was dense enough to eat all eggs settling in deep water, so that a new age-class could become established only in alternate years after the two-year-old larvae had finally emerged. At intermediate depths the life cycles were variable.

Population Control

In the laboratory, cannibalistic behavior may increase population stability and persistence. One of the four species of sheep blowfly studied by Ullyett (103), *Chrysomyia albiceps*, was cannibalistic and it was the only species to persist when food was scarce because the few surviving larvae were heavy enough to produce viable pupae. The noncannibalistic species produced small individuals unable to maintain their populations successfully with intense competition. For *C. albiceps* cannibalism was a mechanism of interference competition that reduced the exploitation pressures on the food resources and maintained the physiological quality of the few surviving individuals. The predatory mite *Blattisocius tarsalis* and its lepidopteran prey *Anagasta kühniella* are both cannibalistic (106). The mites eat eggs, larvae, and other adults of their own species, while *Anagasta* eats its own eggs. Although the details of these complex interactions have not been analyzed, the authors suggest that cannibalism increases stability in this predator-prey system by damping the magnitude of population fluctuations. Cannibalism seems to be an inherent part of the predator-prey dynamics of walleye, *Stizostedion vitreum*, and their major prey, yellow perch, *Perca flavescens* (40), in which Lotka-Volterra-type fluctuations have been observed. Reductions in predator population size, however, were caused by increased cannibalism, not starvation, following a reduction of prey. In most years, cannibalism was the major mortality factor affecting young walleye, and its action was density-dependent, with cannibalism rates increasing during years of food scarcity.

Paine's observations of growth and feeding in the opisthobranch mollusc *Navanax inermis* showed that cannibalism was density-dependent; he suggested that it may be the major factor regulating intertidal populations of *Navanax* (86). Cannibalism occurred especially when food was scarce, when the size distribution was sufficiently heterogeneous for some large individuals to engulf smaller ones, and when the population density was high enough for frequent contacts. *Navanax* was also capable of migrating to subtidal areas or to more favorable intertidal stretches, but since Paine gave no data on the magnitude of these alternatives, regulation by cannibalism was not firmly established.

It was easier to demonstrate the effects of cannibalistic behavior in populations of the freshwater backswimmer *Notonecta hoffmanni*, which were restricted to small pools in a stream with little migration between pools (42, 43). Notonectids ate each other over most of the growing season, but the period of heaviest cannibalism, two weeks in early summer, coincided with a sudden decrease in the availability of other food as well as a reduction in spatial refuges for the young nymphs. This mortality determined population sizes and age structures for the rest of the season and ensured that the first nymphs to hatch each season were the ones most likely to survive and reproduce successfully. Field and laboratory experiments demonstrated that the mortality of young nymphs was not caused by starvation, but by cannibalism as a response to reduced food supplies: nymphs survived well at low natural food densities without older notonectids, and also in the presence of older individuals provided with extra food. However, cannibalism in notonectids was not just a response to food, and occurred whenever vulnerable age-classes were present, even with abundant alternative prey. Thus cannibalism was a predictable part of the life history of this species, acting as a mechanism of population control that rapidly decreased the numbers of intraspecific competitors as food became scarce. There is evidence to suggest that a few well-fed adults would produce more young than many malnourished individuals (41).

In some species cannibalism seems to remove only those individuals that are not immune, in some way, from attack. It has been suggested, without supporting data, that all first instars of the predatory mite *Typhlodromus caudiglans* are eaten unless they have secure hiding places in tree bark (93). Perhaps the most dramatic examples are found in mosquitoes in the genera *Toxorhynchites* (22) and *Megarhinus* (57), which are cannibalistic whenever there is a sufficient size difference among larvae. Usually one individual of a particular species is found in each tree hole, although a few individuals of equal size may occasionally be present. *Toxorhynchites* also has a "killing frenzy" for a few days before pupation, in which all other animals (of any species) are killed, but not eaten (22). This seems to be a mechanism to avoid predation when the individual becomes a vulnerable pupa, since *Toxorhynchites* larvae placed in the same container readily attacked pupae.

Finally, cannibalism is associated with complex life history patterns of some of the larger predaceous rotifers (7, 45, 46, 53). In *Asplanchna sieboldi* there are three morphological forms of asexually reproducing females that differ in their cannibalistic tendencies, as well as in their tendencies to transform directly to the sexual stage. Cannibalistic forms may be induced by high population densities, by eating large food items, or taking in vitamin E. The presence of cannibalistic morphs involves extremely complex interactions that may increase the range of food items available to the species as a whole, reduce population density, change the relative proportions of the different morphological forms present at any time, and delay sexual reproduction.

SELECTION

Tribolium strains show heritable differences in voracity (88). There may be a genetic basis for cannibalism among different morphological forms in flatworms (4), rotifers (7), and spadefoot toads (50), although the expression of these genotypes may require induction by some environmental factor. Some poeciliid fishes form interspecies hybrids that not only have expected patterns of morphological and biochemical variation, but also have cannibalistic tendencies intermediate between those of parental forms. Thibault (101) crossed two species, one cannibalistic (*Poeciliopsis monacha*) and one not (*P. lucida*), and measured cannibalism by adult females on newly hatched young. The proportion of young fish eaten by these hybrids was 0.74, while the two parental species ate 0.95 and 0.0 respectively; back-crosses of the hybrids to each parental type produced individuals with cannibalistic tendencies between those of the hybrid and the original parent species (0.88, 0.12). These differences indicate polygenic inheritance for the trait. Moore & McKay (81) suggested that overall population sizes of these poeciliid species complexes, as well as the proportions of hybrid and parental forms, are controlled by cannibalism.

In spadefoot tadpoles (*Scaphiopus*) cannibalism is also associated with different morphological forms (10, 11), and breeding experiments with *S. holbrooki* imply polygenic inheritance (50). The common form is primarily a scavenger, but a larger morph, having a notched beak and larger, more variable jaw and tooth structures, and a morphologically intermediate type, are both cannibalistic and predaceous. These aggressive morphotypes usually occur in the drier and hotter parts of the ranges of several spadefoot species, but the presence of these forms varies both spatially and temporally among adjacent pools.

The restriction of cannibalism to particular life-history stages of a single species, and the differences in cannibalistic propensities between closely related species, provide further evidence that cannibalism is genetically determined and responsive to selection. Females of *Tribolium* primarily cannibalize eggs, while males concentrate on pupae: in *T. castaneum,* females were 19 times as voracious on eggs as males, but males were 4 times as voracious as females on pupae; in *T. confusum* the equivalent ratios were 7 and 14.5 (79). Similarly, in two species of intertidal oyster drills (*Eupleura caudata* and *Urosalpinx cinerea*), only the females (72), and, in crows, only mature birds (108), were cannibals. Particular age-classes of coccinellid beetles may be either cannibal or victim depending on species; some species, although closely related to those showing strong cannibalistic tendencies, may not cannibalize at all (51, 55, 59). The latter situation has also been demonstrated for some species of *Megarhinus* mosquito larvae (57, 84), and three species of whirligig beetle larvae display a range of responses when offered the same amount of food (56). Finally, geographical differences in cannibalism have been found: populations of the predatory mite *Typhlodromus occidentalis,* sampled from Washington and Utah, showed lower cannibalism rates than those from southern California, although there were no differences in most other life-history characteristics (24).

There are obvious advantages for the survivor of a cannibalistic encounter. The cannibal gains a meal at the same time that it eliminates a potential competitor and perhaps a potential conspecific predator as well. Because population size is reduced, more food will be available to each survivor, enhancing its chances for further survival and rapid growth. As they grow larger, individuals may themselves become less vulnerable to cannibalistic attacks. For instance, many fish grow faster on larger food items, and their own species may be the largest and most available prey (21). Larger size may also be advantageous because the smallest fish in a cohort may be those most vulnerable to mortality factors (including cannibalism by adults) over the winter (17, 62). Similar increases in growth rates are found among cannibalistic morphotypes of spadefoot toads, which develop to larger sizes and may metamophose more successfully (10–12). Cannibalism can also confer direct nutritional advantages that may be expressed in greater reproductive success: *Tribolium castaneum* that fed on eggs produced one more egg per female per day than individuals not provided with eggs, although *T. confusum* did not show a similar increase (52). In the rotifer *Asplanchna sieboldi,* the largest and most cannibalistic morphotype produces more asexual eggs than the other morphs; when these asexual females eventually transform into sexual stages they may produce twice as many resting eggs (53). Less directly, preying on the smaller individuals of its own species allows different sources of food and energy harvested by the young animals to be retained by the species, as may be the case for some fish in which the young eat small items, such as zooplankton, that the adults do not eat (85, 89).

In some species of coccinellid beetles, egg eating may more than double the survival time of instar I larvae when no further food is provided, and 2 or 3 eggs may suffice to allow molting to the next instar (5, 15). Hungry individuals become more aggressive, but larvae that have previously cannibalized may continue to attack other larvae more readily (59). Egg eating may reduce searching activity for about two days and make cannibals less successful at finding other prey (90).

However, the searching ability of these larvae is at best very limited (9), and the increased survival may improve the beetles' chances of finding other prey even at times of high food density.

Cannibalism becomes disadvantageous when individuals become too aggressive. If the cannibal destroys its own progeny or genotype, either completely or faster than those of its conspecific competitors, or if it reduces its own chances of successful reproduction by eliminating suitable mates, cannibalistic behavior will be selected against. The advantages and disadvantages of cannibalistic tendencies must be balanced against other factors that may affect survival. In some situations the disadvantages of cannibalism may be less severe than the consequences of starvation or reproductive failure caused by inadequate nutrition. For example, the growth rates of notonectids maintained at low food levels were significantly reduced by the second instar, but higher mortality did not occur until the fourth and fifth instars; a noncannibalistic population would be very slow to respond to inadequate food supplies by a reduction of its population size (43). A relatively large cohort would persist and continue to reduce resources until catastrophic mortality was caused by starvation, and even the survivors would probably be stressed physiologically (e.g. 102). By contrast, in a cannibalistic population the numbers of competitors are reduced at an early age so that per capita food supplies remain high for the survivors. Therefore, when comparing two populations reduced to the same extent, one by starvation and one by cannibalism, each individual from the cannibalistic population will make a greater contribution to future generations because it was better fed as a juvenile and is likely to grow faster, survive better, breed earlier, and/or produce more young.

Evolutionary arguments about upper limits to the severity of a socially selfish trait such as cannibalism have been based on two major points of view: interdeme selection (107) and individual selection (48, 49). Hamilton stressed that the evolutionary effect of a cannibalistic encounter depends on the degree of relationship between the individuals involved (48). As the genetic relationship between cannibal and victim becomes more distant, a smaller advantage for the cannibal is required before the frequency of this selfish trait increases in the population. If the victim were only a distant cousin, even a small advantage to the cannibal would be favored, but if the victim were a full sibling or one of its own young, then a much greater increase in fitness of the cannibal would be necessary before natural selection would favor the trait. Under some circumstances an equilibrium frequency for the selfish trait should be reached, but the existence and stability of the equilibrium and its actual value will depend on the extent to which mating and the effects of aggressive behavior are distributed nonrandomly (49). King & Dawson (61) suggested that individual preferences for feeding and egg laying in different microhabitats reduce the probability of cannibalism on close relatives in *Tribolium*. Interdeme selection, on the other hand, may oppose the increase of selfish behavior in the species as a whole because the chances of extinction of excessively cannibalistic subpopulations will be high (80, 107). The two arguments are not mutually exclusive, since nonrandom mating among very aggressive individuals could also lead to extinction. The interdeme arguments imply that local extinctions and/or migration between populations will maintain an average upper limit for the species, whereas Hamilton's model suggests that subtle selective pressures may maintain an equilibrium level for each population in response to its own environment.

Eickwort has calculated the increase in individual fitness needed to maintain egg cannibalism on full and half siblings of newly hatched chrysomelid beetle larvae, and has shown this to be surprisingly small (35). An increase in the probability of survival of an instar I beetle by 0.005, from 0.01 to 0.015, will maintain the selective advantage even if full siblings are eaten. Similarly, a decrease of two days in time required to reach maturity also will maintain the trait. The lower the original fitness, the smaller the incremental gain necessary to select for increased aggression; therefore, selection for selfish behavior might be most advantageous in the life stages with the lowest probabilities of survival, such as very young individuals. Thus, despite many of the examples described above, selection for cannibalism does not necessarily demand large increases in individual fitness.

CANNIBALISM VS PREDATION

The occurrence and intensity of cannibalism varies considerably both within and between species; because much of this variability arises from differences in the factors influencing local populations, the extent of cannibalism may not be predictable for particular situations and the proximal factors influencing the evolution of such behavior may be quite obscure. Generalizations about cannibalism must allow for at least two other aspects of individual responses that tend to complicate the picture. First, some species may respond to resource limitation by dispersal, diapause, modifications of physiological characteristics, or interference competition by other means. Second, some animals may be unable to cannibalize because they are physically incapable of handling prey as large, as aggressive, or with such effective escape mechanisms as their own species.

When several species share the same resources, interactions among individuals of competing species further complicate selection for cannibalistic behavior. There are numerous examples of species that prey on potential competitors (3, 9, 10, 25, 28, 38, 41, 59, 66, 69, 88, 94, 103), and in many cases these species are also cannibalistic; the balance between cannibalism and predation influences the coexistence patterns of the competing species. Some of the possible interactions among aggressive species are well illustrated by comparing the responses to food shortage of the larvae of three species of whirligig beetles (Gyrinidae, 56). As food supplies were reduced in laboratory experiments, the developmental rates of *Dineutes nigrior* larvae declined, but there was no cannibalism until food became particularly scarce; *D. horni* became more cannibalistic as food decreased but the larvae maintained the same developmental rates over a wide range of food regimes. *D. assimilis* was very cannibalistic at all food levels and its survivorship did not strongly reflect differences in food supply. When these species were paired in competition experiments the relative amounts of cannibalism and predation, and hence the eventual exclusion or persistence of the less aggressive one, were determined by their different aggressive tendencies at different food levels and by the initial relative abundances of the species. In this example predation was reciprocal, but cases of unilateral predation also occur: the only cannibalistic species of the four sheep blowflies studied by Ullyett (103) was always successful in competition for food because it also was the only one that preyed on individuals of other species.

If food is in short supply for the entire assemblage of competing species, it is not surprising to find predation among both conspecific and heterospecific competitors. The selective advantages of predation upon competitors are similar to those discussed for cannibalism, and, at least in well established systems, the frequencies of cannibalism and predation will be related to the relative advantages to be gained from each type of encounter. This is particularly so if there is an evolutionary cost to being aggressive towards other aggressive individuals. Figure 1 summarizes the

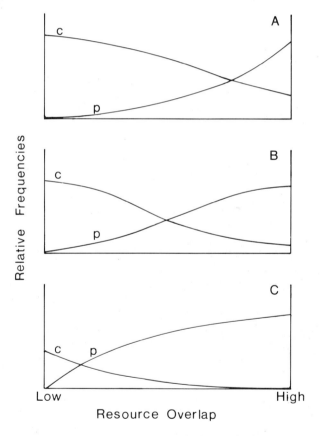

Figure 1:   Relative proportions of cannibalism (*c*) and predation on potential competitors (*p*) by an aggressive species, based on the relative abundances of the species and the degree to which the scarce resources are shared. *A*, the aggressive species is much more common than its competitors; *B*, the species are equally abundant; *C*, the aggressive species is relatively rare.

ways in which the frequencies of cannibalism and predation may change if one assumes that the advantages accruing to successful individuals depend on the relative abundances of competing species and the extent of overlap of the shared resources. Whatever their relative abundances, cannibalism rates should be highest for species without close competitors, i.e. situations where the level of resource overlap is low. In absolute terms, however, rare species may be less cannibalistic than common ones simply because fewer contacts of rare individuals are with their own species. As a larger proportion of the spectrum of a species' resources is shared, interspecific events will increase, but the relative amounts of cannibalism and predation will reflect the chances of encountering other individuals. Common species may always cannibalize to some extent even when interspecific competition is high, although the proportion of cannibalistic events in this case will be lower than if there were less overlap. On the other hand, there may be no cannibalistic events at all in the diets of rare species unless there is little overlap in the use of resources with other species.

These arguments lead to several testable predictions about the occurrence of cannibalism and predation in different situations. First, if natural selection influences a species' predatory behavior, individuals may show different feeding preferences (82) among potential competitors and the hierarchy of these preferences may directly reflect the amount of overlap between the species. For instance, individuals of rare species may have higher preferences for feeding on aggressive competitors than will individuals of dominant species, while intraspecific encounters may more often result in cannibalism in the commoner species. Unfortunately, I know of no data that test these suggestions.

A second set of predictions from Figure 1 is that predation should be particularly common among species with generalized habits and with qualitatively similar resources. These interactions have been described for several different communities including freshwater fish (66, 69), copepods (3) and insects (41, 42, 56), deep sea deposit feeders (25), and some terrestrial insects (28). Among freshwater fish, for instance, Larkin (66) noted that ". . . many species have a relatively wide tolerance of habitat type, a flexibility of feeding habits and in general share many resources of their environment with several other species of fish. Cannibalism and mutual predation are common complications of competitive relationships between species."

The heights and specific shapes of the curves in Figure 1 obviously depend on the inherent aggressiveness of the individuals and on alternative responses to limiting conditions available to the particular species. These, in turn, depend on many other aspects of the environment such as the predictability of the available resources. A fluctuating resource supply may favor the selection of individual behavior that results in a less rapid population response (83), because if there is only a short-term lack of food then species that did not cannibalize immediately, such as the larvae of *Dineutes nigrior* (56) described above, would still be abundant when food supply increased. A rapidly inducible interference mechanism, such as cannibalism, may be particularly advantageous if resource limitation is predictable: California populations of a predatory mite living in a Mediterranean-type climate with dry summers and predictable crashes of insect prey (24) are more cannibalistic than populations from the Pacific northwest (26), and those morphs of spadefoot toads that were both cannibalistic and predaceous were more common in the drier parts of the species' range with less permanent habitats (10, 12).

CONCLUSIONS

Cannibalism is not an aberrant behavior limited to confined or highly stressed populations, but is a normal response to many environmental factors. Since cannibalistic tendencies are often sensitive to selection by local conditions, they may vary considerably among different populations or genetic strains of the same species, and among closely related species. The relative availability of alternative food, the density and behavior of both cannibal and victim, and, more rarely, extreme food deprivation or crowding may influence the intensity of this behavior. Cannibalism is an intraspecific predator-prey interaction that may also function as a means of interference competition, limiting population size before the resource itself becomes limiting. In common with some other self-regulatory mechanisms, cannibalism is a sensitive method of preserving the competitive abilities of the successful individuals while maintaining the reproductive outputs of the survivors. Cannibalism is an age-specific process frequently causing rapid changes in age distribution and population size.

Newly hatched young that eat siblings before dispersing are not sampling their own resource state before they become predators. In these cases the intensity of cannibalism is a function of the numbers of eggs laid by the previous generation, and cannibalism may act as a delayed regulatory mechanism or as a method of providing a guaranteed nutritional source to the young animals. Such situations occur especially among terrestrial insects in which the young larvae do not disperse far from egg batches before searching for food. Egg cannibalism may be inherently different from cannibalism by older individuals, but they both reduce the numbers of potential competitors that would have shared resources as the animals grew larger. Population extinctions are unlikely because cannibalism is usually a short-term event: cannibalism rates decrease as resources become relatively more available to each survivor, and, as vulnerable individuals become scarcer and harder to find, cannibals may switch to other types of prey.

The ecological consequences of predation among trophically similar species are analogous to those of cannibalism. When the same resources are in short supply for the entire assemblage of competing species, both cannibalism and predation among competitors are likely events. In this paper I have stressed examples of these behavioral traits and discussed interactions between them that may influence the populations and perhaps the community as a whole. Cannibalism may result in population self-regulation, while predation among individuals of competing species tends to "regulate" the total biomass of competing individuals. Both interactions change the system from one that might have been superficially described as exploitation competition into one of interference competition mediated through predator-prey interactions. In these systems trophic distinctions become blurred and the complexity of even the most simple food web is greatly increased.

ACKNOWLEDGMENTS

I would like to thank Dan Dykhuizen, Peter Frank, Dave Mertz, Trice Morrow, and Don Potts for their pains in pushing and pulling me through this manuscript at all the right times. Many others at the University of California at Santa Barbara and in Canberra provided very useful discussion and criticism of ideas in this paper.

Literature Cited   (Literature search ended September 30, 1974)

1. Abdel-Salam, von F., El-Lakwak, F. 1973. Über den Kannibalismus bei Larven der Baumwollblatteule *Spodoptera littoralis* (Boisd.) (Lepidoptera, Noctuidae). *Z. Angew. Entomol.* 74:356–61
2. Alm, G. 1952. Year class fluctuations and span of life of perch. *Rep. Inst. Freshwater Res. Drottningholm* 33: 17–38
3. Anderson, R. S. 1970. Predator-prey relationships and predation rates for crustacean zooplankters from some lakes in western Canada. *Can. J. Zool.* 48: 1229–40
4. Armstrong, J. T. 1964. The population dynamics of the planarian, *Dugesia tigrina. Ecology* 45:361–65
5. Banks, C. J. 1956. Observations on the behaviour and mortality in coccinellidae before dispersal from the egg shells. *Proc. R. Entomol. Soc. London A* 31:56–60
6. Beaver, R. Z. 1974. Intraspecific competition among bark beetle larvae (Coleoptera: Scolytidae). *J. Anim. Ecol.* 43:455–67
7. Birky, C. W. Jr. 1969. The developmental genetics of polymorphism in the rotifer *Asplanchna.* III. Quantitative modification of developmental responses to vitamin E, by the genome, physiological state, and population density of responding females. *J. Exp. Zool.* 170:437–48
8. Boice, R. 1972. Some behavioral tests of domestication in Norway rats. *Behaviour* 42:198–231
9. Boldyrev, M. I., Wilde, W. H. A. 1969. Food seeking and survival in predaceous coccinellid larvae. *Can. Entomol.* 101:1218–22
10. Bragg, A. N. 1962. Predator-prey relationships in two species of spadefoot tadpoles with notes on some other features of their behavior. *Wasmann J. Biol.* 20:81–97
11. Bragg, A. N. 1964. Further study of predation and cannibalism in spadefoot tadpoles. *Herpetologica* 20:17–24
12. Bragg, A. N. 1965. *Gnomes of the Night.* Philadelphia: Univ. Pa. Press. 127 pp.
13. Brinkhurst, R. O. 1966. Population dynamics of the large pond-skater *Gerris najas* Degeer (Hemiptera-Heteroptera). *J. Anim. Ecol.* 35:13–25
14. Brower, L. P. 1961. Experimental analyses of egg cannibalism in the Monarch and Queen butterflies, *Danaus plexippus* and *D. gilippus* berenice. *Physiol. Zool.* 34:287–96
15. Brown, H. D. 1972. The behaviour of newly hatched coccinellid larvae (Coleoptera: Coccinellidae). *J. Entomol. Soc. South Afr.* 35:149–57
16. Bulkley, R. V. 1970. Feeding interaction between adult bass and their offspring. *Trans. Am. Fish. Soc.* 99:732–38
17. Chevalier, J. R. 1973. Cannibalism as a factor in first year survival of walleye in Oneida Lake. *Trans. Am. Fish. Soc.* 102:739–44
18. Clark, D. P. 1967. A population study of *Phaulacridium vittatum* Sjöst (Acrididae). *Aust. J. Zool.* 15:799–872
19. Clark, L. R. 1963. The influence of predation by *Syrphus* sp. on the numbers of *Cardiaspina albitextura* (Psyllidae). *Aust. J. Zool.* 11:470–87
20. Colinvaux, P. A. 1973. *Introduction to Ecology.* New York: Wiley. 621 pp.
21. Cooper, G. C. 1936. Food habits, rate of growth and cannibalism of young largemouth bass (*Aplites salmoides*) in state-operated rearing ponds in Michigan during 1935. *Trans. Am. Fish. Soc.* 66:242–66
22. Corbet, P. S., Griffiths, A. 1963. Observations on the aquatic stages of two species of *Toxorhynchites* (Diptera: Culicidae) in Uganda. *Proc. R. Entomol. Soc. London A* 38:125–35
23. Crisp, D. T. 1962. Estimates of the annual production of *Corixa germari* (Fieb.) in an upland reservoir. *Arch. Hydrobiol.* 58:210–23
24. Croft, B. A., McMurtry, J. A. 1971. Comparative studies on four strains of *Typhlodromus occidentalis* Nesbitt

(Acarina: Phytoseiidae). IV. Life-history studies. *Acarologia* 13:460–70

25. Dayton, P. K., Hessler, R. R. 1972. Role of biological disturbance in maintaining diversity in the deep sea. *Deep Sea Res.* 19:199–208

26. Dement, W. A., Mooney, H. A. 1974. Seasonal variation in the production of tannins and cyanogenic glucosides in the chaparral shrub, *Heteromeles arbutifolia. Oecologia* 15:65–76

27. Dempster, J. P. 1963. The natural prey of three species of *Anthocoris* (Heteroptera: Anthocoridae) living on broom (*Sarothamnus scoparius* L.). *Entomol. Exp. Appl.* 6:149–55

28. Dempster, J. P. 1966. Arthropod predators of the Miridae (Heteroptera) living on broom (*Sarothamnus scoparius*). *Entomol. Exp. Appl.* 9:405–12

29. Dixon, A. F. G. 1959. An experimental study of the searching behaviour of the predatory coccinellid beetle *Adalia decempunctata* (L.). *J. Anim. Ecol.* 28:259–81

30. Dornstreich, M. D., Morren, G. E. B. 1974. Does New Guinea cannibalism have nutritional value? *Hum. Ecol.* 2:1–12

31. Eason, E. H. 1964. *Centipedes of the British Isles.* London: Warne. 294 pp.

32. Edgar, W. D. 1969. Prey and predators of the wolf spider *Lycosa lugubris. J. Zool.* 159:405–11

33. Edgar, W. D. 1971. Seasonal weight changes, age-structure, natality and mortality in the wolf spider *Pardosa lugubris* Walck in central Scotland. *Oikos* 22:84–92

34. Edgar, W. D. 1971. The life cycle, abundance and seasonal movement of the wolf spider *Lycosa lugubris* (*Pardosa*) in central Scotland. *J. Anim. Ecol.* 40:303–22

35. Eickwort, K. R. 1973. Cannibalism and kin selection in *Labidomera clivicollis* (Coleoptera: Chrysomelidae). *Am. Natur.* 107:452–3

36. El-Kifl, A. H., Nasr, El-S. A., Moawad, G. M. 1972. On the cannibalistic habit of the black cutworm, *Agrotis ipsilon* (Hufn. Lepidoptera: Noctuidae). *Bull. Soc. Entomol. Egypte* 61:123–26

37. Eliot, J. M. 1973. The diel activity pattern, drifting and food of the leech *Erpobdella octoculata* (L.) (Hirundinea: Erpobdellidae) in a Lake District stream. *J. Anim. Ecol.* 42:449–59

38. Fahy, E. 1972. The feeding behaviour of some common lotic insect species in two

streams of differing detrital content. *J. Zool.* 167:337–50

39. Fischer, Z. 1960. Cannibalism among the larvae of the dragonfly *Lestes nympha* Selys. *Ekol. Pol. Ser. B* 7:33–39

40. Forney, J. L. 1974. Interactions between yellow perch abundance, walleye predation, and survival of alternate prey in Oneida Lake, New York. *Trans. Am. Fish. Soc.* 103:15–24

41. Fox, L. R. 1973. *Food limitation, cannibalism and interactions among predators: effects on populations and communities of aquatic insects.* PhD thesis. Univ. California, Santa Barbara. 209 pp.

42. Fox, L. R. 1975. Some demographic consequences of food shortage for the predator, *Notonecta hoffmanni. Ecology.* In press

43. Fox, L. R. 1975. Factors influencing cannibalism, a mechanism of population limitation in the predator *Notonecta hoffmanni. Ecology.* In press

44. Frost, W. E. 1954. The food of pike, *Esox lucius,* in Windermere. *J. Anim. Ecol.* 23:339–60

45. Gilbert, J. J. 1973. Induction and ecological significance of gigantism in the rotifer *Asplanchna sieboldi. Science* 181:63–66

46. Gilbert, J. J. 1973. The adaptive significance of polymorphism in the rotifer *Asplanchna.* Humps in males and females. *Oecologia* 13:135–46

47. Hallander, H. 1970. Prey, cannibalism and microhabitat selection in the wolf spiders *Pardosa chelata* (O. F. Müller) and *P. pullata* (Clerck). *Oikos* 21: 337–40

48. Hamilton, W. D. 1964. The genetical evolution of social behaviour. I. *J. Theor. Biol.* 7:1–16

49. Hamilton, W. D. 1970. Selection of selfish and altruistic behavior in some extreme models. *Man and Beast; Comparative Social Behavior,* ed. J. F. Eisenberg, W. S. Dillon, 59–91. Smithsonian Third Int. Symp. Washington DC: Smithsonian Inst.

50. Hampton, S. H., Volpe, E. P. 1963. Development and interpopulation variability of the mouthparts of *Scaphiopus holbrooki. Am. Midl. Natur.* 70:319–28

51. Hawkes, O. A. M. 1920. Observations on the life-history, biology, and genetics of the ladybird beetle *Adalia bipunctata* (Mulsant). *Proc. Zool. Soc. London*, 475–90

52. Ho, F. K., Dawson, P. S. 1966. Egg

cannibalism by *Tribolium* larvae. *Ecology* 47:318–22

53. Hurlbert, S. H., Mulla, M. S., Willson, H. R. 1972. Effects of an organophosphorus insecticide on the phytoplankton, zooplankton and insect populations of fresh-water ponds. *Ecol. Monogr.* 42:269–99

54. Hyman, L. H. 1940. *The Invertebrates: Protozoa through Ctenophora.* New York: McGraw. 726 pp.

55. Iperti, G. 1966. Some components of efficiency in aphidophagous coccinellids. *Ecology of Aphidophagous Insects,* ed. I. Hodek, 253. Prague: Academia

56. Istock, C. A. 1966. Distribution, coexistence and competition of whirligig beetles. *Evolution* 20:211–34

57. Jenkins, D. W., Carpenter, S. J. 1946. Ecology of the tree hole breeding mosquitoes of nearctic North America. *Ecol. Monogr.* 16:31–47

58. Jónasson, P. M. 1971. Population studies on *Chironomus anthracinus. Dynamics of Populations,* ed. P. J. den Boer, G. R. Gradwell, 220–31. Wageningen: Pudoc

59. Kaddou, I. K. 1960. The feeding behavior of *Hippodamia quinquesignata* (Kirby) larvae. *Univ. Calif. Publ. Entomol.* 16:181–232

60. King, C. E., Dawson, P. S. 1972. Population biology and the *Tribolium* model. *Evol. Biol.* 5:133–227

61. King, C. E., Dawson, P. S. 1973. Habitat selection by flour beetles in complex environments. *Physiol. Zool.* 46:297–309

62. Kipling, C., Frost, W. E. 1970. A study of the mortality, population numbers, year class strengths, production and food consumption of pike, *Esox lucius,* in Windermere from 1944–1962. *J. Anim. Ecol.* 39:115–58

63. Kirkpatrick, T. W. 1957. *Insect Life in the Tropics.* London: Longmans. 311 pp.

64. Kraus, R. 1963. Food habits of the yellow bass, *Roccus mississippiensis,* Clear Lake, Iowa, summer 1962. *Proc. Iowa Acad. Sci.* 70:209–15

65. Kruuk, H. 1972. *The Spotted Hyena.* Chicago: Univ. Chicago Press. 335 pp.

66. Larkin, P. A. 1956. Interspecific competition and population control in freshwater fish. *J. Fish. Res. Bd. Can.* 13:327–42

67. Lawton, J. H. 1970. Feeding and food energy assimilation in larvae of the damselfly *Pyrrhosoma nymphula* (Sulz.) (Odonata: Zygoptera). *J. Anim. Ecol.* 39:669–89

68. Le Cren, E. D. 1955. Year to year variation in the year-class strength of *Perca vluviatilis. Verh. Int. Ver. Limnol.* 12:187–92

69. Le Cren, E. D. 1965. Some factors regulating the size of populations of freshwater fish. *Mitt. Int. Ver. Theor. Angew. Limnol.* 13:88–105

70. Macan, T. T. 1965. Predation as a factor in the ecology of water bugs. *J. Anim. Ecol.* 34:691–98

71. MacLellan, C. R. 1973. Natural enemies of the light brown apple moth, *Epiphyas postvittana,* in the Australian Capital Territory. *Can. Entomol.* 105:681–700

72. Manzi, J. J. 1970. Combined effects of salinity and temperature on the feeding, reproductive, and survival rates of *Eupleura caudata* (Say) and *Urosalpinx cinerea* (Say) (*Prosobranchia*: Muricidae). *Biol. Bull.* 138:35–46

73. McCormack, J. C. 1965. Observations on the perch population of Ullswater. *J. Anim. Ecol.* 34:463–78

74. McCormack, J. C. 1970. Observations on the food of the perch (*Perca fluviatilis,* L.) in Windermere. *J. Anim. Ecol.* 39:255–68

75. McQueen, D. J. 1969. Reduction of zooplankton standing stocks by predaceous *Cyclops bicuspidatus thomasi* in Marion Lake, British Columbia. *J. Fish. Res. Bd. Can.* 26:1605–18

76. Mecom, J. O. 1972. Feeding habits of Trichoptera in a mountain stream. *Oikos* 23:401–7

77. Mertz, D. B. 1969. Age-distribution and abundance in populations of flour beetles. I. Experimental studies. *Ecol. Monogr.* 39:1–31

78. Mertz, D. B. 1972. The *Tribolium* model and the mathematics of population growth. *Ann. Rev. Ecol. Syst.* 3:51–78

79. Mertz, D. B., Cawthon, D. A. 1973. Sex differences in the cannibalistic roles of adult flour beetles. *Ecology* 54:1400–1

80. Mertz, D. B., Robertson, J. R. 1970. Some developmental consequences of handling, egg-eating, and population density for flour beetle larvae. *Ecology* 51:989–98

81. Moore, W. S., McKay, F. E. 1971. Coexistence in unisexual-bisexual species complexes of *Poeciliopsis* (Pisces: Poeciliidae). *Ecology* 52:791–99

82. Murdoch, W. W. 1969. Switching in general predators: experiments on

predator specificity and stability of prey populations. *Ecol. Monogr.* 39:335–54
83. Murdoch, W. W. 1970. Population regulation and population inertia. *Ecology* 51:497–502
84. Muspratt, J. 1951. The bionomics of an African *Megarhinus* (Dipt., Culicidae) and its possible use in biological control. *Bull. Entomol. Res.* 42:355–70
85. Nikolskii, G. V. 1969. *Theory of Fish Population Dynamics.* Edinburgh: Oliver and Boyd. 323 pp.
86. Paine, R. T. 1965. Natural history, limiting factors and energetics of the opisthobranch *Navanax inermis. Ecology* 46:603–19
87. Pajunen, V. I. 1971. Adaptation of *Arctocorisa carinata* (Sahlb.) and *Callocorixa producta* (Reut.) populations to a rock pool environment. See Ref. 58, 148–58
88. Park, T., Mertz, D. B., Grodzinski, W., Prus, T. 1965. Cannibalistic predation in populations of flour beetles. *Physiol. Zool.* 38:289–321
89. Patriquin, D. G. 1968. Biology of *Gadus morhua* in Ogac Lake, a land locked fiord on Baffin Island. *J. Fish. Res. Bd. Can.* 24:2573–94
90. Pienkowski, R. L. 1965. The incidence and effect of egg cannibalism in first-instar *Coleomegilla maculata lengi* (Coleoptera: Coccinellidae). *Ann. Entomol. Soc. Am.* 58:150–53
91. Poulson, T. L. 1963. Cave adaptation in amblyopsid fishes. *Am. Midl. Nat.* 70:257–90
92. Pritchard, G. 1964. The prey of dragonfly larvae in ponds in northern Alberta. *Can. J. Zool.* 42:785–800
93. Putnam, W. L., Herne, D. H. C. 1964. Relations between *Typhlodromus caudiglans* Schuster (Acarina: Phytoseiidae) and phytophagous mites in Ontario peach orchards. *Can. Entomol.* 96:925–43
94. Rosenzweig, M. L. 1966. Community structure in sympatric carnivora. *J. Mammal.* 47:602–12
95. Schaller, G. B. 1972. *The Serengeti Lion.* Chicago: Univ. Chicago Press. 480 pp.
96. Shankman, P. 1969. Le rôti et le bouilli: Levi-Strauss' theory of cannibalism. *Am. Anthropol.* 71:54–69
97. Smyly, W. J. P. 1952. Observations on the food of the fry of perch (*Perca fluviatilis* Linn.) in Windermere. *Proc. Zool. Soc. London* 122:407–16
98. Somchoudhury, A. K., Mukherjee, A. B. 1971. Biology of *Acaropsis docta* (Berlese) a predator on eggs of pests of stored grains. *Bull. Grain Technol.* 9:203–6
99. Southwick, C. H. 1955. Regulatory mechanisms of house mouse populations: social behavior affecting litter survival. *Ecology* 36:627–34
100. Strawinski, K. 1964. Zoophagism of terrestrial Hemiptera-Heteroptera in Poland. *Ekol. Pol. Ser. A* 12:428–52
101. Thibault, R. E. 1974. Genetics of cannibalism in a viviparous fish and its relationship to population density. *Nature* 251:138–40
102. Twine, P. H. 1971. Cannibalistic behaviour of *Heliothis armigera* (Hubn.). *Queensl. J. Agric. Anim. Sci.* 28:153–57
103. Ullyett, G. C. 1950. Competition for food and allied phenomena in sheep blowfly populations. *Phil. Trans. R. Soc. London* 234:77–174
104. Watson, A., Moss, R. 1970. Dominance, spacing behavior and aggression in relation to population limitation in vertebrates. *Animal Populations in Relation to Their Food Resources,* ed. A. Watson, 167–220. Oxford: Blackwell
105. Way, M. J. 1966. Significance of self-regulatory dispersal in control of aphids by natural enemies. See Ref. 55, 149–50
106. White, E. G., Huffaker, C. B. 1969. Regulatory processes and population cyclicity in laboratory populations of *Anagasta kühniella* (Zeller) (Lepidoptera: Phycitidae). I. Competition for food and predation. *Res. Popul. Ecol. Kyoto* 11:57–83
107. Wright, S. 1960. Physiological genetics, ecology of populations, and natural selection. *Evolution after Darwin,* ed. S. Tax, 429–75. Chicago: Univ. Chicago Press
108. Yom-Tov, Y. 1974. The effect of food and predation on breeding density and success, clutch size and laying date of the crow (*Corvus corone* L.). *J. Anim. Ecol.* 43:479–98
109. Young, A. M. 1970. Predation and abundance in populations of flour beetles. *Ecology* 51:602–19

DEMOGRAPHIC CONSEQUENCES   ❖4089
OF INFANTICIDE IN MAN

Mildred Dickeman

Department of Anthropology, California State College, Sonoma,
Rohnert Park, California 94928

> *For the reproduction of the race, there are two instincts needed, the sexual and the parental, and the way these two are organized is to say the least curious.*
>
> Charles Galton Darwin

INTRODUCTION

The practice of infanticide has been widespread in human cultures, yet, perhaps more than any other means of population limitation, it has been neglected by anthropologist, historian, and demographer. As literature on the subject makes clear, a universal diffidence on the part of practitioners, often coupled with attitudinal bias on the part of the social scientist, has served as justification for inattention. Consideration of specific forms of population control must precede a more sophisticated understanding of the alternative control strategies available to human populations in specific ecological contexts. This review underlines the great potential impact of infanticide as a means of population regulation with social structural implications. Ignoring exceptional and idiosyncratic forms of the practice, I focus here on those cases in which cultural norms raise it to demographically significant frequencies. After reviewing the few existing surveys, I explore potential demographic and social effects on the basis of a few examples and indicate some avenues for future research. If what follows is more speculative and hortatory than conclusive, I must plead the paucity of previous analyses.

The destruction of offspring, far from being a uniquely human phenomenon, is widespread in living organisms. In plants, biochemical inhibitors may prevent germination and normal growth of conspecific seedlings except under propitious conditions; in a variety of invertebrates and vertebrates, egg cannibalism, aggressive neglect of young, and cannibalism of injured and dead have been reported as density-dependent behaviors. In mammals, aggressive neglect and cannibalism have

107

been produced in the house mouse and the laboratory rat as responses to overcrowding, and in the house mouse as a result of circadian stress as well; they have been observed also in the wild in the muskrat. These cases are paralleled in nonhuman primates by laboratory experiments on the tree shrew (5) and by Zuckerman's famous observations on hamadryas baboons in the overcrowded conditions of the London Zoo (114). (For examples and general discussion, see 47, 69, 88, 110.) Many of these examples were experimentally induced through prevention of dispersal. In social animals, laboratory overcrowding results in social and reproductive chaos, with destruction of young as an unintentional consequence of proximity in conditions of stress. Death occurs without regard to the identity of kinship of the offspring. These are, then, mere versions of aggressive neglect.

Statistically, the correlation of proximity with kinship may result in selection against such behavior by removal of shared genes, or may select for aggressive neglect and cannibalism under some conditions through rapid population reduction, reproductive compensation, and even protein conservation (41). While the operation of aggressive neglect is not irrelevant to man, it does highlight some great contrasts in the quality of removal strategies in *Homo sapiens*. There are reports of infanticide and cannibalism in nonhuman primates under apparently normal natural conditions, for example among hanuman langurs and chimpanzees (90, 91, 113), where infants were killed by dominant males. This is not, apparently, a regular practice in all troops of either species.

Cannibalism of infants in time of famine is of course known in some human societies, as among the Eskimos and Australians. Consumption of newborns, fetuses, dead infants, and even, in some districts, regular consumption of all firstborn has a scattered distribution in Australia (46, 102). Infant cannibalism is known to have existed in Pleistocene man. But these instances, like the Aztec consumption of human sacrifices as a regular part of the upper-class diet (21, 75), are exceptional. The mass of human infanticides is never consumed.

What is distinctive in the human species is selective removal on the basis of conscious intent. The rationale varies from case to case, but the usual act of human infanticide is a product of individual or group decision in a healthily functioning social system, and not at all the mere product of accidental proximity in conditions of social disorganization. It is a normative, culturally sanctioned behavior. This truism needs emphasis in the context of the current, often muddy, dialogue regarding human population increase. It reveals, bluntly, the irrelevance of the facile analogies of Calhoun (16) and others to an understanding of human responses to increased density (35, 106). The human act requires the capacity to discriminate between individual offspring, both one's own and others, in terms of a variety of cognitive categories. This capacity for selective removal in response to qualities both of offspring and of ecological and social environments may well be a significant part of the biobehavioral definition of *Homo sapiens*. As Neel (64, p. 358) has suggested:

... man is to my knowledge the only [mammal] who regularly and without "external" provocation purposely and knowingly commits infanticide. ... There have been many attempts to define that point at which our primate ancestors crossed the threshold to "true" man. Most definitions involve tool-making or speech. I suggest that an equally sound definition is the point at which parental care evolved to the level of permitting rapid population increase, with the concomitant recognition of the necessity to limit natural fecundity.

Note that this capacity depends not only on cognitive discrimination, but on the relaxation of innate maternal attachment or imprinting mechanisms that allowed abandonment only in situations of extreme stress. This relaxation may also be a significant criterion of humanity.

DEFINITIONAL PROBLEMS

A precise definition of the practice of infanticide is elusive for reasons both demographic and cultural. Since infanticide is not a permitted registration category in modern urban nations, the practice, where it occurs, must be either misreported under other categories, thus inflating rates of abortion, stillbirth, or postnatal natural death, or join many events in these same classes as unreported. Underreporting in census and registration is most serious at the beginning of life; verbal taboos and negative sanctions only aggravate it. Unreliability is compounded by disagreement among censusing bodies regarding the boundaries of various perinatal categories. Thus an infant dying within 24 hours after birth may be classified in some countries as neither a livebirth nor a death, but as an abortion; in other countries, any infant dying prior to birth registration will be recorded as a stillbirth. The opportunity for covert infanticide in such cases is apparent (25). If such inconsistencies exist in industrial nations, difficulties are increased in societies with minimal or no registration. These inconsistencies are part of a larger taxonomic difficulty: life and death are cultural events, defined differently by every human society. Cultural definitions should correlate with practices such as infanticide, as well as with rates of natural infant mortality, yet we lack a cross-cultural study of the meanings of life and death.

Thus the researcher is confronted with the necessity of imposing arbitrary limits to the subject of his investigation, of which the term "infant" is already one. These may prove inhibiting if the intent is to investigate relative frequencies of different forms of mortality. If infanticide is an alternative to natural mortality, or to death through aggressive neglect or child abuse, then frequencies of directly postnatal events must be compared with those occurring long afterward in the life of the child or even adult. Abandonment likewise presents problems. Exposure is clearly a form of intentional killing, but may have other functions, as in the Netsilik Eskimo (6) where the exposed infant announces by its cries that it is available for adoption by less fertile couples. In Western Europe until the mid-nineteenth century, abandonment, country wet-nursing, and foundling-home donation constituted systematic, if covert, means of infanticide (48, 96). In the literature on infanticide, attention to these definitional problems is minimal. Taeuber (93) has used the term "effective fertility" to refer to livebirths minus intentional removals. In referring to the Yanomama practice of abortion in the sixth or seventh month of gestation, Neel (64) remarks, "The practice may better be described as early induction of labor with infanticide if necessary than as abortion," thus emphasizing the way in which these categories intergrade. The cases under review here involve either exposure or physical trauma of clear intent immediately after birth, such that we can speak of "normal natal human infanticide." But the imprecision of available data will have to be corrected before reliable estimates of the impact of infanticide on rates of natural increase, and of its effectiveness relative to other forms of limitation, can be made.

HISTORY AND LITERATURE

Past treatment of infanticide can best be understood in terms of the larger longstanding dialogue, of which it forms a part, concerning the existence and effectiveness of internal self-regulating mechanisms in man. Some comments on the history of the subject in these terms may clarify the peculiar quality of selective inattention characterizing both demographic and anthropological literature. Early students of population, as educated eighteenth and nineteenth century Europeans, were familiar with the practice of infanticide from classical literature. Malthus (57) reported its existence in Australia, the Pacific, North and South America, Central Asia, India, and China, in addition to ancient Greece and Rome. He emphasized its occurrence in nonurban societies and Asian civilizations, however, and neglected the more covert European forms. Yet even in the Greek and Roman cultures, the importance of infanticide was lost sight of in Malthus's more general treatments, such as his 1830 summary (58), in part because he classified it with epidemics, famine, and warfare as a "positive check." Moral bias caused him to confound control of population growth with control of natality, preventing recognition of infanticide as a preventive check, along with delayed marriage and reduced coitus. In addition, his uncertainty about its effectiveness led him to conclude that in the "ancient nations and less civilized parts of the world, . . . war and violent disease were the predominant checks to . . . population" (58, p. 41).

Thus began a long neglect of cultural controls on the part of demographers, for whom "savage" society was typified by maximal natality and maximal mortality, resulting in moderate or low levels of increase. This of course is the primary stage in the unilinear theory of the demographic transition. Developing in the context of the growth of national censusing and of sociology, demography focused increasingly on mass societies, whether "developed" or "undeveloped." Lacking research interest in the hypothetical initial condition of man, it adopted one of two versions of the "primitive" available in Western thought, that which saw uncivilized and early man in nineteenth-century evolutionary terms, lacking the capacity for rational thought and planning that distinguishes civilized man, and hence subject to the control of external calamities only. Though there are of course exceptions, this intellectual stance is by no means dead. Thus, in a recent introductory text on demography, a long list of prenatal and postnatal mechanisms affecting fertility makes no mention of infanticide (39). Even such an exceptional attempt to conjoin the insights of anthropology and demography as Lorimer's *Culture and Human Fertility* (54), while recognizing the relation of social structure to demography, is criticized by the anthropologist Kluckhohn (43) as underplaying "rational (or, at any rate, articulate) fertility control among 'primitives.' " (For more balanced demographic and anthropological comments on this general problem, see 38, 76, 80.)

Carr-Saunders was an early notable exception to this general view. In his first work he devoted much space to nonurban and non-Western societies, concluding that infanticide, abortion, and abstention from intercourse were widespread in human culture: "There is no indication of the correlation of any one practice with any one economic stage." His notion of "optimum balance" was a theory of internal regulation culturally achieved; thus "the evidence shows that there is even among the most primitive races at times at least some deliberation as to whether a child shall be allowed to live"; ". . . some semi-conscious adjustment of numbers . . ." (18, pp. 213–23). Human societies maintained themselves at or below carrying capacity of their environments, whether by conscious or unconscious means, although Carr-Saunders recognized that all were not equally effective and proposed the operation of some sort of cultural selection. But all men were distinguished from other animals by the restriction of fecundity (potential fertility) through artificial means: "reproduction among men is never 'mechanical' " (18, p. 54). However, Carr-Saunders's later work moved away from a concern with non-Western societies, and the impact of his first book was regrettably slight.

Anthropology took another path. As student of, and hence spokesman for, otherwise neglected cultures, it emphasized the functional and moral validity, if not always the conscious rationality, of "primitive" life. At the same time, attention to many behaviors that had been common content of eighteenth- and nineteenth-century accounts tended to diminish. Until recently, such matters as cannibalism, warfare, human sacrifice, wife capture, torture, slavery, sexual mutilation, and infanticide were either largely neglected or became the subjects of embarrassed dispute. Anthropology became, and to some degree remains, the twentieth century purveyor of the myth of the noble savage.

In recent years, several biologists interested in human ecology have come to positions similar to those of Carr-Saunders and most anthropologists. Cook long ago pointed to the minimal role of disease in regulating precontact North American populations (22), whereas Bates (7) emphasized the likely antiquity of infanticide and abortion. Most influential at the moment is of course Wynne-Edwards (110–112), who, in his search for homeostatic mechansims, has declared himself the intellectual heir of Carr-Saunders. For him, primitive man's cultural customs, "conscious or not, kept the population density nicely balanced against the feeding capacity of the hunting range." The agricultural revolution, for some reason, resulted in the breakdown of human homeostasis. (Note that this contradicts Carr-Saunders.) A paradoxical shift has occurred. "Primitive" man is once again seen in terms of a biological model, but one no longer red in tooth and claw, derived rather from modern population biology, and permitting a role to formerly embarrassing customs. In Eden, Adam and Eve infanticidal lay down with the territorial lion and the socially hierarchical lamb. Expulsion from the hunting-and-gathering paradise, with the invention of agriculture, marked an end to homeostatic innocence of which Cain and Abel were the first fruits.

Not surprisingly, anthropologists most receptive to the Wynne-Edwardian view are among those concerned with the study of hunter-gatherers. Lest some readers think the above account fictional, I quote an extreme example (9, p. 326): "Most demographers agree that functional relationships between the normal birth rate and other requirements (for example, the mobility of the female) favor abortion, lactation taboos, etc. These practices have the effect of homeostatically keeping population size below the point at which diminishing returns from the local habitat would come into play. (See Carr-Saunders 1922; Wynne-Edwards 1962, 1964; Birdsell 1958, 1968; Deevey 1960; Hainline 1965; Dumond 1965; and Halbwachs 1960.)" The reader will note that most of the authors cited are not demographers at all; demographers agree on no such thing.

Recently, the geneticist Neel, on the basis of a long-term study of small Amazonian societies, has come to view "primitives" in somewhat the same light. "It is obvious," he stated, "that in groups such as this there is a deep commitment to regulating the entry of infants into the population" (65, p. 358). Pointing to the intermediate infant mortality characterizing "relatively uncontacted primitive man," and the use of a variety of cultural methods of birth control, he emphasized the "generally harmonious relationship with his ecosystem that characterizes primitive man. There is an identification with and respect for the natural world, beautifully described by Radcliffe-Brown, Redfield, and Levi-Strauss, among others . . . ," supported by a "religion that regards man as a part of a system . . . " (65). Here I cry caution. For the human ecologist, interdisciplinary approaches are necessary, and I do not underestimate the difficulties of cross-disciplinary communication. The entry of biological theory into anthropology is long overdue. Still, with anthropologists uncritically assimilating a population model by no means universally accepted by biologists, and biologists responding unquestioningly to the romantic typologies that have plagued anthropology, I wonder whether we are not in danger of creating a theoretical superstructure that has more scholarly apparatus than biocultural fact as its foundation. Neither available data nor the sample of societies involved justify the assertions quoted above.

I doubt whether there is any such entity as "hunter-gatherers" as a socioeconomically meaningful class of human societies. If we refer to the totality of human adaptations prior to the expansion of agriculture and urban life, the majority of human population was surely concentrated in grassland, open woodland, riverine, and coastal habitats, precisely those areas now monopolized by agrarian and urban groups. Adaptations to the Kalahari or central Australian deserts, or to primary tropical forest, are no more guides to life in optimal environments than are those adaptations of the polar Eskimos. These are all examples of human adaptation at the limits of physiological tolerance. The Kwakiutl, Pomo, prehorse Blackfoot, Magdalenians, and Maglemosians were hunter-gatherers. Generalizations about the range of human adaptations without agriculture must be weighted in the direction of such societies, rather than derived from small bands of relict Bushmen, Australians, and Pygmies, who do not even represent their own known historic ranges— hence Birdsell's exclusion of the Northwest Coast, California, and similar groups from his studies of size, density, and structure in "generalized hunting and collecting populations" (11). That our limited sample of living hunter-gatherers was given us as a result of the expansive vigor of agricultural and pastoral systems in all but the most severe environments must give pause to those concerned with human population. Surely we may use that sample in a search for models of human adaptation without being enslaved by the biases implicit in it.

I turn now to a review of the literature on infanticide, deferring studies of specific societies for later sections. In the context of his general analysis of human population control, Carr-Saunders was the first to review evidence for infanticide on a world-wide basis. Available data on the small family size of "primitive races" led him to explore, as possible explanations, lower fecundity as compared with civilized groups, mortality through disease and warfare, customary abstention from intercourse, abortion, and infanticide. Reviewing a mass of governmental, travellers', and ethnographic accounts, he concluded that the primary means of limitation were the latter three practices. ". . . Everywhere among primitive races either abortion, infanticide or prolonged abstention from intercourse are practiced in such a degree and in such a manner as to have as their primary result the restriction of increase" (18, p. 292). Unable to distinguish between hunter-gatherers and agriculturalists in this regard, he was puzzled by the overpopulation of traditional civilizations such as India and China. Carr-Saunders proposed that cultural checks had ceased to be employed systematically in these societies, being used only in response to famine and other crises, and that this cultural breakdown was in some way a correlate of political oppression, aggravated by the influence of Christianity, resulting in a state of "apathy and listlessness" (18, pp. 276–77). He noted the need to restrict numbers for nursing and mobility among nomads, and the widespread removal of defectives as a "quality" control; he recognized the frequency of female infanticide, but offered no explanation.

Although no detailed analysis of forms and motives is given, Carr-Saunder's list of societies in which infanticide occurs is impressive. He believed the practice dated from the Upper Paleolithic. Only for two societies are quantitative data given, drawn from other sources. A study of 160 Chinese women over 50, who had borne 631 sons and 538 daughters, revealed that 158 of the females but none of the sons had been destroyed. Another calculation was that 20–30% of all female livebirths in two Chinese provinces were removed. For India, Carr-Saunders repeated estimates of the annual destruction of 33,000 livebirths. Neither estimate is discordant with more recent ones. An attempt has been made to dismiss the evidence marshalled by Carr-Saunders as "out of date" and the argument as "very naive" (26). Yet his sources are often detailed and his estimates of frequency and impact are confirmed by our few modern studies. Although a thorough review of early sources is long overdue, Carr-Saunders was probably correct in assuming that the positive bias of observers and the influence of Western contact had resulted in an underrepresentation of the practice in existing literature.

Aptekar's (4) review of infanticide, abortion, and contraception is a rejoinder to Pearl's emphasis on the logistic curve in human population growth (75a, b), and hence a defense both of the impact of cultural practices and of primitive rationality. His treatment of infanticide does not go beyond Carr-Saunders in use of sources, nor is he able to deal with the then-popular problems of depopulation and of the relation between polyandry and infanticide. Aptekar concluded that although societies at all stages of economic development engage in abortion, infanticide, and abstention from sex, infanticide is the "primitive population check," due to a lack of precise knowledge of the fertilization process and to the lack of effective alternatives. Carr-Saunders's contention that primitive societies experienced lower fecundity than civilized ones was challenged in detail by Wolfe in 1933 (107). Wolfe regarded the low fertility of "the primitive hunter" as a consequence of the "low state of culture and . . . high mortality," rather than culturally induced. Cultural checks, he believed, appeared only with the advent of agriculture. Carr-Saunders's evidence for infanticide among hunter-gatherers was ignored. Adherence either to the anthropological or to the biological-demographic model seems to have more to do with these interpretations than do available data.

The first significant contribution to anthropological demography after Carr-Saunders is Krzywicki's *Primitive Society and its Vital Statistics* (46). Although the theoretical framework is strongly unilinear, that perspective has little impact on his painstaking demographic analyses. Although he deals with group size, warfare, depopulation, dependency ratios, and marriage rates, his most valuable contribution is an extended analysis of female fertility, childhood mortality, and family size in hunter-gatherers and agriculturalists. The data on family size go far beyond Carr-Saunders, but confirm the latter's original contention: on the average, from two to three children are reared by hunter-gatherers, from three to four by North American agriculturalists, and from two to three by African agriculturalists. (Of course, effective natality, completed family size, and survivorship are often conflated in early sources, but their consistency is striking.)

Discussing the means of limitation, Krzywicki offered the first functional analysis of infanticide, based on the imperatives of long lactation and mobility in hunter-gatherers, and of long lactation even in sedentary groups. In so doing he discriminated between systematic infanticide for child spacing and those more occasional and demographically less significant forms such as removal of defectives, twins, and illegitimates, and idiosyncratic motives. Analysis of Australian sources is especially detailed. Regional differences, for which we still lack explanation, emerge clearly in preferences for one form of limitation over another: abortion in North and South America (except Eskimo) and infanticide in Australia; the rarity of both customs in agricultural Africa, except in the case of twins and defectives, is apparent. Krzywicki also gave the first serious attention to distorted sex ratios in these societies, but was unable to offer either causal or functional explanation. In some ways this work is unique: use of sources is cautious and demographically informed, and the appendix is a valuable entré to early materials. Krzywicki's appreciation of the role of conscious economic motives in population control is important, as is his ability to discriminate between society's ideological imperatives regarding limitation and the emotions of the individual actors who must implement them.

A cross-cultural sample of 64 societies, later enlarged to 200, was used by Ford in his 1945 analysis of human reproduction and birth control (31, 32). The two publications form a useful introduction to cultural behaviors and attitudes, although they do not provide a functional or demographic analysis, nor are the data handled quantitatively. Ford's discussion of the apparently contradictory nature of attitudes toward infanticide, one of the sources of Aptekar's confusion, is useful. While it may be said that infanticide is both universally practiced and universally condemned, these generalizations refer to several distinct ideological domains. Both abortion and infanticide are practiced occasionally in every society, but may be strongly disapproved deviant acts: of Ford's early sample of 64 societies, 19 expressly forbade infanticide. Yet even where allowed, ". . . it is also clear that strong social pressures are brought to bear against any excessive indulgence in abortion and infanticide. Most societies permit the riddance of a child under specially defined circumstances only. Generally . . . both abortion and infanticide come under rather strict social control" (32, p. 765). Ford offered two explanations, the first concerning the regulation of homicide: ". . . infanticide is most readily condoned if it occurs before the infant is named and has been accepted as a *bona fide* member of its society. It seems that the primary and fundamental restriction in most societies is the taboo on murder, i.e. killing a member of the ingroup. The less eligible a child is for membership in the group, the less seriously the act of killing the child is viewed" (32, p. 765).

Ford's other explanation is of greater import to human biology. He stated (31, pp. 73, 47):

> The widespread insistence upon not killing newborn infants suggests a tendency on the part of mothers to do away with their own babies. Were this not the case it would be difficult to explain the very general development of sanctions directed against infanticide. Nor, indeed, does it seem unlikely that a woman should occasionally wish to kill her offspring . . . this would be especially true after she had fulfilled the obligations imposed by society and had already borne and reared a number of children.

> The tendency to practice infanticide does not contradict the assumption that women instinctively desire to rear and protect their young. The facts do support the view, however, that the maternal instinct, if indeed there be such an instinct for human beings, is not nearly strong enough to counteract unaided the tendency to destroy unwanted infants.

Nag's (62) study of human fertility is an extremely valuable synthesis of physiological and ethnographic materials, including a cross-cultural sample of 61 societies. However, he does not discuss infanticide in detail, perhaps from lack of confidence in Carr-Saunders's and Krzywicki's sources, but more likely because his taxonomy of factors influencing fertility is derived from the work of sociological demographers, who naturally do not include it. Nag concluded: "The circumstances under which infanticide is or was practiced generally in these societies suggest that the mortality due to this factor is negligible in comparison with the mortality due to natural causes."

Incidental data on infanticide are found in Whiting's article "The effects of climate on certain cultural practices" (105), an explanatory hypothesis for the tropical distribution of circumcision, based on cross-cultural correlations with such factors as polygyny, patrilocality, protein deficiency, and postpartum sex avoidance. Whiting's assumption is that postpartum abstention is a means of preserving both quantity and quality of milk supply to the nursing infant. "Infanticide is as efficient as a postpartum sex taboo for solving the problem posed by nomadism, but only the latter solves the problem of protein deficiency." Consequently, infanticide is not involved further in the hypothesis; however, a tabulation is provided. Of 88 societies that could be coded for it, 15 lacked infanticide, 8 practiced it (although it was unascertainable whether for child spacing or not), 54 practiced it but not for child spacing, and only 9 used it as a means of spacing. Whiting recognized that his sample is skewed in favor of Africa and North America. Granting sampling weaknesses, this distribution suggests the necessity of a search for ecological and societal correlates of infanticide in terms of the specific functions that it serves, rather than as a unitary phenomenon.

An initial step in this direction is Grantzberg's treatment of twin infanticide (36). Pointing to the inadequacy of explanations based on ideology, Grantzberg tested the hypothesis that twin infanticide is a response to the burden of caring for two infants simultaneously. He used a sample of 70 societies rated in terms both of availability of childrearing assistance to the mother, as inferred from family size and settlement patterns, and of the mother's freedom from work responsibilities, as measured by subsistence duties and childcare tasks such as nursing and carrying. A significant correlation with absence of twin infanticide emerges; indeed, only 2 of the 33 societies scored as having no difficult conditions actually practice twin infanticide. However, of those that do suffer "conditions that would cause a mother to disrupt social patterns if she attempted to rear two children at once," only 16 out of 37 practice twin infanticide. Clearly Grantzberg's results support the contention that maternal workload is a significant causal factor, if they cannot yet explain how some other societies manage maternal burdens by other means. "Contrary to the common practice of treating infanticide as one monolithic institution, . . . [it] is . . . rather a heterogeneous phenomenon consisting of different practices, each aimed at solving different problems. . . ." The analysis confirms Carr-Saunders's view that infanticide is a response to the "difficulty of transporting and suckling," but demonstrates, as did Krzywicki's studies, that these burdens and their resolution have much wider application than merely to nomadic societies. We must therefore demur from Sussman's (92) contention that "once human groups became sedentary, social spacing mechanisms were no longer a necessity."

The inadequacies and inconsistencies of existing cross-cultural literature relating infanticide to ecological or demographic factors are patent. It is useful at this point to specify those functions that the practice may be expected to serve; that is, general reduction in population numbers (including twin removal), removal of defectives, elimination of social "illegitimates" (i.e. offspring whose existence violates social group boundaries), response to loss of the nursing mother, control of dependency ratio, manipulation of sex ratio, and, finally, use as a backstop to other methods when those fail. These admittedly overlapping categories may give some notion of possible research attacks. Some of the above, such as removal of defectives, are redundantly cited in the literature; others, such as the adjustment of dependency ratios, have received no attention.

In the following sections, I discuss only three of the above, namely, the removal of defectives, control of numbers, and manipulation of the sex ratio. Results demonstrate that the value of much of the existing discussion is vitiated by a reliance on informants' verbal statements. Emphasis on word rather than deed is a bias implicit in much of anthropology and history; unfortunately, ideology is not a reliable guide to the ecological functions of human customs. The rationales that societies provide for their practices are designed for in-house consumption; the function of ideology is to maintain, not to explain, behavior. A scientific understanding of behavioral functions must be sought in attention to linked ecological, economic, and sociopolitical attributes of total societal systems.

REMOVAL OF DEFECTIVES

Maintenance of the genetic health of a population through removal of congenital malformations at birth would appear on first thought to be a likely function of infanticide. Many sources report such a motivation; Ford (31), for example, stated that such removal prevents the development of useless dependents in societies without specialist roles for them and "discourages the transmission of hereditary strains toward certain abnormalities." However, further consideration suggests that the role of infanticide as a means of artificial selection is equivocal at best. It is estimated that 5–6% of all births have defects partially or wholly of genetic origin, and that 5% of all conceptions result in major congenital defects. Incidence of major malformations identifiable at birth, based on Caucasian, Japanese, North American Indian, and American Black series, is about 1%, ranging from 0.63 to 2.44 (63, 82, 109). Such rates would appear to present adequate opportunity for effective artificial selection.

Unfortunately, the relevance of these figures is not great. Series are to some degree noncomparable, since they may rely on registration data, examination at birth, or hospital nursery records. The first are of course subject to underreporting, while examination at birth, as in Neel's Japanese series, includes both stillbirths and livebirths. Only nursery records, as in Woolf & Turner's Utah study (109), exclude stillbirths, which are irrelevant to infanticide. Thoroughness of examination and methods of coding also vary from study to study. These figures are all products of examination by personnel trained in Western medicine, whereas any assessment of general potential for selection must refer to malformations detectable at birth by an indigenous midwife or experienced mother without benefit of Western training. An inspection of the classification of malformations devised by Neel (63) suggests that the majority of medically major defects is not readily identifiable by the layman. Most musculoskeletal and nervous system complications would be superficially recognizable, while only some of those involving digestive and urogenital systems would be; the greater number of abnormalities of the respiratory, cardiovascular, and lymphatic systems, and of the sense organs, would not be recognizable at all. Additional testimony to the assumption that the majority of defects is not available for normal human intervention at birth is the number of additional defectives uncovered both through autopsy of neonatal deaths and nine-month follow-up of livebirths in Neel's study.

Neel identified that "group of malformations readily diagnosable at birth under almost any conditions" and "sufficiently frequent that their incidence may with meaning be compared with the findings in other extensive series" as: anencephaly, spina bifida manifesta, harelip–cleft palate, atresia ani, anophthalmos–microphthalmos, and polydactyly. Rates of incidence for these six major defects range from Neel's 0.49 (Japanese) to McIntosh's 1.23 (New York City, as recalculated by Neel) (63). A Brazilian study reported a frequency of 0.5% for the "severest cases," listed as "harelip-cleft palate, club foot, limb bone aplasias, congenital blindness, polydactyly, mental retardation, albinism, etc." (82). (Mental retardation and blindness are, however, rarely identifiable by the layman at birth.) With the exception of harelip–cleft palate, polydactyly, clubfoot, and albinism, much of this readily observable group will not live beyond the first year; the likelihood of perinatal death in Neel's six categories is extremely high. And even for such traits as cleft palate, survival without modern medical intervention may be difficult.

Populations do of course vary, both in the incidence of specific congenital defects and in the resulting totals. We would expect a higher incidence of defectives both presented and removed in populations with high inbreeding coefficients. Consanguinity does seem to increase the occurrence of defectives and hence the frequency of multiple malformations, but this only elevates the level of lethality on the average, thus making artificial selection less necessary. In any case, the effect of consanguinity (other than that involving primary kin) in controlled studies is small (63, 67, 86).

Available demographic studies do not help us much in regard to this question, but they do suggest that infanticide has limited effects in small populations. Thus, in the highly endogamous Xavante (69, 83), out of a total of 287 individuals of all ages in two villages, seven congenital defectives were identified (one with polydactyly, one with clubfoot, three mental defectives, and two or three with possible congenital heart disease) (101). Although infanticide is likely in this group, if defectives are already being removed by this means, the rate at birth must be unusually high. Similarly, of 274 Yanomama, one achondroplastic dwarf and six cases of congenital heart disease were observed. Yet these defectives are believed to be a "residue of a relatively high proportion of defective children" removed by infanticide or early death, in a society which admits removal of defectives. Mission birth records for other Yanomama villages likewise contained five major congenital defects out of 93 births (64, 66).

Another small society, the Buang of New Guinea (56), which formerly practiced infanticide against defectives, contained eight living dwarfs in a population of 9000. The authors reported, "although it was difficult to obtain a clear and coherent description of dwarfs during childhood, the people stated that the condition was in most cases apparent at birth, both due to the associated deformities and the small size." It is interesting that such individuals, although fully accepted members of society, are not permitted to marry. This injunction may, however, reflect society's assessment regarding the defective's ability to assume the normal socioeconomic sex role, rather than an attempt to prevent genetic transmission.

Woolf & Dukepoo's study of Hopi albinism (108) explains the high rate of this condition as a result of high extramarital fertility of affected males who spend their days as craftsmen in the village, in a society which does not recognize a hereditary basis for the condition, and which, while it does not find these individuals acceptable marriage mates, does find them attractive as sexual partners. Among the Peruvian Cashinahua, albinism is reported to be intentionally removed as spiritually dangerous. One albino woman in her forties was identified in a demographic study of five villages totalling 206 people. Of 15 pregnancies, she had terminated 7 in abortion or infanticide. (This scarcely suggests avoidance of the spiritually dangerous lady.) The authors further observed two infants, one with spina bifida and one with syndactyly, which would have been killed at birth in the past but were saved as the result of Western influence (42). Our difficulty in assessing these few suggestive reports is of course compounded by Western impact, both ideological and medical, making extremely difficult the reconstruction both of past malformation rates and of the impact of infanticide on them.

As stated, populations differ in the incidence of specific malformations. But, while the Ashanti may have spared hunchbacks to become court jesters (31) and the Hopi male albino had a role in the village as a weaver, it is difficult to see any relationship between the present incidence values in most populations and previous selective infanticide. High rates of anencephaly and spina bifida in the British Isles, of polydactyly in US Blacks, or of harelip–cleft palate in Japanese (64) make no sense as artifacts of cultural selection, at least at present. Finally, the relation of this puzzling situation to reproductive compensation and the dependency ratio may be mentioned. If a defective infant, with short life expectancy, lowered economic efficiency, and lowered fertility, is preserved through investment of nursing and general childcare, we may speak in theoretical terms not of fetal but of "live wastage." This is especially so in societies that observe postpartum sex avoidance, in which case only two years of life for a defective child would significantly delay the return of the mother to reproductive status. One would anticipate that the burden of numbers of defectives in subsistence societies would be too great to bear, but such does not appear to be the case. Weinstein, Neel & Salzano (101) have remarked that, given the relative infrequency of chronic and degenerative diseases in small societies, "congenital defects may play a relatively greater role in the medical burden of the community than in contemporary, more civilized groups." We need much better measures of dependency ratios and energy investments in human economic systems. But a consideration of the data reviewed above leads one to wonder whether what appears dysfunctional on the surface may not in fact be adaptive. The maintenance functions of genetic systems are not reducible to simple input-output models. On the genetic level, Neel (63, p. 435) proposed:

> The similarity in malformation frequencies in such diverse populations as Japanese and European thus finds an explanation in the fact that there is a malformation frequency representing the optimum balance between, on the one hand, fetal loss and physical handicap from congenital defect, and, on the other, population gain from those very same

genes which in certain combinations may sometimes result in congenital defect. The differences between populations as regards the frequencies of specific defects would seem to indicate that within the framework of this optimum figure, different populations have evolved genetic systems differing significantly in their details.

On the level of cultural systems, a parallel proposal may be made, if somewhat hesitantly. Perhaps some level of "live wastage," some negative cost-benefit ratios in the production of young, are tolerated as a means of reducing overall natural increase. Insofar as a population may possess excess reproductive potential in relation to available resources, some of those scarce resources might be channeled into the production and support of individuals of low potential, and the devotion of productive capacity to their support might have some effect in reducing rates of natural increase. In regard to congenital malformations, then, most infanticide would appear to be no more than the disposal of some excess population of low productivity, most of which is in any case destined for early mortality. Available demographic and ethnographic data are not likely to assist us much in testing this rather surprising hypothesis. However, on the basis of what we now know, the impact of infanticide on the incidence of congenital malformations appears to be, in most cases, insignificant.

REGULATION OF NUMBERS

The human potential for excess reproduction, a property of all successful species, is of interest both in terms of the history of human expansion from an early African homeland and of the more general question of the evolution of self-regulating mechanisms. Estimates of human growth potential have been based on the physiological maxima of fertility, observed natality rates, and maximal known rates of increase. Lorimer's (54) fecundity model produced a maximum of 8.32 births per completed reproductive span, but this is too modest, as the hypothetical potential of the human female is about 15 children, even allowing for the effect of lactation on ovulation. Recent natality studies give mean rates of the number of children ever born from 7 to 10, the most famous being that of the North American Hutterites (27, 95), which is 10.6 in spite of a mean female age at marriage of 20.9 years. However, these modern populations enjoy at least some benefits of modern medicine, and so do not provide us with the best notion of past human rates of increase. That human growth can be explosive is well demonstrated by Birdsell's (10) analysis of the rates of modern subsistence societies in unoccupied areas. His four cases (Tristan da Cunha, Pitcairn, Bass Straits Islands, and the Australian Nanja Maraura) underwent one-generational doubling in their early histories; more recently Birdsell (12) has revised this to a tripling within a thirty-year generational period. These recent estimates support Malthus's original proposal, based largely on North American colonial data, of a potential human doubling time of 10–25 years (57, 58). Such growth must soon initiate either internal or external limitation.

Australians

It is natural in the historical context discussed above that living hunter-gatherers should have received the most attention as possible models for early human response to the limits of growth. Krzywicki's extended analysis of Australian infanticide, mentioned earlier, includes a computation of the contribution of infanticide to total infant mortality from early sources, noting correctly that much infanticide affects offspring who would in any case die in infancy of natural causes, and is thus substitutive. Australian societies removed on the average every child above the desired number of three per family. Based on estimates of natality, spacing, and the short reproductive span of Australian women (natality being rare after 35), the infanticide rate is 20–40% of livebirths. Any revision of Krzywicki's assumption of a short reproductive span of 15–20 years would result in an increase in this proportion. Assuming natural infant mortality in such conditions as 35–40% without the intervention of infanticide, systematic removal for child spacing would raise total infant mortality to about 45–50% of livebirths, according to Krzywicki. However, this overlooks the role of child spacing in reducing some mortality by improving conditions in infancy.

Birdsell also emphasized the importance of infanticide in unspecialized hunter-gatherers and its role as a response to infant transport problems, based on extensive research on aboriginal Australia (12, 13). In a recent summary, he reported the adult sex ratio in precontact genealogies as 150:100; this, and the evidence of early sources, led him to estimate a 15–50% infanticide rate, imposed through three-year child spacing. (Throughout, I have reported sex ratios as given, rather than converting to more useful percentages.) "There is reason to believe that the limiting forces involved in the necessary spacing of children to be saved would apply broadly to all hunting groups, although sex preferences might vary" (12, pp. 236–37). The problems of mobility and lactation are induced by a complex of environmental and technological factors preventing both sedentism and early weaning. Infanticide is thus an indirect response to environmental limitations, culturally mediated. If so, then we might well posit moderate levels of increase for all mobile hunter-gatherers. The populations from which estimates of maximal human natality are derived are sedentary, with the exception of Birdsell's Maraura case. This means that explosive colonization based on their high rates may not be the best model for the spread of early man, although it may well be appropriate for early populations in areas of abundance that allowed sedentary collecting, at least until other limiting factors intervened.

Bushmen

Recent field investigations of Kalahari Bushmen have raised more questions than they have answered. (I am indebted here to a critique of sources by my student, D. Clavaud.) The most thorough demographic study (40) involved census of 840 Dobe !Kung Bushmen and interviews of all women over 15. On the basis of childhood life tables, Howell estimated a life expectancy of 32 years at birth and an annual growth rate at 0.5%. Her study agrees with others regarding late menarche (average 15.5 years), and even later age at first livebirth (19.5 years); she also reports a long nursing period of 4–6 years (other reports range from 2½–4 years) (51, 60, 85, 87) and a high female marriage rate, only two mental defectives being unmarried. However, Lee's 1969 !Kung sample (50) had an unusual and unexplained sex ratio in all age groups favoring females. Some sort of birth-spacing mechanism may well be involved in producing resulting low rates of completed fertility (4.6), but what it may be is unclear. Both Howell (40) and Lee (50, 52) maintain that infanticide is uncommon, practiced only on defectives, on the death of the husband, or "sometimes when one birth follows another too closely" (40); in Howell's population, 6 out of a total of 500 livebirths had been destroyed. She believed that neither disease nor malnutrition were significantly involved. Consequently, explanations based on unusually long suppression of ovulation through lactation (51) or control of ovulatory cycle through fat intake (34, 45) have been proposed (45, 51).

However, other Bushman ethnographies suggest a different interpretation. Although I take the liberty of conflating data on several bands, to do so raises valid questions regarding the Dobe area. Marshall (60) reported that Nyae-Nyae !Kung space their children 2–4 years apart through infanticide to ensure lactation during that period; Schapera likewise reported systematic infanticide for reasons of nursing and mobility, but not on defectives, removing one or two out of every three children "without exception" (85, p. 116). Early sources in Krzywicki concur. This implies an infanticide rate of at least 50%. The evidence of postpartum abstinence is less consistent, varying from none (85), to one year's abstinence (51), to abstinence until weaning (87). Questions may also be raised regarding disease and nutrition. I ignore the latter here, but assertions about general health do not accord with the diseases reported, including tuberculosis, leprosy, and endemic malaria. Lee (52) estimated childhood mortality at 30–50%, which accords somewhat with Howell's figure of 30%. Most important, he reported that gonorrhea is the major epidemic disease, while Howell found that 35 out of 840 women interviewed believed they were infected. The fertility rate of these 35 was 2.4, compared with 5.1 for women not so self-identified. I scarcely need remark that self-diagnosis, especially in women, results in gross underreporting of this disease. Howell's tables on completed fertility by cohort reveal a consistent decline from the cohort 65 years and over (5.3) to the cohort aged 45–49 (4.1), and she herself attributes the drop in fertility in younger cohorts to gonorrhea.

I believe that resolution of these contradictions depends on the recognition that they refer to differing points on a century-long chronology of increasing contact with, and submission to, surrounding dense populations. If levels of effective natality were traditionally controlled through infanticide and perhaps postpartum abstinence, then the decline in both these practices in the Dobe area in the late 1960s may be a response to declining fertility, due in part to introduced disease. Hence the requests for fertility medicines that Howell received while in the field. Better data on disease, nutrition, and reproduction are needed to test this hypothesis. If confirmed, it will again reinforce the dangers of generalizing from small groups of embattled marginal hunters, on the verge of cultural extinction, to man's preagricultural past. At present, the Bushmen fail to provide us with any understanding of population regulation among hunter-gatherers.

Pomo

I have found no useful demographic studies of hunter-gatherers or pastoralists in less marginal environments. The sedentary Central Pomo of Northern California practiced infanticide, in addition to abortion, against all excess births over the family ideal of two or three, as well as against twins and illegitimates. The everpresent fear of food shortage as a motivating factor is clear from an early study; unfortunately it provides no quantitative data (1).

Tikopia

Horticultural societies have also received little demographic treatment until recently. Still in many ways most valuable is Firth and co-workers' study of Tikopia, based on fieldwork in 1928–1929 and again in 1952–1953 (14, 28–30, 89). I emphasize in this summary both the number of control methods employed and the locus of decisions, aspects highlighted by the observations of Firth and Spillius on the society's response to severe famine in 1952–1953. The setting is an island of less than three square miles with an economy supplying vegetable carbohydrates and marine protein. Migration was not practical due to distances involved and the prior settlement of nearby islands by culturally distinct groups. Density in 1929 was estimated at over 400 per square mile. In Tikopia, the ideal completed family contained two to three children. Systematic controls were coitus interruptus, abortion, infanticide, male and female celibacy, suicide (including overseas voyaging), and, on two occasions in remembered history, warfare between kin groups and mass expulsion.

There are no quantitative data on coitus interruptus, used by the unmarried against illegitimates and by the married for child spacing. Its significance is clear from the fact that it was part of the traditional moral code, publicly inculcated by the chiefs in religious settings (28). Abortion was least important, serving primarily as an alternative to coitus interruptus and infanticide for unmarried women. According to informants, infanticide was a major family means: "A male and a female are born; they are allowed to live; two only. . . . If another child is born, it is buried in the earth and covered with stones; it is killed. If another child is born also, it is buried in the earth. And two only are left, corresponding to the scarcity of food" (28, 30). Firth noted only one case in 1929, with about half of the island Christianized; his informants were contradictory on the matter of sex preference. Borrie's analysis of sex ratios by cohort (14) showed distortions as great as 153:100 in 1929 (ranging from 110–153:100 for cohorts under age 27), and persisting though less extreme in 1952 (106–125:100). As Borrie stated, this persistent imbalance implies female infanticide in most years; a rate of 10–30% of female livebirths is not unreasonable. During the famine, Spillius (89) recorded 62 pregnancies, of which 14 failed to terminate in livebirths. He believed the majority of these were deliberately destroyed by induced abortion or infanticide. Four certain cases of abortion-infanticide were identified.

The role of celibacy is clear from census data (14). In 1929, 93% of males in the age cohort 38–47 were married, but only 81% of females (this cohort had the maximum incidence of marriage for each sex). In 1952, percentages for the same cohort were 87.8 for males and 84.1 for females, a removal of 15–20% of females from reproduction. These data are from the period of Western contact, but Firth's account makes clear their traditional nature. Younger sons, especially in large families, were enjoined by the family elder not to marry, as the offspring of the oldest son would require all available family resources. The feasibility of such voluntary celibacy, both for males and females, depends on the practice of coitus interruptus, abortion, and infanticide in a society that places high value on sexual activity for all adults.

Mortality data for 1929–1952 include 16 female suicides out of a total female mortality of 187. More significant is the practice of overseas voyaging by males, regarded by the Tikopia as tantamount to suicide, and engaged in for adventure, from personal shame, rejection, or as a legal banishment imposed by the chiefs. Genealogies of chiefly houses prior to 1929 contained 23 males lost at sea out of a total of 69 male deaths. In the period 1929–1952, 30 such voyages occurred, with a loss of 81 and only about 20 survivals. This survival rate would surely have been lower before government pacification and government return of stranded sailors. Most of these men were unmarried and under 30 years of age. An increase in the practice during 1941–1950 (49 deaths in the decade) probably reflects growing population pressure. During the famine, 5 voyages were attempted in a period of a few months. Although all were unsuccessful as suicides, Spillius, then present, felt that these were "dress rehearsals for . . . expulsion of large groups." Yet this significant loss in the period of record was only enough to balance the disproportionate sex ratio in the young (14, 30).

Turning to the decisional aspect of population limitation, responsibility for perception of carrying capacity and for specific acts was lodged at several levels of the social structure, and limitation was in large part conscious. Tikopia referred to total population by the term *fenua* ("land") ("the land is many"), and to resources as *te kai* ("the food"). Many informants' statements reflect a concern for the relation between the two and the need for population control: "If the 'land' becomes very many, where will the food be found?" The conscious use of celibacy, contraception, and infanticide was *fakatau ki te kai,* that is, "measured according to the 'food' as they put it" (30, pp. 39, 43–46). Elders were interested in the result of Firth's island census, and commoners and chiefs expressed the necessity for voyaging and celibacy. In 1929, when Firth foresaw overpopulation, elderly and responsible Tikopia were also concerned at the growing number of children, and asked his advice. ". . . An appeal was actually made to me by one of the leading men of Tikopia that on my return to Tulagi I should persuade the Government to pass a law enjoining infanticide after a married pair had had four or five children, in order that the food supply might not be overburdened" (28, p. 417). In the past, the village council had as one of its functions the ritual public injunction of celibacy, continence, and the conservation of resources. At least one chief saw the decline of this institution as a cause of overpopulation. Thus responsibility for population control devolved at several levels on mothers and fathers, family elders, and chiefs and their assistants.

If the causes of the 1952–1953 famine were in part new, response was largely traditional. In assessing the length of the recovery period after the devastating hurricane, "the estimates of the most responsible Tikopia as to the gravity of the situation were proven in the upshot to have been more precise than those of Europeans. . . . " (29, p. 57). During the crisis, the village council was revived increasingly as a means of social control. A law was promulgated there that married couples should limit sexual intercourse to once a week, and suggestions were made that families without food should either accept death or be put to sea (89, p. 13). This is not to say that traditional means were always successful. There had been famines and droughts in the past. And in 1952, the decay of the traditional authority structure, the extent of overpopulation, and the advent of a second devastating hurricane all placed the situation and the resulting social disorganization beyond the control of chiefs and village council. But it is apparent that traditional roles involved cognitive capacities, responsibility, and authority necessary for effective population control in more normal times. While it is unlikely that complete homeostasis was achieved by the above means in precontact days, it is probable that such a system of conscious social controls greatly damped the oscillations in population density that would have occurred without their intervention.

It has been asserted that the Tikopia are "certainly very unusual, if not unique" (54, 100). I contend that they are unique in our ethnographic corpus, rather than in fact. Conscious understanding of the need for limitation, and the involvement of senior kin group members, shamans, and others in decision making, are reflected in many sources. What we lack, with the exception of recent studies of ritual regulation (78, 79), are analyses of the information network by which relevant assessments of resources flow through the social system, and the relation of authority structures to specific acts of control.

Recently, Bayliss-Smith (8) reanalyzed data on Tikopia and other Polynesian outliers. [For other treatments of population limitation on Pacific islands, see (17) and (100).] Computing maximal human carrying capacity, he concluded that available census data reflected levels of population density maintained at 70–80% of carrying capacity. His analysis emphasized the role of conscious checks. Although he extended his homeostatic hypothesis beyond outliers and beyond island cultures, he did so only to marginal habitats with extreme environments. Our willingness to generalize to this extent, or beyond it, must depend not only on the kind of societal analysis stressed above, but on a classification of human societies that reflects ecological factors, such as the oscillatory or continuous nature of natural controls, degree of isolation of the ecosystem, including limitations on migration and energy transport, and capacity for economic and technological growth. If we are to move beyond the simplistic question of whether human societies regulate their numbers to an inquiry into when and by what means they do so, a meaningful answer will depend heavily on our ability to construct ecologically relevant classifications.

Japan

Demographic literature on nonindustrial agrarian societies is immense, but infanticide, though widespread, has been little studied. Japan has received both ethnographic and demographic treatment, especially in Taeuber's classic demography (93). Her anthropologically informed work can be read as a critique of the more simple versions of demographic stereotypy regarding the Asian family and the demographic transition. In addition, Japan presents some instructive comparisons with Tikopia.

National censuses for early periods of Japanese history indicate slow growth from the thirteenth to sixteenth century, as arable lands were increasingly brought into production, but political instability and recurrent earthquakes, epidemics, and famines resulted in high mortalities. Population increased rapidly in the early Tokugawa period, under a more unified government, then remained stationary for about 150 years, from 1700 to 1852, varying, in national censuses introduced at this time, little more than 10%. Several constraints may be identified: isolation was imposed not only by distance from a hostile mainland, but by political exclusion of most foreign agents and by the prohibition of emigration. Thus neither external outlet nor energy inflow through trade or conquest were available. By about 1750, limits of arable land had been reached. With heavy rice taxation and loss of lands to the growing merchant class, levels of farm debt rose. Escape to the cities was an alternative only for few, and mortalities there were high. Malnutrition and famine were common (15, 93).

In this situation, a variety of techniques of population limitation was used, varying by region and class. Contraception seems never to have been important. Celibacy was not significant in a society where marital rates were high in all classes, but delay in age of marriage was important for some, especially the samurai. Childbirth was generally disapproved of for women over 40 or for those who had daughters-in-law. Abortion, both chemical and mechanical, was widespread. This practice was favored by noble and samurai classes and by urbanites, probably as much to destroy illegitimates as for limitation per se, although the increasingly impoverished samurai were often forced to prevent excess births. Abandonment was also common, but its extent is unknown. For peasants, 90% of the population, the preferred means was infanticide. This practice is old in Japan, with references occurring as early as the ninth century, but its use apparently increased in late Tokugawa. Peasant preference rested on several considerations. Abortion was expensive or risky, and illegitimates could be socially assimilated, so concern for concealment was less. Infanticide allowed removal of defectives and twins. Most importantly, it permitted manipulation of the sex ratio to guarantee patrilineal perpetuation of the localized peasant family. The entrenched nature of the practice is evident from linguistic usage. Bowles (15) gives the common euphemism, *mabiki* ("thinning," as in rice seedlings), as well as other regional indirections. The midwife, if present, inquired if the baby was "to be left" or "to be returned"; traditional burial was near a shrine or in the household mortar yard.

These evidences of longstanding tradition are not quantitative. Although Taeuber cautiously refrained from a quantitative estimate, I have rashly attempted one here, using both her data and Bowles's. It is admittedly tentative; a better could probably be derived from primary sources. Evidence as to size of completed family or number of livebirths preserved is of some help. According to Japanese sources, villagers retained an average of 5–6 offspring until the eighteenth century, but this number was reduced to 1–3 during the period of stationary population. Sex composition of this number was ideally two sons and one daughter, the second son serving as insurance against the death of the first, and the daughter being exchanged for the son's wife.

Both my sources feel that historical materials exaggerate the extent of infanticide; e.g., these materials report that two out of five infants were killed in parts of Kyushu, whereas in parts of Shikoku only one boy and two girls were saved, and in Hyuga all but the firstborn were killed. However, Taeuber's summary of government pronatalist policies, which began in the 1660s and became national campaigns in the next century, suggests that these estimates may not be greatly in error. These policies included inducements to additional births in the form of child allowances and rice rations, and generally were graduated to award greater compensation to higher birth orders. Of those listed by Taeuber, one government policy awarded allowances to all births, five awarded them to each child after the firstborn, two to the third or higher, and one to each family with five or more. At least local governmental agencies perceived that only the firstborn (son?) was certain of preservation. Allowing for the then current but erroneous governmental belief that the island's population was declining, and for local variation in custom, a mean retention of 2–3 seems plausible.

Taeuber's view is based in part on her estimate of child mortality. Assuming a life expectancy at birth of approximately 25 years, she proposed a crude birth rate of at least 40 per thousand to maintain stationary population in the face of recurrent famine and epidemic, probably resulting in an average mortality of over 30 per thousand (93, pp. 29, 32–33). Part of the difficulty resides in the estimation of lifespan, and hence the population model on which her calculations are based. In addition, however, Taeuber's analysis overlooks the fact that infanticide is an alternative to much natural infant mortality, as indicated previously. Thus, whereas she maintained that "when the heavy infant mortality of the period is considered, this would mean that the practical decision as to life or death for a child was not required until four, five, or even six children had been born" (93, p. 31), this neglects the function of infanticide as a spacing mechanism for lactation, to adjust to economic fluctuations, and to control sex ratio. Rather than first bearing five or six infants and accepting natural mortality, the peasant family would substitute cultural mortality increasingly, beginning with the second child or, in the case of a female, even the first.

Evidence from the Meiji period contributes to an estimate as well. The 1950 census revealed an average number of children ever born (which I take to be effective natality, removals unreported) to ever-married women born in 1890 or earlier as 4.7. Overall mean children ever born to farmers were 4.98, to fishermen 4.04 (93, p. 268). These means reflect a rate of retention far below maximum potential. Limitations, as Taeuber showed, occurred increasingly later in the reproductive span, as we would expect. Statistics on completed fertility in the rural Aomori area clearly show cessation of childbearing after age 44. Thus for the female cohort 40–44 years, mean number of children ever born was 5.7; for the cohort 50–54, 5.8; for the cohort 60 and over, 5.5 (93, pp. 265–66). A recent study of a fishing village (44) gave the mean completed fertility for the cohort of women born in 1896–1900 as 2.7; the mean for all parous women in the cohort was 4.9. While the sample of 177 is admittedly small, results are consistent with the above. In evaluating these more recent data, it should be noted that the Japanese crude birth rate, after increasing rapidly from 1852 to

1920, underwent a decline from 1920 to about 1940. Women born in 1890 would be in midreproductive span in 1920; we would expect them to show slightly greater fertility than their pre-Meiji grandmothers.

Sex ratios also reveal the extent of infanticide to some degree. Censuses from the seventh to seventeenth century are grossly distorted in favor of females, probably as a result of underreporting of males to evade taxation. More reliable censuses from 1750 on, however, show a clear distortion in favor of males (876 females per thousand males in 1750), declining gradually in the next two centuries but not achieving parity until 1950 (15, 93). Taeuber's analysis of sex ratios of stillbirths and livebirths for recent censuses shows an increasingly disproportion of males in higher birth orders, reaching an extreme (1947–1949) of 168:100 for women 40 and over bearing their twelfth or higher order child. Bowles cited a "recent" study of one mountain village in which nearly two thirds of the families had a boy as eldest child, which could scarcely be due entirely to the natural influence of birth order on sex. Throughout Japan, rural areas show the greatest masculinity ratios. Finally, Taeuber referred to a 1940s village study that revealed evidence of infanticide through high rates of stillbirth and neonatal death, though no sex ratios were given.

In sum, these data indicate a mean number of children retained somewhere between 4 and 6. Assuming a reproductive span of 25 years (ages 15 to 40), this would imply limitation, by whatever means, of almost 50% of potential livebirths, out of a maximum potential of about 8. (This crude estimate could be refined by analysis of age-specific female mortalities.) We must then assume an even lower completed family size for the late Tokugawa period. Sex ratios imply maximum removal of 50% of all female livebirths, but an overall mean of 10–30% of females. Remembering that some males were destroyed, and that infanticide was the primary means for the majority of the population, a late Tokugawa infanticide rate of 10–25% of livebirths is not unreasonable. Historical assertions that 30,000 children were removed annually by 100,000 households in Chiba Prefecture, or that two out of five livebirths were destroyed in Kyushu, become more believable.

Although this estimate is my own, its significance is well expressed by Taeuber (93, p. 33):

> The necessary conclusion from the Japanese experience is that the role of family limitation in premodern societies may have been underestimated and the motivating factors oversimplified. . . . In Japan, family limitation was not only consistent with, but occurred in the service of, the ancestor-oriented family system. It was not a fundamental deviation from the family values of the great Eastern cultures.

In this agrarian society, reliance on conscious checks was probably a result of factors similar to those operating on Tikopia. Japan remained a closed system until at least 1900. "A family system that functioned to expel young existing within a society that barred migration in theory, and in fact offered few opportunities to migrants, was eminently suited to the development of mores of family limitation" (93, p. 31). However, differences are also apparent. Though old, limitation in Japan did not become sufficiently frequent to achieve nationwide equilibrium until the limits to growth had been reached, judging from the frequency of natural mortalities through epidemic and famine before and during Tokugawa times.

A second contrast may give a clue to the first. As cultural checks intensified on the village level in response to peasant perception of local economic realities, government, both local and national, increasingly opposed such practices, though without success. In this stratified society, leadership did not guide and encourage the adjustment of people to land as in Tikopia. This implies a discrepancy between "cognized environments" (77) of different sociopolitical levels, underlining again the need for attention to information flow in systems of population regulation. Whatever the reason in this case, the consequence is part of history. Once preexisting governmental ideology found support in industrialization and overseas expansion, the largely masculine excess that was preserved in response provided the labor and military force for an episode in modern imperialism.

A final comment on ideology concerns the impact of the West. Japan was distinctive in her successful resistance to Western colonization and missionization. In consequence, infanticide and other forms of population control continued, in memory and in some cases in fact, into modern times, rather than, as in colonial societies, being suppressed, rejected, and denied. Japan retains an understanding of the necessity for often painful acceptance of survival imperatives lost by most modern nations. (See for example Kinoshita's powerful 1958 film, "The Ballad of Narayama," depicting customary senilicide in a pre-Meiji mountain village.) That heritage surely played some role in one of the most dramatic national reductions in population increase in modern history.

MANIPULATION OF SEX RATIO

The studies reviewed above reveal the role of infanticide not only in control of numbers, but in the production of distortions in sex ratios as well. Preferential male infanticide is extremely rare in the ethnographic record, though it occurred in Carthage (103) and in some East African groups (26). Obviously it merits further investigation. Female infanticide on the other hand, is very common, as Carr-Saunders and Krzywicki have attested, resulting in significant sex ratio distortions in early age groups. Similar distortions occur in some Pleistocene human populations (99). Its structural importance is dramatically demonstrated in the following cases. Eskimo hunter-gatherers from Alaska to Greenland practiced preferential female infanticide at rates resulting in childhood sex ratios up to 200:100, averaging around 125:100. These rates were offset by heavy adult male mortality from hunting accidents, homicide, and suicide, resulting in most cases in parity by middle age, and a disproportion of females in older age groups. Moreover, optional polygyny intensified the shortage of females, and encouraged conflict between males, wife stealing, and prepubertal betrothal (6, 102). Among the horticultural Yanomama and Xavante of South America, sex ratios in the 0–14 year cohort are 128.6 and 123.7 respectively; in the Peruvian Cashinahua they reach 148:100. Infanticide is known for Yanomama and Cashinahua and suspected for the Xavante. Yanomama male aggression, both inter- and intragroup, produces 24% of all male mortality, resulting in a more balanced total sex ratio of 110.5 (Xavante 115.1). Again, competition for mates is aggravated by polygyny and encourages child betrothal and wife capture (19, 20, 24, 42, 64, 68, 81). Similarly, the horticultural Bena Bena of New Guinea suffer a shortage of females (no sex ratios given) resulting from female infanticide and polygyny, encouraging wife capture and a 30% rate of prepubertal betrothal (49).

Preferential female infanticide in agrarian societies is represented not only by the Japanese case. The Jhareja Rajputs of northwestern India removed virtually 100% of female livebirths prior to British suppression of the practice. This practice occurred in the context of a hypergynous marriage system within the endogamous Rajput caste, which was composed of ranked exogamous subcastes, of which the Jharejas were the highest. Destruction of Jhareja females guaranteed the upward movement of both brides and dowries from lower ranking subcastes. Again, female shortage was aggravated by polygyny in the higher classes. The practice of child betrothal is well known throughout India. I have as yet no data on the fate of the excess of males thus produced, but the fact that this Rajput group, members of the military Kshatriya varn, had imposed their rule by conquest is suggestive (70–74). (Thanks to G. Berreman for useful discussion of hypergyny.) A similar hypergynous system probably occurred at least in parts of preindustrial Europe, judging from a recent analysis of fifteenth century Florentine materials (96). There, sex ratio distortions were greatest in the upper class (124.5), and clear evidence of selective aggressive neglect and abandonment, if not outright infanticide, is provided. The importance of the dowry in traditional Europe is well known.

These cases could undoubtedly be multiplied by reanalysis of primary historical and ethnographic sources. Their significance in terms of the model of male competitive aggression and reproductive success proposed by Trivers and others (3, 23, 33, 97, 98) is apparent. Female infanticide guarantees the operation of that model; in stratified societies it magnifies the natural discrepancy in primary sex ratios between upper and lower classes that is a result of the influence of socioeconomic status and birth order (94). Granting the noncomparability of these materials in some respects, and the large numbers of currently unanswerable questions they raise, they do confirm that preferential female infanticide operates in a variety of human socioeconomic systems as a significant contributor to the maintenance of social structure at rates ranging from 10–100% of female livebirths per social unit.

CONCLUDING THOUGHTS

Evidently, the role of systematic infanticide in birth spacing, family completion, and adjustment of sex ratios can be significant. At rates of 5–50% of all livebirths, it occurs in hunter-gatherers, horticulturalists, and stratified agrarian societies. As one of several alternative control methods, its choice must depend both on ecological and structural imperatives and on the specific physiological consequences of each method. Better data are needed in both these areas before we can relate a particular cluster of control methods to a specific ecosystem, but a few comments concerning this relation can be made.

I agree with Nag (62) that traditional oral contraceptives are rarely effective, and abortion is ordinarily of little demographic importance due to the high risk of mortality or sterility to the subject. In most societies, the effective means of abortion are direct physical trauma to the cervix and external trauma resulting in expulsion of the fetus as a stillbirth at or very near term. In the former, risks of sepsis (with mortality rates as high as 25%), tetanus (mortalities of approximately 50%) (2), and massive hemorrhage are great. As for the use of external trauma, little is gained over natural parturition except increased maternal risk. In consequence, abortion serves mainly for the concealment of illegitimates, except in societies with strong prohibitions against infanticide. Truly effective methods of systematic control other than lactation effect are thus limited to abstinence through postpartum avoidance or chaste celibacy, incomplete coitus, and infanticide, grading into abandonment and early aggressive neglect.

What is the relation of these systematic methods to other social and survival imperatives? Previous research strongly suggests a correlation of long postpartum abstention with protein deficiencies during lactation (84, 105). Lifelong celibacy must depend on the society's capacity to provide productive supplementary roles to celibates. Where social values preclude sexual activity by such individuals, this alternative must be accompanied by segregative institutions such as militia and monasteries; where sexual activity is allowed, abortion, coitus interruptus, infanticide, or abandonment will be condoned. I have indicated the unique significance of infanticide in adjusting sex ratios, which no other method can do. Yet this alternative is probably a less adequate response to low nutritional levels than postpartum abstention or coitus interruptus.

Beyond this there is unclarity. Both infanticide and postpartum avoidance occur in polygynous societies, either in combination or separately. Female infanticide occurs both where women contribute little to primary production and where they contribute much. Likewise, systems of hypergyny may depend on the practice of differential infanticide in upper classes, or merely on socioeconomic differences in sex ratio and perhaps in female celibacy; they may provide productive alternate roles to males denied mates, or they may produce large numbers of unproductive lower status males who must be controlled by repression. Our explanatory difficulties arise not so much from the inadequacy of existing data as from the fact that differing socioeconomic necessities may produce similar adaptive responses, and vice versa.

Population control practices in man exist, moreover, in culturally variant complexes of cognition and value, as Rappaport's exploration of the cybernetic functions of ritual systems in relation to population dynamics implies (78, 79). Douglas's (26) critique of Carr-Saunders and Wynne-Edwards emphasized the role of prestige consumption in motivating population control, but failed to specify the nature of the link between the two. Extending Rappaport's approach, the exchange and consumption of prestige items probably plays, as does ritual, an informational and regulatory role, relating the level of available resources to competitive and other demographic processes through socially significant symbols. Meggitt (61) has demonstrated the operation of such a system for the Mae Enga, while Lindenbaum's (53) analysis of the idea of female pollution in New Guinea defined a network of ideological components, from sorcery to attitudes toward sex, within which the practice of female infanticide is enmeshed, and proposed a relation between intensity of the ideological complex and population density. These analyses indicate a direction for future research: attention to the density-dependent functions of systems of exchange, ritual, and warfare must now be integrated with data on indirect and direct means of population control. If human societies are adaptive, all these must be parts of a single overarching cybernetic system, operating on both conscious and unconscious levels.

In terms of future research, I call attention to one ideological correlate of infanticide that creates acute problems for those using the ethnographic corpus, especially when attempting cross-cultural quantification. Ideological disapproval and even shame and secrecy are by no means incompatible with significant rates of infanticide. In consequence, ethnographic accounts may report either public ideology or private tolerance, and will not be at all comparable cross-culturally, especially when treatment is superficial. Many examples from the use of cross-cultural files could be cited. While Ford is surely correct in suggesting that disapproval encourages fertility and regulates the occurrence of homicide, I believe there is a biologically more significant reason for these apparent contradictions. If ideological, that is, learned behavioral, systems have adaptive value as contrasted with innate behavioral systems, they do so because of their flexibility, their rapidity of response to changing conditions. Thus the ambivalence and inconsistency that characterize all human thought are probably powerfully adaptive, providing immediate response to the needs of the moment by permitting appeal to alternate available rationales. Tolerance of infanticide provides a rationale in situations of familial economic stress, whereas its disapproval supports and encourages natality in improved economic conditions. It is no surprise that the more public, authoritative ideological statement is always that which encourages fertility: species success as measured by dispersal and colonization depends upon it. But the inherent flexibility of inconsistent ideologies allows fine tuning in relation to external variables without requiring readjustment in the ideological structure itself.

Finally, concern with the cybernetic nature of human control systems relates directly to the humanly critical question of the decline in traditional means of population limitation. The Tikopia case is instructive. Firth's (29) data revealed the impact of several social changes other than the introduction of Christianity. A variety of new tools and root crops appeared to allow increased densities; at the same time a series of epidemics, some with heavy mortalities, had resulted from 100 years of Western contact. The overall meaning of these changes must have been difficult for the Tikopia to assess. Monthly tabulation of the use of major staples during the 1952 famine makes clear that reliance on introduced manioc and sweet potato during traditional seasons of food shortage was implicated in overpopulation. Thus the Tikopia case seems not so much a consequence of the imposition of Western ideology as an overload and breakdown in information acquisition and flow, aggravated by decline in the traditional authority structure. Without outside intervention, one would predict resultant severe population oscillations until a new cognitive structure, based on experience with the introduced variables, had evolved.

This analysis may have more general relevance. The historical demography of the Pacific (55) suggests a correlation between the experience of severe introduced epidemics and the abandonment of infanticide, an interpretation in which Bayliss-Smith (8) concurs. The role of introduced ideology may be different from what we have assumed.

In larger, stratified societies, a similar interpretation seems possible, although other factors intervene. All unifying political systems, whether of expanding conquerors or nationalizing elites, apparently encourage fertility and discourage traditional means of limitation. Muhammad attempted to suppress female infanticide in Arabia, as did the Mogul emperor Jahangir in India two hundred years before the British, and Tokugawa and Meiji rulers in Japan, employing Buddhist sanctions. But such efforts are uniformly unsuccessful unless the reforming authority effects a complete disruption of the traditional socioeconomic system. This disruption is accompanied by, and indeed dependent upon, rapid and fluctuating changes in agricultural productivity, labor demand, urbanization, emigration, and military expansion, often coupled with introduced epidemic disease. The result is an inability of the traditional social system to perceive and process the real consequences of new births. In a discordant sociopolitical system, decisions regarding population limitation or growth must fall to the most constricted, local unit of cognition and the lowest level of social organization, namely the family, which must operate without benefit of any wider information flow. Readers will recognize what I propose here as a variant on Hardin's "tragedy of the commons" (37). I suggest, however, that the inability to relate personal short-term advantage to group long-term advantage is not a universal characteristic of humanity, but a specific consequence of the breakdown in social integration, indeed a failure in one of the most fundamental ecological functions of human social systems.

In this, I follow Mamdani's analysis of Indian family planning (59) and White's interpretation of population growth in colonial Java (104) in recognizing the role of labor demand as an important variable in family decision making. This hypothesis means, moreover, that that "stage" in the theory of demographic transition characterized by high fertility and declining mortality says nothing about traditional agrarian societies. Rather it is an expression of the overwhelming contribution Western colonialism has made to the current world population problem. If so, it may follow that world population control will not be achieved until stable, integrated sociopolitical systems, providing adequate cognitive assessment of moderately stable ecological and socioeconomic variables, have evolved. Meanwhile, the role of Western ideology, now widely adopted by formerly colonial societies, seems to be to prevent a return to traditional methods, even when they are perceived on the local level as necessary, and to demand instead the use of less efficient means, especially unsupervised abortion, aggressive neglect, child abuse, and of course starvation, with a consequent increase not only in socially disruptive adult mortalities but in the totality of human suffering as well.

ACKNOWLEDGMENTS

Thanks are due to G. Berreman, M. Datz, K. Pakrasi, A. Wexler, and J. Wind, and to the Anthropology Library, University of California, Berkeley, for assistance with sources, to M. Rebhan for editorial advice and encouragement, and to R. F. Johnston for encouragement.

Literature Cited

1. Aginsky, B. W. 1939. Population control in the Shanel (Pomo) tribe. *Am. Sociol. Rev.* 4(2):209–16
2. Akinla, O. 1970. The problem of abortion in Lagos. *Proc. World Congr. Fert. Steril. 6th,* 113–20
3. Alexander, R. D. 1974. The evolution of social behavior. *Ann. Rev. Ecol. Syst.* 5:325–83
4. Aptekar, H. 1931. *Anjea: Infanticide, Abortion and Contraception in Savage Society.* New York: Godwin
5. Autrum, H., von Holst, D. 1968. Sozialer "Stress" bei Tupajas (*Tupaia glis*) und seine Wirkung auf Wachstum, Körpergewicht und Fortpflanzung. *Z. Vgl. Physiol.* 58:347–55
6. Balikci, A. 1967. Female infanticide on the Arctic coast. *Man* 2:615–25
7. Bates, M. 1955. *The Prevalence of People.* New York: Scribners
8. Bayliss-Smith, T. 1974. Constraints on population growth: the case of the Polynesian outlier atolls in the precontact period. *Hum. Ecol.* 2(4):259–95
9. Binford, L. R. 1968. Post-Pleistocene adaptations. In *New Perspectives in Archaeology,* ed. S. R. Binford, L. R. Binford, 313–41. Chicago: Aldine
10. Birdsell, J. B. 1957. Some population problems involving Pleistocene man. *Cold Spring Harbor Symp. Quant. Biol.* 22:47–69
11. Birdsell, J. B. 1958. On population structure in generalized hunting and collecting populations. *Evolution* 12(2):189–205
12. Birdsell, J. B. 1968. Some predictions for the Pleistocene based in equilibrium systems among recent hunter-gatherers. In *Man the Hunter,* ed. R. B. Lee, I. DeVore, 229–40. Chicago: Aldine
13. Birdsell, J. B. 1972. *Human Evolution: an Introduction to the New Physical Anthropology.* Chicago: Rand-McNally
14. Borrie, W. D., Firth, R., Spillius, J. 1957. The population of Tikopia, 1929 and 1952. *Popul. Stud. London* 10(3):229–52
15. Bowles, G. T. 1953. Population control and the family in feudal and post-restoration Japan. *Kroeber Anthropol. Soc. Pap.* 8–9:1–19
16. Calhoun, J. B. 1962. Population density and social pathology. *Sci. Am.* 206(2):1399–1408
17. Carroll, V., ed. 1975. *Pacific Atoll Populations.* Honolulu: Univ. Hawaii Press. In press
18. Carr-Saunders, A. M. 1922. *The Population Problem: a Study in Human Evolution.* Oxford: Clarendon
19. Chagnon, N. A. 1968. Yanomamo social organization and warfare. In *War, the Anthropology of Armed Conflict and Aggression,* ed. M. Fried, M. Harris, R. Murphy, 109–59. Garden City, NJ: Natural History
20. Chagnon, N. A. 1968. *Yanomamo, The Fierce People.* New York: Holt, Rinehart & Winston
21. Cook, S. F. 1946. Human sacrifice and warfare as factors in the demography of pre-colonial Mexico. *Hum. Biol.* 18:81–102
22. Cook, S. F. 1947. Survivorship in aboriginal populations. *Hum. Biol.* 19(2):83–89
23. Crook, J. H. 1972. Sexual selection, dimorphism, and social organization in the primates. In *Sexual Selection and the Descent of Man, 1871–1971,* ed. B. Campbell, 231–81. Chicago: Aldine
24. De Oliveira, A. E., Salzano, F. M. 1969. Genetic implications of the demography of Brazilian Juruna Indians. *Soc. Biol.* 16(3):209–15
25. Dorn, H. F. 1959. Mortality. In *The Study of Population,* ed. P. M. Hauser, O. D. Duncan. Chicago: Univ. Chicago Press
26. Douglas, M. 1966. Population control in primitive groups. *Br. J. Sociol.* 17(3):263–73
27. Eaton, J., Mayer, A. J. 1953. The social biology of very high fertility among the Hutterites: the demography of a unique population. *Hum. Biol.* (25)3:206–64
28. Firth, R. 1957. *We, the Tikopia.* London: Allen & Unwin. 2d ed.

29. Firth, R. 1959. *Social Change in Tikopia.* London: Allen & Unwin
30. Firth, R. 1965. *Primitive Polynesian Economy.* London: Routledge & Kegan Paul. 2d ed.
31. Ford, C. 1945. A comparative study of human reproduction. *Yale Univ. Publ. Anthropol.* Vol. 3. New Haven: Yale Univ. Press
32. Ford, C. 1952. Control of contraception in cross-cultural perspective. *Ann. NY Acad. Sci.* 54(5):763–68
33. Fox, R. 1972. Alliance and constraint: sexual selection in the evolution of human kinship systems. See Ref. 23, 282–331
34. Frisch, R. E., McArthur, J. W. 1974. Menstrual cycles: fatness as a determinant of minimum wight for height necessary for their maintenance or onset. *Science* 185:949–51
35. Galle, O. R., Gove, W. R., McPherson, J. M. 1972. Population density and pathology: what are the relations for man? *Science* 176(4030):23–30
36. Grantzberg, G. 1973. Twin infanticide —a cross-cultural test of a materialistic explanation. *Ethos* 1(4):405–12
37. Hardin, G. 1968. The tragedy of the commons. *Science* 162(3859):1243–48
38. Hauser, P. M., Duncan, O. D., eds. 1959. *The Study of Population.* Chicago: Univ. Chicago Press
39. Heer, D. M. 1968. *Society and Population.* Englewood Cliffs, NJ: Prentice-Hall
40. Howell, N. 1975. The population of the Dobe area !Kung (Zun/Wasi). In *Kalahari Hunter-Gatherers,* ed. R. Lee, I. DeVore. Cambridge: Harvard Univ. Press. In press
41. Hutchinson, G. E., MacArthur, R. H. 1959. On the theoretical significance of aggressive neglect in interspecific competition. *Am. Natur.* 93(869):33–34
42. Johnston, F. E., Kensinger, K. M., Jantz, R. L., Walker, G. F. 1969. The population structure of the Peruvian Cashinahua: demographic, genetic and cultural interrelationships. *Hum. Biol.* 41(1):29–41
43. Kluckhohn, C. 1956. Review of Lorimer, F. et al 1954. *Am. J. Phys. Anthropol.* 14(3):527–32
44. Kobayashi, K. 1969. Changing patterns of differential fertility in the population of Japan. *Proc. Int. Congr. Anthropol. Ethnol. Sci.. 8th,* 1:345–347
45. Kolata, G. B. 1974. !Kung hunter-gatherers: feminism, diet, and birth-control. *Science* 185:932–34
46. Krzywicki, L. 1934. *Primitive Society and its Vital Statistics.* London: Macmillan
47. Lack, D. 1954. *The Natural Regulation of Animal Numbers.* Oxford: Clarendon
48. Langer, W. L. 1974. Infanticide: a historical survey. *Hist. Child. Quart.* 1(3): 353–65
49. Langness, L. L. 1967. Sexual antagonism in the New Guinea Highlands: a Bena Bena example. *Oceania* 37(3): 161–77
50. Lee, R. B. 1969. !Kung Bushmen subsistence: an input-output analysis. In *Environment and Cultural Behavior,* ed. A. P. Vayda, 47–79. New York: Natural History
51. Lee, R. B. 1972. Population growth and the beginnings of sedentary life among the Kung Bushmen. In *Population Growth: Anthropological Implications,* ed. B. Spooner, 329–42. Cambridge: MIT Press
52. Lee, R. B. 1972. The !Kung Bushmen of Botswana. In *Hunters and Gatherers Today,* ed. M. Bicchieri, 327–68. New York: Holt, Rinehart & Winston
53. Lindenbaum, S. 1972. Sorcerers, ghosts, and polluting women: an analysis of religious belief and population control. *Ethnology* 11(3):241–53
54. Lorimer, F. et al 1954. *Culture and Human Fertility.* Zurich: UNESCO
55. McArthur, N. 1968. *Island Populations of the Pacific.* Honolulu: Univ. Hawaii Press
56. Malcolm, L. A., Zimmerman, L. 1973. Dwarfism amongst the Buang of Papua New Guinea. *Hum. Biol.* 45(2):181–93
57. Malthus, T. R. 1807. *An Essay on the Principle of Population.* 2 vols. London: Johnson. 4th ed.
58. Malthus, T. R. 1960. A summary view of the principle of population (1830). In *On Population: Three Essays,* ed. F. Notestein, 13–59. New York: New American Library
59. Mamdani, M. 1972. *The Myth of Population Control: Family, Caste and Class in an Indian Village.* New York: Monthly Review
60. Marshall, L. 1960. !Kung Bushmen bands. *Africa* 30(4):325–55
61. Meggitt, M. 1972. System and subsystem: the Te exchange cycle among the Mae Enga. *Hum. Ecol.* 1(2):111–23
62. Nag, M. 1962. Factors affecting human fertility in nonindustrial societies: a cross-cultural study. New Haven: *Yale Univ. Publ. Anthropol.* Vol. 66

63. Neel, J. V. 1958. A study of major congenital defects in Japanese infants. *Am. J. Hum. Genet.* 10(4):398–445

64. Neel, J. V. 1969. Some aspects of differential fertility in two American Indian tribes. *Proc. Int. Congr. Anthropol. Ethnol. Sci., 8th,* 1:356–61

65. Neel, J. V. 1970. Lessons from a "primitive" people. *Science* 170(3960):815–22

66. Neel, J. V. 1971. Genetic aspects of the ecology of disease in the American Indian. In *The Ongoing Evolution of Latin American Populations,* ed. F. M. Salzano, 561–90. Springfield,: Ill. Thomas

67. Neel, J. V., Salzano, F. M. 1967. Further studies on the Xavante Indians. X. Some hypotheses-generalizations resulting from these studies. *Am. J. Hum. Genet.* 19(4):554–74

68. Neel, J. V., Salzano, F. M., Junquiera, P. C., Keiter, F., Maybury-Lewis, D. 1964. Studies on the Xavante Indians of the Brazilian Mato Grosso. *Am. J. Hum. Genet.* 16(1):52–140

69. Odum, E. P. 1971. *Fundamentals of Ecology.* Philadelphia: Saunders. 3rd ed.

70. Pakrasi, K. B. 1968. Infanticide in India —a century ago. *Bull. Socio-Econ. Res. Inst. Calcutta* 2(2):21–30

71. Pakrasi, K. B. 1970. *Female Infanticide in India.* Calcutta: Editions Indian

72. Pakrasi, K. B. 1970. The genesis of female infanticide. *Humanist Rev.* 2(3): 255–81

73. Pakrasi, K. B. 1972. On the antecedents of Infanticide Act of 1870 in India. *Bull. Cult. Res. Inst. Calcutta* 9(1–2):20–30

74. Pakrasi, K. B., Sasmal, B. 1971. Infanticide and variation of sex-ratio in a caste population of India. *Acta Medica Auxol. Italy* 3(3):217–28

75. Padden, R. C. 1967. *The Hummingbird and the Hawk: Conquest and Sovereignty in the Valley of Mexico, 1503–1541.* New York: Harper & Row

75a. Pearl, R. 1922. *The Biology of Death.* Philadelphia: Lippincott

75b. Pearl, R. 1924. *The Curve of Population Growth. Proc. Am. Phil. Soc.* 63:10–17

76. Polgar, S. 1972. Population history and population policies from an anthropological perspective. *Curr. Anthropol.* 13(2):203–11

77. Rappaport, R. A. 1963. Aspects of man's influence upon island ecosystems: alternation and control. In *Man's Place in the Island Ecosystem,* ed. F. R. Fosberg, 155–70. Honolulu: Bishop Museum

78. Rappaport, R. A. 1968. *Pigs for the Ancestors: Ritual in the Ecology of a New Guinea People.* New Haven: Yale Univ. Press

79. Rappaport, R. A. 1971. The sacred in human evolution. *Ann. Rev. Ecol. Syst.* 2:23–44

80. Ryder, N. B. 1957. The conceptualization of the transition in fertility. *Cold Spring Harbor Symp. Quant. Biol.* 22:91–96

81. Salzano, F. M., Cardoso de Oliveira, R. 1970. Genetic aspects of the demography of Brazilian Terena Indians. *Soc. Biol.* 17(3):217–23

82. Salzano, F. M., Freire-Maia, N. 1970. *Problems in Human Biology: A Study of Brazilian Populations.* Detroit: Wayne State Univ. Press

83. Salzano, F. M., Neel, J. V., Maybury-Lewis, D. 1967. Further studies on the Xavante Indians. I. Demographic data on two additional villages; genetic structure of the tribe. *Am. J. Hum. Genet.* 19(4):463–89

84. Saucier, J. F. 1972. Correlates of the long postpartum taboo: a cross-cultural study. *Curr. Anthropol.* 13(2):238–49

85. Schapera, I. 1930. *The Khoisan Peoples of South Africa: Bushmen and Hottentot.* London: Routledge

86. Schull, W. J., Neel, J. V. 1966. Some further observations on the effect of inbreeding mortality in Kure, Japan. *Am. J. Hum. Genet.* 18(2):144–52

87. Silberbauer, G. B. 1972. The G/Wi Bushmen. See Ref. 52, 271–326

88. Slobodkin, L. B. 1961. *Growth and Regulation of Animal Populations.* New York: Holt, Rinehart & Winston

89. Spillius, J. 1957. Natural disaster and political crisis in a Polynesian society: an exploration of operational research. *Hum. Relat.* 10(1):3–27;10(2):113–25

90. Sugiyama, Y. 1967. Social organization of hanuman langurs. In *Social Communication Among Primates,* ed. S. A. Altmann, 221–36. Chicago: Univ. Chicago Press

91. Sugiyama, Y. 1973. The social structure of wild chimpanzees: a review of field studies. In *Comparative Ecology and Behaviour of Primates.* ed. R. P. Michael, J. H. Crook, 375–410. London: Academic

92. Sussman, R. W. 1972. Child transport, family size, and increase in human population during the Neolithic. *Curr. Anthropol.* 13(2):258–59

93. Taeuber, I. 1958. *The Population of Japan.* Princeton: Princeton Univ. Press

94. Teitelbaum, M. S., Mantel, N., Starr, C. R. 1971. Limited dependence of the human sex ratio on birth order and parental ages. *Am. J. Hum. Genet.* 23(3): 271–80

95. Tietze, C. 1957. Reproductive span and rate of reproduction among Hutterite women. *Fert. Steril.* 8(1):89–97

96. Trexler, R. C. 1973. Infanticide in Florence: new sources and first results. *Hist. Child. Quart.* 1(1):98–116

97. Trivers, R. L. 1972. Parental investment and sexual selection. See Ref. 23, 136–79

98. Trivers, R. L., Willard, D. E. 1973. Natural selection of parental ability to vary the sex ratio of offspring. *Science* 179:90–92

99. Vallois, H. V. 1961. The social life of early man: the evidence of skeletons. In *Social Life of Early Man,* ed. S. L. Washburn, 214–35. Chicago: Aldine

100. Vayda, A. P., Rappaport, R. A. 1963. Island cultures. See Ref. 77, 133–42

101. Weinstein, E. D., Neel, J. V., Salzano, F. M. 1967. Further studies on the Xavante Indians. VI. The physical status of the Xavantes of Simões Lopes. *Am. J. Hum. Genet.* 19(4):532–42

102. Weyer, E. M. 1932. *The Eskimos: Their Environment and Folkways.* New Haven: Yale Univ. Press

103. Weyl, N. 1968. Some possible genetic implications of Carthagenian child sacrifice. *Perspect. Biol. Med.* 12(1):69–78

104. White, B. 1973. Demand for labor and population growth in colonial Java. *Hum. Ecol.* 1(3):217–36

105. Whiting, J. W. 1969. Effects of climate on certain cultural practices. See Ref. 50, 416–50

106. Wohlwill, J. F. 1974. Human adaptation to levels of environmental stimulation. *Hum. Ecol.* 2(2):127–47

107. Wolfe, A. B. 1933. The fecundity and fertility of early man. *Hum. Biol.* 5(1): 35–60

108. Woolf, C. M., Dukepoo, F. C. 1959. Hopi Indians, inbreeding, and albinism. *Science* 164:30–37

109. Woolf, C. M., Turner, J. A. 1969. Incidence of congenital malformations among live births in Salt Lake City, Utah, 1951–1961. *Soc. Biol.* 16(4): 270–79

110. Wynne-Edwards, V. C. 1962. *Animal Dispersion in Relation to Social Behaviour.* Edinburgh: Oliver & Boyd

111. Wynne-Edwards, V. C. 1964. Population control in animals. *Sci. Am.* 211(2):68–74

112. Wynne-Edwards, V. C. 1965. Self-regulating mechanisms in populations of animals. *Science* 147(3665):1543–48

113. Yoshiba, K. 1968. Local and intertroop variability in ecology and social behavior of common Indian langurs. In *Primates: Studies in Adaptation and Variability,* ed. P. C. Jay, 217–42. New York: Holt, Rinehart & Winston

114. Zuckerman, S. 1932. *The Social Life of Monkeys and Apes.* London: Kegan Paul

ENERGETICS OF POLLINATION

♦4090

Bernd Heinrich

Department of Entomological Sciences, University of California, Berkeley, California 94720

INTRODUCTION

Nectar from flowers provides nourishment for animals ranging in size from mites (55) to man. However, few of the organisms that can use these food resources are pollinators. One of the factors affecting whether or not an animal can be a dependable flower visitor is the relationship between its energy demands and the quantity of food it can harvest from the flowers (53, 107, 131, 133, 136, 255). This provides perhaps the most common basis upon which mutual adaptations for pollination have evolved. With some well-known exceptions [including the scents collected by male bees, possibly for territorial marking (64, 284) and as sex attractants (164, 232)], most attractants to flowers are food. The food quantity provided in relation to the energy demand of the flower visitor influences the amount of flower to flower and plant to plant movement. If the food rewards are too great, a flower visitor could restrict its movements, and become a "nectar thief" rather than a pollinator. If the food rewards are too small, the flower visitor learns to avoid that plant species and forages from another. It is probable that the optimal strategy of both foragers and plants would evolve (54) along the lines of an existentialist "game" (246). The "rules" of this game evolve from the foragers' strategy, where time and energy are used to optimal advantage (71, 241). The foragers' behavior is analogous to that of a predator (212, 247), but the evolutionary response of the "prey" is to maximize rather than minimize discovery and exploitation. Numerous theoretical (e.g. 10, 18, 101, 131, 195, 252, 257) and practical aspects (35, 47, 88, 207) of the pollinator-plant evolution have been reviewed.

From the evolutionary perspective it follows that those animals that are the most frequent vagile flower visitors to a given species will be the most dependable pollinators. Thus they are the ones that would most likely determine the evolution of flower signals to increase the frequency of visits to the flower and to increase the percentage of the pollination events per visit. Concurrently, features should evolve to exclude "nectar thieves" (151).

139

It is likely that natural selection would tend to produce those quantities of food reward that result in the most cross-pollination at the least cost in nectar. Plants probably do not direct more energy to food production for pollinators than is necessary for the flowers to be adequately pollinated: flowers must provide sufficient food to be attractive, yet if the rewards are too rich, the potential pollinator may be restricted to a single plant. Selective pressures would tend to produce enough food reward for optimal pollinators and, at the same time, to provide exclusion mechanisms for poor pollinators and robbers. However, the exclusion of ineffective pollinators is but the first step in a complex pattern of interrelationships that acts to achieve energy balance between specific pollinators and specific plant species.

During the last two centuries, the focus of studies on pollination has been on morphological coadaptations of flowers and their pollinators (9, 60, 73, 103, 151, 155, 157, 210, 211, 229, 230, 233, 251). These adaptations are astonishing in variety and complexity. The hidden food rewards, in contrast, might seem relatively unvarying and superficially uninteresting. However, they determine whether or not a plant is visited in preference to a neighbor, and whether or not the visitor moves between plants. The ecological significance of food rewards and energetics has not received much scrutiny until quite recently. I address myself here to the question of how, in the ecological context of numerous competing plant species, the floral rewards could affect the movements of flower visitors and hence the evolution of flowers.

The topic encompasses several disciplines, each with its own rich historical background. The information that is relevant and available is enormous. It is far beyond the aims of this essay, and my abilities, to review the subject comprehensively. Rather, I am forced to be synoptic, hoping to provide an overview to indicate general patterns and the possible scope, complexity, and nature of ongoing work.

MEASUREMENTS OF POLLINATOR ENERGETICS

The energy costs of foraging can be allocated to different categories. Energy intake during foraging must exceed the energy expenditure of harvesting. Moreover, the profit during the time available for foraging must be sufficient for long-term energy balance, which in obligate flower visitors includes the energy demands for reproduction. *Energy balance* thus may mean different things depending on the activities and duration in question. Suitable methods for measuring the different rates of energy expenditure vary accordingly.

Direct measurements of food consumption, particularly when foods are chemically defined as are sugars from nectar, can be reliable indicators of total energy expenditure. However, since ingested sugars may be stored or converted to lipid or glycogen, it is necessary to continue the measurements for time periods sufficient to achieve steady-state conditions. In hummingbirds and bees that are presumably not accumulating fat reserves, rapid utilization of sugars has made it possible to compute 24-hr energy budgets on the basis of food intake. However, such data by themselves give no indication of the time or conditions under which the food calories are expended.

The standard and probably most reliable indicator of energy expenditure is the rate of either oxygen consumption or carbon dioxide emission. For most animals feeding on nectar sugar, it can be assumed that the respiratory quotient is close to 1.0, and every milliliter O_2 consumed or milliliter of CO_2 liberated is equivalent to an expenditure of 5.0 calories. Nectars containing lipid (15, 16), as well as pure lipid (which yields approximately 9.0 calories per milligram) produced by "elaiophors," have been reported (271), and 1 ml O_2 consumed during lipid utilization represents an expenditure of 4.7 calories. By far the greatest bulk of foodstuff in nectar is sugar, which yields about 4 calories per milligram. Carbohydrate is a poor substrate for long-distance travel because its weight/energy ratio is about one-eighth that of lipid, and birds and some insects (including Orthoptera and Lepidoptera) convert carbohydrate to fat (122, 280). Hymenoptera and Diptera have a respiratory quotient (RQ) of 1.0, indicating that their flight muscles use primarily carbohydrate.

Despite its precision, the greatest drawback of using metabolic rate for energy budgets is its limitation to specific and often experimentally controlled activity states such as locomotion, maintenance metabolism, or temperature regulation. The animals are generally confined to a respirometer jar, in which their activity patterns and metabolic rates may not correspond with those in natural conditions. However, when combined with careful field observations, the measurements can be a powerful tool to infer energy budgets. For example, from timed observations of flight durations, perching, and body temperature during torpor, in conjunction with the corresponding rates of oxygen consumption during these activity states, accurate energy budgets corresponding to the observed rates of food consumption (221, 254, 287) have been calculated for hummingbirds.

Although the extrapolations from laboratory-derived metabolic rates have proven highly useful in calculating energy budgets of free-living hummingbirds, they are less suitable for some other animals. An endothermic insect, for example, may vary its metabolic rate by an order of magnitude in a few minutes while perching (117, 122, 149), yet display no outward sign of this process. Thus observations of discrete activity states such as perching or flying cannot be used in such cases for extrapolations to energy budgets.

A third, though costly, technique of circumventing the above difficulties in determining energy budgets is the use of isotope-labeled water. Monitoring the amount of isotope in the blood after a given amount has been injected gives an indication of the amount of energy expended during the time between injection and sampling (169, 209, 265). Social insects, trap-lining bats (132), and territorial hummingbirds (183, 255, 287) may be ideal candidates for the technique because these animals can be recaptured in the field at given intervals. To my knowledge, however, the method has yet to be used to measure energy budgets of free-living pollinators. Although ideal for long-term measurements of energy expenditure, the method cannot be used to determine the metabolic rate at any one time.

Body temperature is possibly the most reliable indicator of "instantaneous" energy expenditure of free-living animals in which discrete activity states are not apparent by visual inspection. At least 80% of an animal's energy expenditure is degraded to heat, due to inefficiency at the biochemical and mechanical levels of organization (282). An increase in heat production, usually accompanied by an increase in body temperature (19, 122, 281), thus closely parallels an increase in energy expenditure. If the metabolic rate is large enough, as it is for most pollinators, it can be calculated from body temperature when mass and cooling rates are known (19, 123, 128). The method, though restricted to those conditions where radiant heat input and active cooling are at a minimum, has been useful in examining the energy expenditure of flying grasshoppers (281) and bumblebees (120).

ENERGY EXPENDITURE BY POLLINATORS

Pollinators conform to energy relationships similar to those of other animals. However, many of them are small, highly mobile, and must restrict foraging activity to the sometimes relatively short periods during the day when their host flowers present nectar. Flower-visiting insects are probably the most extravagant utilizers of energy on a weight-specific basis (118, 120, 131, 136, 278). However, large expenditures are often required to make small profits.

One of the generalities applicable to vertebrate animals is that resting metabolism is inversely related to body size (19, 50, 154, 166, 202). The metabolic rates of passerines and nonpasserines have been reviewed on several occasions (154). Both groups of birds have metabolic rates close to those of homeothermic mammals (166). The energy expenditure of some insects while thermoregulating depends similarly on their size. On the basis of whole body weight, the metabolic rate of a bumblebee while incubating is 170 cal $(g\ hr)^{-1}$ at $0°C$. A hummingbird weighing 10 times more than the bee has a weight-specific respiratory rate 2.4 times less than that of the bee, and a bat weighing 10 times more than the bird has a respiratory rate at the same temperature that is 2.8 times less than that of the bird (121). The smaller the animal, the greater the energetic barrier to activity at low ambient temperature. As Bartholomew has pointed out, "as long as they [small animals] maintain high body temperatures, they are never more than a few hours from death by starvation, particularly at low ambient temperatures" (19, p. 348).

The basal metabolic rate (BMR) of homeothermic animals is measured at temperatures where no energy is expended for thermoregulation. Departure from thermoneutrality results in marked changes in metabolic rate, particularly in small animals. For example, hummingbirds weighing 8 g may increase their metabolic rate from about 9.0 cal $(g\ hr)^{-1}$ at $33°C$ to 65 cal $(g\ hr)^{-1}$ at $0°C$ (105). A stationary bumblebee weighing approximately 0.5 g increases the metabolic rate of its thoracic muscles from 85 cal $(g\ hr)^{-1}$ to 850 cal $(g\ hr)^{-1}$ over the same range of ambient temperature while incubating brood (121).

Since many flowers bloom only for short durations, the small high-energy pollinators could face severe energy problems. However, they have evolved a solution to the diurnal fluctuations of food availability—periods of torpidity. Some social insects avoid this torpidity by storing food energy in the nest. A queen bumblebee may use the entire contents of her honeypot in a single night (121, 122). When all available food has been utilized, the bee enters torpor (117). When at $0°C$, a torpid bumblebee has a metabolic rate 1000–2000 times less than when it is regulating its body temperature (121, 149). An important difference between the torpor of a hummingbird and that of an insect is that the bird regulates its body temperature at a lower set-point (288), but the insect does not regulate it at all. Energy saving by torpidity in hummingbirds was first discussed by Pearson (220, 221), who calculated that a male Anna hummingbird (*Calypte anna*) expends 10.3 Kcal during 24 hr as opposed to only 7.6 Kcal when torpid, an energy saving equivalent to the nectar contents of 370 *Fuchsia* blossoms. Wolf & Hainsworth have made similar calculations of the time and energy economy of torpor in tropical hummingbirds (287). The hummingbird *Selasphorus flammula* must visit 313 *Salvia* flowers to match calculated energy expenditure for 1 hr (107).

The rates of increase in body temperature during warm-up are strongly size-dependent and impose severe limits on the feasibility of hypothermia as an energy-saving strategy for larger animals. A bumblebee weighing 0.6 g may warm up at 12°C/min (122), but an animal weighing 300 g warms up about 120 times less rapidly (128). In addition, the energy costs of warm-up are clearly unfavorable for larger animals (19, 222). It costs a 0.5 g bumblebee 7.5 cal to warm up from 13.5°C – 38.0°C (123), equivalent to the energy expenditure during 3.0 min of flight. A sphinx moth weighing 2.0 g requires 30 cal to warm up from 15°C (126), equivalent to the energy expended during approximately 3.7 min of flight (116, 129). A small bat or hummingbird expends about 114 cal during a warm-up from 10°C, corresponding to approximately 1.2% of the total energy budget for 24 hr. In contrast, a 200 kg bear would need as much energy to warm up as it uses during an entire 24-hr activity period (222).

Other than thermoregulation, the highest energy costs are those of locomotion. Flight, particularly hovering (262, 282), is metabolically the most expensive mode of locomotion, although for a given distance of travel, it can be energetically less costly than walking (263). For insects and birds, the energetic cost of flight has been shown to vary markedly with load (26, 123) and flight speed (262), but it is relatively independent of ambient temperature (25, 27, 113, 123, 129). The above generalities, however, are insufficient to allow the preparation of accurate energy budgets for specific animal species. The following discussions concern energy expenditures in common classes of pollinators.

Bats

The pollinators of over 500 Neotropical plants are Microchiroptera (272). These animals may at times rely on fruit, using flowers as a secondary source of food (17). Some of the smaller species of nectivors, e.g. Australasian Megachiroptera, are known to enter torpor at night (20). Few data are available on the energetic costs of bat flight, but those available (260) indicate that it is similar to bird flight, which agrees with theoretical considerations (264).

Birds

The most common bird pollinators include the honeycreepers (Drepanididae) of Hawaii, the sunbirds (Nectarinidae) of Africa and Asia, some parrots (lorikeets) from Australia, some honeyeaters (Melliphagidae) of Oceanea and Australia, and hummingbirds (Trochilidae) from America. Except for the latter group, the birds perch while visiting flowers, and their energy relations during foraging, flight, and temperature regulation are not known to differ in any significant way from those of other birds whose energetics have been recently reviewed (50, 154). Except lorikeets, which may derive large portions of their energy supplies from pollen (53), most of the birds [including hummingbirds (254, 273)] also feed on insects.

When breeding, both male and female hummingbirds require relatively large amounts of food energy. The males require time and energy to defend territories (254, 287). The females do not enter torpor at night while incubating (49), and they must have sufficient time for insect collecting to feed their young.

Hummingbirds range in weight from approximately 2.5 to 12 g. The weight-specific metabolic rates that have been measured during flight are close to 43 ml O$_2$ (g hr)$^{-1}$, or 215 cal (g hr)$^{-1}$, regardless of body size (108, 165), but metabolic rate during flight is related to the weight-relative wing area (72).

Bees and Other Insects

Except for some social Hymenoptera, most insects enter torpor at relatively frequent intervals, usually arousing only when preparing to fly (122). When inactive, their metabolic rate continues to decline with decreasing temperature. The metabolic rate in a torpid bumblebee at 10°C is near 0.5 ml O$_2$ (g hr)$^{-1}$, and about four times this rate at 20°C (149). These rates are similar to those observed in other insects (150). The great range of metabolic rates of "resting" insects is undoubtedly due, in part, to occasional or persistent endothermy.

The metabolic costs of rest and walking have not been differentiated in insects; however, it is probable that the metabolic costs of walking, at least at slow speeds, are very near those during rest, relative to the costs of flight. The metabolic cost of flight varies markedly between different types of flight, such as gliding and hovering. Sotavalta suggested that the rate of fuel consumption varies with the 1.4 power of body weight (250), but a comparison of the energy expenditure of a few insects (using various types of flight) weighing from several milligrams to several grams showed no dependence of energy expenditure on body weight (126). Rates of power output during flight in objects ranging in size from aphids to pigeons (156) and up to DC-8 jet transport planes have been compared (263, 264). Honeybees (*Apis mellifera*) utilize 10–11 mg sugar/hr of flight (30, 31, 240, 250), corresponding to a metabolic rate of 77 ml O$_2$ (g hr)$^{-1}$, or 385 cal (g hr)$^{-1}$, near that of flying bumblebees (*Bombus* sp.) (123). Sphinx moths (Sphingidae) have a metabolic rate during flight that is near 60 ml O$_2$ (g hr)$^{-1}$, slightly lower than that of flying bumblebees, (116, 129). However, as they range in weight from ≥ 100 mg to over 6 g (128), their total fuel consumption per hour of flight corresponds to about 8–480 mg of sugar.

The metabolic rate of butterflies (Papilionidae) has not been measured in free flight or flight at 100% lift, but those with large wings, permitting low wing-loading should, by extrapolation from moths (116), locusts (281), and birds (72) have low metabolic rates during flight. Many species of butterflies bask (119, 269), which reduces or eliminates the need for energy expenditure by endothermic warm-up (274). Not having to rapidly accumulate large energy profits to feed larvae, they often wait for sunshine before initiating activity. Moreover, unlike bees, they take the time to bask in the sun rather than foraging without pause.

Except for fruitflies and blowflies, few data on the metabolic rates of Diptera during flight are available (51, 59, 293). Fruitflies tend to have a weight-specific metabolic rate about one-third that of bees. Hovering syrphid flies, on the basis of endothermy, probably have a metabolic rate at least as high as that of bees. Although endothermic by shivering, syrphids practice considerable energy economy by basking (130). In the Arctic, small flies have been observed to bask in heliotropic flowers (139).

The numerous energy-saving mechanisms observed in many nectivorous animals suggest that energy supplies have historically been, and are, sometimes limiting to survival.

FOOD REWARDS IN FLOWERS

With the rare exception of lipid in the nectars of some flowers (16), by far the largest dry weight of nectar is represented in sugar. The sugars are primarily the monosaccharides glucose and fructose and the disaccharide sucrose. Sucrose predominates in most flowers with tubular corollas and its hydrolysates, glucose and fructose, in open flowers (8, 109, 227, 256, 291). Nectar also contains amino acids and other components (15, 16). While these may be of great significance in nutrition, they are probably not a significant source of food energy, for they usually comprise less than 0.03% of the total dry weight of the nectar (H. G. Baker, personal communication).

Various methods of nectar analysis are available. The volume of nectar per flower may be determined by centrifuging individual flowers or by withdrawing the nectar, using capillary tubes of various sizes (16, 29, 32). The concentration of sugars in samples of several microliters of nectar can usually be measured with a pocket refractometer (107, 109, 118, 255, 292). The biochemical composition of the nectar is usually detected using chromatographic techniques (291) and a variety of other detection methods, depending on whether sugars (227) or amino acids (15, 16) are the components of interest.

It is assumed that 1 mg of sugar, regardless of type, yields about 4.0 cal., probably a reasonable estimate for most ecological questions. However, in honeybees, for example, there are physiological differences in the ability to taste (93), live on (270), and utilize in flight (187) various sugars, some of which are found in nectars. Hummingbirds void essentially no sucrose from the cloaca, implying that nearly all ingested is fully utilized (110). Generally, the sugars found in nectars are the ones for which honeybees have the greatest preference (290, 291) and the ones most readily utilized (187, 270). To honeybees, mixtures of glucose, sucrose, and fructose are more attractive than the individual sugars (290), but hummingbirds prefer pure sucrose (256).

The total caloric rewards available in flowers vary greatly. For example, nectar available in flowers of different plants in Central America (106, 107, 132, 255) varies from less than 0.03 mg to approximately 1800 mg in *Ochroma lagapus,* a range of 60,000 times (P. Opler, personal communication). Most Holarctic flowers of the north temperate (29, 33, 74, 93, 125, 152, 228, 244) and Arctic regions (137) contain <1 mg sugar per floret and are visited by bees. The amount of sugar in "bird flowers" (12, 109, 255) is considerably larger than that in "bee flowers," although it overlaps with them. "Bat flowers" contain some of the largest amounts of sugar. Up to 15 ml of nectar is produced per flower per night by some bat-pollinated flowers in West Africa (17) and Costa Rica (132). As is discussed later, the amount of nectar per floret is undoubtedly only one of several variables affecting the profits that a flower visitor can obtain from a given plant.

The concentration of the nectar of open flowers is highly variable, ranging from less than 10% to near 80% (33). In part, this range is known to be caused by environmental conditions that foster desiccation or dilution (216, 242). However, the concentration of nectar in flowers with tubular corollas is much more independent of environmental conditions. In bird, bat, and "butterfly flowers" (275), the nectar is usually dilute (15–25%), whereas that of bee flowers is often more than 50% sugar. Environmental factors affecting nectar secretion have been discussed (142, 245).

Whether or not a given caloric reward is presented as dilute or as concentrated solution is important in the energetics of foraging. Sugar presented in dilute solution sets an absolute limit on the amount that can be taken at any one time. Since endotherms require more food energy at low temperatures than at high, it is of interest that nectar from high elevation hummingbird flowers tends to be more dilute than that from low elevations (106). Baker recently suggested a possible functional significance for this difference (14). The rate of nectar uptake by hummingbirds is markedly dependent on its viscosity (106), which is temperature-dependent. The lower temperatures in the highlands increase the viscosity of the nectar, but the lower concentrations produced counteract this effect so that the rate of uptake can remain high.

There is some evidence to suggest that alpine plants produce more nectar than those of lower elevations (245, 261), and that those plants residing north of the Arctic circle produce more than those growing at lower latitudes (137). Nevertheless, honeybees do not accumulate much honey north of the Arctic circle (138), presumably because low temperature either greatly restricts their activity or requires them to consume honey for temperature regulation as fast as it is collected. Bumblebees, on the other hand, because of their prodigious endothermy, are able to forage for 24 hr a day in the Arctic, even at ambient temperatures below the freezing point of water (43, 237), and they have been observed foraging in rain and snow (286). Some minimum amount of nectar must obviously be present to attract pollinators, but the range of nectar amounts within which pollination is optimal may be narrow and variable from one locality to the next.

Several attempts have been made to relate seed with nectar production in genetic strains of plants varying in nectar production, but the results have not been clear-cut (223, 224). However, recently F. L. Carpenter and R. E. MacMillen (personal communication) have measured seed-set in Hawaiian Ohia trees (*Metrosideros col-*

lina) as a function of nectar availability, and found that seed-set declines significantly above and below some optimal degree of nectar availability.

Pollen is an important food reward in many flowers. With some exceptions (53), however, its importance is not in its energy content, but in its protein, which is used for egg maturation and larval growth. Pollen is probably a relatively more important food item for solitary bees than for social bees. A considerable portion of the energy resources collected by social bees is used in heating the nest (89, 117, 121), which accelerates brood development. This energy is derived from sugar. The solitary bees, which dispense with such energy expenditure, accumulate only enough food reserves in the nest to feed the larvae, and should thus have a much smaller need for food energy than do social bees.

For social bees, the demands for food energy are often so great relative to protein that pollen often appears to be collected only incidentally to nectar collecting. It is often discarded, and it thus has been concluded that nectar is the primary attractant for honeybees to flowers (226, 228). Free (81) observed that honeybees collect pollen only when there is no nectar, but Gary et al disagreed (97). It is probable that no generalities can be gleaned from isolated examples, since the preferred foods depend on needs in the hive (84), which vary greatly from one instance to the next.

The total amounts of pollen produced by some flowers have been measured (53, 173, 225, 266). Honeybees take approximately 1000 pollen grains from a flower of *Trifolium pratense,* visiting on the average 284 flowers per load in 24 min (44). *Colias* butterflies visiting *Phlox glabberima* unintentionally take a similar number of pollen grains from each flower (173).

The labor required to collect a load of pollen is often less than that required to collect a load of nectar. A pollen-gathering bee, for example, may only visit 7–120 apple blossoms per trip; one collecting nectar visits 250–1446 (234). But bees collecting pollen pollinate a greater percentage of the flowers they visit (83). Honeybees foraging from vetch (*Viccia* sp.) visit the same number of flowers per unit time whether they are collecting only nectar or both nectar and pollen (277). Weaver (279) has shown that 1 lb of white clover honey represents approximately 17,330 foraging trips. Since bees visit about 500 flowers during an average foraging trip of 25 min, each pound of the honey represents the food rewards from approximately 8.7 million flowers, and 7221 hr of bee labor. As long as energy supplies are limiting, as they often are to social bees, the supply of pollen is usually secondarily limited.

FORAGING PROFITS

The foraging profits of individuals are ultimately related to reproductive or hive success; several studies have linked the two. For example, time-labor factors of individual honeybee foragers on selected crops (219, 279) and as a function of flight distances (30, 31, 68, 69, 236, 237) have been translated to total honey production per hive. More recently, time-labor factors of flower visiting by birds in relation to nesting have been investigated (53, 254).

Factors affecting foraging profits of individuals are examined more specifically below. Foraging economics of honeybees (276, 277, 279), bumblebees (118, 120,

125), and hummingbirds (107, 221, 254) on different plant species have been calculated. It is not necessary that a forager be rewarded at every flower it visits. Honeybees visit and pollinate the flowers of a nectarless variety of muskmelon, but only as long as those flowers are intermingled with others bearing nectar (37). The precise mechanisms whereby bees assess the suitability of flowers are not known. However, bees can learn to associate a scent (158) or a color in a single reward during the 2 sec before feeding (204). The information resides initially only in the short-term memory, from which it fades within a few minutes (205). It can be assumed that the rate at which the reward is presented or becomes available must not drop below a minimum dictated by short-term memory in order for conditioning to occur. (Long-term memory may last for a month.) The food rewards that a given plant species yields must ultimately be assessed in terms of the quantity that can be collected per unit time. This quantity is a function of the distance between florets and the speed with which food can be extracted from them. For example, bumblebees usually visit the flowers of Chelone glabra at only 2.8 per min, in part because up to 30 sec may be required to enter a single blossom (127). The relatively great effort required to search for and enter the widely distributed blossoms may be energetically worthwhile since each blossom (foragers excluded) contains on the average 3.3 mg of sugar. In contrast, Trifolium pratense may contain only 0.05 mg of sugar per floret, but the florets, arranged into inflorescences, are probed at a sustained rate near 40 per min by the long-tongued Bombus fervidus. While foraging from capitula of Hieracium sp. (Compositae) growing in patches, the sustained rate of probing florets by short-tongued bees (Bombus terricola) averages 110 per min. The amounts of nectar per floret of Hieracium are minute (usually not visible to the unaided eye). The high rates of probing, probably necessary for an energetic profit, might not be possible if the florets were not in dense inflorescences and the plants in relatively dense colonies.

The absolute distance between flowers and the ease of entry into them may not always be the only relevant variables to foragers. The effective distance is related to the time of flight from one flower to the next, and larger objects are more attractive from a distance and are visited by a more direct line of flight than small ones (161). It is of critical energetic importance that flower signals be conspicuous, so that search times are minimized and the line of flight from one flower to the next is direct and unimpeded. Ideally flowers should be located outside the foliage and marked by color patterns that contrast sharply with the background. As Lovell has indicated, conspicuousness has a profound effect on the rate of food discovery (189–191). Thus, if the rewarding flower is not conspicuous, the forager may fail to detect it and continue foraging from less rewarding flowers.

Crowding of flowers into inflorescences, or the presence of large petals, makes targets more conspicuous to potential flower visitors from a distance, thus aiding discovery (286) and shortening the flight path between successive flowers visited. However, a large target at close range obscures the "bullseye"—the nectar or pollen source. Additional time and energy is saved by close-in signals. The honey guides (58, 199, 251) and scent guides (7, 182) act to direct the movements to the nectar without delay. Since the forager can visit more flowers when its time at each is decreased, the flowers could produce less nectar, and the pollinator could visit more flowers while maintaining a similar profit margin.

The rate at which florets can be manipulated, which could make the difference between profit and loss (125), depends also on various morphological features of the flowers and foragers. Hummingbirds, for example, are able to extract nectar from flowers of a wide range in length of corolla tube. However, the rate at which the nectar can be extracted decreases markedly with increasing corolla length for a given forager (109). Similarly, the rate at which different species of bumblebees manipulate the florets of *Trifolium pratense* is directly related to tongue length (see 140 for review). Rates of visitation of short-tongued bumblebees to campanulate flowers of *Uvularia sessifolia* may be sufficiently low so that they are energetically "excluded," although physically only impeded (125).

Hummingbirds need only to maintain a daily energy balance at which input equals output. During the breeding season, however, the net profit must be sufficient, after individual energy requirements are met, to leave time for territorial defense by males and for insect catching to feed the young by females. Total energy expenditure is approximately the same in the breeding and nonbreeding seasons, but the time allocated for different activities shifts (253, 254).

For Lepidopterous pollinators, some foods required for reproduction are drawn from energy stores derived from leaves on which the larvae have fed. Some moths rely on these reserves for their entire food supply. Other moths and butterflies (275) may lay a few eggs without feeding, but their life spans and reproductive potentials are greatly reduced unless they feed (99).

Heliconius butterflies have lifespans possibly longer than 6 months (98), during which food reserves accumulated from the larval stage are exhausted and the insects rely nearly exclusively on the food derived from flowers, both as a protein source for egg production and as their energy source. These butterflies provide an interesting contrast in their relations with their hosts compared to high-energy hymenopterous pollinators. Because of their slow gliding flight and their small energy investment to offspring, spread out over a long time, their rates of energy intake and expenditure are very low. The butterflies readily meet these requirements by visiting less than a dozen flowers a day, whereas a bumblebee may have to visit some flowers in its habitat at a rate of 10–20 per min to make an acceptable energy profit (125).

A social bee worker must collect food energy in great excess of what it expends (95), and is usually also under the rigid constraints of time, particularly where the environment is marked by seasonality. Such a time limit may be particularly severe in bumblebees, which must bring in sufficient profit to allow rapid buildup of the colony, followed by the production of reproductives, in a single season (231). The foraging speed in different bees may be a reflection of selective pressures on foraging profits. *Bombus* visits flowers at nearly twice the rate of honeybees (41, 66, 86), and the rate of foraging of a variety of solitary bees (289) is at least half that of honeybees and often less than one-fourth that of a *Bombus* (127). Although the parasitic bumblebee *Psithyrus* is similar in size to eusocial bumblebees, its rate of flower visitation is several times lower than that of *Bombus* on the same species of flowers (B. Heinrich, unpublished). *Psithyrus* does not need to collect a large profit since it does not feed its own larvae.

Because of their large size, some sphinx moths, birds, and bats have very high net rates of energy expenditure relative to the food energy of most flowers. One of these pollinators weighing 3 g, for example, expends approximately 11 cal per minute of flight, which is energetically equivalent to the sugar contained in 15 μl of a 20% sugar solution. Due to the high rate of energy expenditure during locomotion, the heavier animals must necessarily have high rates of food intake; this can be accomplished by visiting only flowers with high food rewards or visiting flowers at a very rapid rate.

Hovering greatly accelerates the rate at which flowers can be visited. The ruby-throated hummingbird (*Archilochus colubris*), for example, may visit *Impatiens biflora* blossoms at a rate of 37 per min, whereas bumblebees foraging from the same flowers at the same place and time visit them at 10 per min (127). However, although hovering permits the making of a rapid profit when food rewards per flower are ample, it is also the most rapid means of accumulating an energetic debt when they are not.

When flowers are tightly clumped, as on compact inflorescences, a forager that perches presumably can visit as many florets per unit time as a hoverer, but without incurring the high energetic costs. Thus the hoverer necessarily reduces the spectrum of flowers from which it can forage, but it has an energetic advantage over others. While foraging from large inflorescences, bees have the option of reducing both the percentage of flight time and their body temperature, and energy expenditure could drop nearly an order of magnitude while nectar gathering on a restricted group of flowers continues (118). Torpor during foraging (118) appears to be a reserve mechanism observed only if high-energy flowers are no longer available in the habitat. These mechanisms may permit the animals to maintain an energy balance, but probably preclude them from making a rapid profit. Analogous low energy foraging behavior is observed on a continuous basis in ants foraging from *Polygonum cascadense* (Polygoniaceae) (133).

Foraging profits by social insects depend markedly on the distance of the food source from the hive. The energy expenditure in flight to and from the hive, though great, is often negligible in comparison to the food energy that could be collected during the same time (30, 32, 277). For example, a bumblebee visiting 40 clover blossoms per minute (66) potentially collects enough sugar (2 mg) in this minute for 6 min of flight, that is, for one round trip to a food source nearly 1 km from the hive at a speed of 18 km/hr (61). However, a foraging distance of 1 km costs the bee 6.7 min of foraging time equivalent to the nectar content of 267 clover blossoms. Thus at 1 km from the hive, the clover blossoms are worth much less energetically than they are at 0 distance from the hive. Eckert (68) has calculated that when a food source is 2–3 km from the hive, a honeybee can make 20 round trips per hour, but at a distance of 14 km it can make only 1 trip per hour. Beütler (30) reached similar conclusions and Hamilton & Watt analyzed more general aspects of resources and foraging distance (112).

With regard to increasing sophistication of control, four levels of geographic orientation can be recognized in terrestrial animals, two for insects and two for birds. Minimal geographic orientation control is exhibited by migratory locusts, which ascend and descend in and out of trade winds that carry and guide them over vast distances (119). Cabbage butterflies (*Pieris brassicae* and *P. rapae*) in England do better in controlling their seasonal migrations by flying directly toward the sun in the fall and away from the sun in spring, thus simply utilizing basic directional light orientation (positive and negative phototaxis) (9). In their end result, such migrations are biologically meaningful, though not optimal because of the lack of linearity.

Really straight migrations may be controlled by visual celestial compass or magnetic compass orientation, as in the extrinsic component of geographic vector orientation in birds. [70] The associated distance orientation, as in the topographic vector orientation of bees and ants, is under intrinsic control (54). Navigation, finally, is the still-unexplained ability of birds with some experience to reach from an unknown location a geographic goal over medium to great distances. Several reviews of long distance bird orientation are available (2, 53, 91, 105).

Many fish migrations, especially up and down river systems, may be considered as extensions of zonal orientation. An important component of the geographic orientation of fish is following an odor-marked current of water, a form of guideline orientation (28, 62, 138).

CONCLUDING REMARKS

For orientation ecology to flourish it is deemed essential that the still predominating classification of taxes and kineses be supplemented or replaced by terms with strong ecological connotations. The proposed unified theory and terminology of orientation ecology serves two main purposes: it organizes multitudinous phenomena from bacteria to man, thus rendering them more readily accessible, and it brings numerous details within reach of prediction by means of general rules and principles. A total of seventy such generalizations is marked by square brackets in this review, and only space limited the establishment of many more. From this, I venture to say that no other area of ecology or ethology of comparable scope is presently understood in such depth and breadth as orientation ecology, and this statement definitely holds for the interface between ecology and ethology. Apart from this, orientation ecology derives its importance from the fact that all motile organisms, while interacting with their environments, spend more time orienting than in any other similarly complex behavioral activity. In addition, orientation is the only universal behavior of all motile organisms. It is also of general importance that cropping paths, a subject of orientation ecology, are the best preserved complex behavior in the fossil record.

ACKNOWLEDGMENTS

I am indebted to Mr. E. M. Barrows, Dr. R. F. Johnston, Dr. C. D. Michener, Dr. W. J. O'Brien, Mrs. M. O'Brien, and Dr. O. Taylor for helpful comments on various parts of the draft. Most of my work on animal orientation had been supported by the "Deutsche Forschungsgemeinschaft"; it is presently supported by NSF Grant BMS 72-02575.

Literature Cited

1. Abel, E. 1953. Lichtrückenreflex eines Fisches in der blauen Grotte. Österr. Zool. Z. 4:397–401
2. Adler, H. 1970. Ontogeny and phylogeny of orientation. In *Development and Evolution of Behavior: Essays in Memory of T. C. Schneirla*, ed. L. R. Aronson, E. Tobach, D. S. Lehrman, J. S. Rosenblatt, 303–36. San Francisco: Freeman
3. Adler, J. 1973. Chemotaxis in *Escherichia coli*. In *Behavior of Micro-Organisms*, ed. A. Perez-Miraveta, 1–15. NY: Plenum
4. Adler, J., Tso, W.-W. 1974. "Decision"-making in bacteria: chemotactic response of *Escherichia coli*. *Science* 184:1292–94
5. Alcock, J. 1972. Observations on the behaviour of the grasshopper *Taeniopoda eques* (Burmeister), Orthoptera, Acrididae. *Anim. Behav.* 20:237–42
6. Alexander, R. D., Otte, D. 1967. The evolution of genitalia and mating behavior in crickets (Gryllidae) and other Orthoptera. *Misc. Publ. Mus. Zool. Univ. Mich.* No. 133
7. Armitage, K. B. 1962. Social behaviour of a colony of the yellow bellied marmot (*Marmota flaviventris*). *Anim. Behav.* 10:319–31
8. Bainbridge, R. 1961. Migrations. In *The Physiology of Crustacea*, ed. T. H. Waterman, 2:431–63. N.Y.: Academic
9. Baker, R. R. 1968. A possible method of evolution of the migratory habit in butterflies. *Phil. Trans. R. Soc. London B* 253:309–41
10. Banse, K. 1964. On the vertical distribution of zooplankton in the sea. *Prog. Oceanogr.* 2:52–125
11. Barlow, G. W. 1964. Ethology of the Asian teleost *Badis badis*. V. Dynamics of fanning and other parental activities, with comments on the behavior of the larvae and postlarvae. *Z. Tierpsychol.* 21:99–123
12. Barth, R. H. 1970. The mating behavior of *Periplaneta americana* (Linnaeus) and *Blatta orientalis* (Linnaeus) (Blattaria, Blattinae), with notes on three additional species of Periplaneta and interspecific action of female sex pheromones. *Z. Tierpsychol.* 27:722–48

13. Bässler, U. 1962. Zum Einfluss von Licht und Schwerkraft auf die Ruhestellung der Stabheuschrecke (*Carausius mososus*). *Z. Naturforsch. B* 17:477–80
14. Beukema, J. J. 1968. Predation by the three-spined stickleback (*Gasterosteus aculeatus* L.). The influence of hunger and experience. *Behaviour* 31:1–126
15. Bischof, N., Scheerer, E. 1970. Systemanalyse der optischvestibulären Interaction bei der Wahrnehmung der Vertikalen. *Psychol. Forsch.* 34:99–181
16. Blaxter, J. H. S., Holliday, F. G. T. 1963. The behaviour and physiology of herring and other clupeids. *Adv. Mar. Biol.* 1:262–393
17. Bourne, D. W., Heezen, B. C. 1965. A wandering enteropneust from the abyssal Pacific, and the distribution of spiral tracks on the sea floor. *Science* 150:60–63
18. Boycott, B. B. 1954. Learning in *Octopus vulgaris* and other cephalopods. *Pubbl. Stn. Zool. Napoli* 25:67–93
19. Budelmann, B. U., Wolf, H. G. 1973. Gravity response and angular acceleration receptors in *Octopus vulgaris*. *J. Comp. Physiol.* 85:283–90
20. Bünning, E. 1973. *The Physiological Clock. Circadian Rhythms and Biological Chronometry*. NY: Springer
21. Burger, J. 1974. Breeding adaptations of Franklin's gull (*Larus pipixan*) to a marsh habitat. *Anim. Behav.* 22:521–69
22. Burger, M. C. 1971. Zum Mechanismus der Gegenwendung nach mechanisch aufgezwungener Richtungsänderung bei *Schizophyllum sabulosum* (Julidae, Diplopoda). *Z. Vergl. Physiol.* 71:219–54
23. Calow, P. 1974. Some observations on locomotor strategies and their metabolic effects in two species of freshwater gastropods, *Ancylus fluviatilis* Mull. and *Planorbis contortus* Linn. *Oecologia Berlin* 16:149–61
24. Campos, J. J., Langer, A., Krowitz, A. 1970. Cardiac responses on the visual cliff in prelocomotor human infants. *Science* 170:196–97
25. Carthy, J. D. 1958. *An Introduction to the Behaviour of Invertebrates*. London: George, Allen and Unwin

26. Clayton, R. K. 1964. Phototaxis in Micro-organisms. In *Photophysiology,* ed. A. C. Giese, 2:50–77. NY: Academic

27. Cody, M. L. 1974. Optimization in ecology. *Science* 183:1156–64

28. Cooper, J. C., Hasler, A. D. 1974. Electroencephalographic evidence for retention of olfactory cues in homing coho salmon. *Science* 183:336–38

29. Cott, H. B. 1940. *Adaptive Coloration in Animals.* London: Methuen

30. Croze, H. J. 1970. Searching images in carrion crows. *Z. Tierpsychol. Beih.* 5:1–85

31. Crumpacker, D. W., Williams, J. S. 1973. Density, dispersion, and population structure in *Drosophila pseusoobscura. Ecol. Monogr.* 43:499–538

32. Dawkins, M. 1974. Behavioural analysis of coordinated feeding movements in the gastropod Lymnaea stagnalis (L.). *J. Comp. Physiol.* 92:255–71

33. Delius, J. D., Vollrath, F. W. 1973. Rotation compensating reflexes independent of the labyrinth. *J. Comp. Physiol.* 83:123–34

34. Dethier, V. G. 1957. Communication by insects: physiology of dancing. *Science* 125:331–36

35. Dijkgraaf, S. 1963. Nystagmus and related phenomena in *Sepia officinalis. Experientia* 19:29

36. Downes, J. A. 1969. The swarming and mating flight of Diptera. *Ann Rev. Entomol.* 14:271–98

37. Easter, St. S., Johnes, D. D., Heckenlively, D. 1974. Horizontal compensatory eye movements in goldfish (*Carausius morosus*). I. The normal animal. *J. Comp. Physiol.* 92:23–35

38. Emlen, J. M. 1966. The role of time and energy in food preferences. *Am. Nat.* 100:611–17

39. Emlen, S. T. 1971. Celestial rotation and stellar orientation in migratory warblers. *Science* 173:460–61

40. Enright, J. P. 1970. Ecological aspects of endogenous rhythmicity. *Ann. Rev. Ecol. Syst.* 1:221–38

41. Ercolini, A., Scapini, F. 1974. Sun compass and shore slope in the orientation of littoral amphipodes (*Talitrus saltator* Montagu). *Monit. Zool. Ital.* (N.S.) 8:85–115

42. Farkas, S. R., Shorey, H. H., Gaston, L. K. 1974. Sex pheromones of Lepidoptera and visual cues on aerial odor-trail following by males of *Pectinophora gossypiella. Ann. Entomol. Soc. Am.* 67:633–38

43. Ferguson, D. E. 1971. The sensory basis of amphibian orientation *Ann. NY Acad. Sci.* 188:30–36

44. Forbes, L., Seward, M. J. B., Crisp, D. J. 1971. Orientation to light and the shading response in barnacles. In *Fourth European Marine Biology Symposium,* ed. D. J. Crisp, 539–58. Cambridge, Engl.: Cambridge Univ. Press

45. Fossey, D. 1974. Observations on the home range of one group of mountain gorillas (*Gorilla gorilla* Beringei). *Anim. Behav.* 22:568–81

46. Fraenkel, G. 1930. Die Orientierung von *Schistocerca gregaria* zur strahlenden Wärme. *Z. Vergl. Physiol.* 13:300–13

47. Fraenkel, G., Gunn, D. L. 1940. *The Orientation of Animals.* Oxford: Monogr. Anim. Behav.

48. Franck, D., Hailman, E. 1972. Orientierung des Prozessionsspinners *Thaumetopoea pityocampa* Schilf. auf den Verpuppungswanderungen (Lep., Notodontidae). *Entomol. Mitt. Zool. Mus. Hamburg* 4:299–301

49. Frisch, K. V. 1965. *Tanzsprache und Orientierung der Bienen* Berlin-Heidelberg. NY: Springer; Engl. ed. *The Dance Language and Orientation of Bees.* Cambridge, Mass.: Harvard Univ. Press

50. Goodman, L. J. 1965. The role of certain optomotor reactions in regulating stability in the rolling plane during flight in the desert locust *Schistocerca gregaria. J. Exp. Biol.* 42:385–407

51. Görner, P. 1973. Beispiele einer Orientierung ohne richtende Aussenreize. *Fortsch. Zool.* 21:20–45

52. Götz, K. G. 1968. Flight control in *Drosophila* by visual perception of motion. *Kybernetik* 4:199–208

53. Gwinner, E. 1971. Orientierung. In *Grundriss der Vogelzugkunde,* ed. E. Schütz, 299–348. Berlin-Hamburg: Parey

54. Gwinner, E. 1974. Endogenous temporal control of migratory restlessness in warblers. *Naturwissenschaften* 61:405

55. Haaker, U. 1967. Tagesrhythmische Vertikalbewegungen bei Tausendfüsslern (Myriopoda, Diplopoda). *Naturwissenschaften* 54:346–47

56. Häntschel, W. 1975. Trace fossils and problematica. In *Treatise on Invertebrate Paleontology,* ed. C. Teichert, Pt. W., Suppl. 1. Boulder, Colorado: Geol. Soc. Am.; Lawrence, Kansas: Univ. Kansas Press

57. Hagen, H.-O. 1967. Nachweis einer kinästhetischen Orientierung bei *Uca rapax*. *Z. Morphol. Oekol. Tiere* 58:3Q1–20
58. Hainsworth, F. R., Wolf, L. L. 1972. Energetics of nectar extraction in a small, high altitude, tropical hummingbird, *Gelaphorus flammula*. *J. Comp. Physiol.* 80:377–87
59. Hall, K. R. L. 1965. Ecology and behavior of baboons, patas, and vervet monkeys in Uganda. In *The Baboon in Medical Research,* ed. H. Vagteborg, 43–61. Austin, Texas: Texas Univ. Press
60. Hamilton, W. J. III. Watt, K. E. E. 1970. Refuging. *Ann. Rev. Ecol. Syst.* 1:263–86
61. Harden-Jones, F. R. 1963. The reaction of fish to moving backgrounds. *J. Exp. Biol.* 40:437–46
62. Hasler, A. D. 1966. *Underwater Guideposts.* Madison: Univ. Wisconsin Press
63. Heatwole, H. 1970. Thermal ecology of the desert dragon *Amphibolurus inermis. Ecol. Monogr.* 40:425–57
64. Heinrich, B. 1974. Thermoregulation in endothermic insects. *Science* 185: 747–56
65. Heran, H., Lindauer, M. 1963. Windkompensation und Seitenwindkorrektur der Bienen beim Flug über Wasser. *Z. Vergl. Physiol.* 47:39–55
66. Herrnkind, W. F. 1972. Orientation in shore-living arthropods, especially the sand fiddler crab. *Behavior of Marine Animals,* 1:1–59. NY: Plenum
67. Hisada, M., Tamasige, M., Suzuki, N. 1965. Control of the flight of the dragonfly *Sympetrum darwinianum* Seleys. I. Dorsophotic response. *J. Fac. Sci. Hokkaido Imp. Univ. 6: Zool.* 15: 568–77
68. Holst, E. V. 1950. Quantitative Messungen von Stimmungen im Verhalten der Fische. *Symp. Soc. Exp. Biol.* 4: 143–72
69. Horridge, G. A. 1971. Primitive examples of gravity receptors and their evolution. In *Gravity and the Organisms,* ed. S. A. Gordon, M. J. Cohen, 203–21. Chicago: Univ. Chicago Press
70. Humphries, D. A., Driver, P. M. 1970. Protean defense by prey animals. *Oecologia Berlin* 5:285–302
71. Hutchinson, G. E. 1953. The concept of pattern in ecology. *Proc. Acad. Nat. Sci. Philadelphia* 105:1–12
72. Hutchinson, G. E. 1967. *A Treatise on Limnology,* Vol. II. NY: Plenum
73. Jacobs, W. 1952. Vergleichende Verhaltensstudien an Feldheuschrecken (Orthoptera, Acrididae) und einigen anderen Insekten. *Verh. Dtsch. Zool. Ges.* 1952:115–38
74. Jahn, Th. 1960. Optische Gleichgewichtsregelung und zentrale Kompensation bei Amphibien, insbesondere der Erdkröte (*Bufo bufo* L.). *Z. Vergl. Physiol.* 43:119–40
75. Jander, R. 1957. Die optische Richtungsorientierung der Roten Waldameise (*Formica rufa* L.). *Z. Vergl. Physiol.* 40:162–263
76. Jander, R. 1962. The swimming plane of the crustacean *Mysidium gracile* (Dana). *Biol. Bull.* 122:380–90
77. Jander, R. 1963. Grundleistungen der Licht- und Schwerkraftorientierung von Insekten. *Z. Vergl. Physiol.* 47:381–430
78. Jander, R. 1965. Die Phylogenie von Orientierungsmechanismen der Arthropoden. *Verh. Dtsch. Zool. Ges. Jena* 1965:266–306
79. Jander, R. 1968. Über die Ethometrie von Schlüsselreizen, die Theorie der telotaktischen Wahlhandlung und das Potenzprinzip der terminalen Cumulation bei Arthropoden. *Z. Vergl. Physiol.* 59:319–56
80. Jander, R. 1970. Ein Ansatz für die moderne Elementarbeschreibung der Orientierungshandlung. *Z. Tierpsychol.* 27:771–78
81. Jander, R. 1975. Interaction of light and gravity orientation in *Daphnia pulex. Fortsch. Zool.* 23:174–84
82. Jander, R., Daumer, K. 1974. Guideline and gravity orientation of blind termites foraging in the open (Termitidae: Macrotermes, Hospitalitermes). *Insectes Soc.* 21:45–69
83. Jander, R., Horn, E., Hoffmann, M. 1970. Die Bedeutung der Gelenkrezeptoren in den Beinen für die Geotaxis der höheren Insekten (Pterygota). *Z. Vergl. Physiol.* 66:326–42
84. Jander, R., Jander, U. 1970. Über die Phylogenie der Geotaxis innerhalb der Bienen (Apoidea). *Z. Vergl. Physiol.* 66:355–68
85. Jander, R., Quadagno, D. 1974. An interval scale for measuring visual pattern discrimination of a mammal (*Rattus norvegicus*). *Behav. Biol.* 12: 93–99
86. Jander, R., Waterman, T. H. 1960. Sensory discrimination between polarized light and light intensity patterns by arthropods. *J. Cell. Comp. Physiol.* 56:137–360
87. Johnson, C. G. 1969. *Migration and*

Dispersal of Insects by Flight. London: Methuen

88. Kähling, J. 1961. Untersuchungen über den Lichtsinn und dessen Lokalisation bei dem Höhlenfisch *Anoptichthys jordani* Hubbs and Inns (Characidae). *Biol. Zentralbl.* 80:439–51

89. Kaiser, H. 1974. Verhaltensgefüge und Territorialverhalten der Libelle *Aeschna cyanea. Z. Tierpsychol.* 34:398–429

90. Kaufman, J. H. 1974. Social ecology of the whiptail wallaby, *Macropodus parryi,* in northeastern New South Wales. *Anim. Behav.* 22:281–369

91. Keeton, W. 1974. The orientational and navigational basis of homing in birds. *Adv. Study Behav.* 5:48–132

92. Kennedy, J. J., Marish, D. 1974. Pheromone-regulated anemotaxis in flying moths. *Science* 184:999–1001

93. Klopfer, P. H. 1973. *Behavioral Aspects of Ecology.* Englewood Cliffs: Prentice-Hall

94. Knight-Jones, E. W. 1954. Relations between metachronism and the direction of ciliary beat in metazoa. *Q. J. Microsc. Sci.* 95:503–52

95. Krebs, J. R. 1973. Behavioral aspects of predation. In *Perspectives in Ecology,* ed. P. P. G. Bateson, P. H. Klopfer. NY: Plenum

96. Kühn, A. 1919. *Die Orientierung der Tiere im Raum.* Jena: Fischer

97. Land, M. F., Collett, T. S. 1974. Chasing behaviour of houseflies (*Fannia canicularis*). *J. Comp. Physiol.* 89:331–57

98. Lindauer, M. 1961. *Communication Among Social Bees.* Cambridge, Mass.: Harvard Univ. Press

99. Linsenmair, K. E. 1969. Anemotaktische Orientierung bei Tenebrioniden und Mistkäfern (Insecta, Coleoptera). *Z. Vergl. Physiol.* 64:154–211

100. Linsenmair-Ziegler, Ch. 1970. Vergleichende Untersuchungen zum photogeotaktischen Winkeltransponieren pterygoter Insekten. *Z. Vergl. Physiol.* 68:229–62

101. Ludwig, W. 1929. Untersuchungen über die Schraubenbahnen niederer Organismen. *Z. Vergl. Physiol.* 9:734–801

102. MacArthur, R. H., Connell, J. H. 1966. *The Biology of Populations.* NY: Wiley

103. MacArthur, R. H., Pianka, E. R. 1966. On optimal use of patchy environment. *Am. Natur.* 100:603–9

104. Mackinnon, J. 1974. The behaviour and ecology of wild orang-utan. *Anim. Behav.* 22:3–74

105. Matthews, G. V. T. 1968. *Bird Navigation.* Cambridge, Engl.: Cambridge Univ. Press

106. Merkel, F. W., Fischer-Klein, K. 1973. Winkel-kompensation bei Zwergwachteln (*Exhaltoria chinensis*). *Vogelwarte* 27:39–50

107. Milborrow, B. V. 1974. The chemistry and physiology of abscissic acid. *Ann. Rev. Plant Physiol.* 25:259–307

108. Mills, J. N., ed. 1973. *Biological Aspects of Circadian Rhythms.* NY: Plenum

109. Milne, L. J., Milne, M. 1965. Stabilization of the visual field. *Biol. Bull.* 128:285–96

110. Mittelstaedt, H. 1950. Physiologie des Gleichgewichtes bei fliegenden Libellen. *Z. Vergl. Physiol.* 32:422–63

111. Mohr, H. 1972. *Lectures on Photomorphogenesis.* NY: Springer

112. Mrosovsky, N. 1972. The water finding ability of sea turtles. *Brain Behav. Evol.* 5:202–25

113. Müller, K. 1974. Stream drift as a chronobiological phenomenon in running water. *Ann. Rev. Syst. Ecol.* 5:309–323

114. Pardi, L. 1960. Innate components in the solar orientation of littoral amphipods. *Cold Spring Harbor Symp. Quant. Biol.* 25:395–401

115. Pearre, S. 1973. Vertical migration and feeding in *Sagitta elegans. Ecology* 54:300–14

116. Pick, E. E., Lea, A. 1970. Field observations on spontaneous movements of solitary hoppers of the brown locust *Locusta pardalina* (Walker) and behavioural differences between various colour forms. *Phytophylactica* 2:203–310

117. Pollard, E. C. 1971. Physical determinants of receptor mechanisms. In *Gravity and the Organisms,* ed. S. A. Gordon, J. Cohen, 25–34. Chicago & London: Univ. Chicago Press

118. Pringle, J. W. S. 1948. The gyroscopic mechanism of the halteres of Diptera. *Phil. Trans. R. Soc. London B* 233:347–84

119. Rainey, R. C. 1974. Biometerology and insect flight. Some aspects of energy exchange. *Ann. Rev. Entomol.* 19:407–39

120. Rensch, B. 1959. Trends towards progress of brains and sense organs. *Cold Spring Harbor Symp. Quant. Biol.* 24:291–303

121. Rensch, B. 1972. *Neuere Probleme der Abstammungslehre. Die Transpezifische Evolution.* Stuttgart: F. Enke Verlag

122. Richter, G. 1973. Field and laboratory observations on the diurnal vertical mi-

gration of marine gastropod larvae. *Neth. J. Sea Res.* 7:126–34

123. Roberts, A. M. 1970. Geotaxis in motile micro-organisms. *J. Exp. Biol.* 53: 687–99

124. Robinson, M. H. 1973. The evolution of cryptic postures in insects with special reference to some New Guinea Tettigoniids (Orthoptera). *Psyche* 80:159–65

125. Rock, J. 1974. *Orientation and Form.* NY: Academic

126. Rosenberg, N. J. 1974. *Microclimate. The Biological Environment.* Chichester, Engl.: Wiley

127. Ruiter, L. de 1955. Countershading in caterpillars. *Arch. Neerl. Zool.* 11:285–341

128. Sandeman, D. C. 1975. Dynamic receptors in the statocyst of crabs. *Fortsch. Zool.* 23:185–91

129. Schaefer, K. P., Schott, D., Meyer, D. L. 1975. On the organization of neuronal circuits in the generation of the orientation response (visual grasp reflex). *Fortsch. Zool.* 23:199–212

130. Schneider, G. 1953. Die Halteren der Schmeissfliege (*Calliphora*) als Sinnesorgane und als mechanische Flugstabilisatoren. *Z. Vergl. Physiol.* 35: 416–58

131. Schoener, T. W. 1971. Theory of feeding strategies. *Ann. Rev. Ecol. Syst.* 2:369–404

132. Schöne, H. 1959. Die Lageorientierung mit Statolithenorganen und Augen. *Ergeb. Biol.* 21:161–209

133. Seifarth, E., Barth, F. G. 1972. Compound slit sense organs on the spider leg: mechanoreceptors involved in kinesthetic orientation. *J. Comp. Physiol.* 78:76–191

134. Silver, P. H. 1974. Photoptic spectral sensitivity of the neon tetra (*Paracheirodon innesi,* Myers) found by the use of a dorsal light reaction. *Vision Res.* 24:329–34

135. Smith, J. N. M. 1974. The food searching behaviour of two European thrushes. II. The adaptiveness of the search pattern. *Behaviour* 49:1–61

136. Southwood, T. R. E. 1966. *Ecological Methods.* London: Methuen

137. Stang, G. 1972. Die zentralnervöse Verrechnung optischer Afferenzen bei der Gleichgewichtshaltung von Fischen. *J. Comp. Physiol.* 80:95–118

138. Stasko, A. B. 1971. Review of field studies on fish orientation. *Proc. NY Acad. Sci.* 188:12–28

139. Stein, A., Schöne, H. 1972. Über das Zusammenspiel von Schwereorientierung und Orientierung zur Unterlage beim Flusskrebs. *Verh. Dtsch. Zool. Ges.* 65:225–28

140. Sudd, J. H. 1967. *An Introduction to the Behaviour of Ants.* London: Arnold

141. Tsang, N., Macnab, R., Koshland, D. E. 1973. Common mechanisms for repellents and attractants in bacterial chemotaxis. *Science* 181:60–63

142. Vielmetter, W. 1958. Physiologie des Verhaltens zur Sonnenstrahlung bei dem Tagfalter *Argynnis paphia* L. I. Untersuchungen im Freiland. *Insect Physiol.* 2:13–37

143. Walk, R. D., Walter, C. P. 1973. Effect of visual deprivation on depth discrimination of hooded rats. *J. Comp. Physiol. Psychol.* 85:559–63

144. Wallraff, H. G. 1972. An approach toward an analysis of pattern recognition involved in the stellar orientation of birds. In *Animal Orientation and Navigation,* ed. S. R. Galler, K. Schmidt-Koenig, G. J. Jacobs, R. E. Belleville, 211–22. Washington DC: NASA SP-262

145. Walls, G. L. 1962. The evolutionary history of eye movements. *Vision Res.* 2:69–80

146. Warburton, K. 1973. Solar orientation in the snail *Nerita plicata* (Prosobranchia: Neritacea) on a beach near Watamu, Kenya. *Mar. Biol.* 23:93–100

147. Weiss, P. 1925. Tierisches Verhalten als "Systemreaktion." Die Orientierung der Ruhestellung von Schmetterlingen (*Vanessa*) gegen Licht- und Schwerkraft. *Biol. Gen.* 1:167–248. Transl. 1959. *Yearb. Soc. Gen. Syst. Res.* 4:1–44

148. Werner, E. E., Hall, D. J. 1974. Optimal foraging and size selection of prey by blue gill sunfish (*Lepomis macrochirus*). *Ecology* 55:1042–52

149. Wickler, W., Seibt, U. 1972. Zur Ethologie der Stielaugenfliegen (Diptera, Diopsidae). *Z. Tierpsychol.* 31: 113–30

150. Wiltschko, H., Gwinner, E. 1974. Evidence for innate magnetic compass in garden warblers. *Naturwissenschaften* 61:406

151. Youthed, G. J., Moran, V. C. 1969. Pit construction by myrmeleontid larvae. *J. Insect Physiol.* 15:867–75

SPECTRAL ANALYSIS IN ECOLOGY

❖4092

Trevor Platt and Kenneth L. Denman

Marine Ecology Laboratory, Bedford Institute of Oceanography, Dartmouth, Nova Scotia, Canada

PREAMBLE

Ecologists, like other biologists, are involved with the study of living systems, and they, again like other biologists, are handicapped in their work by the inadequacy of the conceptual framework to which they must relate their quantitative results, and within which they must generate new and testable hypotheses. One way in which the weakness of this structure manifests itself is in the lack of a theoretical approach consistently applicable to a wide variety of living systems, and able to lead to powerful predictions of high generality. Various such approaches have been tried and found wanting; for example, the most recent candidate, the systems analysis method, requires colossal labor and expense to produce a solution that is self-consistent for a particular data set, but for which is guaranteed neither the uniqueness in the original data set nor the self-consistency in any other data set that may be drawn from the same system ensemble, regardless of the boundary conditions. In other words, both the predictive power and the generality of this inelegant approach are very low.

A new foundation for a theoretical biology has been proposed based on nonlinear statistical mechanics (49). This method, which is neither strictly holism nor wholly reductionism, is emerging from attempts to find a general mathematical approach that would account for living as well as nonliving phenomena. It identifies as the most important characteristic of complex systems the property that the differential equations describing the functional relationships between the system components be of the nonlinear kind.

A crucial characteristic of nonlinear systems is their disposition toward *periodic* behavior, even for nonperiodic boundary conditions. Therefore, they tend to a periodic (cyclic) organization in time, in space, or in both. According to this view, the biosystem is seen as an ensemble of nonlinear oscillators, coupled together in various functional configurations at each hierarchic level of system description. The

189

basic element of temporal organization in such a system is the cycle: a complete description of the system would include a list of the frequencies of all of its dominant cycles. That is to say, one way of specifying a system might be to specify its spectrum, since a spectrum is merely a representation of possible frequencies of action of a process arranged according to their relative importance or magnitude.

Much recent work suggests that the nonlinear statistical mechanical description of biological systems is a plausible one (35). New structures may arise from instabilities in living systems; they are a direct consequence of the nonlinear processes acting on random fluctuations. In other words, there is a response at the macroscopic level of description (emergence of new organization) resulting from microscopic excitations (fluctuations). In general, the new structures may be periodic in the spatiotemporal domain (27). It is important to note that these features may be present when the system is far from thermodynamic equilibrium, and that their maintenance requires a steady supply of energy; it is therefore appropriate to refer to them as *dissipative structures*. The earliest biological example of dissipative structure dealt with the chemical basis of morphogenesis (46). Most of the subsequent work has treated biochemical oscillators (11, 19). An example from ecological theory has been provided (38).

If it is realistic to adopt a statistical mechanical representation of living systems in which nonlinear effects are predominant, then the approach should certainly be valid at the ecosystem level of description. The incidence of fluctuations increases with the number of degrees of freedom of the system, and we would therefore anticipate that the approach would be applicable *a fortiori* to ecosystems where the number of degrees of freedom is immense. This suggests that ecosystems should exhibit periodic behavior in time and space, and that the identification and interpretation of the characteristic frequencies or periodicities should be one of the central goals of the discipline of ecology.

Unfortunately, the historical tendency among ecologists has been to design their sampling programs to minimize the apparent variability in time and space, and to be less interested in the variations about the means than in the means themselves. Such an approach to ecological sampling makes it unlikely that periodicities, if present, will be discovered in ecosystem behavior, and it may well be that ecologists, in seeking to minimize the noise so that they might better detect the signal, have been overlooking the real signal of interest.

How will we recognize these fundamental modes of ecosystem behavior? In general, we expect that there will be simultaneous response at several or many frequencies; to resolve these from each other in the presence of random effects can be a formidable task. The only known applicable mathematical technique is the method of spectral analysis, and it is the use of this technique in the ecological context that we deal with in this review.

We distinguish two broad classes of periodic or organized behavior. The first, discussed above, depends on the nonlinear properties of the system itself: the energy supply that maintains the structure might be constant or aperiodic (or random). In this case the modes of behavior of the system are independent of the external conditions. Wiener (47) has shown that when a nonlinear system of low dissipation rate is excited by a random (white noise) field of high energy and small characteristic geometrical scale, the oscillations of the system are confined to a few sharp frequency bands (a "line spectrum"). Second, periodic behavior in the system may result from periodicity in the external energy supply, that is, from periodic forcing. In this case, the modes of excitation of the system might be simple functions of the frequency of forcing (10).

Two particular cases of the second class have received considerable attention from ecologists in the past. They are daily and seasonal cycles that are induced, respectively, in response to the frequency of the earth's rotation about its own axis and to the frequency of rotation of the earth about the sun. Very little attention has been given in ecology to periodic behavior of the first kind. In general, nonlinear oscillations may exist in a much larger number of possible modes than those excited by geophysical forcing at discrete frequencies. As we show below, in any one series of measurements it is not possible to examine the whole spectrum of possible frequencies of behavior; it is far better to make a detailed examination of a selected portion, or "observational window," of the available spectrum. For this reason, sampling for circadian or annual ecological rhythms would not necessarily be the best way to establish the existence of a spectrum of oscillations excited by nonlinear interactions. The daily and seasonal cycles are but two possible modes of ecosystem response; we should not be insensitive to the infinite variety of other possibilities.

THE PRINCIPLES AND METHODS OF SPECTRAL ANALYSIS

Introduction

In this section, we present a summary of the theory of power spectral analysis together with a discussion of the various practical considerations worth keeping in mind while planning for collection of time-series data. Detailed treatments of time-series analysis are available (3, 7, 23). Three articles in *Technometrics,* Volume 3 (22, 28, 45), provide a useful comparison of theoretical treatments. An excellent short review on time-series analysis is given by Hinich & Clay (21).

A *time series* is a sequence of numbers representing a progression of some parameter in time, i.e. it is a function $x(t)$ of time t. The characteristic feature of a time series is that it is a random or nondeterministic function of time: its future behavior cannot be predicted exactly. Rather, we can make predictions only about statistical or average properties of the series; hence the notion of a stochastic process arises. We use the term *time series* loosely to include cases where, for example, space rather than time is the independent variable.

Power spectral analysis is a form of analysis of variance of a time series in which the variance of the series of numbers about their mean is partitioned into contributions at frequencies that are harmonics of the length of the data set. Unlike harmonic analysis, however, spectral analysis takes a statistical approach in which the time series is assumed to be no more than a single realization of a stochastic process.

Theory of the Power Spectrum

THE AUTOCOVARIANCE FUNCTION   Most ecologists are familiar with the autocovariance function of a data series, thus it is a convenient starting point in the theoretical development of power spectral analysis. Consider first a finite series of N equally spaced data points x_t, $t = 1, \ldots, N$, where the mean,

$$x_m = N^{-1}\sum\nolimits_{t=1}^{N} x_t,$$

has been set equal to zero for simplicity. Also, assume that the series obeys a Gaussian or normal probability distribution and is stationary (by which we mean intuitively that subsets of the total data set have statistically the same mean and variance).

The sample autocovariance function is defined as

$$c_{xx}(k) = N^{-1}\sum\nolimits_{t=1}^{N-k} x_t \, x_{t+k}, \quad k = 0,1,2, \ldots, M-1, \qquad 1.$$

where k is the lag number and $(M-1)$ is the maximum number of lags. In words, the autocovariance function is the operation of sliding the data set along beside a copy of itself and summing products of adjacent data points. It provides information about how neighboring points are correlated. When $k = 0$, $c_{xx}(0) = N^{-1}\sum\nolimits_{t=1}^{N} x_t x_t$, which is, of course, just the sample variance s_N^2. As the lag k increases, the statistical uncertainty in the estimate $c_{xx}(k)$ also increases because the series overlaps with itself over a smaller fraction of the total time series, and $c_{xx}(k)$ is formed from the sums of fewer and fewer products.

The quantities we have defined so far $[c_{xx}(k), s_N^2 = c_{xx}(0)]$ are sample estimates of the true corresponding values in the infinite population $[\gamma_{xx}(k), \sigma_{xx}^2 = \gamma_{xx}(0)]$.

THE PERIODOGRAM   A fundamental result of Fourier analysis is that one can decompose a finite time series into a sum of sine and cosine functions of different frequencies. The periodogram then is a simultaneous least squares fit of a finite number of sine and cosine functions of different frequencies to a time series of given length where one attempts to minimize the error ϵ_t in the approximation to the time series x_t, $t = 1, \ldots, N$:

$$x_t = \alpha_0 + \sum\nolimits_{i=1}^{q} (\alpha_i \cos 2\pi\Delta f_i t + \beta_i \sin 2\pi\Delta f_i t) + \epsilon_t, \qquad 2.$$

where $f_i = i/N\Delta$ is the ith harmonic of the fundamental frequency $1/N\Delta$ and Δ is the time interval. If the least squares estimates of the coefficients α_0 and (α_i, β_i) are a_0 and (a_i, b_i), the periodogram consists of the $q = N/2$ values,

$$I(f_i) = \tfrac{1}{2}N(a_i^2 + b_i^2),$$
$$i = 1, 2, \ldots, q-1, \qquad 3.$$
and
$$I(f_q) = I(0.5/\Delta) = Na_q^2,$$

where $I(f_i)$ is the intensity at the frequency f_i. The periodogram estimates then are defined for equispaced frequencies ranging from $1/(2N\Delta)$ up to $f_q = 1/(2\Delta)$, often called the Nyquist frequency. These two frequencies define the lower and upper boundaries of a frequency band that we refer to as the *observational window*. By our choice of the length of the series N and of the sampling interval Δ, we have excluded the possibility of obtaining information about our system at all frequencies outside this band.

If we allow the periodogram intensity to be a continuous function defined at all frequencies from zero to the Nyquist frequency f_q, we obtain the sample spectrum

$$S_{xx}(f) = \tfrac{1}{2}N(a_f^2 + b_f^2); \qquad 0 \leq f \leq f_q. \tag{4.}$$

A fundamental result of Fourier analysis is that the sample spectrum is the Fourier cosine transform of the estimate of the autocovariance function:

$$S_{xx}(f) = 2\Delta[c_{xx}(0) + 2\sum_{k=1}^{M-1} c_{xx}(k) \cos 2\pi\Delta fk]; \qquad 0 \leq f \leq f_q. \tag{5.}$$

If we take the expected or average value of the sample spectrum as N becomes infinitely large, we get the true power spectrum $\Gamma_{xx}(f) = \lim_{N\to\infty} E[S_{xx}(f)]$.

Since the power spectrum is the Fourier cosine transform of the autocovariance function, knowledge of the autocovariance function is mathematically equivalent to knowledge of the power spectrum, and vice versa.

Another important result linking the power spectrum to the true variance is that $\sigma_{xx}^2 = \int_0^{f_q} \Gamma_{xx}(f)df$. In words, the integral of the power spectrum over all frequencies is equal to the total variance of the original data series.

The power spectrum, then, gives the variance of a stochastic process as a function of frequency. From a finite time series representing a single realization of that stochastic process, one can obtain a sample spectrum either by taking a Fourier cosine transform of the estimated autocovariance function or by a periodogram analysis. In the next section, we evaluate these methods according to the following criterion: how well does our sample spectrum estimate the true power spectrum, and how can we improve this estimation?

Estimating the Power Spectrum

THE NECESSITY FOR SMOOTHING   Although the sample spectrum $S_{xx}(f)$ is an unbiased estimate of the true power spectrum $\Gamma_{xx}(f)$, the variance of the sample spectrum does not go to zero as N becomes large, but rather remains finite: $\lim_{N\to\infty}\mathrm{Var}[S_{xx}(f)] = \Gamma_{xx}(f)$. Thus the raw spectral estimator obtained from the autocovariance function or from a periodogram analysis, has a standard error of 100% and, as such, is not a consistent estimator of $\Gamma_{xx}(f)$. Obviously, some form of correction is necessary if the method of power spectral estimation is to be of any use.

BARTLETT'S PROCEDURE, LAG WINDOWS, SPECTRAL WINDOWS   In Bartlett's procedure, the earliest method for improving spectral estimators (1, 2), the series of length N is split into p series each of length $m = N/p$. The smoothed spectral estimate

$$S_{xx}^{*} (f_i)$$

is just the average of the spectral estimates calculated for each of the subseries at frequency f_i. The variance in the smoothed spectral estimate is reduced then by a factor of approximately m/N.

This procedure can be shown to be equivalent (except for minor corrections at the end points) to the following weighting or averaging of the autocovariance function entering into the spectral estimates obtained by the Fourier cosine transform (Equation 5):

$$S_{xx}^{*} (f) = 2\Delta[c_{xx} (0) + 2\sum_{k=1}^{N-1} w(k)c_{xx}(k) \cos2\pi\Delta fk]; \quad (0 \leq f \leq f_q), \quad 6.$$

with $w(k) = 1 - k/m$ ($k \leq m$) and $w(k) = 0$ ($k > m$). In Equation 6, we weighted the covariances being transformed by a lag window, $w(k)$, which decreases linearly from one at zero lag ($k = 0$) to zero at a lag m ($k = m$), where m is the length of each of the subseries used in Bartlett's procedure.

Using properties of the Fourier convolution theorem, one can show that Equation 6 is equivalent to a smoothing of the spectral estimates themselves (whether obtained by periodogram analysis, Equations 2 and 3, or by a Fourier transform of the unweighted autocovariance function, Equations 1 and 5):

$$S_{xx}^{*}(f) = \int_0^{f_q} S_{xx}(f) W(f-G)dg, \qquad 7.$$

where $W(f') = m(\sin\pi\Delta f'm/\pi\Delta f'm)^2$; $-\infty \leq f' \leq \infty$ in the specific case of Bartlett's procedure. The function $W(f')$ is called a *spectral window*. The advantage of smoothing with either a lag window, $w(k)$, or a spectral window, $W(f)$, is that they can be generalized to provide a variety of recipes for improving spectral estimates according to specific data requirements.

One can see that the Bartlett lag window neglects the lagged products beyond a maximum lag m; obviously the maximum lag M in Equation 1 should be the same as m. Thus smoothing according to Bartlett's procedure not only reduces the variance of the spectral estimates, but also reduces the number of lags for which the autocovariance function need be calculated. In practice, the maximum lag M should be chosen between $N/4$ and $N/10$. With a maximum lag M, there will be only $M/2$ independent spectral estimates. However, such a wide spacing is not sufficient for proper smoothing, and it is therefore also advisable to compute the estimates in Equation 6 at the $M + 1$ discrete frequencies $f_i = if_q/M$, where $i = 0, 1, 2, \dots, M$.

Although the smoothed spectral estimate

$$S^*_{xx}(f_i)$$

has a reduced variance, the power or variance in the elementary frequency band of width $1/(2\Delta M)$ centered at f_i has been "smudged" over a range of frequencies by the spectral window $W(f)$. In the common terminology, there is leakage of power between the smoothed spectral estimates. If the true power spectrum has peaks or a range of frequencies over which the spectrum is varying rapidly, then this leakage will cause the smoothed estimates in that region to be biased: their expected values will not equal the true power spectrum.

For minimum bias then, the smoothing procedure should not spread the power at f_i appreciably beyond the elementary band $1/(2\Delta M)$; i.e. the bandwidth should be kept small. For minimum variance, however, a broad bandwidth is desirable. In fact, this apparent inverse relation between bandwidth and variance can be shown to be a general result:

Variance $\times$ Bandwidth = Constant.   8.

In choosing a suitable spectral window, Equation 8 implies the necessity of a compromise between variance and bias. Which is more important—small variance or small bias—will depend on the shape of the spectrum. If the spectrum is reasonably flat or smooth, leakage will not introduce much bias and one should probably choose a wider window that minimizes the variance of the smoothed spectral estimates. However, if the spectrum is expected to have sharp peaks or steep slopes, as in the cases of ocean tides and ocean waves, then the leakage should be minimized at the expense of variance.

According to detailed comparisons of several windows (7, 23), a good compromise seems to be the Hanning or Tukey window. If spectral estimates are calculated at the standard frequency intervals of $1/2\Delta M$, then the Tukey window can be written simply as a moving average over three adjacent raw estimates with weights 1/4, 1/2, 1/4.

THE BLACKMAN-TUKEY METHOD VERSUS THE FAST FOURIER TRANSFORM METHOD   The general technique of estimating power spectra by calculating the autocovariance function, weighting with a lag window, and then operating with a Fourier cosine transform was first presented in (7); the most complete and rigorous treatment is that of (23). The advantage of this technique lies in the use of the lag window, which reduces both the variance in the spectral estimates and the number of calculations required to compute them (by a factor of about 1/4 to 1/10).

However, in 1965 an algorithm called the fast Fourier transform (FFT) was introduced (13), which dramatically reduced the time required to compute power spectral estimates (12, 23, 48).

For a series of N points, periodogram analysis takes about N^2 operations and the FFT takes $2 \log_2 N$. For $N = 1024$, the FFT would take about 2% as long. Thus for any long series (N greater than 10^4), all the periodogram estimates can now be calculated in a much shorter time. In practice, the desired reliability can be obtained much more effectively by smoothing the linear coefficients, a_i and b_i, before they are squared and added to form the quadratic estimates $S_{xx}(f)$ defined in Equation 4, [see (6)].

The relative merits of the Blackman-Tukey and the FFT methods depend on the characteristics of the series involved. A comparison of five sets of spectra computed by the two techniques for corresponding time series (16) showed that the FFT resulted in more leakage, and thus more bias, where there were steep slopes in the data.

For a short historical series, the type usually available in ecology, the Blackman-Tukey approach is as economical as, and produces better resolution of lines and peaks in the spectrum than, the FFT (23). Another feature ecologists may find advantageous is that the autocorrelation function, a necessary product in the Blackman-Tukey method, is often required input for the development of stochastic models (8).

However, the FFT method is usually preferred in geophysical and turbulent fluid dynamics applications because it preserves all the phase information from the original series, whereas all phase information is lost in the Blackman-Tukey method. The FFT method is also the only computationally feasible approach for more than about 10^4 points.

Practical Considerations

ALIASING   Aliasing may be defined as the use of a sampling interval Δ, which is too large to resolve the shortest fluctuations present in the data. The effect on the spectrum is that this high-frequency energy appears erroneously in the spectral estimates for some lower frequency below the Nyquist frequency ($f_q = 1/2\Delta$). This is an especially serious problem in ecology, where each data point often represents a discrete sample, as in bottle station sampling in marine ecology. Compare such a procedure with continuous sampling where each data point would represent an average over a whole or part of the sampling interval.

As an example, consider a small bay with a large semidiurnal tide (period 13 hr) in which the biological processes are strongly influenced by the tidal currents. If a single station were sampled once a day, only an apparent cycle with a period of 6.5 days would show up in the data, as shown by Figure 1. There would be no indication of the actual 13 hr periodicity. Sampling once every 12 hr would yield the same spurious result. Even sampling once every 6 hr, or more than twice each tidal cycle, still modulates the semidiurnal periodicity with a period of 6.5 days.

In practice, the sampling rate should be such that there are at least four samples per cycle for the shortest period fluctuation that may be present. If such a high sampling rate is impractical, then some averaging or "low-pass" filtering process must be applied before the sampling is done. Often the averaging may be accomplished by means of an instrument with a sufficiently slow response time. For

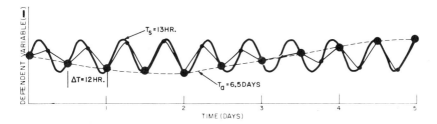

Figure 1   An example of aliased sampling. The heavy solid line represents the signal to be sampled with a period (T_s) of 13 hr. The large dots represent discrete sampling at 12-hr intervals, and the dashed line represents the apparent 6.5 day (157 hr) periodicity. The thin line connecting the large and small dots represents the apparent signal for a 6-hr sampling interval.

example, if one wants to sample the temperature in a culture at 1-min intervals (where there may be unwanted fluctuations with periods shorter than 1 min), a temperature probe should be chosen with a response time of at least 1 or 2 min. Such a probe would effectively average the temperature fluctuations over 1 or 2 min. The same effect of averaging could also be obtained electronically by means of a low pass filter circuit or a "sample and hold" circuit.

To summarize, aliasing should not be a serious problem provided that it is properly considered in the design of an experiment or an observational program.

EQUISPACED DATA   The techniques presented for calculating power spectral estimates require equispaced data points. However, in marine ecology or in any type of long term field experiment, the time series may have unequal spacings. In the case of uneven spacing, a new data set at equal intervals may be obtained from the original data set by interpolation.

In an ecological example (42), a fourth-order polynomial interpolation was used to form an equispaced time series with a 15-day interval from an original series that had points at approximately 14-day intervals. In addition, a moving average was used to filter out high frequency oscillations to enhance the dominant fluctuation having a periodicity of several months. It could then be stated with confidence that this periodicity was not the result of aliasing, illustrating the one advantage offered by a nonuniform sampling interval: namely the almost complete elimination of the possibility of serious aliasing (44).

NON-GAUSSIAN DATA   The theoretical results concerning power spectral analysis are strictly valid only for time series distributed in a Gaussian or normal manner about their mean. Sometimes non-normal data can be transformed to give a new time series which is normally distributed (e.g. the lognormal distribution). Even if they cannot, however, non-normality does not seem to be a great problem in practice (45). Provided that repeated estimates of the spectrum are consistent, non-normal data yield valid, useful information despite the fact that the estimated spectrum is not a complete description of the original series.

STATIONARITY   A strict definition of stationarity for a stochastic process states
that all the moments of the probability distribution calculated for any realization
or time series of that process be constant (in a statistical sense) from series to series.
In practice, a sufficient condition is that of reduced stationarity in which the mean
and variance of a time series depend only on the length of that time series and not
on the absolute time. Sometimes a process exhibits stationarity properties if examined over sufficiently short time series. Usually, however, very low frequency oscillations will be present in the data as long term trends, which can cause a serious
problem by overemphasizing the relative importance of the lowest frequency estimates. To reduce these effects, one should, in most cases, eliminate the very low
frequency fluctuations either by least squares trend removal or by prewhitening with
a difference filter (7, 15, 23).

Nonstationarity of the variance in certain frequency bands will cause large variations in the spectral estimates at those frequencies. In the extreme case, estimates
of the spectrum for different segments of data will be totally inconsistent. However,
where the spectrum is varying in some uniform manner, useful information can be
obtained concerning changes in the behavior of the system or stochastic process
itself.

Multiple Time Series

The spectrum of a single variable is often of limited use or interest; rather, one is
usually more interested in comparing the spectra for different variables, especially
if the object is to determine causal relationships between variables. In such cases,
one would use cross-spectral analysis techniques where the covariation is estimated
between pairs of variables as a function of frequency.

CROSS-COVARIANCE FUNCTION   The sample cross-covariance function (ccf) for
two series of N points, x_t and y_t, with zero means, is defined in a manner completely
analogous to the autocovariance function (acf):

$$c_{xy}(k) = N^{-1}\sum_{t=1}^{N-k} x_t y_{t+k}$$
$$c_{xy}(-k) = N^{-1}\sum_{t=1}^{N-k} x_t y_{t-k}$$
$$0 \leq k \leq (M-1), \qquad\qquad 9.$$

where k is the lag number and $(M-1)$ is the maximum lag. Unlike the acf, the ccf
is not an even function of lag, nor is it necessarily a maximum at zero lag. As an
example, consider interacting populations of phytoplankton and zooplankton in a
bay or fjord. If, as is often the case, the "blooms" in zooplankton lag those in the
phytoplankton by about one month, then a sample ccf between two time series of
their biomasses would also have its maximum at a lag of one month.

CROSS-SPECTRA AND COHERENCE   The simple cross-covariance function
$c_{xy}(k)$ gives no information about the correlation between two series as a function
of frequency. However, the cross-spectrum between two series can be calculated in

a manner analogous to that for power spectra. The coherence spectrum and the phase spectrum, derived from the cross-spectrum, correspond to knowledge of both the maximum ccf and the corresponding lag for each frequency in the cross-spectrum.

Suppose that we have calculated the smoothed power spectral estimates,

$$S^{\bullet}_{xx}(f_i) \text{ and } S^{\bullet}_{yy}(f_i), (f_i = i/(2\Delta M), i = 0, 1, \ldots, M),$$

for the two series x_t and y_t, and also the cross-covariance function at positive and negative lags. Even and odd functions can be formed from the cross-covariance function:

$$l_{xy}(k) = \tfrac{1}{2}[c_{xy}(k) + c_{xy}(-k)] \text{ and}$$

$$q_{xy}(k) = \tfrac{1}{2}[c_{xy}(k) - c_{xy}(-k)].$$

We can now Fourier transform $l_{xy}(k)$ and $q_{xy}(k)$ to give the smoothed cospectral estimate $L^{\bullet}_{xy}(f)$ and the smoothed quadrature spectral estimate $Q^{\bullet}_{xy}(f)$:

$$L^{\bullet}_{xy}(f) = 2\Delta[l_{xy}(0) + 2\sum_{k=1}^{M-1} l_{xy}(k)w(k)\cos 2\pi\Delta fk]; \qquad 0 \le f \le f_q,$$

$$Q^{\bullet}_{xy}(f) = 4\Delta\sum_{k=1}^{M-1} q_{xy}(k)w(k)\sin 2\pi\Delta fk; \qquad 1/(2\Delta M) \le f \le f_q, \qquad 10.$$

$$Q^{\bullet}_{xy}(0) = Q_{xy}(f_q) = 0.$$

The cospectrum gives the in-phase correlation at a given frequency between two series, and the quadrature spectrum gives the correlation at a given frequency between the two series where the time axis of one has been shifted by a quarter of a wavelength. As an example, the sine and cosine functions are in perfect quadrature.

Finally, the squared coherency estimate

$$K^{\bullet}_{xy}(f)^2$$

and the smoothed phase estimate

$$F^{\bullet}_{xy}(f)$$

can be formed from the cospectral and quadrature spectral estimates,

$$K^{\bullet}_{xy}(f)^2 = [L^{\bullet}_{xy}(f)^2 + Q^{\bullet}_{xy}(f)^2]/[S^{\bullet}_{xx}(f)S^{\bullet}_{yy}(f)]$$

and   11.

$$F^{\bullet}_{xy}(f) = \arctan[-Q^{\bullet}_{xy}(f)/L^{\bullet}_{xy}(f)], \text{ for } 0 \le f \le f_q.$$

The phase is for the Fourier component of the y-series relative to that of the x-series. As in the case of the simple cross-correlation function, the significance level must be determined with care (4, 18, 23). In practice, the behavior of the phase function can usually be used to indicate the significance of the coherence-squared function. Where the phase is a smooth function of frequency, the squared coherence is usually significantly different from zero; where the phase function oscillates rapidly with frequency, the squared coherence is usually low and not significant.

The purpose of cross-spectral analysis is sometimes to determine cause and effect relationships for model purposes. In such instances, it is preferable to form a gain or transfer function from the cross spectrum rather than calculate the squared coherence (8).

EXAMPLES OF SPECTRAL ANALYSIS IN ECOLOGY

Analysis of Natural Population Dynamics

INSECTS AND THEIR PARASITES   Bigger (5) has examined by Fourier analysis the records from six years of daily suction trap catches of the coffee leaf-miners *Leucoptera meyricki* and *L. caffeina*. Some 88% of the variance in his original data may be explained by a combination of processes acting at four principal frequencies. Analysis of the simultaneous counts of leaf-miners and some half-dozen hymenopterous parasites showed that the amplitudes and phase relationships of their cycles were consistent with the hypothesis that parasitism was a key factor controlling the abundance of the leaf-miners. On the basis of these results, a simple analytical model of the host-parasite interaction was developed which was useful in evaluating strategies of pest control. Interpretation of the biological significance of the separate harmonic terms revealed that over a ten-year period the species composition of the parasite complex had been readjusting from a state of presumed imbalance induced by the use of DDT in the 1950s. Various other insect interactions were noted in which the same method might be usefully employed.

BLUE-GREEN ALGAE IN THE TROPICAL OCEAN   Steven & Glombitza (42) examined the records of plankton samples taken at 2-week intervals over 3 yr from a station 9 km west of Barbados in 460 m water. In the surface layer (5 m), most of the chlorophyll was associated with the blue-green alga *Trichodesmium thiebaudii*. Inspection of the data indicated periodicity in the record, and autocorrelation analysis confirmed this, fixing the periodicity at 120 days; variations in chlorophyll and in numbers of filaments of *Trichodesmium* were strictly in phase. It was considered that the cycle was the free-running type, being independent of the annual solar cycle, since the annual variations of day length, surface temperature, and salinity are small at the station studied; the blooms at the surface were thought to be regenerated from a seed population near the base of the euphotic zone, which had more or less permanent access to an essential growth factor.

SHORT-TERM RESPONSES OF PHYTOPLANKTON POPULATIONS   Zlobin (50) attempted to find evidence for oscillations in the responses of natural phytoplankton population on time scales shorter than one day. Short-term fluctuations in oxygen concentration and temperature in the upper layer of the ocean have been treated by Fourier analysis. Spectral lines at 6- and 12-hr periods were attributed to variations in solar illumination; a 4-hr peak was attributed to temperature, but a peak at 2.4-hr was thought to be due to a strictly biological oscillation.

The Spatial Organization of Phytoplankton Populations

The first ecological application of spectral analysis to the determination of scales of organization in space was probably that involving the spatial structure of phytoplankton populations. The development of this work, in which chlorophyll concentration was used as an index of phytoplankton abundance, relied heavily on technological progress in automatic sampling.

Spatial heterogeneity is increasingly recognized as a crucial factor in controlling the stability of populations (41); for many organisms (including commercial fish species), with a planktonic larval stage, it is thought that survival of the larvae is critically dependent on their finding a higher than average food concentration at some key phase of their life history (24). Attempts to relate the observed structure in the biological populations to the physical structure of the environment in which they live have been particularly fruitful.

Early attempts (32) used point sampling at a single depth (5 m) in the coastal zone. The number of data points in each time series (80) was marginal for a spectral analysis, and only a narrow observational window (350 m to 30 km) was accessible with this experimental design, which was optimal for the resources available at the time. In addition to chlorophyll concentration (measured on filtered samples in the laboratory), time series were also collected for temperature and salinity as properties of the physical environment.

Both the precision and resolution of the statistical estimates were improved when an automatic chlorophyll measuring system suitable for this application became available (29). The width of the observational window was also increased. By use of this apparatus, the fluctuations in chlorophyll concentration at a fixed point in the Gaspé current, Gulf of St. Lawrence, were analyzed (29). Under certain conditions, it is possible to consider variations in time as equivalent to variations in space (e.g. Taylor's frozen field hypothesis). If this is permissible for the experiment in question, the spectral window examined was from 10–1000 m for each of three time series having 600–700 data points. The slope of the variance spectrum of chlorophyll concentration corresponded exactly to that which would be predicted for a variable under the control of turbulence homogeneous and isotropic in three dimensions according to the theory of Kolmogorov (25), although the assumptions on which the theory is based are not strictly justified in the present case.

An important point is illustrated here: spectral analysis can be used to test the predictions of theoretical analyses, particularly those dealing with the relationship between physical and biological structure in aquatic media. Many of the theoretical treatments in physical oceanography are based on dimensional arguments, and these often yield suggestions about spectral shapes as their primary predictions. Theoretical marine ecology, pursuing the same line, can end up with similar predictions; spectral analysis is the only relevant method of verification.

A coherence analysis (14) between temperature and chlorophyll for the same experiment showed significant correlation between these two quantities for most scales examined, which may be taken as a further indication of the control of the phytoplankton concentration by the turbulence. The coherence dropped off rapidly at length-scales of a few meters, where the variance in both parameters decreased to insignificant values. In another study (30), transects 50–80 km long were made in the Gulf of St. Lawrence to study, in particular, the low frequency component of the variability in chlorophyll concentration. Coherence between temperature and chlorophyll was reduced below significance for scales longer than $\sim$5 km (14).

The studies mentioned so far have been limited in that they were made at only a single depth. In 1973, a transect 16-km long was run in the Gulf of St. Lawrence with time series of chlorophyll and temperature recorded at two depths (5 and 9 m) simultaneously (14). For this work, the response time of the automatic measuring equipment was decreased to give a sampling resolution of 3.2 m, and attention focused on the higher frequency variability. Inspection of the raw time series (Figure 2) shows that although chlorophyll and temperature are correlated at each depth, there is no apparent correlation between signals at the two depths.

In both the temperature and chlorophyll spectra (Figure 2), there is a change of slope at a scale $\sim$100 m, the slope being steeper for scales smaller than this. The coherence between temperature and chlorophyll became insignificant at about the same length-scale, and the phase relation between the two time series also became random at the same scale (14).

In our studies on the spatial structure of phytoplankton populations, we have used various experimental designs, each one giving particular information about a part of the possible spectrum of variations in chlorophyll concentrations. However, to date, using purely shipboard measurements, we have investigated but a narrow band of the spectrum of structures that may exist. We have used spectral analysis to relate biological and physical structures, and we make the tentative conclusion that there exists a range of length-scales (below $\sim$5 km) for which chlorophyll behaves as a passive scalar whose local abundance is controlled by physical transport processes; for length-scales longer than $\sim$5 km it appears that the role of phytoplankton growth in promoting spatial heterogeneity dominates over that of the physical transport processes in eroding it.

Between $\sim$5 km and $\sim$100 m, where the coherences between chlorophyll and temperature are high, we believe that the observed variations in both signals result primarily from physical transport processes, in particular from motions associated with the internal wave field. Beyond $\sim$100 m, where both spectra decrease quickly and the coherence becomes very low, fluctuations in both parameters are totally controlled and damped out by the turbulent diffusion [note also that, based on theoretical arguments (31), the smallest patch of phytoplankton that could maintain itself against diffusion would have a length-scale of $\sim$50 m]. That the coherence in this region is small is not important, because variance-conserving plots $[fS^* \ (f)$ versus $\ln f]$ of either parameter would show that the fraction of the total variance occurring beyond about 50 m is insignificantly small.

Spatial spectra of chlorophyll and the spectra of current fluctuations from a stationary meter mounted beneath a subsurface buoy have also been measured (34) in Lake Tahoe, California (500 km^2 in area and 500 m in depth). These spectra are thought to be free from contamination by internal waves. For length-scales less than

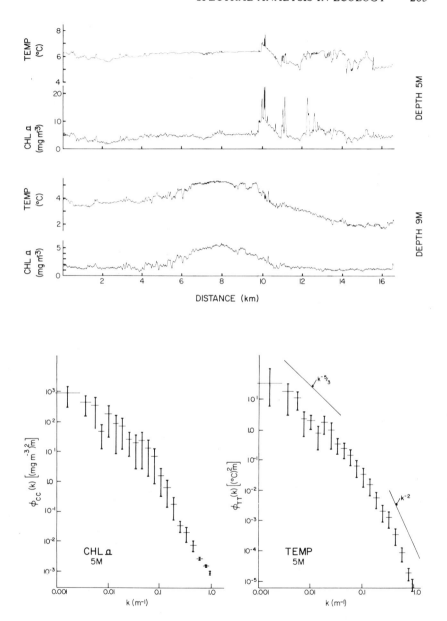

Figure 2 Upper panel: Series of chlorophyll and temperature obtained simultaneously at depths of 5 and 9 m over a 16.5 km transect. The sampling interval was 1 sec or 3.2 m. *Lower panel:* Variance or power spectra for two series at 5 m plotted on logarithmic axes. The $k^{-5/3}$ and k^{-2} lines are not fitted, but are presented only for reference. From Denman & Platt (14).

100 m, both chlorophyll and current spectra show similar negative power-law dependence with slope of approximately $-5/3$ (Figure 3). For length-scales greater than 100 m, the spectra are quite different. Variance of chlorophyll continues to increase with decreasing wavenumber. Current fluctuations, on the other hand, show a tendency to level off for longer wavelengths. For length-scales below 100 m, the distribution of phytoplankton seems to be controlled directly by turbulence. The divergence between the two spectra for scales more than 100 m is taken to indicate that in this region different processes contribute to the variance of phytoplankton and momentum, i.e. that spatial variability in the biological parameters such as cell growth and community structure have an important role in shaping the chlorophyll distribution. The requisite spatial inhomogeneities have already been observed for phytoplankton community structure in lakes (36) and for turnover time of phytoplankton in the sea (33).

Finally, in this section we mention the recent calculation of variance spectra of individual phytoplankton species abundance in Lake Tahoe (37). For the nannoplanktonic diatom *Cyclotella stelligera,* for example, a characteristic length-scale of ∽220 m was found.

Terrestrial Plant Communities

Time-series analysis finds an indirect application to the measurement of the productivity (photosynthetic rate) of plants on land. The method is to estimate the actual flux of carbon dioxide from the air into the plants. A knowledge is required of the vertical gradient of CO_2 in the air above the plants and also of the coefficient of turbulent diffusion of CO_2. Spectral analysis of the turbulent fluctuations in CO_2 concentration could be used to estimate the diffusion coefficient if a sufficiently fast sensor were available, but usually it is assumed that the diffusion of CO_2 is equal to that of momentum, and fluctuations in wind speed are measured to deduce the momentum transfer. This method has been applied extensively to both agriculture and forestry (26, 39).

For terrestrial plant communities, Hill (20) has given a useful comparison of several methods (including spectral analysis) for determining the intensity of spatial pattern.

Spectral Analysis and Perturbed Ecosystems

Eisner (17) suggested a method, based on spectral analysis, for doing experiments on natural ecosystems. Experiments are difficult to make on natural systems because too many variables are outside the control of the experimenter and because the controlled variable has to be changed by an unrealistically large amount to produce a measurable change that can be related unambiguously to the experimental perturbation. Eisner pointed out that measurable changes should be induced in the system with a rather modest stimulus, if the stimulus is applied periodically. In other words, he suggested perturbing the ecosystem with a forcing function that has a narrow Fourier spectrum with peak frequency chosen to be a frequency unrepresented in the undisturbed system. The applicability of such a technique is much enhanced for systems in which the fluctuations are strongly non-Gaussian, i.e. for which the

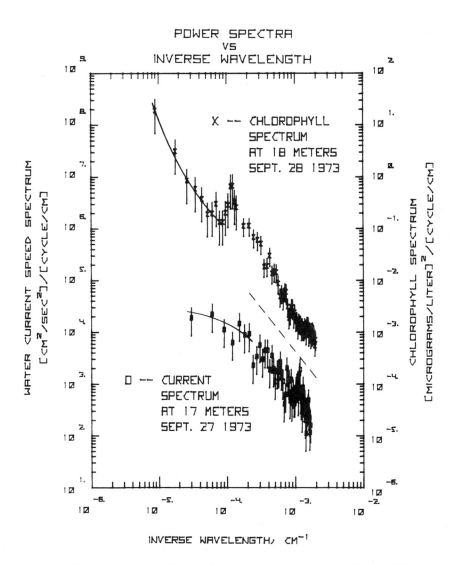

Figure 3   Variance spectra for chlorophyll and current speed from Lake Tahoe, California. Error bars are 80% confidence limits. The broken line has the slope –5/3. From Powell et al (34).

variance spectrum is not flat. This would certainly be true of living systems if we accept a representation of them as an assemblage of nonlinear oscillators. In fact, Eisner claimed that his approach could be applied to judge the degree of nonlinearity in the system under study. The method is interesting but, as yet, awaits experimental trial.

Historical Ecology and Paleoecology

A particularly fascinating application of spectral analysis in ecology deals with fluctuations in the growth of trees as measured by relative thickness of tree rings. In such studies, time series of up to 3000 yr can be established for the relative annual growth of individual trees (9). For a given relative width of tree ring, the probability distribution of corresponding paleoclimate can be constructed (43), and fluctuations in tree ring growth correlate well with fluctuations in annual precipitation and annual mean temperature.

Variance spectra for tree ring width can be computed for groups of trees in a particular region and comparison made between the spectra for different parts of the world. These comparisons reveal several interesting features (9). First, there is a relatively universal peak with a period of from 2–3 yr, present on nearly every tree ring spectrum computed, including one for Miocene trees. The widespread persistence of this peak suggests that it is associated with some external causative factor present everywhere on the earth. One possibility is the annual variation in the earth's magnetic field. Second, significant subsidiary peaks sometimes occur. Their periods vary from area to area over an interval of 5–40 yr. These peaks are thought to be essentially ecological in origin, and may represent the locally possible modes of response to a (perhaps random) external forcing function. In specimens of sequoia from California, the importance of these subsidiary peaks diminishes in the second millenium of life relative to the first millenium. This leads to a possible interpretation of the "ecological" peaks as representing competitive interactions between the tree and other organisms in its ecosystem, based on the argument that after age 1000, a tree should be sufficiently well established to be essentially independent of influence of competitors, herbivores, or parasites that might do considerable damage to younger trees. Third, there is no tendency for the occurrence of a spectral peak at 11 yr, which might be associated with annual variations in sunspot activity. Indeed, tree ring chronologies represent one of the few natural time series long enough to permit a test for a possible peak at the sunspot frequency to be made with any precision. Again, for a 155 yr period for which a continuous series of direct measurements of solar activity was available, no significant 11 yr peak could be found in the cross-spectrum of sunspots versus tree rings (sequoia in California).

Stochastic Models and Time Series in Ecology

Time-series data may be collected in ecology for the express purpose of constructing stochastic models of the process under consideration, that is, of constructing models that predict the probability that a future value of the measured variate will fall between two specified limits. We thus approach the important problem of prediction in the ecosystem, but from a probabilistic rather than a deterministic viewpoint. In this case, the time series of observations is regarded as but one realization from an

infinite ensemble of corresponding time series that could be generated by the stochastic process.

A central objective of such models is the elucidation of the dynamic system
response (transfer function) which translates the time series of the forcing variable
(input) into that of the variable of interest (output). The determination of the
transfer function in the presence of noise is a difficult problem whose solution
requires the application of the methods of spectral analysis. Once the transfer
function between two variables is known, past values of both the input and output
series may be used to predict future values of the output variable, improving the
precision of the forecasting (8).

An ecological application of these techniques is being conducted by Ford, Milne,
and Deans (Unit of Tree Biology, Institute of Terrestrial Ecology, Midlothian,
Scotland). The effect of the physical environment on the growth of a young Sitka
spruce forest, considered as a dynamic control system, is being investigated, as is
the general utility of correlation (time domain) and spectral (frequency domain)
techniques in forest growth modeling. Correlation functions are being used to test
the proposal that day to day (and shorter) weather and growth patterns can be
modeled as stochastic variables in time. In addition, the cross-correlation function
has been used to measure phase lags between the annual growth patterns of shoots
on the same tree and on different trees.

Potential Use of Spectral Analysis as an Objective Test of Population Models

Manipulation of laboratory populations of an organism is a common way to study
its population dynamics. In recent years, such laboratory studies have often been
accompanied by a simulation model which purportedly describes the numerical
fluctuations of the manipulated population. Interpretation of these simulation models is rather subjective, and their impact is somewhat blunted by the lack of an
objective criterion to test their validity. It would be desirable to have a quantitative
index by which simulations could be judged and through which a choice could be
made between alternative versions of the model; because the simulation output
is usually in the form of a time series of some population parameter, it seems
natural to suppose that we might look to spectral analysis to provide our test
statistic.

In an interesting paper on laboratory populations of *Drosophila,* Shorrocks (40)
used autocorrelation analysis to find evidence for regular fluctuations in abundance
that were independent of regular change in any environmental variable. Having
shown that this periodicity depends at least partly on an inherited factor related to
the differential fertility of adults raised in crowded or uncrowded conditions, he used
this conclusion to construct a simple simulation model, the output of which is
compared with the original experimental time series. But he was content to note the
"remarkable similarity" between the two series, stopping short even of finding the
dominant period in the reconstructed one.

We suggest that a direct, objective way to compare theory and experiment in such
cases might be by means of a cross-correlation analysis between the two series,
checking for the strength of the correlation at the frequencies of interest and also

for the phase relationships if these are thought to be important. At the very least, this procedure seems superior to the tests available at present.

THE FUTURE OF SPECTRAL ANALYSIS IN ECOLOGY

It has been shown above that ecosystems may be represented as ensembles of coupled nonlinear oscillators and that their typical response to excitation or perturbation, whether random or periodic, is to execute oscillations in both temporal and spatial domains. By virtue of its own intrinsic resonant frequencies, and through dynamic interaction with other oscillating components of the system, the variable of interest may show simultaneous fluctuations at many frequencies. If we wish to decompose the data record of this variable to discover its constituent periodicities, be they spatial, temporal, or both, we must have recourse to spectral analysis.

Given the fundamental importance of the concepts involved, and the richness of the potential reward, one might ask why spectral analysis has not been, and is not being, applied with more intensity to ecological problems. One reason is the relative unfamiliarity of the method among ecologists; after all, spectral analysis has been developed from scratch in communications theory only within the last 20 yr. A second reason is the basic difficulty of the subject and the lack of textbooks that do not assume a high level of mathematical sophistication in the reader. Again, most of the effort in ecology in the past 15 yr has been dissipated in the cause of reductionism. There has been an aversion among ecologists to take a holistic view of the systems they study, and a tendency to suppress variability in favor of the measurements of mean concentrations and fluxes to put into their narcissan system models. Only in the last few years has this trend begun to reverse itself with the gradual realization of the inutility of systems models and the appearance of works treating living systems as nonlinear thermodynamic systems (19, 49).

We hope that these obstacles to progress, if not completely removed, have at least been moderated by our review. But there remains another problem, and that is the difficulty of obtaining suitable data records of ecological variables for the time-series analysis. That this is not impossible is demonstrated by the variety of successful studies we have quoted as examples of the application of spectral analysis in ecology. It is a question of choosing carefully the variables to measure and of designing the sampling program with the final analysis in mind, rather than trying post facto to interpret an inadequate set of data. The sampling will become easier as automatic measuring and recording devices become more available for more variables of ecological interest. As has been illustrated above for the studies on the spatial structure of phytoplankton populations, progress has gone hand in hand with technical refinements in measuring.

While a considerable variety of problems has already been tackled, the number of published applications is still very small; the surface has barely been scratched. We expect with confidence that the next 10 or 15 years will see the development of the field of spectroscopy of the ecosystem: the identification and interpretation of periodicities in ecosystem behavior, the characterization of ecosystems according to their fundamental scales in space and time, and the investigation of the implications for perturbed systems by the mathematical method of normal mode analysis. The rate of expansion will depend on whether, and how fast, the nonlinear oscillator representation of living systems is accepted as a new paradigm for theoretical biology (49).

Literature Cited

1. Bartlett, M. S. 1948. Smoothing periodograms from time-series with continuous spectra. *Nature* 161:686–87
2. Bartlett, M. S. 1953. *An Introduction to Stochastic Processes with Special Reference to Methods and Applications.* Cambridge, Engl.: Cambridge Univ. Press. 362 pp.
3. Bendat, J. S., Piersol, A. G. 1966. *Measurement and Analysis of Random Data.* New York: Wiley. 390 pp.
4. Benignus, V. A. 1969. Estimation of the coherence spectrum and its confidence interval using the Fast Fourier Transform. *IEEE Trans. Audio Electroacoust.* Au-17(2):145–50
5. Bigger, M. 1973. An investigation by Fourier analysis into the interaction between coffee leaf-miners and their larval parasites. *J. Anim. Ecol.* 42:417–34
6. Bingham, C., Godfrey, M. D., Tukey, J. W. 1967. Modern techniques of power spectrum estimation. *IEEE Trans. Audio Electroacoust.* Au-15(2):56–66
7. Blackman, R. B., Tukey, J. W. 1958. *The Measurement of Power Spectra From the Point of View of Communications Engineering.* New York: Dover. 190 pp.
8. Box, G. P., Jenkins, G. M. 1970. *Time Series Analysis–Forecasting and Control.* San Francisco: Holden-Day. 542 pp.
9. Bryson, R. A., Dutton, J. A. 1961. Some aspects of the variance spectra of tree rings and varves. *Ann. NY Acad. Sci.* 95:580–604
10. Cardon, S. Z., Iberall, A. S. 1970. Oscillations in biological systems. *Curr. Mod. Biol.* 3:237–49
11. Chance, B., Pye, E. K., Ghosh, A. K., Hess, B., eds. 1973. *Biological and Biochemical Oscillators.* Johnson Research Foundation Colloquium. New York: Academic. 534 pp.
12. Cochran, W. T. et al 1967. What is the fast Fourier transform? *IEEE Trans. Audio Electroacoust.* Au-15:45–55
13. Cooley, J. W., Tukey, J. W. 1965. An algorithm for the machine calculation of complex Fourier series. *Math. Comput.* 19:297–301
14. Denman, K. L., Platt, T. 1974. Coherence in the horizontal distributions of phytoplankton and temperature in the upper ocean. *Proceedings of Sixth Liège Colloquium on Ocean Hydrodynamics, Mémoires de la Société Royale des Sciences de Liège,* 6 sér. 8:19–30
15. Durbin, J. 1962. Trend elimination by moving-average and variate-difference filters. *Bull. Int. Statist. Inst.* 39:131–41
16. Edge, B. L., Liu, P. C. 1970. Comparing power spectra computed by Blackman-Tukey and fast Fourier transform. *Water Resour. Res.* 6(6):1601–10
17. Eisner, E. 1971. Experiments in ecology: a problem in signal extraction. In *Statistical Ecology Volume 2: Sampling and Modeling Biological Populations and Population Dynamics,* ed. G. P. Patil, E. C. Pielou, W. E. Waters, 237–51. Based on the Proceedings of the International Symposium on Statistical Ecology, New Haven, Connecticut, August 1969. Philadelphia: Pa. State Univ. Press. 420 pp.
18. Enochson, L. D., Goodman, N. R. 1965. Gaussian approximations to the distribution of sample coherence. *US Air Force Flight Dynamics Lab. Tech. Rep.,* 65–57. Wright-Patterson Air Force Base, Ohio. 27 pp.
19. Glansdorff, P., Prigogine, I. 1971. *Thermodynamic Theory of Structure, Stability and Fluctuations.* New York: Wiley. 306 pp.
20. Hill, M. O. 1973. The intensity of spatial pattern in plant communities. *J. Ecol.* 61:225–35
21. Hinich, M. J., Clay, C. S. 1968. The application of the discrete Fourier transform in the estimation of power spectra, coherence, and bispectra of geophysical data. *Rev. Geophys.* 6(3):347–63
22. Jenkins, G. M. 1961. General considerations in the analysis of spectra. *Technometrics* 3(2):133–66
23. Jenkins, G. M., Watts, D. G. 1968. *Spectral Analysis and Its Applications.* San Francisco: Holden-Day. 525 pp.
24. Jones, R. 1973. Density dependent regulation of the numbers of cod and had-

dock. *Rapp. P. V. Reun. Cons. Perm. Int. Explor. Mer* 164:156–73

25. Kolmogorov, A. N. 1941. The local structure of turbulence in an incompressible viscous fluid for very large Reynolds number (In Russian). *Dokl. Akad. Nauk USSR* 30: 299–303

26. Monteith, J. L. 1968. Analysis of the photosynthesis and respiration of field crops from vertical fluxes of carbon dioxide. In *Functioning of Terrestrial Ecosystems at the Primary Production Level*, 349–58. *Proceedings of UNESCO Symposium (Copenhagen)*, UNESCO (Belgium). 516 pp.

27. Nicolis, G., Auchmuty, J. F. G. 1974. Dissipative structures, catastrophes and pattern formation: a bifurcation analysis. *Proc. Nat. Acad. Sci. USA* 71: 2748–51

28. Parzen, E. 1961. Mathematical considerations in the estimation of spectra. *Technometrics* 3(2):167–90

29. Platt, T. 1972. Local phytoplankton abundance and turbulence. *Deep Sea Res.* 19:183–87

30. Platt, T. 1972. The feasibility of mapping the chlorophyll distribution in the Gulf of St. Lawrence. *Fish. Res. Bd. Can. Tech. Rep. 332*, 8 pp. + 20 fig.

31. Platt, T., Denman, K. L. 1974. A general equation for the mesoscale distribution of phytoplankton in the sea. See Ref. 14, 31–42

32. Platt, T., Dickie, L. M., Trites, R. W. 1970. Spatial heterogeneity of phytoplankton in a near-shore environment. *J. Fish. Res. Bd. Can.* 27(8):1453–73

33. Platt, T., Filion, C. 1973. Spatial variability of the productivity: Biomass ratio for phytoplankton in a small marine basin. *Limnol. Oceanogr.* 18(5): 743–49

34. Powell, T. M. et al 1975. Spatial scales of current speed and phytoplankton biomass fluctuations in Lake Tahoe. *Science.* In press

35. Prigogine, I., Nicolis, G. 1971. Biological order, structure and instabilities. *Q. Rev. Biophys.* 4:107–48

36. Richerson, P., Armstrong, R., Goldman, C. R. 1970. Contemporaneous disequilibrium, a new hypothesis to explain the "paradox of the plankton." *Proc. Nat. Acad. Sci. USA* 67(4): 1710–14

37. Richerson, P. J., Dozier, B. J., Maeda, B. R. 1975. The structure of phytoplankton associations in Lake Tahoe

38. Segel, L. A., Jackson, J. L. 1972. Dissipative structure: An explanation and an ecological example. *J. Theor. Biol.* 37:545–59

39. Sestak, Z., Catsky, J., Jarvis, P. J. 1971. *Plant Photosynthetic Production. Manual of Methods.* The Hague. Junk. 818 pp.

40. Shorrocks, B. 1970. Population fluctuations in the fruit fly (*Drosophila melanogaster*) maintained in the laboratory. *J. Anim. Ecol.* 39:229–53

41. Steele, J. H. 1974. Spatial heterogeneity and population stability. *Nature* 248 (5443):83

42. Steven, D. M., Glombitza, R. 1972. Oscillatory variation of a phytoplankton population in a tropical ocean. *Nature* 237(5350):105–7

43. Stockton, C. W., Fritts, H. C. 1971. Conditional probability of occurrence for variations in climate based on width of annual tree rings. *Tree Ring Bull.* 31:3–24

44. Thompson, R. 1971. Topographical Rossby waves at a site north of the Gulf Stream. *Deep Sea Res.* 18(1):1–20

45. Tukey, J. W. 1961. Discussion, emphasizing the connection between analysis of variance and spectrum analysis. *Technometrics* 3(2):191–219

46. Turing, A. M. 1952. The chemical basis of morphogenesis. *Proc. R. Soc. London B* 237:37–72

47. Wiener, N. 1964. On the oscillations of non-linear systems. In *Stochastic Models in Medicine and Biology*, ed. J. Gurland, 167–74. *Mathematics Research Center Publications*, 10:393. Madison: Univ. Wisconsin Press

48. Welch, P. D. 1967. The use of fast Fourier transform for the estimation of power spectra: a method based on time averages over short, modified periodograms. *IEEE Trans. Audio Electroacoust.* Au-15:70–73

49. Yates, F. E., Marsh, D. J., Iberall, A. S. 1972. Integration of the whole organism—A foundation for a theoretical biology. In *Challenging Biological Problems: Directions Towards Their Solutions*, 25th Anniversary Volume. The American Institute of Biological Sciences, 110–32. New York: Oxford Univ. Press. 502 pp.

50. Zlobin, V. S. 1973. *Osnovi Prognozirovaniya Pervichnoi Produktivnosti Foticheskovo Sloya Okeana.* Murmansk. 515 pp. (In Russian)

THE POPULATION BIOLOGY OF CORAL REEF FISHES

❖4093

Paul R. Ehrlich

Department of Biological Sciences, Stanford University, Stanford, California 94305

INTRODUCTION

The development of self-contained underwater breathing apparatus (scuba) has, in the past few decades, provided population biologists with access to the shallow waters of the sea. The availability of scuba has also created an enormous sportdiving industry, with much attention paid to tropical areas.

A popular interest in the spectacular fish faunas of coral reefs, analogous to that shown in birds and butterflies, seems to be developing. This interest will have advantages for the professional scientist. Many observations and photographs will be made in field and aquarium by amateurs, and large markets will permit publication of what otherwise would have been unprofitable literature. For instance, an excellent guide to the common fishes of Caribbean reefs, designed for underwater use, has recently appeared (51).

Coral reefs provide diving biologists with a stunning diversity of organisms accessible in three dimensions; a diver is, in a sense, able to do the rough equivalent of flying over, through, and along the edge of a tropical rain forest. Most of the hermatypic (reef-building) corals live in less than 45 m of water (311), a depth that, coincidentally, is almost exactly the maximum recommended by the U.S. Navy for scuba diving with compressed air. Because their symbiotic zooxanthellae require light, no hermatypic corals live below 90 m (311), and even that depth is occasionally approached by scuba divers with more courage than good sense.

Coral reefs occur through the warm waters of the Indo-Pacific and the western Atlantic (primarily Caribbean) seas. Associated with them to one degree or another are the perhaps 6000–8000 species of fishes that are the subject of this review (41, 59).

Scientific work on the biology of reef fishes began relatively recently. Before World War II, pioneering work was done by a few biologists, especially William H. Longley, who used a diving helmet, took pictures of fishes with a camera enclosed in a watertight glass box with front and rear windows of plate glass, and took notes

211

on a wax-covered slate. The high quality of his work can be appreciated by examining the text and plates of Longley's "Systematic Catalogue of the Fishes of Tortugas, Florida," posthumously completed and edited by Samuel T. Hildebrand of the U.S. Fish and Wildlife Service (153).

After the war, a relatively small but dedicated group of marine biologists using scuba apparatus began to make the observations upon which this review is based. The bibliography indicates the rapid expansion of this field in the 1960s. I would like to single out one man, however, whose taxonomic and ecological work still forms a substantial portion of the significant literature, namely, John E. Randall of the Bernice P. Bishop Museum of Honolulu. Among diving biologists, Jack Randall is a legend in his own time, and the others will understand why I have chosen to dedicate this review to him.

SOME NOTES ON TAXONOMY AND DIVERSITY

The teleostean fishes of coral reefs are sometimes compared with butterflies; indeed many of the most common and beautiful chaetodontids are called butterflyfishes. There are certain similarities between butterflies and reef fishes, but there are striking differences as well. One of these differences lies in the taxonomic structure of the two groups. In general, reef fishes tend to show much more phenetic, ecological, and behavioral diversity than do butterflies, but within any type they show much less differentiation into arrays of closely related genera and species. While these differences to some degree may be due to divergent views of taxonomists working with invertebrates and vertebrates, in general they are reflections of the different patterns of phenetic similarities within the groups. The teleostean fishes comprise perhaps 20,000 species (59), while the butterflies are a somewhat smaller group with 12,000–15,000 (82). The butterflies, a superfamily of the order Lepidoptera, are divided into five families (78), or an average of almost 3000 species per family. A recent work (120) divided the teleosts into 30 orders and 413 families (some 50 species per family).

The taxonomic structure of the teleosts shows up clearly in the diversity of reef fish faunas. For instance, in a study of the fishes of small section of Tague Bay Reef in the Virgin Islands, Ogden (177) found 125 species belonging to 84 genera and 44 families (1.5 species per genus, 1.9 genera per family). The genus most heavily represented was *Haemulon,* the grunts (Pomadasyidae), with six species; the wrasse genus *Halichoeres,* parrotfishes of the genus *Sparisoma,* and pomacentrids of the genus *Eupomacentrus* had five species each. Five other genera were represented by three species, and all the remaining by two or one.

In the Capricorn group at the southern end of the Great Barrier Reef (by no means the richest faunistic area of that reef) the diversity of fishes is much greater (116; B. C. Russell and F. H. Talbot, personal communication). Some 850 species are present, representing about 297 genera and 84 families (2.4 species per genus; 3.5 genera per family). Several genera contain a dozen or more species; *Chaetodon,* 25; *Scarus,* ±24; *Apogon,* ±23; *Pomacentrus,* 19; *Acanthurus* 13; *Halichoeres,* 12.

In contrast, a sample from a rich butterfly fauna in northwest India (260) contained 87 species belonging to 46 genera and four families (1.9 species per genus, 11.5 genera per family). One genus was represented by five species, four genera by four species, and seven genera by three species. Thus, while the numbers of congeneric species in the three faunas are not strikingly disparate, the butterflies have many more phenetically similar groups living sympatrically than do the fishes. The probable reasons for this difference are discussed below.

Reef fishes, like most organisms, include groups considered taxonomically difficult. Considerable confusion in the past has been caused by the dramatic changes in form and color that occur during the development of many of them. For instance, individuals of the bluehead wrasse, *Thalassoma bifasciatum,* are among the most common reef fishes of the Caribbean. The vast majority of individuals are small (<5 cm), with yellow backs and white bellies divided by a black lateral stripe. A second phase of development is comprised of larger individuals that have the black stripe lightened and broadened into a series of rectangular blotches; many in this phase reach 12 cm in length. A third phase is totally different. It consists of large males, generally 10 cm or longer, that have blue heads followed by two vertical black bands and separated by a narrow white or pale blue band just behind the pectorals. The rear half of the body varies in color from yellow-green to blue, and the tail develops dorsal and ventral lobes not seen in the other phases. All fishes of this bluehead phase are males, and they make up only a few percent of individuals seen on reefs. Although the relationship of these forms was long suspected (150, 151) it was not confirmed until Stoll (281) showed that injection with methyl testosterone converted yellowphases to blueheads.

Fishes of all three phases may be sexually mature; individuals less than 4 cm long may spawn. The yellowphase spawns communally; bluehead males spawn with single females (208, 216). The unwary scientist, accustomed to the less varied sexual habits of terrestrial organisms, could well be misled into believing that he or she was dealing with populations of two quite distinct species. Indeed, complex color changes in the parrotfishes has led to the terminal phase ("termphase") male being described as a species distinct from the "midphase" fishes—smaller adult females and males that are very similar in appearance (204, 227, 248, 249, 306, 308). Different phases have been repeatedly given different specific names in other reef fish groups, for example in the wrasse genus *Halichoeres* (210).

While careful taxonomic work has removed much of the confusion caused by phase change in the scarids and labrids, other groups remain taxonomically refractory. One such group is the genus of the hamlets, *Hypoplectrus* (Serranidae), in which the species are "poorly defined" [(29); G. W. Barlow, manuscript]. These small relatives of the groupers are frequently found on reefs and usually occur singly; many individuals are encountered that do not "fit" descriptions in the literature. Whether this is a reflection of complex relationships among the populations with a lack of phenetically distinct species (as it appears to be in certain butterflies such as the *machaon* group of the genus *Papilio*) or whether it simply reflects inadequate taxonomic work cannot currently be determined.

POPULATION ECOLOGY

Despite some taxonomic confusion and occasional difficulty in making accurate specific identifications without killing individuals, reef fishes, in general, make excellent material for ecological investigations. After they have reached maturity, most species are easily distinguished visually from their nearest relatives; the young of many can also be identified (although usually with greater difficulty). Visual censusing and close observation is possible by day or night to a degree seldom possible in investigations of terrestrial organisms, and tagging can be done with relative ease, although wounds from tags may cause heavy mortality in some groups and impede growth in others (198). Promising new techniques for marking individuals with "dayglo" paints are now under development (J. C. Ogden, personal communication). It is even possible to use sonic tags to follow the movements of individuals, but at present this technique is satisfactory only for the largest reef fishes.

Population Structure

In spite of these advantages, there have been relatively few studies that throw light on the structure of reef fish populations. Extensive tagging programs have been carried out with diverse species on reefs off Bermuda (16), the Virgin Islands (198, 203), and Florida (270), leading to the general conclusion that most reef species are sedentary. These and other (220) results coincide with the impressions of divers who visit the same area repeatedly and learn to expect to see a certain bigeye (*Priacanthus cruentatus*) under the same staghorn coral, a familiar butterflyfish in the vicinity of certain holes in the reef (104), or the same group of cleaner wrasse (*Labroides phthirophagus*) at the same station day after day. Stories of pet groupers (Serranidae) living in certain holes on the reef are legion in popular diving magazines.

This conclusion is hardly surpising. Reef fishes are normally utterly dependent upon the reef for food and shelter from predators, and thus selection against wandering must be great. Many strongly territorial species, such as various members of the damselfish genera *Pomacentrus* and *Eupomacentrus*, defend very restricted areas of the reef (219). This close attachment to defended substrate, often casually observed, has been studied in detail in the Great Barrier Reef species *Pomacentrus flavicauda* (160). Individuals were observed to remain in areas roughly 2 m² over a five month study period. These home ranges were defended in their entirety not just against conspecifics, but against a wide variety of herbivorous and ominivorous species. Low was able to adduce considerable evidence that this interspecific territoriality functions to assure individuals of *P. flavicauda* an ample supply of their algal food. In *Eupomacentrus planifrons*, the size of the defended territory varied with the *species* of the intruder, but not with the size of the intruding individual (172); this led to the suspicion that different sizes of territories secure different resources.

In contrast, the interspecific territoriality of *Pomacentrus jenkinsi* in Hawaiian waters is reportedly applied indiscriminately to all other fish species (218), and *Eupomacentrus* spp. about 10 cm in length have vigorously attacked me in the shallow waters off St. Croix, Virgin Islands. The bicolor damselfish, *Eupomacentrus partitus,* also defends its territory against selected species, but the basis of its discriminations is less clear (169). The territorial behavior of these, and presumably other interspecifically territorial fishes, would appear to have a considerable influence on the home ranges of other reef fishes.

In some pomacentrids, such as *Eupomacentrus leucostictus* (46), the young are highly territorial. In others, such as *Abudefduf zonatus,* they are not, and adult territoriality may periodically be abandoned for clustering behavior (144). Pomacentrid fishes of the genus *Amphiprion* (7) live in an intimate mutualistic relationship with sea anemones and show territoriality in aquaria (97, 163, 183). In nature, however, territoriality may be either present or absent (7, 88, 164). In either case, the home range of these fishes, which appear to spend their entire lives associated with a sessile anemone, is extremely restricted. Allen (7) reported *Amphiprion* living more than 30 months with the same anemone. Spawning takes place in a substrate nest immediately adjacent to the anemone (7).

Other pomacentrids that are not territorial nonetheless have highly restricted home ranges on the reef. Sale's careful studies of the wide-ranging Indo-Pacific "banded humbug" *Dascyllus aruanus* (236, 237, 239, 240) showed that in less than 5% of observations were these fish found more than 1 m from an original capture site. Although these fishes are not territorial, agonistic behavior occurs within groups taking shelter in branched corals. The intensity of this behavior varies both with group size and the amount of shelter (and there is a significant interaction between them). This behavior influences patterns of movement among suitable areas of coral and tends to space individuals efficiently where suitable coral colonies are in short supply.

In contrast to the relatively straightforward home range and territorial arrangements of certain pomacentrids (e.g. *Pomacentrus jenkinsi* defending a fixed feeding territory both as juveniles and adults), other reef fishes show a bewildering variety of behavior in their utilization of space. Some Acanthuridae and Pomacentridae establish fixed feeding territories as young, but abandon them as they mature, sometimes forming pairs that defend temporary feeding territories (218). Other groups, such as the chaetodontids, do not normally appear to hold feeding territories (83, contrary to 218).

Surgeonfishes (23) may either hold small permanent or semipermanent territories (5–20 m²), hold temporary grazing territories, or be nonterritorial and school. Adults of the same species may show territorial behavior at one time or place and not at another. Juvenile acanthurids generally appear to be territorial (23, 183), with the possible exception of *Acanthurus triostegus* where the juveniles are reported both to be territorial (22) and nonterritorial (199, 234). In Hawaii, juveniles of this species are also unique in that they occupy intertidal pools (119).

The significance of different patterns of territoriality and schooling in acanthurids has been discussed (23, 24). The Acanthuridae are, along with the parrotfishes (Scaridae), some pomacentrids, balistids, kyphosids, and blenniids (129, 208), the major grazers on the reef. For instance, one guild grazes on the sand adjacent to the reef and is made up of relatively large, mobile species (*Acanthurus dussumieri, A. xanthopterus, A. mata*). They exploit an almost unlimited food resource, sifting diatoms and algae from the sand (142). Individuals of the sand grazer's guild are not territorial, but feed in heterotypic schools, quite likely as a predator defense mechanism (80). In contrast, the guild of small reef browsers, such as the spectacular *Acanthurus achilles* of the surge zone, has some highly territorial members. Presumably the reef browsers are defending an area necessary to provide them with an adequate supply of the multicellular algae that grow on the rock or coral substrate (142).

Most interesting is Barlow's (24) interpretation of solitary and schooling behavior in the manini (*Acanthurus triostegus*), the most common surgeonfish in the shallow waters of Hawaii. The "scalpel" at the base of the caudal fin that gives the surgeonfishes their name is very weakly developed in *A. triostegus,* presumably putting it at a disadvantage in social encounters (32). While in some situations manini schooling is an antipredator defense, in others it is a mechanism for gaining access to feeding surfaces defended by the dominant lavender tang, *Acanthurus nigrofuscus.* In essence, a schooling stream of manini swamps the tang's defenses. Barlow showed that schooling reduced the number of attacks on each manini by at least 30% per unit time spent grazing on the substrate. At Lizard Island on the Great Barrier Reef, *A. triostegus* has been observed using this technique to swamp the defenses of territorial pomacentrids (G. Anderson, P. Ehrlich, F. Talbot, and B. Russell, unpublished). Similarly, *Pomacentrus lividus,* defending its feeding territory in the Red Sea, can have its defenses swamped by invading schools of *Acanthurus sohal* (295).

The other prominent group of reef grazers, the parrotfishes (Scaridae), also show a spectrum of spatial behavior that varies between and within species. G. W. Barlow (manuscript) found two species of the genus *Sparisoma* more solitary than two species of *Scarus,* confirming earlier observations of generic difference in social behavior (308). Ogden & Buckman (180), working on the Caribbean coast of Panama, found stationary-territorial, stationary-nonterritorial, and moving flock fishes of *Scarus croicensis* in shallow water (<3 m). While midphase fishes may be territorial in shallow water, they are not in deeper water (at least not at Barlow's Puerto Rico study site in 15–30 m of water).

In shallow water, Ogden & Buckman (180) recorded *Scarus croicensis* foraging in heterospecific schools with one or two species of *Acanthurus,* the barred hamlet (*Hypoplectrus puella*), and, less frequently, the spotted goatfish (*Pseudupeneus maculatus*); I observed similar assemblages in the Grenadines and Virgin Islands. The scarids and acanthurids presumably gain predator protection from increased school size. The hamlets apparently benefit by enhanced feeding opportunities; they snap up small invertebrates stirred up by the milling herd. The goatfishes may gain some protection from predators, but probably do not gain a feeding benefit since they probe the substrate for invertebrates with their barbels and are often themselves followed by small fishes feeding on what they stir up.

In Panama, *Scarus croicensis* spends the days in the shallow waters of the reef and nights in deeper waters (180). Morning and evening migrations occur over narrowly defined routes and thus provide an excellent opportunity to census the population. This behavior is strikingly complementary to that of various nocturnally feeding grunts (primarily *Haemulon flavolineatum* and *H. plumieri*), goatfishes (*Mulloidichthys martinicus*), and snappers (*Lutjanus*), which spend their days resting in heterotypic schools on Caribbean reefs. These fishes leave the reef in a ritualistic series of movements along narrow routes just at last light. They reverse the process at dawn (J. C. Ogden and P. R. Ehrlich, unpublished). There is some evidence that the same fishes rest in the same schools day after day, and may feed together over sand flats at night. Fishes in these schools are of medium size; nothing is known of the distribution and movements of larger fishes that have outgrown the schools. Studies of the margate, *Haemulon album* (65), indicate similar behavior, with tagged fishes returning to the same daytime positions. Again, parrotfishes that spend their nights in offshore caves at Bermuda migrate daily to feeding grounds along the shore, using sun-compass orientation (309).

Although various aspects of the population structure of a few groups of reef fishes are being gradually elucidated, a great deal remains to be learned. Attempts should be made to tag and follow individuals over major portions of their lives on the reef, and spatial patterns in more groups should be carefully investigated. For instance, final interpretation of the spectacular patterns of butterflyfishes, so abundant on reefs in the Indo-Pacific, will probably have to await more detailed knowledge of their population structure.

Essentially nothing is known about the population structure of such prominent predators on other reef fishes as the carangids (jacks), aulostomids (trumpetfish), synodontids (lizardfish), and lutjanids (snappers). We are similarly ignorant of the activities of the predaceous fishes, such as the holocentrids (squirrelfishes), apogonids (cardinalfishes), and moray eels (Muraenidae), which spend their days sequestered in cavities in the reef (132), or of the abundant bottom-dwelling blennies (Blenniidae) and gobies (Gobiidae) that observations (264, 275) seem to indicate are quite sedentary. We also lack information on the often strangely shaped fishes of the order Tetraodontiformes (Plectognathi), so familiar to divers on reefs. These include the trigger and filefishes (Balistidae), trunkfishes (Ostraciontidae), puffers (Tetraodontidae), and porcupinefishes (Diodontidae), which are predators on invertebrates and, with the exception of the balistids, are relatively slow swimmers and presumably quite amenable to capture-recapture experiments.

Population Dynamics and Genetics

Results of tagging studies have not been used as the basis for estimates of population size, probably wisely so because of the small fraction of the populations marked and the probable violation of basic Lincoln Index assumptions (randomization, no traphappiness). Other methods of population estimation have been proposed (44), but none seem to have been employed systematically. Indeed, even anecdotal observations of changes in population size are infrequent. Ogden & Buckman (180), with the benefit of unusually accurate census data, recorded a decline in the population of *Scarus croicensis* coincident with the onset of the dry season. They hypothesized that this was related to a reduction in food supply resulting from a lowered level of sedimentation.

In any event, without more knowledge of the population structure of reef fishes, any observed change in numbers is difficult to interpret. From the point of view of the factors controlling both the size of populations and their genetic composition, it is essential that population units be recognized (83). If, for instance, in every generation, pelagic larvae of a species are widely dispersed over hundreds of miles of reef by currents, then essentially panmictic populations may extend over enormous areas (71). Whether this occurs frequently is questionable, for in many reef areas prevailing currents would at least tend to make movements of larvae more or less unidirectional. Thus the Hawaiian Islands populations of reef fishes with pelagic eggs and larvae must suffer enormous losses as their propagules are swept from the relatively localized coastal current system into the major western current set up by the trade winds (142).

One would expect, furthermore, that populations of such fishes around small oceanic islands would have difficulty maintaining themselves in the face of this attrition. Could, for instance, the acanthurid populations of isolated Johnston Island, "downstream" from the Hawaiian groups (117, 119), be maintained by a steady input of larvae from Hawaiian populations to balance the losses? The evidence seems to be against this (199). The Johnston population of manini, the most common Hawaiian acanthurid, is differentiated from that of Hawaii (194), and so is that of *Ctenochaetus strigosus,* the kole, another common Hawaiian tang (193).

Probably other factors conserve the Johnston fauna. Jones (142) suggested that, as Boden (28) found in Bermuda, there may be temperature and density factors that interact to create patterns of circulation tending to keep planktonic larvae from being carried out to sea. There is some evidence that local gyrals may trap small larvae that, when they are partially grown, may be able to escape and return to islands (235). In addition, "coast countercurrents, long back swirls and eddies" keep many California lobster larvae in the area of the adult distribution (141). Having pelagic larvae does not necessarily mean having huge, virtually panmictic, populations.

Many, if not most, dynamic units in reef fish populations are probably quite large, with a large pool of very young fishes available to reoccupy any segment of reef where larger fishes have suffered a catastrophic decline from storms, predation, or temporary exhaustion of food supply. On undisturbed large reefs these populations are likely to show the kind of stability of size often predicted and occasionally demonstrated (81) for tropical forest animal populations. And, indeed, Smith & Tyler (264) claimed there is no evidence of large fluctuations in abundance of reef fishes throughout the year. As discussed below, however, Australian biologists have observed dramatic fluctuations in species composition (and thus, of course, in the populations of different species) in small areas of reef.

Very recently, the work of M. Soulé (personal communication) and his collaborators has begun to give us a glimpse of the population genetics of reef fishes. Using the techniques of starch gel electrophoresis, he has examined 21–25 loci in populations of five pomacentrids—*Dascyllus aruanus, D. melanurus, Pomacentrus wardi, P. flavicauda,* and *Acanthrochromis polyacanthus*—in the southwest Pacific. *A. polyacanthus* is unusual in that it has no pelagic larval stage (D. R. Robertson, manuscript) and demonstrates parental care of its young on the reef. It shows a high degree of geographic phenetic differentiation, and Soulé has shown that this is accompanied by a high degree of genetic differentiation. Not only are gene frequencies different from population to population, but many alleles are also unique to one or a few populations.

In contrast, Soulé found that the widespread *D. aruanus,* which has pelagic larvae, shows virtually idential allele frequencies over a 3000 km transect from the southern part of the Great Barrier Reef of Australia to Madang, Papua New Guinea. In addition, a single sample of *D. melanurus* from Madang showed at most only one fixed allele difference from *D. aruanus. Pomacentrus wardi,* which also has pelagic larvae, appears to show a pattern similar to *D. aruanus,* with no fixed allelic differences among populations. Between the two species of *Pomacentrus* there is just one fixed allelic difference.

Soulé's data also indicate that geographically and ecologically marginal populations at the southern end of the Great Barrier Reef tend to be less polymorphic than the equatorial populations, a pattern consistent with what little is known for other vertebrates (268).

Soulé (268) and others are in the process of studying allozyme variation in the sergeant major, *Abudefduf saxatilis (sensu lato)* a circumtropical reef fish. With the exception of one or two esterase loci, Bermuda and Panama Atlantic populations seem virtually identical at about 28 loci. The eastern Pacific form, *A. troischelii,* virtually indistinguishable morphologically from *A. saxatilis,* differs from it by six fixed alleles. Like *A. saxatilis,* however, *A. troischelii* shows almost no genetic differentiation between the two sampled populations in the Revillagigedo Islands and the Pacific coast of Panama, more than three thousand kilometers apart. *A. "saxatilis"* from Madang, New Guinea is almost a perfect intermediate between *A. troischelii* from the Pacific and *A. saxatilis* from the Atlantic.

These data raise many questions about the role of gene flow in preventing the differentiation of populations. The following comments have arisen from discussions with Soulé. Except for *Acanthochromis,* which apparently lacks a between-reef dispersal stage, the pattern in the pomacentrids is consistent; unless there is complete isolation by emergent land, very little allelic differentiation appears in populations separated by as much as 3000–5000 km.

In sharp contrast, *Acanthochromis* populations on different parts of the Great Barrier Reef and the islands to the north show striking phenetic and genetic differentiation. In fact, there are much greater genetic differences among the *Acanthochromis* populations than between sympatric congeners in the genera *Dascyllus* and *Pomacentrus* and among populations of many terrestrial animals (83). Soulé contended that the relatively low rate of migration and gene flow in *Acanthochromis* probably facilitates geographic differentiation, although without a thorough ecological analysis we cannot rule out the possibility that selection pressures unique to *Acanthochromis* are causing its striking diversification.

Not only should patterns of differentiation among other reef fish populations prove of great interest, but so should comparisons of the level of genetic variability in these ancient (174) "climatically" stable communities with those of other communities—especially the tropical rain forest. Somero & Soulé (267) predicted that populations on reefs, in deep seas, and in rain forests should show more heterozygosity than those in less climactically stable regions, and they have presented some data to support this hypothesis in a comparison of the genetic variation in samples of fishes, including three reef species, *Dascyllus reticularus, Amphiprion clarkii* (both pomacentrids), and *Halichoeres* sp. (Labridae). These species seem to be more variable than those from areas (Antarctic, temperate estuarine) where one might expect strong directional selection. That hypothesis should be thoroughly tested in the future.

Understanding the genetics of reef fishes will require more, however, than assaying samples for allozyme variation. As indicated above, knowledge of levels of gene flow will be essential, and this means evaluating both the movements of individuals and the probabilities of migrant individuals successfully reproducing (83). Studies relating the frequency with which larvae are found in the oceanic plankton between populations to the differentiation of those populations (244) would be most useful. And, of course, thorough knowledge of gene flow levels will also involve understanding of the genetic systems of various reef species. Fundamental to both of these is spawning behavior and the fate of the eggs and newly hatched young. We have some information about both for a small proportion of reef fishes.

Pomacentrid fishes, both benthic feeders and those that forage in the water column, spawn in pairs on the substrate (3, 6, 43, 57, 72, 97, 99, 110, 111, 126, 144, 148, 170, 232, 238, 279, 285, 292; D. R. Robertson, manuscript). Their spawning behavior is quite similar to that of the closely related freshwater cichlids (301), familiar to home aquarists. The males of some species (e.g. *Abudefduf saxatilis, Chromis multilineata*) spawn consecutively with several females (98, 170), often in a lek system similar to that of certain birds (146). Curiously, this system has not led to the evolution of visually prominent males in many pomacentrid species, or to a terminal phase male as it apparently has in many scarids and labrids.

While the known larval pomacentrids, except *Acanthochromis polyacanthus,* are all pelagic, in some species, at least, they remain closer to the reef than do the young of reef species that are not substrate spawners. The hatchlings of *Abudefduf saxatilis* are reported to assemble in depressions on the reef, and juvenile fishes are found along the outer reef wall (98), an observation that needs confirmation since the morphology of pomacentrid larvae indicates they would be incapable of avoiding being carried by currents (P. F. Sale, personal communication).

The spawning behavior of blennies and gobies seems to be quite similar to that of pomacentrids (38, 219), although most of our information is based on observations of species that do not occur on coral reefs (e.g. 77, 290, 296).

At the opposite end of a behavioral spectrum from the pomacentrids, blennies, and gobies are the acanthurids, scarids, labrids, and at least one goatfish (*Pseudupeneus maculatus*) (216). These fishes do not anchor their eggs to the substrate and then attempt to defend them from numerous substrate predators. Instead they spawn in a frantic dash towards the surface, launching the fertilized eggs into the relatively egg-predator-free upper waters (G. W. Barlow, manuscript; 48, 94, 199, 200, 308). The reproductive behavior of these fishes is extremely varied. Spawning has been observed in only three species of acanthurids, all Indo-Pacific (199, 200). In all three cases, small groups of fishes left a much larger school for the upward spawning rush.

Randall, who made these pioneering observations, at first thought that the upward movement served to confuse possible predators, but eventually decided that its main function was to aid the ejection of the sperm and eggs. The rapid dash to the surface expands the swimbladder and increases pressure on the gonads; a quick flexing of the body as the fish rapidly changes direction and turns back downward may help to release sperm and eggs and mix them. As Jones (142) suggests, however, Randall's first reason may be more important. Fishes that lay demersal eggs must expend a large amount of energy defending them from the abundant predators of the reef substrate—wrasses, butterflyfishes, many pomacentrids, and plectognaths. In contrast, the waters near the surface are relatively free of potential egg predators, many of which are tied to the reef for shelter. As many fishes in both salt and fresh water are able to spawn successfully without such an upward dash, I am tempted to speculate that the ultimate selective value of "gamete-launching" is that relative freedom from predation more than compensates for the loss of pelagic eggs and larvae from currents. The dash also minimizes the exposure of the spawning fishes to predators such as carangids, lutjanids, and serranids.

In scarids and labrids (227), the situation is more complex. Many species have two adult phases. In the midphase, males and females are similar in appearance and spawning is communal, in much the manner of acanthurids (93, 227). The other phase consists of large, brightly colored primary or terminal phase (termphase) males that spawn singly with midphase females in a dash to the surface.

In spawning aggregations of the yellowtail parrotfish, *Sparisoma rubripinne*, over 75% of the fishes were males and each spawning group seemed to consist of a single female and 3–12 males (216). This species also spawns in termphase-midphase pairs, although much more rarely. The abundant striped parrotfish of the Caribbean, *Scarus croicensis*, shows, similar patterns in deep waters (216; G. W. Barlow, manuscript), as does the princess parrotfish, *Scarus taeniopterus* (Barlow, manuscript), and the redband parrotfish, *Sparisoma viride* (216, 308). In shallow waters, *S. croicensis* shows a different social system in which females are the prime territory holders (48).

Barlow (manuscript) speculated that the lek society evolved as males contested for the best spawning locations (which are in short supply). He argued logically that this system could, through sexual selection, explain the evolution of the conspicuous termphase males. The evolution of the shallow water *S. croicensis* system remains unexplained. Barlow also addressed the problem of the retention of midphase group spawning and concluded that in different ecological situations (e.g. different population densities) one or the other system would be favored and that overall selection would favor males that can breed both as midphase and termphase.

In scarids and labrids there is another interesting reproductive phenomenon in addition to two-stage spawning. In many of these fishes, there is protogynous hermaphroditism—functional females changing into males (223–225). The best known case is that of the famous cleaner wrasse *Labroides dimidiatus*, in which Robertson (224, 225) was able to show social control of sex reversal. Social groups of these fishes (in which the sexes are similar in appearance and there is no male termphase) consist of a single male and a harem of females. The tendency of females to turn into males is suppressed by the presence of the dominant male. The death or removal of that male ordinarily permits the dominant individual in the female pecking order to change rapidly into a male. These fishes are territorial, however, and the death of a male may result in a neighboring male occupying the dead fish's territory and incorporating the dead fish's females into its own harem.

A very similar situation was reported (99) in a serranid (anthiid) fish, *Anthias squamipinnis*, that inhabits reefs in the Indo-Pacific. The system was considered highly adaptive because males are only produced if there is need for them, that is if their density in the population decreases. The breeding ecology of this species is complex and may involve simultaneous group spawning of territorial and nonterritorial males (187).

In *Labroides dimidiatus* there are only secondary males—those derived from females. In other labrids, such as *Thalassoma lunare,* there are both secondary males and primary males, that is, those born male (226). The complexities of the selective systems leading to these patterns in labroids are discussed in detail by Robertson & Choat (226) and expanded upon by them in a discussion of the extremely flexible ontogeny of the parrotfishes (J. H. Choat and D. R. Robertson, unpublished). In the latter work, the sexual ontogeny and ecology of nine Great Barrier Reef scarids are described in detail. One species has only secondary males, others have varying mixtures of primary and secondary males.

Other serranid fishes are synchronous hermaphrodites (can produce eggs and sperm simultaneously), and still others have completely separate sexes (53, 54, 262). Some porgies (Sparidae) are also hermaphroditic (53, 221). Part of the selective value of hermaphroditism may well be increased fecundity, but, other factors must be considered (224, 225), in particular the degree of inbreeding produced by these genetic systems [interestingly, synchronous hermaphrodites have been observed spawning in pairs (53)].

There is a scattering of information on reproduction of other groups of reef fishes, albeit often garnered from nonreef species of the same family (38). Most families appear to have pelagic eggs: Carangidae (jacks), Chaetodontidae (butterfly and angelfishes), Kyphosidae (chubs), Lutjanidae (snappers), Mullidae (goatfishes), Muraenidae (morays), Pomadasyidae (grunts), and Sparidae (porgies). Many cardinalfishes (Apogonidae), often the prominent nocturnal group on a reef, are mouthbrooders (184). Many of the plectognath fishes, including various Tetraodontidae (puffers), Diodontidae (porcupinefishes), and perhaps many Balistidae (triggerfishes) have demersal eggs. The Serranidae (groupers, hamlets, basslets, etc) seem to have species with both demersal and pelagic eggs. A few groups, such as some cliniids and brotulids, are viviparous (29). In these small fishes, populations are highly differentiated, perhaps because the livebearing habit keeps the young close to home (229; J. E. McCosker, personal communication).

Before a reasonably coherent picture of the population genetics of reef fishes can be assembled, several kinds of knowledge are necessary. Further data on allozyme variation must be gathered and evaluated as to the nature of the sample of loci obtained. Enough crossing work should be done to give reasonable assurance that the variation investigated has a genetic basis. In some cases, it should be possible to do this with more readily bred freshwater fishes closely related to reef fishes, e.g.

cichlids instead of pomacentrids. More information needs to be gathered on population structure, especially on the potential for long range gene flow. This is probably the most challenging task. It is one that might be tackled by careful observations of the spread of transplanted fishes into areas where they did not occur before. Transfers have been done and can "take" without causing disruption in the local fauna, as has been demonstrated by the importation of reef game fishes (groupers and snappers) into the Hawaiian Islands (197, 214). Such transplants should only be attempted under very special circumstances, if ever, but when they are carried out or occur naturally they provide an excellent opportunity to gather information on the movement of propagules. Finally, the role of the diverse genetic systems on the reef, involving such phenomena as the "pseudo-lek" systems and hermaphroditism, must be entered into the equation.

COEVOLUTION

Considering the ancient, relatively stable conditions on coral reefs, it is not surprising that the organisms of the reef community show a bewildering variety of tightly coevolved relationships. Some of these, such as the anemonefish-anemone and cleaner-cleanee relationships have been studied in considerable detail. Others, such as plant-herbivore and predator-prey coevolution are just beginning to be investigated, and studies of food utilization and growth in reef fishes are few and far between (e.g. 19, 167, 168).

Plant-Herbivore Coevolution

Of the impressive diversity of reef fishes, only a few groups are primarily herbivorous (10, 129, 207). By far the most prominent are the parrotfishes and the surgeonfishes, which often graze and browse on the reef in herds, reminiscent of large terrestrial herbivores. The chubs, Kyphosidae, are also all herbivorous, but do not usually form large schools. Other herbivorous species are found scattered through families such as the Pomacanthidae (94), Pomacentridae, Siganidae (295), Blenniidae, and various plectognaths. In spite of the presence of these abundant herbivores, however, benthic algae usually do not form prominent stands on reefs as do plants in most terrestrial areas (14, 70, 201).

That reef herbivores have a large impact on benthic algae is obvious from the results of exclosure experiments (76, 201, 277). In Hawaii two months after an exclosure was set up, the dominant alga inside *(Ectocarpus indicus)* attained a height up to 15 mm, and other species averaged 4–30 mm (201). The algal mat in the area outside of the exclosure averaged 1–2 mm in depth. Randall obtained similar results in the Virgin Islands, where algae grew profusely within a one square yard exclosure despite the presence of 16 individuals of the herbivorous *Diadema* sea urchin, which, on the average, tripled in size during the 11 week period of the experiment. When the exclosure was removed, acanthurids and scarids moved in and immediately began grazing.

Reef fishes have also been shown to be important in producing bare "halos" (Randall zones) around reefs; these are areas devoid of the sea grasses (primarily *Thalassia testudinum* and *Cymodocea manatorum*) that cover large areas of the sea bottom in shallow waters of the Caribbean (206). Apparently, the fishes will only venture a short distance from the shelter of the reef to feed and this produces the Randall zone, an area of cleared sand that fringes the reefs, and is roughly 10 m wide. The phenomenon is quite similar to the production of cropped areas in grassland adjacent to chaparral in California (25). Small herbivores appear to be wary of exposing themselves to predation both above and below water!

However, it is difficult to generalize about the impact of grazing fishes and of grazing invertebrates on the reef flora. For instance, at St. Croix, and possibly in other areas of the Caribbean, the sea urchin *Diadema antillarum* seems to be a more important grazer on algae and sea grasses than are reef fishes, and may be primarily responsible for the reef-edge Randall zone (178, 179). Indeed, on Caribbean reefs where it is abundant, *Diadema* may also add its impact to that of browsing fishes on populations of sessile benthic macro-invertebrates (242). When a reef is cleared of *Diadema* there is an increase in the biomass both of herbivorous fishes and of carnivores, especially labrids that prey on invertebrates associated with the benthic flora (181).

Recent work on the coevolution of terrestrial herbivores and their food plants indicates that the selective impact of the herbivores is great, that major elements of the plants' defenses tend to be biochemical (39, 82, 102), that among closely related plants some are more subject to attack than others (40), and that differential susceptibility is related to biochemical defense strategy (73). Considering the heavy level of grazing on reef algae, one would expect that this flora, too, would show differential susceptibility to grazing; and this expectation is met. On a heavily *Diadema*-grazed patch reef some reef algae did attain leafy growth—*Laurencia obtusa, Caulerpa* spp., and *Dictyota* spp. (178). Experimental exposure of these and "control" algae to fish predation showed that the former three were not eaten as readily as the controls (J. C. Ogden, unpublished).

These algae, as well as blue-green algae that are often avoided by fishes (202) probably have some sort of chemical defense, although low caloric value or structural defenses cannot be eliminated (185). *Caulerpa* contains caulerpin and caulepicin (75, 243), which presumably have a defensive function.

Further effort to isolate possible defensive compounds from algae would seem highly worthwhile. Not only will knowledge of the evolution of algal defenses help in our understanding of the evolution of reef ecosystems, but it may also have highly practical applications in mariculture and elsewhere. There is, for instance, a powerful toxin in some large individuals of tropical marine fishes that frequently causes human poisoning. The disease, known as *ciguatera* or tropical fish poisoning, is sometimes fatal (202). The source of this poison may be an algal toxin that concentrates as it moves up marine food chains (195), perhaps originating in pioneer blue-green algae. If so, there should be a correlation of the poisoning with cyclonic disturbances (276), heavy anchoring activity by boats, or other disturbances likely to provide virgin surfaces on the reef.

On the other hand, ciguatera toxin may have its source in toxic invertebrates, which are much more abundant in the tropics than temperate waters (15), or in small poisonous fishes. Or indeed, the disease may be a compound entity with several distinct sources.

Plankton-Planktivore Coevolution

A common feeding niche of reef fishes is that of fishes that forage for plankton in the water column. Various herring-like fishes, small wrasses, occasional lutjanids, and, especially, a wide array of pomacentrids, make their living eating phytoplankton and zooplankton. Although lagoon plankton appear to be largely resident, the feeding by fishes on oceanic plankton as it is swept over the reef seems to represent a substantial import of energy into the reef (91). Relatively little is known of the devices evolved by plankton in response to predation pressures from fishes, although there are signs of plankton schooling in caves and crevices that provide shelter from predators (91).

Predator-Prey Coevolution

PREDATION ON INVERTEBRATES   Marine ecosystems differ from terrestrial ones in having what might be considered a "pseudoflora" of sessile invertebrates potentially subject to grazing by fishes and other mobile predators. Hermatypic corals are consumed by "herbivores," such as scarids, especially in the Pacific where coral cover is more extensive and marine phanerogams are rare in comparison with Caribbean reefs (12, 114, 284). Grazing pressure from fishes seems to account for the relative paucity of sessile invertebrates on the exposed surfaces of the reef (70, 92, 175). Although the upper surfaces of coral slabs at Fanning Island (10) were nearly devoid of benthic organisms, sponges and algae grew on the bottoms. These were immediately attacked by acanthurids, wrasses, scarids, and goatfishes when slabs were turned over. These observations should be repeated since wrasses and goatfishes have not been reported to feed on sponges elsewhere, nor are there records of scarids intentionally feeding on them. In addition, while grazers such as acanthurids and scarids may limit the distribution of some invertebrates, they encourage others by creating suitable settlement surfaces by removing green filamentous algae (295).

Bakus (10, 11) has suggested that grazing constitutes a major selective pressure on benthic marine invertebrates. This view has been supported by the demonstration that many species of sponges and holothurians are toxic, and that toxicity may occur in 100% of coral reef holothurian species. In addition, the bright colors of many unarmored nudibranch molluscs strongly indicate aposematic coloration, and the stinging ability of sea anemones and their relatives is obviously a deterrent to many potential piscine predators. It also seems a logical evolutionary parallel to terrestrial systems where, as mentioned above, plants defend themselves with noxious chemicals and arthropods also commonly have chemical defenses (e.g. 89, 90, 230).

Some fishes, most especially Caribbean grunts (Pomadasyidae), rest on the reef during the day and at night forage off the reef on a wide variety of invertebrates (68). Nothing is known of their coevolution with their prey, but as grunts are eaten by reef piscivores, which deposit much of their feces on the reef, they constitute a channel for energy flow into the reef system.

Sessile or relatively sessile invertebrates are not, of course, limited to chemical defenses. Many have developed mechanical defenses against predation—the shells of many molluscs and the spines of sea urchins are obvious examples. To overcome the latter, the triggerfishes have evolved appropriate teeth, setback eyes, and behavior that permits them to make fast work of urchins as well as other armored invertebrates (107, 129, 207, 208), while other fishes adopt other successful modes of attack, including the smashing of urchins against rocks by labrids of the genus *Cheilinus* (107). Some small clingfishes and gobies appear to behave as ectoparasites on the urchins with which they are associated, feeding on the urchin's tube feet (289).

CO-OPTION OF INVERTEBRATE DEFENSES   All of the fishes of the reef, except the larger sharks and rays, must contend with predators that hunt them. The protective devices they have evolved in coevolutionary races with their predators are extremely varied. Some, for instance, co-opt the defenses of invertebrates. Small apogonids (cardinalfishes) cluster among the spines of sea urchins and crown-of-thorns starfish, presumably gaining protection there when they are not feeding (1, 2, 145, 161, 217, 282). Many of these fishes have longitudinal stripes that help them blend in with the urchin. Fishes of other families also shelter in sea urchins (279, 282), including shrimpfishes (Centriscidae), relatives of the seahorse that hang head downward among the spines (127). The degree of camouflage varies among urchin inquilines; apparently the ability of some predators to hunt among the spines has, in some cases, produced selection for close color resemblance of the fish to the urchin.

Pomacentrid fishes of the genus *Amphiprion* take advantage of the stinging tentacles of sea anemones, much as the apogonids take advantage of sea urchin spines (7). But this is a more perilous symbiosis since sea anemones often use their nematocysts to capture fishes (122). The anemonefish-anemone relationship has thus attracted more attention from biologists (64, 110, 163, 164, 166, 294) than has the fish-urchin relationship. The anemonefishes are not born with immunity to the stings of the anemones, but acquire it gradually by acclimation (67, 162, 165, 245). The fishes are territorial when associated with an anemone. They capture food and bring it to their anemone (86, 115, 163, 166)—although more readily in the aquarium than in nature (7)—so that the relationship is truly mutualistic. Other reef pomacentrids are also sometimes associated with anemones (86, 278, 279), as is at least one wrasse, *Thalassoma amblycephalus* (246), and an apogonid, *Apogon quadrisquamatus* (61). These species, however, do not seem to form as "tight" a relationship with anemones as do the *Amphiprion*.

CHEMICAL DEFENSES   Many reef fishes have acquired toxins of their own. Best known of these are the Scorpaenidae (scorpionfishes, stonefishes, lionfishes), which have poisonous defensive spines (35, 208) capable in some species of killing a man. The fish may use these spines to strike out in active defense (130). Siganids have toxic spines, and toadfishes (Batrachoididae) have "the most highly developed venom apparatus in the fish world" (208). Soapfishes (Grammistidae) produce a protein skin toxin, grammistin (209), which has been known to kill other fishes confined with them and is lethal to mammals, producing in cats symptoms similar to ciguatera (212). Trunkfishes (*Ostracion*) and puffers (*Canthigaster*) also have skin toxins repellent to large predators (J. E. Randall, personal communication). At least one species of flatfish (Pleuronectiformes) produces a toxin lethal to other fishes (56). Some parrotfishes form a mucous envelope in which they pass the night on the reef surface (305), and there is some evidence that this helps to protect them from the nocturnal depredations of moray eels (307). Since these eels use their sense of smell to locate prey (20), the mucous envelope, rather than being distasteful, may simply serve to limit the release of odorant molecules from the resting scarids.

SCHOOLING   Another predator-defense strategy utilized by reef fishes, schooling, is seen in diverse groups: tiny fry, herring-like fishes, reef grazers, daytime resting groups of grunts and snappers, and patrolling groups of large predators such as jacks. Eibl-Eibesfeldt (87) differentiated pelagic schooling prey fish, pelagic schooling predators, semipelagic schoolers, schooling substrate grazers, schooling fishes with a fixed refuge on the reef, and facultative schoolers. In agreement with most biologists before and since who have studied fish schooling (e.g. 37, 45, 80, 124, 131, 132, 173, 188, 189, 192, 250, 269), Eibl-Eibesfeldt considered the school "primarily a protective association," citing, among other things, the behavior of schools in the presence of predators, especially the zig-zag streaming of fleeing fishes and the formation of "vacuoles" around the predator. He also suggested that predatory fishes hunting in schools were able to surround schools of food fishes, split off individuals, and corner them.

The principal means by which schooling seems to benefit the prey is through what might be called the "confusion-effect," or the moving swarm (132)—an effect familiar to those who have attempted to spear a fish in a school or net a single butterfly from a swarm. Schools also may serve to lessen the frequency of predator-prey encounters—the larger the school, the greater the advantage (45). Some fishes have coevolved patterns of heterotypic schooling to permit increase in school size (80). In addition, a tightly packed school may frighten a predator by resembling a very large fish (269).

The development, physiology, dynamics, and adaptive value of schooling have been extensively studied by behaviorists (33, 34, 36, 50, 253–259, 303, 304), but definitive experiments on the question of greatest importance to the population biologist—the adaptive value of schooling—remain to be done. It has not been possible to show unequivocally, for instance, that a solitary grunt resting during the day on a patch reef would be less likely to be eaten if it were in a school. Besides the intrinsic difficulty of doing appropriate experiments in nature, an additional complication is the apparently differential response of schooling fishes to different predators. Jack mackerel (*Trachurus symmetricus*) attacked by scorpionfish or rockfish only reacted individually (58). When they were attacked by a bonito or sea lion, however, the mackerel formed schools and rushed off together. The sea lions and bonitos more commonly attack jack mackerel than do rockfish or scorpionfish, and the former make prolonged high speed attacks from above while the latter lunge from below. Shaw (258) also observed individuals from silversides schools being consumed by a fish attacking from below "without ruffling the school." The final complication is, of course, that schooling undoubtedly serves other functions in addition to predator protection (24, 258).

COLORATION   One of the most interesting unanswered questions about reef fishes is to what extent color plays a role in predator-prey relationships. In some cases, the concealing function of color in predator and prey species is obvious—flounders, synodontids, groupers, and rockfishes that resemble the substrate, countershaded snappers, and so on. The principles involved are discussed in detail by Cott (63) with numerous examples taken from reef fishes. In some otherwise conspicuous reef fishes, countershading may serve not to conceal the fish but to conceal a conspicuous social signal (123), and it has been suggested that such signals help coordinate interspecific movements such as schooling (186). Many reef fishes also resemble other objects or organisms (212, 215, 302) or have patterns that make it difficult for a predator to fix on the fish when moving rapidly (22). Similarly convergent color patterns (mimicry) in heterotypic diurnal resting schools clearly seem a result of selection by predators (80).

Beyond such observations, however, there is considerable dispute over the significance of coloration in reef fishes. The dispute goes at least as far back as the early years of the century when Reighard (221) concluded that conspicuous colors were without biological significance, and Longley (150–152) argued that most reef fish color patterns served for concealment. More recently, Lorenz (154, 155) claimed that the bright, contrasting "poster colors" of many marine fishes are species-specific sign stimuli that release intraspecific aggression and result in intraspecific territoriality. Lorenz also claimed that these fishes do not change color under various circumstances as do many other fishes (300). This poster coloration is also supposed to provide for species recognition, at least over long range, in the Chaetodontidae (313).

Lorenz appears to have been misled by studying behavior in aquaria where it is often dramatically different from that in nature (e.g. 182, 183). His hypothesis is falsified by a variety of observations. One is that among the most aggressive territorial reef fishes are dull-colored pomacentrids of such genera as *Pomacentrus* and *Eupomacentrus*. Similarly, hamlets (*Hypoplectrus*) fit Lorenz's model in that some of them are brightly colored but some of them are relatively dull-colored and, like many pomacentrids, show interspecific as well as intraspecific territoriality (G. W. Barlow, manuscript).

In contrast, many fishes with spectacular poster colors are not territorial. Many chaetodontids and moorish idols (*Zanclus*)—classic examples of poster coloration —often form feeding assemblages of up to a dozen individuals, (P. Ehrlich, F. Talbot, B. Russell, and G. Anderson, unpublished). These fishes only show interspecific aggression under limited circumstances, especially at close range (<1 m) during crepuscular periods. Their response in nature to both conspecific and heterospecific models is also generally one of curious attraction rather than aggression, and field observations seem to indicate that a major function of their striking colors is to permit pairs or larger groups to assemble when appropriate (P. Ehrlich, F. Talbot, B. Russell, and G. Anderson, unpublished).

Fricke (105, 109) suggested that these colors were species recognition signals in the chaetodontids. This seems unlikely to be the only function, however, when one considers the sympatry of such similar species as *Chaetodon auriga* and *C. vagibundus*. Poster coloration probably does not have a single function (22, 87).

Several other explanations for poster coloration are possible. Cott (63) interpreted the striking dark-light pattern of fishes such as *Dascyllus aruanus* and *Heniochus* sp. as disruptive coloration. In *D. aruanus,* this striking pattern is attractive to conspecifics (103), but whether or not the "disruptive pattern" is functional is in some doubt. My personal observations are that it is not very functional, at least not during most of the day. In the high risk crepuscular period (or to the visual systems of some predators) it may be very effective, and P. F. Sale (personal communication) says that it makes individuals very difficult to see when they are sheltering in coral heads.

There remains the possibility that much poster coloration is aposematic. For instance, I have observed rather large (±10 cm) young french angelfish (*Pomacanthus paru*), which are strikingly colored in black and gold, repeatedly dally unattacked within 1 m or less of large synodontids. Their immunity, however, may be a result of the cleaning behavior of the smaller young with the same pattern. If poster colors were warning colors, one might expect that chaetodontids would be relatively rare in the stomachs of most predators. A check of major works containing stomach content analysis (52, 129, 207, 211, 274) revealed that chaetodontids and the closely related pomacanthids (angelfishes) were virtually absent from the array of prey fishes.

If chaetodontid-pomacanthid coloring is, at least in part, aposematic, one might also expect to find other nonchaetodontid flat-bodied fishes mimicking them. Only one such case has been reported in the literature. Randall & Randall (215) discovered an *Acanthurus* that closely resembled a chrome-yellow *Centropyge* (Pomacanthidae) with a blue eye-ring, a blue line under the mouth, a blue margin on the opercle, and an orange patch just above the base of the tail. They made several attempts to test the *Centropyge* for distastefulness by exposing individuals on lines to predators, with negative results. Although their field experiments were not conclusive, they do indicate that the *Centropyge* was not in a class of distastefulness with grammistids. Direct distastefulness, however, is not the only possible basis for warning coloration. Flattened and armed with spines, pomacanthids and chaetodontids may be difficult targets for many predators to catch or engulf and not generally worth the effort and risk of an attack. J. E. Randall (personal communication) pointed out that the *Acanthurus (A. pyroferus)* which mimics *C. flavissimus,* mimics a *Centropyge, C. vrolikii,* of entirely different appearance at Palau where *C. flavissimus* does not occur. Furthermore, *A. pyroferus* loses the resemblance when it gets larger than *C. vrolikii.* The key element in this system may be the extreme familiarity of the Centropyges with every nook and cranny of their home range. This gives them an advantage over predators not shared by the widely ranging (but equally spiny) *Acanthurus,* and roaming predators, mistaking the latter for *Centropyge,* may decide an attack is not worth the effort.

A point in the argument against aposematic coloring in butterflyfishes and angelfishes is, however, the apparent absence of close Mullerian complexes among them, although many Indo-Pacific species have the same general gestalt, and juveniles of the Atlantic *Pomacanthus paru* and *P. arcuatus* are very similar.

The difficulty predators have in attacking chaetodonids may be enhanced by the frequent presence of eye stripes that tend to conceal the eye, accompanied by false eyes on the posterior part of the body. With the squarish shape of the fishes and their general slow movements, it may be quite difficult for some predators to decide which end is forward, especially in dim light. These patterns may also serve to protect against hit-and-run predatory attacks of blennies (*Plagiotremus* sp.), which often tear chunks of flesh from the vicinity of the eye (84, 154, 298). Russell worked extensively with the blennies in the field, and believes that chaetodontids are not normally threatened by *Plagiotremus* (B. C. Russell, personal communication). These blennies tend to attack large fishes such as lethrinids and siganids, and bite them on the hind end.

Finally, the spectacular colors of chaetodontids may be a form of "flash coloration." The fishes display a broad, flat "poster" to a predator that disappears when they turn to flee, showing only their narrow posterior aspect or, in some circumstances, a narrow dorsal aspect (123). Interestingly, butterflies display difficult-to-explain poster colors also, but in butterflies we can be sure that these colors rarely, if ever, serve an aggressive purpose, and frequently function as flash colors to deceive predators. Under different conditions, poster patterns on reef fishes may be social signals at one moment and serve as a predator defense at the next.

MIMICRY   Mimicry itself is a frequent phenomenon on reefs, often described but sometimes difficult to explain (8, 215). Various hamlets (*Hypoplectrus* spp.) appear to mimic pomacentrids, and I have been momentarily fooled at a distance by the resemblance between the yellowtail hamlet *H. chlorurus* and the yellowtail damselfish, *Microspathodon chrysurus*. (The dark-body, light-tail pattern of these fishes is quite common among reef species, as is the accentuated dark front-half, light rear-half pattern typified by the Caribbean bicolor damselfish, *Eupomacentrus partitus,* the Pacific *Chromis margaritifer,* and some populations of *Acanthochromis polyacanthus.*) It is unlikely that predators willing to attack the largely herbivorous *Microspathodon* would be reluctant to tackle the hamlet—most likely the hamlet more readily approaches its prey disguised as a pomacentrid. Other mimetic pairs include the Pacific saddled puffer, *Canthigaster valenti,* and the saddled filefish, *Paraluteres prionurus;* presumably the latter gains protection when it is mistaken for the poisonous puffer (8). The resemblance of a young *Bodianus* (probably *perdito*) to *Chaetodon ephippum* (49) is more difficult to explain.

The use of eye spots to deflect predator attack has already been mentioned, and such eye spots are frequently found near the base of the caudal fin of many reef fishes or on the dorsal fin. When they are spectacularly developed on the dorsal fin, as in various wrasses (e.g. *Coris aygula, Halichoeres centriquadratus, Anampses twisti*) and juvenile *Bolbometopon bicolor* (Scaridae), the eyespot may, in some situations, be suddenly erected and used to startle predators in much the same way as moths use suddenly exposed eye spots to startle birds (27).

J. E. McCosker (personal communication) recently discovered the use to which at least one reef fish puts its rear eyespot. The Indo-Australian *Calloplesiops altivelis* has a spotted pattern closely resembling that of a sympatric moray eel, *Gymnothorax meleagris.* When alarmed, the fish dashes into a hole in the reef, leaving the rear part of its body exposed. It holds its fins in a manner that exposes the eyespot near the base of the rear of the dorsal fin. The eye spot, and the attitude of the fins, produces a remarkable simulation of the head of a predaceous moray! Other moray-patterned fishes with rear eye spots, such as *Anampses meleagrides,* may also mimic morays, but appropriate behavior has not been observed.

ATTACK STRATEGIES   Against this array of defenses, reef predators mount a variety of attacks (139) ranging from the use of lures (251) and the slow herding of prey into corners by lionfishes (*Pterois*) (108) to the headlong dash of synodontids, groupers, and *Priacanthus* (108, 132). A trumpetfish (*Aulostomus*) will often swim along close to another large fish, using it as cover from which to dash out at prey (87), and groupers follow foraging morays apparently in hope of catching fishes forced into the open by the eel (108). And, of course, predator color patterns have evolved in ways that lessen the chance of detection by prey (e.g. Barlow, 21). Giant Queensland groupers (*Promicrops lanceolatus*) that attain a length of 3 m and a weight of more than a quarter ton sometimes hide in caves and suck fishes weighing several pounds into their mouths (190). Much of the predator activity on many reefs seems to take place in the crepuscular periods as the changeover between day-active and night-active faunas occurs (131, 132, 137, 138, 272).

Cleaning

A very specialized form of predation—the eating of ectoparasites of fishes by small cleaner fishes—forms the center of a most interesting coevolutionary complex on the reef (95). Fishes attacked by ectoparasites have evolved special cleaning postures and color changes when they are solicited by cleaners (84, 95, 148, 156, 203). These serve to make the parasites more obvious and accessible. On reefs, there are two general classes of cleaner fishes. One class is the young of species that are rarely cleaners as adults. Cleaning behavior is frequent in young butterflyfishes, angelfishes, pomacentrids, and wrasses, and is probably always facultative (148, 213). Occasionally, adult chaetodontids (213) and adult wrasses (62) are observed cleaning; and adult pipefishes (Syngnathidae) are now also known to be cleaners (128).

In contrast, two groups of fishes invariably clean as adults and are almost certainly obligatory cleaners. One is the cleaner gobies of the Atlantic (30, 31). Five of twelve species of the subgenus *Elacatinus* of *Gobiosoma* are known to be cleaners and "feed mainly, if not exclusively, on gnathiid isopods that are removed from the bodies, fins, gill chambers and mouths of a wide variety of reef species and fishes that visit reef areas" (30). These small gobies (<3 cm), brightly marked with yellow and black stripes, often set up "cleaning stations" on large sponges, brain, or palmate corals, and are visited by the cleanees. They clean a wide range of fishes, including large piscivores, while the sympatric *Thalassoma bifasciatum,* cleaning while juveniles, is largely restricted to nonpiscivores (66). The *Thalassoma* were devoured by predators in tank experiments, while the gobies were generally spared.

In the Indo-Pacific, wrasses of the genus *Labroides* wear a similar livery and play a similar role (196, 312). The Hawaiian species *Labroides phthirophagus* is an obligate cleaner (312), and all other *Labroides* probably are as well. These cleaner wrasses consume mucus, scales, and dermal and epidermal tissues in addition to ectoparasites, and they may supplement this cleaning diet with zooplankton (312). Unlike the gobies, however, at least *Labroides dimidiatus* and especially *L. bicolor* do not remain confined to small cleaning stations but move over the reef. At Bora Bora, they were often observed to travel in pairs or small groups, visiting especially the pomacentrids that hold territories (P. Ehrlich and J. P. Holdren, unpublished) in a manner reminiscent of tropical butterflies that "trap-line" from flower to flower gathering pollen (81, 112).

Various authors (105, 106, 156, 297–299, 302) have studied the nexus of signals that mediate the cleaner-cleanee relationship. This interest has been, in part, stimulated by an additional member of the parasite-host-cleaner coevolutionary complex, the cleaner mimic. This mimic, a saber-toothed blenny, *Aspidontus taeniatus,* imitates the appearance and behavior signals of *Labroides* (85, 298, 299, 302) but instead of cleaning the prospective cleanee, *Aspidontus* bites a chunk out of a fin and eats it (interestingly it does not use the saber-teeth in attacking—they are reserved for intraspecific battling). Apparently, the mimetic resemblance is only good enough to fool young or naive fishes; those with more experience learn to avoid the *Aspidontus* (215). The distinction between the behavior of the cleaner and the mimic is blurred, however. On occasion, cleaners will chase and bite fishes without inspecting for parasites (159; B. C. Russell, personal communication).

Many blennies make their living by snatching chunks off larger fishes in hit and run attacks, and are involved in cases of both aggressive and Batesian mimicry (8, 134, 157, 271). These blennies sometimes school with wrasses, which they resemble, dashing at prey from the cover of the school (134). There are also less specialized cleaners than *Labroides* among adult wrasses (213), and the young of many wrasses are cleaners even though the adults are not. Russell (personal communication) thinks that, evolutionarily, cleaning in the wrasses was a natural extension of foraging, the larger fishes simply being treated as additional substrate. This has been followed by more aggressive behavior on the part of *Labroides,* the *Aspidontus* pattern of mimicry, and, finally, the rapid "dash-from-a-hole, take-a-bite, and retreat-to-the-hole" behavior seen in some members of the blenniid genus *Plagiotremus.*

Many questions remain, however, about the present evolutionary state of the *Labroides-Aspidontus* complex, just as they do about all terrestrial mimetic associations (79). Is the mimic putting any selective pressure on the model by increasing the frequency with which *Labroides* are eaten, or by driving away prospective cleanees? If the former were occurring, one might find that where *Aspidontus* is common, *L. dimidiatus* looks slightly more different from *Aspidontus* than it does where *Aspidontus* is rare. To enhance cleanee discrimination the *Labroides* should evolve away from the *Aspidontus* pattern as rapidly as possible without losing its basic cleaner livery.

There should be plenty of flexibility in the system, since the general appearance of the *L. dimidiatus* (as opposed to its detailed appearance) would seem to be enough to announce "cleaner." A Caribbean grouper in an aquarium recognized *L. dimidiatus* as a cleaner even though it had never seen one (125) and naive Pacific fishes recognize *Gobiosoma* (30). Presumably prior experience with the cleaners in the grouper's home waters was sufficient—indeed recognition of the highly special-ized pattern of the cleaners is probably part of the genetic repertoire of piscivores (66). In theory, *Labroides* might be able to evolve a sufficiently different garb to permit cleanees to discriminate between relatively painless cleaners (*L. dimidiatus*) and painful ones (*Aspidontus*), but there is considerable anecdotal evidence that *Aspidontus* can evolve rapidly enough to "stay even" in the evolutionary race.

COMMUNITY ECOLOGY

One question that remains partly open concerns the effects of cleaning symbioses, both those involving cleaner fishes and those involving cleaner shrimp (121, 228), on the diversity of the fish community. Limbaugh claimed (148) that cleaning symbioses were more common in tropical than in temperate waters, and that remov-ing cleaners led to a rapid decline in the diversity of fishes present on the reef. Hobson (135), however, strongly questioned the assumption that tropical communi-ties have a higher proportion of cleaners, and, indeed, numerous cases of cleaning in temperate marine (136, 148) and freshwater (4) fishes are coming to light.

Limbaugh's experiment of removing the cleaners from a reef has been repeated several times (158, 312) without resulting in a decline in diversity. Partial removal of cleaners did, however, lead to an increase in the cleaning behavior of those remaining, as well as changes in the behavior of the cleanees and the behavior and distribution of the remaining cleaners (158, 159). Interestingly, Losey found no increase in the number of ectoparasites (compared with a control reef) after cleaner removal. Further cleaner removal experiments certainly should be carried out in a variety of reef situations.

Whatever their effect on total reef diversity, the presence of cleaners leads to high "point diversity." Slobodkin & Fishelson (261) found that the larger nonterritorial diurnal fish species of the reef tended to aggregate at cleaning stations. These fish species make up about one fourth of the species found on the reef. Thus cleaners form a biological focus of diversity analogous to physical foci such as deep holes and areas of great geometric complexity.

More general studies of reef fish communities using a wide variety of collecting and observational techniques are numerous (5, 13, 17, 55, 61, 62, 69, 116, 118, 143, 176, 191, 263–265, 284, 291) and have painted a reasonably complete picture of the structure of these communities. Carnivores are much more diverse than herbivores or omnivores, but on large reefs the latter two may dominate both by numbers and weight (17, 129, 176). In contrast, carnivores made up more than 50% of the fauna (by weight) in the Virgin Islands (205) and Tanganyika (284). The problem of partitioning the reef fish fauna into carnivores, omnivores, and herbivores is difficult because of sampling problems (284), and compounded because many fishes take refuge on the reef but feed on adjacent sand flats or in the water column (69, 205, 280).

Perhaps the best general description of the way the reef fish fauna is fitted to the reef is "crammed." One interesting bit of evidence for this is the one-way migration of fish species from the crowded reefs of the Red Sea through the Suez Canal.

Mediterranean fishes have been notably unsuccessful at penetrating the reef fauna (9). There are essentially two shifts of fishes on the reefs, day and night, with a time of great activity in the dawn and dusk turnover period (26, 62, 101, 131–133, 149, 247, 272, 273), as the fishes respond to changing light levels (62) in a species-specific sequence (74, 83). Detailed studies of the sharing of resources by reef fish species (264–266, 293) have led to the conclusion that space itself "may play a major, if not decisive role in maintaining numerical stability in coral reef fish communities" (264). An excellent example of this is the severe limitation of populations of saber-toothed blennies, *Plagiotremus* (*Runula*), by the availability of suitable holes for shelter and nest sites. (B. C. Russell, personal communication). These slender predators have highly specific requirements for the size and shape of holes they occupy. By providing suitable artificial nest blocks, Russell has been able to produce an approximate doubling in the populations of two species (*P. rhynorhynchos, P. tapeinosoma*).

The limitations of space are increased by the inter- and intraspecific territorial behavior of so many species, discussed earlier, and seem fundamental to the evolution of many major features of the reef fish fauna. For example, the guilds of grazers (23, 142, 295) might be thought of as largely a function of the finite area available for the growth of algae.

It is difficult to imagine how more species could enter many reef faunas, especially in the "day shift," which is relatively crowded compared to the night (62). The Shannon-Weaver diversity index for the fish community at the Virgin Islands Tektite site was 3.3154 based on individuals and 3.0056 based on biomass; the lower biomass figure reflects the great size of individuals of a few species (264). If, for instance, the large *Scarus vetula* could be replaced by several smaller herbivores, then the reef fish diversity could be increased.

It is, however, impossible to determine whether or not a reef fish fauna is saturated with species without doing field experiments. Similarly the conjecture that reef transients are generally *r*-selected and residents *k*-selected (264) will have to await further studies, especially of reproductive biology, for confirmation.

Workers in Australia have recently been making rapid progress toward elucidating both the structure of coral reef fish communities and the causes of their extraordinary species diversity. Extensive work on One Tree Island at the southern end of the Great Barrier Reefs (116, 286, 287) has shown striking differences in fish fauna and fish biomass in ecologically different regions of the reef (e.g. windward and leeward slopes, reef crest, transition to inter-reef ocean bottom, etc). In spite of this general site specificity, however, the precise fauna on a given site within a habitat seems governed largely by chance factors (233, 241; P. Sale, manuscript). On artificial reefs (233), for instance, at least at the southern end of the barrier where breeding is highly seasonal, the physical structure of the reef is relatively unimportant in determining at least the initial species composition of the community.

Sale's work (241; manuscript) on a guild of pomacentrids similarly indicated that random processes involved in removal and recruitment tend to determine which of several highly territorial species occupy a given rubble patch on the reef. Indeed, the growing consensus of the "Australian school" seems that the prevalent notion that high diversity on reefs is possible because of the narrow niches of the species is false, and that many reef fish species are generalists. Apparently diversity is maintained in the face of limited space by a pattern of steady high mortality and random recruitment (the latter mostly from pelagic young). Talbot (285) and Russell and Talbot (manuscript) suggested along these lines that high tropical diversity in fish faunas exists largely because there has been a great deal of time during which reefs have remained much the same. Presumably this has permitted the allopatric evolution of numerous species to fill broad niches; species that have been able to become and remain sympatric on high productivity reefs because of the patchwork nature of the environment and the stochastic recruitment-loss processes. This pattern agrees with the theoretical conclusion of Levin (147) that heterogeneity may be introduced in homogeneous environments by random colonization events.

An area of considerable importance to the community ecology of reefs that needs much more investigation is the mode of communication among reef species. Visual communication is common (60) but sound and smell may play an important role in the lives of reef fishes (42, 171, 288, 309). Indeed, considering the chemoreceptive abilities of sharks, it seems likely that the reef community is bound together by a web of pheromones and other chemical cues that have not yet even begun to be unravelled.

Reef fishes have virtually been omitted as sources of data to test mathematical theory of community ecology. The only study that has come to my attention is that of Roughgarden (231), who used Randall's (207, 208) data on food habits of Caribbean fishes to illustrate his models of "species packing."

Not only does the structure of the reef influence the structure of the fish community, but the reverse is also true. Fishes, especially scarids, plectognaths, and some pomacentrids (114), destroy coral during their feeding, and are important in bioerosive processes on the reef. They retard coral growth and open corals to invasion by algae (113). In fact, these fishes are a major factor in converting coral reefs into coral sand (18). Ogden and Gerhard (manuscript), for example, estimated that, in Panama, parrotfishes generated an estimated 0.5 kg m^{-2} yr^{-1} carbonate sediment. Overfishing may dramatically change the flora and benthic fauna of the reef (242; J. Ogden, personal communication). This is just one example of the diverse human activities that threaten these fascinating and productive ecosystems (100, 140).

CONCLUSION—REEF FISHES AND BUTTERFLIES

At this point, the major reason for the dramatically different taxonomic structure of reef fishes and butterflies is clear. The reef fishes are a sample of all teleostean fishes, organisms that play a vast array of ecological roles. Even a cursory survey of their morphology reveals a diversity to match these roles. In contrast, the butterflies are a sample of a huge (±200,000 species) order of insects that is almost entirely herbivorous. They are the major group to penetrate the daytime aerial niche as adults; unlike reef fishes, butterflies have no "night shift." Interesting questions remain, however. Why, for instance, can one find more than two dozen species of chaetodontids sympatrically on the Great Barrier Reef, with more than a dozen mingling together in small areas, when such concentrations of congeners seem very rare in the much smaller butterflies?

Population structure and dynamics are better understood in the butterflies (83) than in the reef fishes, but both groups appear to have relatively sedentary adults and populations subject to rather frequent extinction. In the fishes, however, lack of information on gene flow makes it difficult to state whether disappearance of a species from a section of reef means the end of a mendelian population or merely of a local subunit of such a population. Similarly, although it is now clear that many butterfly populations remain genetically similar because they are under similar selection pressures (83), we do not have enough information to make such statements about the fishes.

Butterflies are known to sequester poisons from their plant food that help protect them from vertebrate predators (47). Little is known about the origins of fish poisons. Although most appear to be endogenous, the ciguatera situation suggests that at least some of the toxins may be derived from the fish's food.

Both groups tend to have bright "poster colors" that have not been entirely explained. Both also have numerous members involved in mimicry—the butterflies apparently more so than the fishes. This may be explained by the frequent use of highly poisonous food plants (82) and the likelihood that, to one degree or another, all butterflies are distasteful (C. L. Remington, personal communication).

At the community level, much more has been done with the fishes than with the butterflies, and the progress made in explaining reef fish diversity suggests that more efforts should be made toward uncovering the determinants of butterfly community composition. The work that has been done on butterflies (252) indicates that, at least in temperate zones, they show much less habitat selection than do the fishes. How can this observation be related to the different phenetic patterns of the two groups? It seems likely that the availability of a diverse array of food plants has presented the butterflies with the evolutionary opportunity to radiate into numerous large complexes of closely related species. But, where any given food plant is widely distributed, adult butterflies can move about their business with few restrictions, since predation pressure on them is generally light. In contrast, much of the behavior of reef fishes seems substrate-oriented because of strong predation pressure. Habitat selection then becomes critical since food and other requisites of life must be found in close proximity to suitable shelter.

ACKNOWLEDGMENTS

I wish to thank Gordon Anderson, George W. Barlow, Anne H. Ehrlich, Richard W. Holm, John E. McCosker, John C. Ogden, John E. Randall, Jonathan Roughgarden, Barry C. Russell, Peter F. Sale, Michael E. Soulé, Frank H. Talbot, J. C. Tyler, and Evelyn Shaw-Wertheim, all of whom have reviewed the manuscript, made most helpful comments, and, in many cases, given me access to their unpublished results. J. Howard Choat was kind enough to supply me with two of his unpublished manuscripts. Margaret Craig and her staff at the Stanford Biology Library were extremely helpful in running down sources, and Darryl Wheye showed great patience in helping me prepare the manuscript.

Literature Cited

1. Abel, E. 1960. Fische zwischen Seeigel-Stacheln. *Natur Volk* 90:33–37
2. Abel, E. 1960. Zur Kenntnis des Verhaltens und der Ökologie von Fischen on Korallenriffen bei Ghardaga (Rotes Meer). *Z. Morph. Oekol. Tiere* 48:430–503
3. Abel, E. 1961. Freiwasserstudien über das Fortpflanzungsverhalten des Mönchfisches *Chromis chromis* Linné, einem Vertreter der Pomacentriden im Mittelmeer. *Z. Tierpsychol.* 18:441–9
4. Abel, E. 1971. Zur Ethologie von Putzsymbiosen unheimischer Süsswasserfische im natürlichen Biotop. *Oecologia* 6:133–51
5. Abel, E. 1972. Problem der Ökologischen definition des "Korallenfisches." *Proc. Symp. Corals Coral Reefs, 1969, Mar. Biol. Assoc. India* 449–56
6. Albrecht, H. 1969. Behaviour of four species of Atlantic damselfishes from Colombia, South America (*Abudefduf saxatilis, A. taurus, Chromis multilineata, C. cyanea;* Pisces, Pomacentridae). *Z. Tierpsychol.* 26:662–76
7. Allen, G. R. 1972. *The Anemonefishes: Their Classification and Biology.* Neptune City, NJ: TFH Publ. 288 pp.
8. Allen, G. R., Russell, B. C., Carlson, B. A., Starck, W. H. 1975. Mimicry in marine fishes. *Trop. Fish Hobbyist.* In press
9. Aron, W. I., Smith, S. H. 1971. Ship canals and aquatic ecosystems. *Science* 174:13–20
10. Bakus, G. J. 1964. The effects of fish grazing on invertebrate evolution in shallow tropical waters. *Allan Hancord Found. Publ. Occas. Pap. No. 27.* 22 pp.
11. Bakus, G. J. 1966. Some relationships of fishes to benthic organisms on coral reefs. *Nature* 210:280–84
12. Bakus, G. J. 1967. The feeding habits of fishes and primary production at Eniwetok, Marshall Islands. *Micronesica* 3:135–49
13. Bakus, G. J. 1969. Energetics and feeding in shallow marine waters. *Int. Rev. Gen. Exp. Zool.* 4:275–369
14. Bakus, G. J. 1972. Effects of the feeding habits of coral reef fishes on the benthic biota. *Proc. Symp. Corals Coral Reefs, 1964, Mar. Biol. Assoc. India* 445–48
15. Bakus, G. J., Green, G. 1974. Toxicity in sponges and holothurians: a geographic pattern. *Science* 185:951–53
16. Bardach, J. E. 1958. On the movements of certain Bermuda reef fishes. *Ecology* 39:139–46
17. Bardach, J. E. 1959. The summer standing crop of fish on a shallow Bermuda reef. *Limnol. Oceonogr.* 4:77–85
18. Bardach, J. E. 1961. Transport of calcareous fragments by reef fishes. *Science* 133:98–99
19. Bardach, J. E., Menzel, D. W. 1956. Field and laboratory observations on the growth of some Bermuda reef fisheries. *Proc. Gulf Carib. Fish. Inst., 9th Ses.* 106–13
20. Bardach, J. E., Winn, H. E., Menzel, D. W. 1959. The role of the senses in the feeding of the nocturnal reef predators *Gymnothorax moringa* and *G. vicinus. Copeia* 1959:133–39
21. Barlow, G. W. 1967. The functional significance of the split-head color pattern as exemplified in a leaf fish, *Polycentrus schomburgkii. Ichthyol./Aquar. J.* April–June: 57–70
22. Barlow, G. W. 1972. The attitude of fish eye-lines in relation to body shape and to stripes and bars. *Copeia* 1972: 4–12
23. Barlow, G. W. 1974. Contrasts in social behavior between Central American cichlid fishes and coral-reef surgeon fishes. *Am. Zool.* 14:9–34
24. Barlow, G. W. 1975. Extraspecific imposition of social grouping among surgeon fishes. *J. Zool.* 174:In press
25. Bartholomew, B. 1970. Bare zone between California shrub and grassland communities: the role of animals. *Science* 170:1210–12
26. Bertram, B. C. R. 1965. The behaviour of Maltese fishes by day and night. *Rep. Underwater Assoc. Malta* 1:39–41
27. Blest, A. D. 1957. The function of eye-spot patterns in the Lepidoptera. *Behaviour* 11:209–56
28. Boden, B. P. 1952. Natural conservation of insular plankton. *Nature* 169:697–99
29. Böhlke, J. E., Chaplin, C. C. G. 1968. *Fishes of the Bahamas and Adjacent Tropical Waters.* Wynnewood: Livingston. 771 pp.
30. Böhlke, J. E., McCosker, J. E. 1973. Two additional West Atlantic gobies (genus *Gobiosoma*) that remove ectoparasites from other fishes. *Copeia* 1973:609–10
31. Böhlke, J. E., Robins, C. R. 1968. Western Atlantic seven-spined gobies, with descriptions of two new species and a new genus, and comments on Pacific

relatives. *Proc. Acad. Nat. Sci. Philadelphia* 120:45–175

32. Breder, C. M. 1948. Observation on coloration in reference to behavior in tide pool and other marine shore fishes. *Bull. Am. Mus. Natur. Hist.* 92:285–311

33. Breder, C. M. 1954. Equations descriptive of fish schools and other animal aggregations. *Ecology* 35:361–70

34. Breder, C. M. 1959. Studies of social groupings in fishes. *Bull. Am. Mus. Natur. Hist.* 117:397–481

35. Breder, C. M. 1963. Defensive behavior venom in *Scorpaena* and *Dactylopterus*. *Copeia* 1963:698–700

36. Breder, C. M. 1965. Vortices and fish schools. *Zoologica* 50:97–114

37. Breder, C. M. 1967. On the survival value of fish schools. *Zoologica* 52:25–40

38. Breder, C. M., Rosen, D. E. 1966. *Modes of Reproduction in Fishes.* New York: Natural History Press. 941 pp.

39. Breedlove, D. E., Ehrlich, P. R. 1968. Plant-herbivore coevolution: lupines and lycaenids. *Science* 162:671–72

40. Breedlove, D. E., Ehrlich, P. R. 1972. Coevolution: patterns of legume predation by a lycaenid butterfly. *Oecologia* 10:99–104

41. Briggs, J. C. 1967. Relationships of the tropical shelf regions. *Stud. Trop. Oceanogr.* 5:569–92

42. Bright, T. J. 1972. Bio-acoustic studies on reef organisms. *Results of the Tektite Program: Ecology of Coral Reef Fishes, Bull. Natur. Hist. Mus. L. A.* 14:45–69

43. Brinley, F. J. 1939. Spawning habits and development of beau-gregory (*Pomacentrus leucostictus*). *Copeia* 1939:185–88

44. Brock, V. E. 1954. A preliminary report on a method of estimating reef fish populations. *J. Wildl. Manage.* 18:297–308

45. Brock, V. E., Riffenburgh, R. H. 1960. Fish schooling: a possible factor in reducing predation. *J. Cons. Perm. Int. Explor. Mer.* 25:307–17

46. Brockmann, H. J. 1973. The function of poster-coloration in the beaugregory, *Eupomacentrus leucostictus* (Pomacentridae, Pisces). *Z. Tierpsychol.* 33:13–34

47. Brower, L. P. 1970. Plant poisons in a terrestrial food chain and implications for mimicry theory. In *Biochemical Coevolution,* ed. K. L. Chambers, 69–82. Corvallis: Univ. Oregon

48. Buckman, N. S., Ogden, J. C. 1973. Territorial behavior of the striped parrotfish *Scarus croicensis* Block (Scaridae) *Ecology* 54:1377–82

49. Burgess, W., Axelrod, H. R. 1971. *Pacific Marine Fishes, Book 2.* Neptune City, NJ: TFH Publ.

50. Cahn, P. H., Shaw, E. 1963. Schooling fishes: the role of sensory factors. *Anim. Behav.* 11:405–6

51. Chaplin, C. G., Scott, P. 1972. *Fishwatchers Guide to West Atlantic Coral Reefs.* Wynnewood: Livingston. 65 pp.

52. Choat, J. H. 1968. Feeding habits and distribution of *Plectropomus maculatus*. *Proc. R. Soc. Queensl.* 80:13–17

53. Clark, E. 1959. Functional hermaphroditism and self-fertilization in a serranid fish. *Science* 129:215–16

54. Clark, E. 1965. Mating of groupers. *Natur. Hist.* 74:22–25

55. Clark, E., Ben-Tuvia, A., Steinitz, H. 1968. Observations on a coastal fish community, Dahlak Archipelago, Red Sea. *Sea Fish. Res. Sta. Haifa Bull.* 49:15–31

56. Clark, E., Chao, S. 1972. A toxic secretion from the Red Sea flatfish *Pardachirus marmoratus* (Lacépède). *Hebr. Univ. Jerusalem, Steinitz Mar. Lab. Sci. Newslett. No. 2,* p. 14

57. Clarke, T. A. 1970. Territorial behavior and population dynamics of a pomacentrid fish, the Garibaldi, *Hypsypops rubicunda. Ecol. Monogr.* 40:189–212

58. Clarke, T. A., Flechsig, A. O., Grigg, R. W. 1967. Ecological studies during Project Sealab II. *Science* 157:1381–39

59. Cohen, D. 1970. How many recent fishes are there? *Proc. Calif. Acad. Sci.* 38:341–46

60. Colin, P. L. 1971. Interspecific relationships of the yellowhead jawfish, *Opistognathus aurifrons* (Pisces, Opistognathidae). *Copeia* 1971:469–73

61. Colin, P. L. 1974. Mini-prowlers of the night reef. *Sea Frontiers* 20:139–45

62. Collette, B. B., Talbot, F. H. 1972. Activity patterns of coral reef fishes with emphasis on nocturnal-diurnal changeover. *Results of the Tektite Program: Ecology of Coral Reef Fishes, Bull. Natur. Hist. Mus. L.A.* 14:98–124

63. Cott, H. B. 1940. *Adaptive Coloration in Animals.* London: Methuen. 508 pp.

64. Crespigny, C. C. de 1869. Notes on the friendship existing between the malacopterygian fish *Premnas biaculeatus* and the *Actinia crassicornis*. *Proc. Zool. Soc. London* 1869:248–49

65. Cummings, W. C., Brahy, B. D., Spires, J. Y. 1966. Sound production, schooling, and feeding habits of the margate, *Haemulon album* Cuvier, off North

Bimini, Bahamas. *Bull. Mar. Sci.* 16:626–40

66. Darcy, G. H., Maisel, E., Ogden, J. C. 1974. Cleaning preferences of the gobies *Gobiosoma evelynae* and *G. prochilos* and the juvenile wrasse *Thalassoma bifasciatum. Copeia* 1974:375–79

67. Davenport, D., Norris, K. S. 1958. Observations on the symbiosis of the sea anemone *Stoichactes* and the pomacentrid fish *Amphiprion percula. Biol. Bull.* 115:397–410

68. Davis, W. P. 1967. Ecological interactions, comparative biology and evolutionary trends of thirteen pomadasyid fishes at Alligator Reef, Florida Keys. PhD dissertation. Univ. Miami, Miami, Fla. 94 pp.

69. Davis, W. P., Birdsong, R. S. 1973. Coral reef fishes which forage in the water column. *Helgol. Wiss. Meeresunter.* 24:292–306

70. Dawson, E. Y., Aleem, A. A., Halstead, B. W. 1955. Marine algae from Palmyra Island with special reference to the feeding habits and toxicology of reef fishes. *Allan Hancock Found. Occas. Pap. No. 17.* 39 pp.

71. Day, J. H. 1963. Complexity in the biotic environment. *Speciation in the Sea, Syst. Assoc. Publ. No. 5,* 31–49

72. Delsman, H. C. 1930. Fish eggs and larvae from the Java Sea, 16. *Amphiprion percula. Treubia* 12:367–70

73. Dolinger, P. M., Ehrlich, P. R., Fitch, W. L., Breedlove, D. E. 1973. Alkaloid and predation patterns in Colorado lupine populations. *Oecologie* 13:141–204

74. Domm, S. B., Domm, A. J. 1973. The sequence of appearance at dawn and disappearance at dusk of some coral reef fishes. *Pac. Sci.* 27:128–35

75. Doty, M. S., Santos, G. A. 1966. Caulerpicin, a toxic constituent of *Caulerpa. Nature* 211–440

76. Earle, S. A. 1972. The influence of herbivores on the marine plants of Great Lameshur Bay, with an annotated list of plants. *Results of the Tektite Program: Ecology of Coral Reef Fishes, Bull. Natur. Hist. Mus. L.A.* 14:17–44

77. Ebert, E. E., Turner, C. H. 1962. The nesting behavior, eggs and larvae of the bluespot goby. *Calif. Fish Game* 48:249–52

78. Ehrlich, P. R. 1958. The comparative morphology, phylogeny and higher classification of the butterflies (Lepidoptera: Papilionoidea). *Univ. Kansas Sci. Bull.* 39:305–70

79. Ehrlich, P. R. 1970. Coevolution and the biology of communities. *Biochemical Coevolution,* ed. K. L. Chambers, 1–11. Corvallis: Oregon State Univ. Press

80. Ehrlich, P. R., Ehrlich, A. H. 1973. Coevolution: heterotypic schooling in Caribbean reef fishes. *Am. Natur.* 107:157–60

81. Ehrlich, P. R., Gilbert, L. E. 1973. Population structure and dynamics of the tropical butterfly *Heliconius ethilla. Biotropica* 5:69–82

82. Ehrlich, P. R., Raven, P. H. 1965. Butterflies and plants: a study in coevolution. *Evolution* 18:586–608

83. Ehrlich, P. R., White, R. R., Singer, M. C. Gilbert, L. E., McKechnie, S. W. 1975. Checkerspot butterflies: a historical perspective. *Science* 188:221–28

84. Eibl-Eibesfeldt, I. 1955. Über Symbiosen, Parasitismus und andere besondere zwischenartliche Beziehungen tropischer Meeresfische. *Z. Tierpsychol.* 12:203–19

85. Eibl-Eibesfeldt, I. 1959. Der Fisch *Aspidontus taeniatus* als Nachahmer des Putzers *Labroides dimidiatus. Z. Tierpsychol.* 16:19–25

86. Eibl-Eibesfeldt, I. 1960. Beobachtungen und Versuche on Anemonenfischen (Amphiprion) der Malediven und der Nicobaren. *Z. Tierpsychol.* 17:1–10

87. Eibl-Eibesfeldt, I. 1962. Freiwasserbeobachtungen zur Deutung des Schwarmverhaltens verschiedener Fische. *Z. Tierpsychol.* 19:165–82

88. Eibl-Eibesfeldt, I. 1965. *Land of a Thousand Atolls.* London: Macgibbon & Kee. 195 pp.

89. Eisner, T. 1970. Chemical defense against predation in arthropods. *Chemical Ecology,* ed. E. Sondheimer, J. B. Simeone, 157–217. New York: Academic

90. Eisner, T., Meinwald, J. 1966. Defensive secretions of arthropods. *Science* 153:1341–50

91. Emery, A. R. 1968. Preliminary observations on coral reef plankton. *Limnol. Oceanogr.* 13:293–303

92. Endean, R., Stephenson, W., Kenny, R. 1956. The ecology and distribution of intertidal organisms on certain islands off the Queensland coast. *Austr. J. Mar. Freshwater Res.* 7:317–42

93. Feddern, H. A. 1965. The spawning, growth, and general behavior of the bluehead wrasse, *Thalassoma bifasciatum* (Pisces: Labridae). *Bull. Mar. Sci.* 15:896–941

94. Feddern, H. A. 1968. *Systematics and ecology of western Atlantic angelfishes, family Chaetodontidae, with an analysis of hybridization in Holacanthus.* PhD dissertation. Univ. Miami, Miami, Fla. 117 pp.

95. Feder, H. M. 1966. Cleaning symbiosis in the marine environment. In *Symbiosis, Vol. 1,* ed. S. M. Henry, 327–80. New York: Academic

96. Fishelson, L. 1963. Observations on the biology and behaviour of Red Sea coral fishes. *Sea Fish. Res. Sta. Haifa Bull.* 37:11–26

97. Fishelson, L. 1965. Observations and experiments on the Red Sea anemones and their symbiotic fish *Amphiprion bicinctus. Bull. Sea Fish. Res. Sta. Haifa. Bull.* 39:1–14

98. Fishelson, L. 1970. Behavior and ecology of a population of *Abudefduf saxatilis* (Pomacentridae, Teleostei) at Eliat (Red Sea). *Anim. Behav.* 18: 225–37

99. Fishelson, L. 1970. Protogynous sex reversal in the fish *Anthias squamipinnis* (Teleostei, Anthiidae) regulated by the presence or absence of a male fish. *Nature* 227:40–41

100. Fishelson, L. 1973. Ecology of coral reefs in the Gulf of Aqaba (Red Sea) influenced by pollution. *Oecologia* 12:55–67

101. Fishelson, L., Popper, D., Ganderman, N. 1971. Diurnal cyclic behaviour of *Pempheris oualensis* Cuv. and Val. (Pempheridae, Teleostei). *J. Natur. Hist.* 5:503–6

102. Fraenkel, G. 1959. The raison d'etre of secondary plant substances. *Science* 129:1466–70

103. Franzisket, L. 1959. Experimentelle Untersuchung über die optische Wirkung der Streifung beim Preussenfisch (*Dascyllus aruanus*). *Behaviour* 15: 77–81

104. Fraser-Brunner, A. 1950. *Holacanthus xanthotis* sp. n. and other chaetodont fishes from the Gulf of Aden. *Proc. Zool. Soc. London* 120:43–48

105. Fricke, H. 1966. Attrappenversuche mit einigen plakatfarbigen Korallenfischen im Roten Meer. *Z. Tierpsychol.* 23:4–7

106. Fricke, H. 1966. Zum Verhalten des Putzerfisches, *Labroides dimidiatus. Z. Tierpsychol.* 23:1–3

107. Fricke, H. 1971. Fische als Feinde tropischer Seeigel. *Mar. Biol.* 9:328–38

108. Fricke, H. 1972. *The Coral Seas.* New York: Putnam. 224 pp.

109. Fricke, H. 1973. Behaviour as part of ecological adaptation—in situ studies in the coral reef. *Helgol. Wiss. Meeresunter.* 24:120–44

110. Garnaud, J. 1951. Nouvelles données sur l'éthologie d'un pomacentride: *Amphiprion percula* Lacépède. *Bull. Inst. Oceanogr. (Monaco).* 48:1–11

111. Garnaud, J. 1957. Ethologie de *Dascyllus trimaculatus* (Ruppell). *Bull. Inst. Oceanogr. Monaco* 54:1–10

112. Gilbert, L. E. 1972. Pollen feeding and reproductive biology of *Heliconius* butterflies. *Proc. Nat. Acad. Sci. USA* 69:1403–7

113. Glynn, P. W., Stewart, R. H., McCosker, J. E. 1972. Pacific coral reefs of Panama: structure, distribution and predators. *Geol. Rundsch.* 61:483–519

114. Glynn, P. W. 1973. Aspects of the ecology of coral reefs in the western Atlantic region. In *Biology and Geology of Coral Reefs,* ed. O. A. Freedman, R. Endean, 271–324. New York: Academic

115. Gohar, H. A. F. 1934. Partnership between fish and anemone. *Nature* 134:291

116. Goldman, B. 1973. *Aspects of the ecology of the coral reef fishes of One Tree Island.* PhD dissertation. Macquarie Univ., Sydney, Australia. 193 pp.

117. Gosline, W. A. 1953. The nature and evolution of the Hawaiian inshore fish fauna. *Proc. Pac. Sci. Cong., 8th Philippines,* 3:347–58

118. Gosline, W. A. 1965. Vertical zonation of inshore fishes in the upper water layers of the Hawaiian Islands. *Ecology* 46:823–31

119. Gosline, W. A., Brock, V. E. 1960. *A Handbook of Hawaiian Fishes.* Honolulu: Univ. Hawaii Press. 377 pp.

120. Greenwood, P. H., Rosen, D. E., Weitzman, S. H., Myers, G. S. 1966. Phyletic studies of the teleostean fishes, with a provisional classification of living forms. *Bull. Am. Mus. Natur. Hist.* 131:341–455

121. Grobe, J. 1960. Putz-Symbiosen zwischen Fischen und Garnelen. *Natur Volk* 90:152–57

122. Gudger, E. W. 1941. Coelenterates as enemies of fishes. IV. Sea anemones and corals as fish eaters. *New England Natur.* 10:1–8

123. Hamilton, W. J. III., Peterman, R. M. 1971. Countershading in the colourful reef fish *Chaetodon lunula*: concealment, communication or both? *Anim. Behav.* 19:357–64

124. Hartline, A. C., Hartline, P. H., Szmant, A. M., Flechsig, A. O. 1972. Escape response in a pomacentrid reef fish, *Chromis cyaneus*. *Results of the Tektite Program: Ecology of Coral Reef Fishes, Bull. Nat. Hist. Mus. L.A.* 14:93–97

125. Hediger, H. 1968. Putzer-fische in Aquarium. *Natur. Mus.* 98:89–96

126. Helfrich, P. 1958. *The early life history and reproductive behavior of the maomao, Abudefduf abdominalis (Quoy and Gaimard).* PhD dissertation. Univ. Hawaii, Honolulu. 207 pp.

127. Herald, E. S. 1961. *Living Fishes of the World.* Garden City, NY: Doubleday. 304 pp.

128. Herald, E. S., Randall, J. E. 1972. Five new Indo-Pacific pipefishes. *Proc. Calif. Acad. Sci.* 39:121–40

129. Hiatt, R. W., Strasburg, D. W. 1960. Ecological relationships of the fish fauna on coral reefs of the Marshall Islands. *Ecol. Monogr.* 30:65–127

130. Hinton, S. 1962. Unusual defense movements in *Scorpaena plumieri mystes.* *Copeia* 1962:842

131. Hobson, E. S. 1965. Diurnal-nocturnal activity of some inshore fishes in the Gulf of California. *Copeia* 1965:291–302

132. Hobson, E. S. 1968. Predatory behavior of some shore fishes in the Gulf of California. *Res. Rep. 73 Bur. Sport Fish. Wildl.,* 1–92

133. Hobson, E. S. 1968. Coloration and activity of fishes, day and night. *Underwater Natur.* Winter:6–11

134. Hobson, E. S. 1969. Possible advantages to the blenny *Runula azalea* in aggregating with the wrasse *Thalassoma lucasanum* in the tropical Western Pacific. *Copeia* 1969:191–93

135. Hobson, E. S. 1969. Comments on certain recent generalizations regarding cleaning symbiosis in fishes. *Pac. Sci.* 23:35–39

136. Hobson, E. S. 1971. Cleaning symbiosis among California inshore fishes. *Fish. Bull.* 69:491–523

137. Hobson, E. S. 1972. Activity of Hawaiian reef fishes during the evening and morning transitions between daylight and darkness. *Fish. Bull.* 70:715–40

138. Hobson, E. S. 1973. Diel feeding migrations in tropical reef fishes. *Helgo. Wiss. Meeresunter.* 24:361–70

139. Hobson, E. S. 1974. Feeding relationships of teleostean fishes on coral reefs in Kona, Hawaii. *Fish. Bull.* 72:915–1031

140. Johannes, R. E. 1972. Coral reefs and pollution. In *Marine Pollution and Sea Life,* ed. M. Ruivo. London: Fishing News Books

141. Johnson, M. W. 1959. The offshore drift of larvae of the California spiny lobster *Panulirus interruptus.* *Calif. Coop. Oceanic Fish. Invest. Rep.* 7:147–61

142. Jones, R. S. 1968. Ecological relationships in Hawaiian and Johnston Island Acanthuridae (surgeonfishes). *Micronesica* 4:309–61

143. Jones, R. S., Randall, R. H., Cheng, Y., Kami, H. T., Mak, S. 1972. A marine biological survey of southern Taiwan with emphasis on corals and fishes. *Inst. Oceanogr. Nat. Taiwan Univ. Spec. Publ. No. 1*

144. Keenleyside, M. H. A. 1972. The behaviour of *Abudefduf zonatus* (Pisces, Pomacentridae) at Heron Island, Great Barrier Reef. *Anim. Behav.* 20:763–74

145. Lachner, E. A. 1955. Inquilinism and a new record for *Paramia bipunctata,* a cardinal fish from the Red Sea. *Copeia* 1955:53–54

146. Lack, D. 1968. *Ecological Adaptations for Breeding in Birds.* London: Methuen. 409 pp.

147. Levin, S. 1974. Dispersion and population interactions. *Am. Natur.* 108:207–28

148. Limbaugh, C. 1964. Notes on the life history of two California pomacentrids: garibaldis, *Hypsypops rubicunda* (Gerard), and blacksmiths, *Chromis punctipinnis* (Copper). *Pac. Sci.* 18:41–50

149. Livingston, R. T. 1971. Circadian rhythms in the respiration of eight species of cardinal fishes (Pisces: Apogonidae); comparative analysis and adaptive significance. *Mar. Biol.* 9:253–66

150. Longley, W. H. 1914. Report upon color of fishes of the Tortugas reefs. *Carnegie Inst. Washington, Yearb.* 13:207–8

151. Longley, W. H. 1915. Coloration of tropical reef fishes. *Carnegie Inst. Washington Yearb.* 14:208–9

152. Longley, W. H. 1916. The significance of the colors of tropical reef fishes. *Carnegie Inst. Washington Yearb.* 15:209–12

153. Longley, W. H., Hildebrand, S. F. 1941. Systematic catalogue of the fishes of Tortugas, Florida with observations in color, habits, and local distribution. *Carnegie Inst. Washington Publ. 535. Pap. Tortugas Lab. 34.* 331 pp.

154. Lorenz, K. 1962. The function of colour

in coral reef fishes. *Proc. R. Inst. G. B.* 39:282–96

155. Lorenz, K. 1963. *Das Sogenannte Böse, Zur Naturgeschichte der Aggression.* Vienna: Barotha-Schoeler. 306 pp.

156. Losey, G. S. 1971. Communication between fishes in cleaning symbiosis. In *Aspects of the Biology of Symbiosis,* ed. T. C. Ching, 45–76. Baltimore: Univ. Park Press

157. Losey, G. S. 1972. Predation protection in the poison-fang blenny, *Meiacanthus atrodorsalis,* and its mimics, *Escenius bicolor* and *Runula laudadus* (Blenniidae). *Pac. Sci.* 26:129–39

158. Losey, G. S. 1972. The ecological importance of cleaning symbiosis. *Copeia* 1972:820–33

159. Losey, G. S. 1972. Behavioural ecology of the "cleaning fish." *Aust. Natur. Hist.* 17:232–38

160. Low, R. M. 1971. Interspecific territoriality in a pomacentrid reef fish *Pomacentrus flavicauda* Whitley. *Ecology* 52:648–54

161. Magus, D. B. 1967. Ecological and ethological studies on echinoderms of the Red Sea. *Stud. Trop. Oceanogr. Miami* 5:635–64

162. Mariscal, R. N. 1969. The protection of the anemone fish, *Amphiprion xanthurus* from the sea anemone, *Stoichactis kenti.* *Experientia* 25:1114

163. Mariscal, R. N. 1970. A field and laboratory study of the symbiotic behavior of fishes and sea anemones from the tropical Indo-Pacific. *Univ. Calif. Berkeley Publ. Zool.* 91:1–33

164. Mariscal, R. N. 1970. The nature of the symbiosis between Indo-Pacific anemone fishes and sea anemone. *Mar. Biol.* 6:58–65

165. Mariscal, R. N. 1970. An experimental analysis of the protection of *Amphiprion xanthurus* Cuvier and Valenciennes and some other anemone fishes from sea anemones. *J. Exp. Mar. Biol. Ecol.* 4:134–49

166. Mariscal, R. N. 1972. Behavior of symbiotic fishes and sea anemones. In *Behavior of Marine Animals,* ed. H. E. Winns, B. L. Olla, 2:2327–60. New York: Plenum

167. Menzel, D. W. 1959. Utilization of algae for growth by the angelfish, *Holacanthus bermudensis.* *J. Cons. Perm. Int. Explor. Mer.* 24:308–13

168. Menzel, D. W. 1960. Utilization of food by a Bermuda reef fish, *Epinephelus gut-*

tatus. *J. Cons. Perm. Int. Explor. Mer.* 25:216–22

169. Myrberg, A. A. Jr. 1972. Social dominance and territoriality in the bicolor damselfish, *Eupomacentrus partitus* (Poey) (Pisces: Pomacentridae). *Behaviour* 41:207–31

170. Myrberg, A. A. Jr., Brahy, B. D., Emery, A. R. 1967. Field observations on reproduction of the damselfish *Chromis multilineata* (Pomacentridae), with additional notes on general behavior. *Copeia* 1967:819–27

171. Myrberg, A. A. Jr., Spites, J. Y. 1972. Sound discrimination by the bicolor damselfish, *Eupomacentrus partitus.* *J. Exp. Biol.* 57:727–35

172. Myrberg, A. A. Jr., Thresher, R. E. 1974. Interspecific aggression and its relevance to the concept of territoriality in reef fishes. *Am. Zool.* 14:81–96

173. Neill, S. R. S. J., Cullen, J. M. 1974. Experiments on whether schooling by their prey affects the hunting behavior of cephalopods and fish predators. *J. Zool. London* 172:549–69

174. Newell, N. D. 1971. An outline history of tropical organic reefs. *Am. Mus. Novi.* 265:1–37

175. Newman, W. A. 1960. On the paucity of intertidal barnacles in the tropical western Pacific. *Veliger* 2:89–94

176. Odum, H. T., Odum, E. P. 1955. Trophic structure and productivity of a windward coral reef community on Eniwetok Atoll. *Ecol. Monogr.* 25:291–320

177. Ogden, J. C. 1972. An ecological study of Tague Bay Reef, St. Croix, U.S.V.I. *Spec. Publ. Mar. Biol. West Indies Lab. Fairleigh Dickenson Univ. No. 1*

178. Ogden, J. C., Abbott, D. P., Abbott, I. 1973. Studies on the activity and food of the echinoid *Diadema antillarum* Philippi on a West Indian patch reef. *Spec. Publ. Mar. Biol. West Indies Lab. Fairleigh Dickenson Univ. No. 2*

179. Ogden, J. C., Brown, R. A., Salesky, N. 1974. Grazing by the echinoid *Diadema antillarum* Philippi: formation of halos around West Indian patch reefs. *Science* 182:715–17

180. Ogden, J. C., Buckman, N. S. 1973. Movements, foraging groups, and diurnal migrations of the striped parrotfish *Scarus croicensis* Block (Scaridae). *Ecology* 54:589–96

181. Ogden, J. C., Sammarco, P. W., Abbott, D. P. 1973. *Diadema antillarum* and plants: a study of shallow water marine

food chain. *Proc. Assoc. Isl. Mar. Labs. Caribb. Mayaguey* (Abstr.)
182. Okuno, R. 1962. Intra- and interspecific relations of saltwater fishes in aquarium. I. Butterfly fishes. *Jpn. J. Ecol.* 12:129–33
183. Okuno, R. 1963. Observations and discussion on the social behavior of marine fishes. *Publ. Seto Mar. Biol. Lab.* 11:281–336
184. Oppenheimer, J. R. 1970. Mouthbreeding in fishes. *Anim. Behav.* 18:493–503
185. Paine, R. T., Vadas, R. L. 1969. The effects of grazing by sea urchins, *Strongylocentrotus* spp. on benthic algal populations. *Limnol. Oceanogr.* 14:710–19
186. Peterman, R. M. 1971. A possible function of coloration in coral reef fishes. *Copeia* 1971:330–31
187. Popper, D., Fishelson, L. 1973. Ecology and behavior and *Anthias squamipinnis* (Peters, 1855) (Anthiidae, Teleostei) in the coral habitat of Eilat (Red Sea). *J. Exp. Zool.* 184:409–24
188. Potts, G. W. 1969. Behaviour of the snapper, *Lutjanus monostigma* around Aldabra. *Underwater Assoc. Rep.* 1969:96–99
189. Potts, G. W. 1970. The schooling ethology of *Lutjanus monostigma* (Pisces) in the shallow reef environments of Aldabra. *J. Zool. London* 161:223–35
190. Power, A. 1969. *The Great Barrier Reef.* London: Hamlyn. 115 pp.
191. Quiguer, J. 1969. Quelques données sur la repartition des poissons des récifs coralliens. *Cah. Pac.* 13:181–85
192. Radakov, D. V. 1973. *Schooling in the Ecology of Fish.* New York: Wiley. 173 pp.
193. Randall, J. E. 1955. A revision of the surgeon fish genus *Ctenochaetus,* family Acanthuridae, with descriptions of five new species. *Zoologica* 40:149–66
194. Randall, J. E. 1956. A revision of the surgeon fish genus *Acanthurus. Pac. Sci.* 10:154–235
195. Randall, J. E. 1958. A review of ciguatera, tropical fish poisoning, with a tentative explanation of its cause. *Bull. Mar. Sci. Gulf Caribb.* 8:236–67
196. Randall, J. E. 1958. A review of the labrid fish genus *Labroides,* with description of two new species and notes on ecology. *Pac. Sci.* 12:327–47
197. Randall, J. E. 1960. New fishes for Hawaii. *Sea Frontiers* 6:31–43
198. Randall, J. E. 1961. Tagging reef fishes in the Virgin Islands. *Proc. Gulf Caribb. Fish. Inst.* 14:201–41

199. Randall, J. E. 1961. A contribution to the biology of the convict surgeonfish of the Hawaiian Islands, *Acanthurus triostegus sandvicensis. Pac. Sci.* 15:215–72
200. Randall, J. E. 1961. Observations on the spawning of surgeonfishes (Acanthuridae) in the Society Islands. *Copeia* 1961:237–38
201. Randall, J. E. 1961. Overgrazing of algae by herbivorous marine fishes. *Ecology* 42:812
202. Randall, J. E. 1961. Ciguatera: tropical fish poisoning. *Sea Frontiers* 7:130–39
203. Randall, J. E. 1962. Fish service stations. *Sea Frontiers* 8:40–47
204. Randall, J. E. 1963. Notes on the systematics of the parrotfishes (Scaridae), with emphasis on sexual dichromatism. *Copeia* 1963:225–37
205. Randall, J. E. 1963. An analysis of the fish populations of artificial and natural reefs in the Virgin Islands. *Caribb. J. Sci.* 3:31–47
206. Randall, J. E. 1965. Grazing effect on sea grasses by herbivorous reef fishes in the West Indies. *Ecology* 46:255–60
207. Randall, J. E. 1967. Food habits of reef fishes of the West Indies. *Stud. Trop. Oceanogr.* 5:665–847
208. Randall, J. E. 1968. *Caribbean Reef Fishes.* Jersey City, NJ: TFH Publ. 318 pp.
209. Randall, J. E. et al 1971. Grammistin, the skin toxin of soapfishes, and its significance in the classification of the Grammistidae. *Publ. Seto Mar. Biol. Lab.* 19:157–90
210. Randall, J. E., Böhlke, J. E. 1965. Review of the Atlantic labrid fishes of the genus *Halichoeres. Proc. Acad. Nat. Sci. Philadelphia* 117:235–59
211. Randall, J. E., Brock, V. E. 1960. Observations on the ecology of epepheline and lutjanid fishes of the Society Islands, with emphasis on food habits. *Trans. Am. Fish. Soc.* 89:9–16
212. Randall, J. E., Emery, A. R. 1971. On the resemblance of the young of the fishes *Platax pinnatus* and *Plectorhynchus chaetodontoides* to flat worms and nudibranchs. *Zoologica* 56:115–19
213. Randall, J. E., Helfman, G. 1972. *Diproctacanthus xanthurus,* a cleaner wrasse from the Palau Islands, with notes on other cleaning fishes. *Trop. Fish Hobbyist* 20:87–95
214. Randall, J. E., Kanayama, R. K. 1972. Hawaiian fish immigrants. *Sea Frontiers* 18:144–53
215. Randall, J. E., Randall, H. A. 1960. Examples of mimicry and protective re-

semblance in tropical marine fishes. *Bull. Mar. Sci. Gulf Caribb.* 10:444–80

216. Randall, J. E., Randall, H. A. 1963. The spawning and early development of the Atlantic parrotfish, *Sparisoma rubripinne*, with notes on other scarid and labrid fishes. *Zoologica* 48:49–60

217. Randall, J. E., Schroeder, R. E., Starck, W. A. 1964. Notes on the biology of *Diadema antillarum*. *Caribb. J. Sci.* 4:421–33

218. Rasa, O. A. E. 1969. Territoriality and the establishment of dominance by means of visual cues in *Pomacentrus jenkinsi* (Pisces: Pomacentridae). *Z. Tierpsychol.* 26:825–45

219. Reese, E. S. 1964. Ethology and marine zoology. *Ann. Rev. Oceanogr. Mar. Biol.* 1964:455–88

220. Reese, E. S. 1973. Duration of residence by coral reef fishes on "home" reefs. *Copeia* 1973:145–49

221. Reighard, J. 1908. An experimental field study of warning coloration in coral-reef fishes. *Carnegie Inst. Washington Pap. Tortugas Lab.* 1:257–325

222. Reinboth, R. 1962. Morphologische und funktionelle Zweigeschlechtlichkeit bei marinen Teleostien (Serranidae, Sparidae, Centracanthidae, Labridae). *Zool. Jahrb. Physiol.* 69:405–80

223. Reinboth, R. 1968. Protogynie bei Papageifischen (Scaridae). *Z. Naturforsch. Teil B.* 23:852–55

224. Robertson, D. R. 1972. Social control of sex reversal in coral-reef fish. *Science* 177:1007–9

225. Robertson, D. R. 1973. Sex changes under the waves. *New Sci.* 31 May: 538–39

226. Robertson, D. R., Choat, J. H. 1974. Protogynous hermaphroditism and social systems in labrid fish. *Proc. Int. Symp. Coral Reefs, 2nd.* 1:217–25

227. Roede, M. J. 1972. *Color as related to size, sex, and behavior in seven Caribbean labrid fish species (genera Thalassoma, Halichoeres and Hemipteronotus)*. The Hague: Nykoff

228. Roessler, C., Post, J. 1972. Prophylactic services of the cleaning shrimp. *Natur. Hist.* May:30–37

229. Rosenblatt, R. H. 1963. Some aspects of speciation in marine shore fishes. *Speciation in the Sea, Syst. Assoc. Publ. No. 5,* 171–80

230. Roth, L. M., Eisner, T. 1962. Chemical defenses of arthropods. *Ann. Rev. Entomol.* 7:107–36

231. Roughgarden, J. 1974. Species packing and the competition function with illustrations from coral reef fish. *Theor. Popul. Biol.* 5:163–86

232. Russell, B. C. 1971. Underwater observations on the reproductive activity of the demoiselle *Chromis dispilus* (Pisces: Pomacentridae). *Mar. Biol.* 10:22–24

233. Russell, B. C., Talbot, F. H., Domm, S. B. 1974. Patterns of colonization of artificial reefs by coral reef fishes. *Proc. Int. Symp. Coral Reefs, 2nd.* 1:207–15

234. Sale, P. F. 1969. Pertinent stimuli for habitat selection by the juvenile manini, *Acanthurus triostigus sandvicensis. Ecology* 50:616–23

235. Sale, P. F. 1970. Distribution of larval Acanthuridae off Hawaii. *Copeia* 1970:765–66

236. Sale, P. F. 1971. Extremely limited home range in a coral reef fish, *Dascyllus aruanus* (Pisces: Pomacentridae) *Copeia* 1971:324–27

237. Sale, P. F. 1971. Apparent effect of prior experience on a habitat preference exhibited by the reef fish, *Dascyllus aruanus* (Pisces:Pomacentridae). *Anim. Behav.* 19:251–56

238. Sale, P. F. 1971. The reproductive behavior of the pomacentrid fish, *Chromis caeruleus. Z. Tierpsychol.* 29:156–64

239. Sale, P. F. 1972. Influence of corals in the dispersion of the pomacentrid fish, *Dascyllus aruanus. Ecology* 53:741–44

240. Sale, P. F. 1972. Effect of cover on agonistic behavior of a reef fish: a possible spacing mechanism. *Ecology* 53:753–58

241. Sale, P. F. 1974. Mechanisms of coexistence in a guild of territorial fishes at Heron Island. *Proc. Int. Symp. Corals and Coral Reefs, 2nd.* 1:193–206

242. Sammarco, P. W., Levinton, J. S., Ogden, J. C. 1974. Grazing and control of coral reef community structure by *Diadema antillarum* Phillipi (Echinodermata: Echinoidae): a preliminary study. *J. Mar. Res.* 32:47–53

243. Santos, G. A., Doty, M. S. 1968. Chemical studies on three species of the marine algal genus *Caulerpa*. In *Drugs From the Sea*, ed. H. D. Frendenthal, Washington: Mar. Tech. Soc.

244. Scheltema, R. S. 1971. Larval dispersal as a means of genetic exchange between geographically separated populations of shallow-water benthic marine gastropods. *Biol. Bull.* 140:284–322

245. Schlichter, D. 1968. Das Zusammenleben von Riffanemonen und Anemonenfischen. *Z. Tierpsychol.* 25:933–54

246. Schlichter, D. 1970. *Thalassoma amblycephalus* ein neuer Anemonefisch. *Typ. Mar. Biol.* 7:269–72

247. Schroeder, R. E., Starck, W. A. II 1964. Photographing the night creatures of Alligator Reef. *Nat. Geogr.* 125:128–54

248. Schultz, L. P. 1958. Review of the parrotfishes, family Scaridae. *Bull. US Nat. Mus.* 214:1–143

249. Schultz, L. P. 1969. The taxonomic status of the controversial genera and species of parrotfishes with a descriptive list (Family Scaridae). *Smithsonian Contrib. Zool.* 17:1–49

250. Seghers, B. H. 1974. Schooling behavior in the guppy (*Poecilia reticulata*): an evolutionary response to predation. *Evolution* 28:486–89

251. Shallenberger, R. J., Madden, W. D. 1973. Luring behavior in the scorpionfish, *Iracundus signifer. Behaviour* 47:33–47

252. Sharp, M. A., Parks, D. R., Ehrlich, P. R. 1974. Plant resources and butterfly habitat selection. *Ecology* 55:870–75

253. Shaw, E. 1960. The development of schooling behavior in fishes. *Physiol. Zool.* 33:263–72

254. Shaw, E. 1961. The development of schooling in fishes. II. *Physiol. Zool.* 34:263–72

255. Shaw, E. 1962. The schooling of fishes. *Sci. Am.* 206:128–38

256. Shaw, E. 1965. The optomotor response and the schooling of fish *Int. Comm. Northwest Atl. Fish. Spec. Publ. No. 6*

257. Shaw, E. 1969. The duration of schooling among fish separated and those not separated by barriers. *Am. Mus. Novit.* 2373:1–13

258. Shaw, E. 1970. Schooling in fishes: critique and review. In *Development and Evolution of Behavior,* ed. R. Aronson et al, 453–80. San Francisco: Freeman

259. Shaw, E., Tucker, A. 1965. The optomotor reaction of schooling carangid fishes. *Anim. Behav.* 13:330–36

260. Shull, E. M. 1962. Over one hundred butterfly species caught in a single day (3rd June 1961) at Mussoorie, India. *J. Lepid. Soc.* 16:143–45

261. Slobodkin, L. B., Fishelson, L. 1974. The effect of the cleaner fish *Labroides dimidiatus* on the point diversity of fishes on the reef front at Eliat. *Am. Natur.* 108:369–76

262. Smith, C. L. 1965. The patterns of sexuality and the classification of serranid fishes. *Am. Mus. Novit.* 2207:1–20

263. Smith, C. L. 1973. Small rotenone stations: a tool for studying coral reef fish communities. *Am. Mus. Novit.* 2512:1–21

264. Smith, C. L., Tyler, J. C. 1972. Space resource sharing in a coral reef fish community. *Results of the Tektite Program: Ecology of Coral Reef Fishes, Bull. Natur. Hist. Mus. L.A.* 14:125–70

265. Smith, C. L., Tyler, J. C. 1973. Population ecology of a Bahamian suprabenthic shore fish assemblage. *Am. Mus. Novit.* 2528:1–38

266. Smith, C. L., Tyler, J. C. 1973. Direct observations of resource sharing in coral reef fish. *Helgol. Wiss. Meeresunter.* 24:264–75

267. Somero, G., Soulé, M. 1974. Genetic variation in marine fishes as a test of the niche-variation hypothesis. *Nature* 249:670–72

268. Soulé, M. 1973. The epistasis cycle: a theory of marginal populations. *Ann. Rev. Ecol. Syst.* 4:165–87

269. Springer, S. 1957. Some observations on the behavior of schools of fishes in the Gulf of Mexico and adjacent waters. *Ecology* 38:166–71

270. Springer, V. G., McErlean, A. J. 1962. A study of the behavior of some tagged south Florida coral reef fishes. *Am. Med. Natur.* 67:386–97

271. Springer, V. G., Smith-Vaniz, W. F. 1972. Mimetic relationships involving fishes of the family Blenniidae. *Smithsonian Contrib. Zool.* 112:1–36

272. Starck, W. A. II, Davis, W. P. 1966. Night habits of fishes of Alligator Reef, Florida. *Ichthyologica* 38:313–56

273. Starck, W. A. II, Schroeder, R. E. 1965. A coral reef at night. *Sea Frontiers* 11:66–74

274. Starck, W. A. II, Schroeder, R. E. 1971. Investigations on the gray snapper, *Lutjanus griseus. Stud. Trop. Oceanogr. Miami* 10:1–224

275. Stephens, J. S., Johnson, R. K., Key, G. S., McCosker, J. E. 1970. The comparative ecology of three sympatric species of California blennies of the genus *Hypsoblennius* Gill (Teleostomi, Blenniidae). *Ecol. Monogr.* 40:213–33

276. Stephenson, W., Endean, R., Bennett, I. 1958. An ecological study of the marine fauna of Low Isles, Queensland. *Austr. J. Mar. Freshwater Res.* 9:262–318

277. Stephenson, W., Searle, R. B. 1960. Experimental studies on the ecology of intertidal environments at Heron Island. I. Exclusion of fish from beach rock. *Austr. J. Mar. Freshwater Res.* 11:241–67

278. Stevenson, R. A. 1963. Behavior of the pomacentrid reef fish *Dascyllus albisella*

Gill in relation to the anemone *Marcanthia cookei. Copeia* 1963:612–14

279. Stevenson, R. A. 1963. *Life history and behavior of Dascyllus albisella Gill, a pomacentrid reef fish.* PhD dissertation. Univ. Hawaii, Honolulu, Hawaii. 221 pp.

280. Stevenson, R. A. 1972. Regulation of feeding behavior of the bicolor damselfish (*Eupomacentrus partitus* Poey) by environmental factors. In *Behavior of Marine Animals,* 2: *Vertebrates,* ed. H. E. Winn, B. L. Olla. New York: Plenum

281. Stoll, L. M. 1955. Hormonal control of sexually dimorphic pigmentation of *Thalassoma bifasciatum. Zoologica* 40:125–31

282. Strasburg, D. W. 1966. Observations on the ecology of four apogonid fishes. *Pac. Sci.* 20:338–41

283. Swerdloff, S. N. 1970. *The comparative biology of two Hawaiian species of the damselfish genus Chromis (Pomacentridae).* PhD dissertation. Univ. Hawaii, Honolulu, Hawaii. 183 pp.

284. Talbot, F. H. 1965. A description of the coral structure of Tutia Reef (Tanganyika Territory, East Africa) and its fish fauna. *Proc. Zool. Soc. London* 145:431–70

285. Talbot, F. H. 1970. The south east Asian area as a centre of marine speciation: an ecological analysis of causes. *Rep. Austr. Acad. Sci.* 12:43–50

286. Talbot, F. H., Goldman, B. 1972. A preliminary report on the diversity and feeding relationships of the reef fishes of One Tree Island, Great Barrier Reef system. *Proc. Symp. Corals Coral Reefs, 1969, Mar. Biol. Assoc. India* 425–44

287. Talbot, F. H., Goldman, B. 1974. The ecology of coral reef fishes. In *The Biology and Geology of Coral Reefs,* ed. O. A. Jones, R. Endean. NY: Academic

288. Tavolga, W. N. 1956. Visual, chemical and sound stimuli as cues in the sex discriminatory behavior of the gobiid fish *Bathygobius soporator. Zoologica* 41:49–64

289. Teytaud, A. R. 1971. Food habits of the goby, *Ginsburgellus novemlineatus,* and the clingfish, *Arcos rubiginosus,* associated with echinoids in the Virgin Islands. *Caribb. J. Sci.* 11:41–45

290. Thomson, J. M., Bennett, A. E. 1953. The oyster blenny, *Omobranchius anolus* (Valenciennes) (Blenniidae). *Austr. J. Mar. Freshwater Res* 4:227–33

291. Thresher, R. E. 1974. Small predators on the reef. *Sea Frontiers* 20:219–27

292. Turner, C. H., Ebert, E. E. 1962. The nesting of *Chromis punctipinnis* (Cooper) and a description of their eggs and larvae. *Calif. Fish Game* 48:243–48

293. Tyler, J. C. 1971. Habitat preferences of the fishes that dwell in shrub corals on the Great Barrier Reef. *Proc. Acad. Nat. Sci. Philadelphia* 123:1–26

294. Verivey, D. J. 1930. Coral reef studies. I. The symbiosis between damselfishes and sea anemones in Batavia Bay. *Treubia* 12:305–53

295. Vine, P. J. 1974. Effects of algal grazing and aggressive behaviour of the fishes *Pomacentrus lividus* and *Acanthurus sohol* on coral-reef ecology. *Mar. Biol.* 24:131–36

296. Wickler, W. 1957. Vergleichende Verhaltensstudien an Grundfischen. I. Beitrage zur Biologie, besonders zur Ethologie von *Blennius fluviatilis* Asso im Vergleich zu einigen anderen Bodenfischen. *Z. Tierpsychol.* 14:393–428

297. Wickler, W. 1960. Aquarienbeobachtungen an *Aspidontus* einem ektoparasitischen Fisch. *Z. Tierpsychol.* 17:277–92

298. Wickler, W. 1961. Uber des Verhalten der Blenniiden *Runula* und *Aspidontus* (Pisces: Blenniidae). *Z. Tierpsychol.* 18:421–40

299. Wickler, W. 1963. Zum Problem der Signalbildung, am Beispiel der Vehaltsenmimikry zwischen *Aspidontus* und *Labroides* (Pisces, Acanthopterygii) *Z. Tierpsychol.* 20:657–79

300. Wickler, W. 1967. Specialization of organs having a signal function in some marine fish. *Stud. Trop. Oceonogr. Miami* 5:539–48

301. Wickler, W. 1967. Vergleich des Ablaichverhaltens einiger paarbildender sowie nicht-paarbildender Pomocentriden und Cichliden (Pisces: Perciformes). *Z. Tierpsychol.* 24:457–70

302. Wickler, W. 1968. *Mimicry in Plants and Animals.* New York: World Univ. Library. 249 pp.

303. Williams, G. C. 1964. Measurement of consocation among fishes and comments on the evolution of schooling. *Publ. Mus. Mich. State Univ.* 2:351–83

304. Williams, M. M., Shaw, E. 1971. Modifiability of schooling behavior in fishes: the role of early experience. *Am. Mus. Novit.* 2448:1–19

305. Winn, H. E. 1955. Formation of a mucous envelope at night by parrot fishes. *Zoologica* 40:145–47

306. Winn, H. E., Bardach, J. E. 1957. Behavior, sexual dichromatism and

species of parrotfishes. *Science* 125: 885–69

307. Winn, H. E., Bardach, J. E. 1959. Differential food selection by moray eels and a possible role of the mucous envelope of parrotfishes in reduction of predation. *Ecology* 40:296–98

308. Winn, H. E., Bardach, J. E. 1960. Some aspects of the comparative biology of parrotfishes at Bermuda. *Zoologica* 45:24–34

309. Winn, H. E., Marshall, J. A., Hazlett, B. 1964. Behavior, diel activities, and stimuli that elicit sound production and reactions to sounds in the longspine

squirrelfish. *Copeia* 1964:413–25

310. Winn, H. E., Salmon, M., Roberts, N. 1964. Sun-compass orientation by parrot fishes. *Z. Tierpsychol.* 21:798–812

311. Yonge, C. M. 1968. Living corals. *Proc. R. Soc. B* 169:329–44

312. Youngbluth, M. J. 1968. Aspects of the ecology and ethology of the cleaning fish, *Labroides phthirophagus* Randall. *Z. Tierpsychol.* 25:915–32

313. Zumpe, D. 1965. Laboratory observations on the aggressive behaviour of some butterfly fishes (Chaetodontidae). *Z. Tierpsychol.* 22:226–36

LATE QUATERNARY CLIMATIC CHANGE IN AFRICA

❖4094

D. A. Livingstone
Department of Zoology, Duke University, Durham, North Carolina 27706

INTRODUCTION

During the past decade, emphasis in ecology has shifted from ecosystem studies to the evolutionary background of natural communities. The community analog of physiology, though far from completely understood, has been supplanted by a community analog of embryology as the most compelling and provocative branch of the subject.

This recognition of the systematic and evolutionary complexity of ecological phenomena follows a period during which ecologists sometimes lost sight of the organisms while in pursuit of transferred calories or cycled geochemicals. By recognizing the complexity of the world we will be in a better position to deal with it realistically. If, at the same time, we can discern some order in the complexity by using or extending evolutionary and genetic theory, so much the better.

By adding evolutionary time to the ecological game, however, we generate a need to know things about the natural geography of our planet that are not part of the standard intellectual equipment of biologists. In particular, an accurate understanding of the nature, amplitude, and timing of changes in the geographical parameters of ecological systems is vital. Some of the fundamental data, such as the rate of movement of tectonic plates, have only recently been established. Facts such as the basic mutability of tropical environments were not taught because they seemed of little importance in biology, even though they have been well known to geographers and geologists for two generations. Ecologists interested in making effective use of an evolutionary temporal perspective must be alert to ways in which the world has changed, ways in which it is changing now, and ways in which it will change. This review summarizes current knowledge in one branch of natural geography, the late-Quaternary paleoclimatology of Africa.

Unlike Darwin, leaders of the new evolutionary movement in ecology have not been especially concerned with environmental change. Most of them were educated in the temperate zone where both the annual march of the seasons and the great Pleistocene climatic changes are expressed vividly. When temperate ecologists go to

249

the tropics they tend to believe, in the absence of clear evidence for great temperature change, that no significant changes of climate have occurred there. In America, at least, the most readily accessible tropical environments have been West Indian islands and the Isthmus of Panama, small land masses with atypical oceanic climates. Brief field experience in such habitats reinforces the expectation that the climate of the tropics has been unchanging.

Some idea of the impact of this misapprehension upon ecology may be drawn from two examples: one in pedagogy, one in research. Krebs (67, p. 513), in one of the best of a recent crop of thoughtful ecology texts, presented without adverse comment the idea that tropical species diversity is high because of the "constant favorable environment" of the tropics. MacArthur & Wilson (76) developed this decade's most fertile body of deductive ecological theory on the assumption of a climatic steady state. It is significant that the most spectacular failure of island biogeographic theory concerns the mammals of the Cordillera (16) to which climatic change is of overriding importance, and that the accuracy of biogeographic theory predictions for the Andean flora can be markedly improved (B. Simpson, personal communication) by recognition of changes in tropical climates.

GEOGRAPHY OF TROPICAL AFRICA

Africa has a high mean elevation but few mountain ranges (Figure 1). Such relief as does occur is due mostly to faulting and associated volcanism rather than to folding, and is limited in area. There are extensive plateau areas in Ethiopia above 2000 m; less extensive high plateaus, with isolated mountains rising above 5000 m, in Uganda, Kenya, and Tanzania; and Mt. Cameroon, rising to 4069 m in the west. The Atlas, a true mountain range, exceeds 4000 m. Plateaus of more modest altitude, but over 1000 m high, are very common, especially in a belt from Ethiopia to South Africa.

The high elevation has prevailed for a long time, and Africa has been little affected by the extensive shallow seas that have often covered great areas of other continents. As a result, calcareous sedimentary rocks are scarce, and much of Africa is composed of crystalline Precambrian rocks overlain in places by Tertiary and Quaternary volcanics. The main weathering reactions involve silicate rocks rather than congruent solution of sediments. Garrels & Mackenzie (49, p. 125) estimate that 38% of the CO_2 consumed by Africa goes into silicate reactions, as opposed to 7% for Europe, a continent rich in limestones and dolomites.

Most of Africa is warm, in places even hot, but few parts are well watered. The Sahara is the most extensive and one of the most severe deserts, but even at lower latitudes rain is often not abundant. The equatorial Zaire (Congo) cuvette and a coastal strip along the Gulf of Guinea in West Africa are consistently moist enough to maintain evergreen tropical forest (Figure 2), and there are small patches of evergreen forest elsewhere, particularly on mountains. The widespread vegetation types of tropical Africa are, however, woodlands, savannas, and grasslands. East Africa, even along the equator, is anomalously dry, and vegetation types approaching desert are to be found at latitudes well within the "humid" tropics.

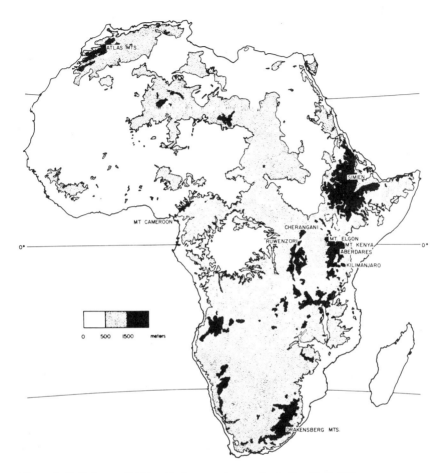

Figure 1 Relief map of Africa. Redrawn and generalized after Clark (26).

Such rain as does fall usually follows the sun; a single dry season and a single wet season is the rule except along the equator, where many stations experience a rainy season associated with each equinox. The pattern of airflow exhibits monsoon effects, and is influenced by the Indian Ocean monsoon of neighboring Asia.

Tropical rain is associated in most equatorial parts of the world with the Inter-tropical Front, or Intertropical Convergence Zone, which shifts its position north and south with the seasons. Thompson (125), in a succinct and informative essay on the climate of Africa, recounted the futile early attempts to forecast African weather on the assumption that the frontal pattern also prevailed there. On climatic maps it is possible to plot a latitudinal convergence zone, especially in West Africa, and the zone does move north and south with the seasons. On synoptic maps, representing the weather at a particular time rather than long-term average condi-

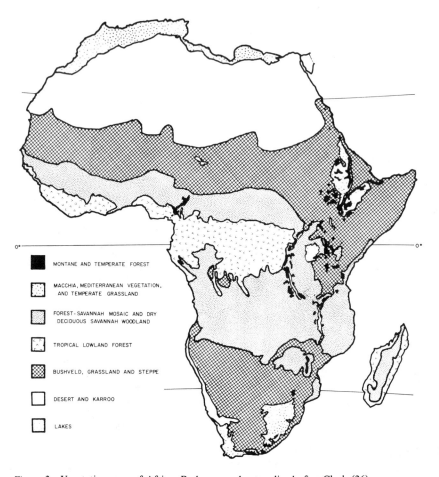

Figure 2   Vegetation map of Africa. Redrawn and generalized after Clark (26).

tions, it is less easy to recognize such a convergence, and the convergence is seldom associated with rain.

Even in western Africa, where it develops best, the convergence is a very shallow feature, and the dominant air flows are divergent, even at quite moderate altitudes. Rain is not associated with frontal systems or with traveling disturbances. Tropical African rainfall is local. A rainstorm develops, rain falls for a while, and then stops. There is no movement of the rainstorm to another locality, and weather, for the most part, must be forecast on the basis of the pressure field rather than on the basis of frontal theory and assumed streamline flows.

Near the equator the horizontal components of the Coriolis' force are weak, and plotting geostrophic wind directions from the pressure field is difficult. The vertical component of the force is strong, and combines with centrifugal effects to give westerly winds a tendency to rise and generate rain, and easterly winds a tendency to sink and produce dry weather. Although this phenomenon may account for part of the aridity of eastern equatorial Africa by preventing much rain from monsoonal flows off the Indian ocean, a general deep divergent flow also seems to be involved. The topographic diversity of eastern Africa produces many local rain shadows and orographic wet spots, but the general pattern is one of remarkably low equatorial rainfall.

The Zairean rain forest and the forest fringe along the Gulf of Guinea, watered by rising westerly flows off the Atlantic, are surrounded by concentric bands of progressively more xeric vegetation (Figure 2): forest-savanna mosaic, woodland, savanna, grassland, and a variety of thickets and thorn scrubs. Large areas of the forest-savanna mosaic are sprinkled with outliers of rain forest, especially in fire-protected localities, and it is reasonable to believe that these regions would be more heavily forested without a large human population.

The flora of equatorial Africa is somewhat poorer than that of South America, and there is a corresponding relative poverty of bird species. However, there is a very rich mammal fauna, particularly of Bovidae, and the ancient Rift Valley lakes support spectacular species flocks of fishes, molluscs, and other animals.

Temperature seems to have little effect on the distribution of most organisms, which cover enormous ranges of latitudes wherever the rainfall is suitable. Many high mountain species, however, are apparently under considerable temperature control, and some montane species are widely disjunct from their nearest relatives in the temperate zones. Butterflies (25) and birds (82–86) show clear affinities between Mt. Cameroon in the west and the highlands and high mountains of East Africa in the east. Although the distance is much shorter between East Africa and the Ethiopian highland to the north, few birds have been able to make the short crossing across very dry country, whereas many have traversed the more humid route to Mt. Cameroon. This pattern has been treated as presumptive evidence for climatic change because a general lowering of climatic belts would permit montane species now isolated in East and West Africa to extend their ranges over the intervening plateaus.

HIGH MOUNTAINS

Glacial Geology

The highlands of Ethiopia and East Africa (Figure 1) provide some of the clearest and most convincing evidence of climatic change on the entire continent. Large termino-lateral moraines festoon the sides of the snow-clad peaks Kilimanjaro, Kenya, and Ruwenzori, as well as the slopes of Mt. Elgon, Satima in the Aberdares and the mountains of Simien in Ethiopia, which do not support glaciers today. Anomalous though an equatorial ice age may seem, there is no doubt that one occurred. Apparently it involved a lowering of mean annual temperature comparable to that of the temperate zone.

The essential facts were summarized by Roccati [(105), cited in Osmaston (100)] and presented in painstaking detail by Nilsson (92–94, 96, 98). Nilsson was able to show that there had been repeated episodes of glacial advance. On all mountains there are numerous small moraines of relatively recent still-stands or readvances, but there were good indications of at least four major episodes on Elgon, Mt. Kenya, and Kilimanjaro, and two on Ruwenzori. In Nilsson's day there were no radiometric methods for dating Quaternary moraines, but the resemblance to the Alpine and American sequences, both of which then seemed to consist of four major ice ages, suggested comparable ages in the temperate and tropical zones. The presence of only two sets of moraines on Ruwenzori was explained in terms of the apparent youth of that mountain.

Ascending the lowest of these moraines is an instructive exercise. They have a local relief of 100–200 m and are covered with a thick rich evergreen forest, wild bananas, and tree-ferns, and crisscrossed by the trails of elephant and Cape buffalo. The traveler toiling up the trail in oppressive tropical heat cannot but wonder at the great climatic change that has occurred since the fresh and well-preserved moraines were formed.

Radiocarbon provides an excellent tool for dating the younger Pleistocene glaciations but has been little used for this purpose in Africa. Livingstone (71) dated a core from Mahoma Lake at 3000 m, just behind the terminal moraines of the last extensive glaciation, and produced a minimum age for deglaciation of 14,750 yr. This is in good accord with temperate data. The older glaciations have not been dated, although on the volcanic mountains, where the till sheets are separated by volcanic ash layers and lavas, this should be possible.

The early glacial geology of Nilsson has been followed by more recent work by Downie and co-workers (39, 40), by Baker (4), and by Osmaston (100). Downie combined till studies on Kilimanjaro with geological examination of the interleaved lava flows and worked out a sequence of glaciations, some of which are obviously of great antiquity. Baker mapped in great detail on Mt. Kenya a long series of retreat stages of the last glaciation and provided evidence for at least one older drift. Apparently Nilsson mistook trachytic agglomerate for old till in some localities and therefore overestimated the number of old glaciations on Mt. Kenya.

Osmaston's conclusions from Ruwenzori rest upon field work spanning ten years, with careful study of aerial photographs, and are worth considering in some detail.

The main glacial geological features on Ruwenzori are the terminal moraines of the Mahoma Lake glaciation. These extend 2400 m below the present lower limit of ice, are very massive, and have been radiocarbon dated at something greater than 14,750 yr. Above the Mahoma Lake moraines it is possible to identify a consistent series of much smaller moraines, the Omurubaho moraines, at 3600–3900 m. These are not well dated, but Osmaston pointed out that the volume of material in them was very much less than the volume of slide-rock in the screes formed more recently. He concluded that they must be close in age to the Mahoma Lake moraines, and Livingstone (73) produced a minimum age for Kitandara Lake, behind the Omorubaho moraines, of 6890 yr.

Below the Mahoma Lake moraines the valleys are U-shaped for some distance, and in some of them there are irregular mounds, apparently the remnants of terminal moraines. This is good evidence of a much older glaciation, judging from the great contrast with the fresh Mahoma Lake moraines. Osmaston gave, as an order of magnitude, an estimate of 100,000 yr for these Rwimi Basin moraines.

A still older glaciation is indicated by old moraines on the flat interfluves between U-valleys of Rwimi Basin age. These, the Katabarua moraines, are morphologically well preserved, for they are well protected from river erosion by their location, but the boulders of which they are composed show deep onionskin weathering. There was a major inversion in the topography of the mountains between the Rwimi Basin and Katabarua glaciations, indicating a great difference in their ages.

Osmaston (100) made a detailed study of the shape and altitudinal distribution of each of the former glaciers and, with a few reasonable assumptions about their general regime, he was able to use iterative electronic computation to estimate precisely the past positions of the firn-line, or annual limit between net ablation and net accumulation:

	Firn line lowering (meters)
Modern conditions	0
Omurubaho glaciation	180
Mahoma Lake glaciation	630
Rwimi Basin glaciation	720
Katabarua glaciation	630
Complete deglaciation	−300

Assuming that the lowering of firn-line was due exclusively to temperature drop, with no increase in precipitation, Omaston concluded that the temperature had been 4°C lower during Mahoma Lake time than it is now. This estimate is conservative; there is independent evidence that the climate was actually dryer during the last glaciation.

Osmaston applied the same iterative method to Kilimanjaro, where he suggested a temperature drop of 5–6°C for full glacial conditions. Again, this figure is likely to be conservative, but agrees reasonably well with the estimates of 7–8° temperature drop for Europe or 4–7° drop for temperate America, indicated by similar glacial geological evidence (46, p. 72).

The modern firn-line of Ruwenzori tilts to the wetter western side, and the fossil firnlines are parallel to it, indicating that there has been no shift in the wind direction and pattern of precipitation on the mountain for a long time. It is watered today by winds blowing from both the Atlantic and the Indian Oceans. If the Ruwenzori has not suffered a major change in airflow, it is unlikely that the mountains closer to the Indian Ocean coast have done so.

Perhaps because they are so similar to the European homelands of most investigators of North African paleoclimatology, the Atlas Mountains have been neglected, and are perhaps paleoclimatically the least well known part of the continent. The High Atlas Mountains in Morocco between 30°45' and 31°25'N and 7°15' and 8°50'W supported glaciers during the Pleistocene (46, p. 698), but not even the age of glaciation seems to be known. This relatively moist glaciated area is probably the only part of northern Africa likely to yield cores organic enough for the absolute pollen frequency work needed to distinguish local temperate tree pollen from long-distance transport.

Pollen Analysis

After exploratory beginnings by Hedberg (62) and Osmaston (99), the first long pollen diagram from the mountains of Africa was provided by van Zinderen Bakker (127, 129) and Coetzee (29) from 2900 m in the Cherangani Hills of Kenya (Figure 3). Cores up to 4.7 m long from the sediments of Kaisungor Swamp, Cherangani, yielded a pollen record for the past 27,750 yr. Interpretation of the diagrams is complicated by the fact that certain elements of importance in the montane vegetation, notably Cyperaceae and *Alchemilla,* a rosaceous genus of herbs and low woody shrubs, are likely to be over-represented because of pollen production on the bog surface. Nevertheless, the record contains clear indications of major vegetational change. The lower 2.5 m yield pollen spectra dominated by elements of the nonforest vegetation, whereas the upper 2.2 m contain more abundant pollen of taxa important in the montane forest including *Olea, Podocarpus, Ilex,* and *Juniperus.* The shift between these two major vegetation types is not well carbon dated. Interpolation suggests that it was about 9000 years ago.

In describing the Kaisungor pollen sequence, van Zinderen Bakker gave a detailed interpretation correlated on a zone-for-zone basis with the standard pollen sequence from Western Europe. This procedure has been criticized by several authors (73, 88), largely because too little was known about the pollen production of modern African vegetation types and about ecological factors, including climatic ones, of importance in determining the vegetation. Van Zinderen Bakker used grass pollen as his main climatic indicator, interpreting it as an indicator of cold conditions. The various grasses cannot be distinguished by pollen morphology, and the family contains genera important from sea level to snowline. It is possible, despite the ubiquity of grasses, that they might be important in the pollen rain only of cold, high altitude vegetations. Van Zinderen Bakker provided no surface spectra to show that this was so, and later authors (60, 66, 73, 89, 90) gave abundant evidence that it is not. Very high percentages of grass grains are to be found in surface spectra from a wide variety of grassland, savanna, woodland, and even some forest types.

Livingstone (73) presented pollen diagrams from several altitudes in Ruwenzori (Figure 3), the most informative from Mahoma Lake at 3000 m, the type locality of the Mahoma Lake glaciation. Although there are differences in detail, the general sequence resembles that from Kaisungor, 600 km to the east. In early postglacial times, from about 14,700 to about 12,500 yr ago, the pollen rain was dominated by grasses, by other herbs, notably Compositae, and by a variety of small heliophilous trees and shrubs, especially *Myrica* and *Hagenia.* Since 12,500 years ago the pollen assemblages have been dominated by a variety of montane forest plants, including Urticaceae, *Olea,* Myrsinaceae, *Celtis, Acalypha,* and *Podocarpus.* During the last few thousand years, forest pollen has declined somewhat and grass has recovered, although not to its original levels.

Figure 3 Map of important paleoclimatic sites in Africa. Base from Clark (26). (A) Aber-dares, (AL) Alexandersfontein, (AN) Aliwal North, (AT) Atlas Mts., (B) Bosumtwi, (BG) Bangueulu Swamps, (BL) Bloemfontein, (BN) Blue Nile, (C) Mt. Cameroon, (CF) Cape Flats, (CH) Cherangani, (D) Drakensberg Mts., (E) Mt. Elgon, (G) Groenvlei, (H) Hoggar Mts., (IAD) L. Idi Amin Dada, (IN) Ishiba Ngandu, (KD) Kalahari Desert, (KE) Mt. Kenya, (KF) Kalambo Falls, (KG) Kigezi, (KI) Kilimanjaro, (LC) L. Chad, (LK) L. Kivu, (LM) L. Manyara, (LN) L. Naivasha, (LR) L. Rudolf, (LT) L. Tanganyika, (LV) L. Victoria, (MSS) L. Mobutu Sese Seko, (N) Niger R., (NEA) Northeast Angola, (O) Orange R., (OV) Omo Valley, (R) Ruwenzori, (RH) S. Rhodesia, (S) Senegal R., (SI) Simien Mts., (SP) Stanley Pool, (T) Tibesti Mts., (TR) Transvaal, (V) Virunga Volcanoes, (VA) Vaal R., (WN) White Nile, (Z) Zaire R., (ZM) Zambezi R.

Livingstone was cautious about attaching climatic significance to the pollen shifts, recognizing that human deforestation, low temperature, or low rainfall all tend to reduce the importance of forest trees and their associates in the vegetation. Land-use is unlikely to have been a major factor 12,500 yr ago, but both temperature and moisture are serious possibilities. A positive indication that conditions were dry in the open vegetation prior to 12,500 BP was provided by *Artemisia* grains of that age. At present, this composite genus is restricted in tropical Africa to highlands dryer than Ruwenzori. Evidence against very cold conditions was provided by the importance of *Myrica* pollen in the early spectra. At present, this small tree is most abundant at altitudes lower than Mahoma Lake. Even on the nearby Virunga Volcanoes, where it grows to tree-line, *Myrica* does not extend to elevations more than a few hundred meters above Mahoma Lake. Livingstone also invoked the raw soil of the newly deglaciated countryside as a possible determinant of the open nature of the early vegetation at Mahoma Lake.

Morrison (87, 89) and Morrison & Hamilton (90) studied a somewhat longer pollen sequence from Kigezi in western Uganda. Morrison's most informative diagram came from Muchoya Swamp at 2500 m. Although today the site is a swamp, it was a lake for much of its history, so the problems of local over-representation are less severe than at Kaisungor. The exact time represented is unknown, but the sedimentation rate suggested by a carbon date of 12,890 yr in the eighth meter, indicates that the bottom of the core near 12 m may be as much as 25,000 years old.

The general vegetational sequence is broadly similar to those of the other mountains, but land use by man seems to have been a dominant factor during the latter part of the swamp's history. Unlike Mahoma, at the altitudinal boundary of bamboo and giant heather forests little touched by man, the vegetation around Muchoya is much affected by cultivation and burning.

The longest and most interesting pollen profile from a tropical African mountain has been provided by Coetzee (28, 29) from Sacred Lake at 2400 m on Mt. Kenya. The oldest carbon date is 33,350 yr, but it is possible, considering the heterogeneous and rather inorganic material in the lower part of the core, that deposition may not have been continuous. There is every reason, however, to expect a complete record in the organic gyttja of the upper 13.3 m. Extrapolation suggests an age of some 25,000 yr for the base of the gyttja.

Prior to 10,600 years ago at Sacred Lake the pollen assemblages were dominated by grasses, *Artemisia,* and other indicators of open vegetation. Then a series of other taxa entered the pollen rain: first *Hagenia,* suggestive of treeline conditions, and finally such montane forest elements as *Podocarpus, Afrocrania, Pygeum, Olea,* and *Macaranga.* The early part of the record is not without changes. *Hagenia, Podocarpus,* and other forest genera were of some prominence between 33,350 and 20,000 years ago.

Coetzee's interpretation of this record is similar to that of van Zinderen Bakker for the shorter section at Kaisungor. The same objections may be raised to the use of grass pollen as an indicator of cold conditions, but Coetzee's interpretation must certainly be correct for at least the glacial-age part of her record, for Osmaston showed with glacial geological evidence that the climate was at least 4° colder during the Mahoma Lake glaciation of East African mountains. The periglacial climate of Mt. Kenya may have been dry as well, particularly in view of the abundance of *Artemisia* pollen. *Artemisia afra,* the sole tropical African species, is still a constituent of the vegetation of Mt. Kenya, but contributes only a trace to the modern pollen rain at Mahoma Lake. Recently deglaciated soil conditions are less likely to have been important than at Mahoma, because Sacred Lake is volcanic in origin and lies many kilometers from the terminal moraines of Mt. Kenya. It is possible that solifluction of the lightly vegetated ice age slopes maintained somewhat more youthful soils than those currently supporting the dense montane forest around Sacred Lake.

Since the original stratigraphic studies were carried out, Hamilton (60) has examined surface spectra from highland Uganda in a variety of vegetational contexts. His new data are particularly important for discriminating between pollen types that are dispersed long distances and those that fall close to the places where they are produced. Unfortunately many of the spectra come from terrestrial sites—lakes at a sufficient variety of altitudes are not available in the Uganda Highlands—and climatic information about the places from which they come is probably insufficient for a multivariate analysis of the relation between pollen assemblages and climate such as Webb & Bryson have provided for parts of temperate America (17, 134–136).

Hamilton (60) attempted to reconcile the interpretations of Livingstone and Morrison with those of Coetzee and van Zinderen Bakker. He believed that the embarrassingly early peak of *Myrica* pollen at Mahoma Lake is an artifact of differential pollen dispersal.

Taking into account the information available for all four highland areas, it is clear that the vegetation during the peak of the Mahoma Lake glaciation was much more open than that of today. The shift to modern vegetation was accompanied more or less closely by increases in temperature and humidity. The radiocarbon control of this shift is not good, but it does not seem to have been simultaneous at the four sites. Rather, it seems to have occurred first on Ruwenzori, the wettest mountain, and last on Cherangani, the driest. If these temporal discrepancies are real, they may be due partly to biogeographic factors as well as to paleoclimatic ones.

The precise timing and nature of climatic changes at high altitudes in equatorial Africa is still unclear. Deglaciation beginning shortly before 14,700 BP is in accord with the timetable of deglaciation in Europe and America, and a major shift of vegetation occurred between 9500 and 12,500 BP at all sites. The montane evidence alone is not enough to place, even to the nearest millenium, the time of greatest climatic change.

Biogeographers have frequently assumed that the vegetation belts of mountains represent stable communities, in which the constituent species have an obligatory association with each other. Pollen stratigraphy shows that this has not been the case. Most clearly on Ruwenzori, but to some extent (especially with respect to the importance of *Podocarpus*) on the other mountains, there has been a tendency for the constituent taxa of the montane forest belts to behave individualistically. There is not as yet any compelling stratigraphic evidence that an appreciable number of the taxa of the modern montane forest found refuge during the last ice age in a belt at any altitude around these mountains, but the density of pollen diagrams from the surrounding plateau is probably not yet great enough to deny categorically the possibility of species seeking refuge at some altitude on the lower slopes. It does seem, however, that in tropical Africa, no less than in Western Europe and temperate North America, we are faced with a problem in determining in just what locations the constituent species of the modern mesic forests survived the ice ages.

LAKES OF THE EAST AFRICAN PLATEAU

Lakes of the Upper Nile Drainage

LAKE VICTORIA   Lake Victoria lies athwart the equator and forms the headwater of the River Nile. It contains a very rich fauna of endemic fish species, indicating some antiquity, but most belong to the single cichlid genus *Haplochromis* and may have evolved more recently than most animal species. The lake is surrounded by a variety of vegetation ranging from savanna to evergreen forest, and includes in its catchment high mountain areas on the Mau escarpment and Mt. Elgon. Despite its great size, it is only 79 m deep. Some parts of the lake basin, particularly Nyanza Gulf in the northeast, are probably much older than the main part of the lake, which was ponded along west-flowing rivers behind a slowly and intermittently rising tectonic dome. Renewed uplift of the dome has generated a series of strandlines tilted up toward the west, and the lake is also surrounded by three younger horizontal strandlines at 18–20, 12–14, and 3–4 m. These young strandlines probably reflect pauses in down-cutting of the Nile outlet at Jinja, Uganda, rather than past episodes of wetter climate (122–124).

Kendall (66) examined several cores from the northern part of Lake Victoria for minerals, pollen grains, and other microfossils. His longest core appears to be virtually complete and represents about 15,000 radiocarbon yr. The section was radiocarbon dated in 28 places, providing unusually accurate determination of the sedimentation rate, and permitting Kendall to express his analyses in terms of microfossils deposited per unit area per unit time. This led to an accuracy in vegetational reconstruction that has not yet been matched for any other site in Africa.

Near the bottom of the section, with a radiocarbon age of 14,730 yr, there was an erosional unconformity. Below the unconformity, water content of the sediment was low, mineral content high, and microfossil preservation very poor. The organic content was modest in this lower member, about 20%. Above the unconformity the sediment was a uniform organic lake mud with excellent preservation of pollen grains and other microfossils. There was, however, some chemical differentiation of this overlying member. Below about 15.4 m, with a radiocarbon age of 12,000 yr, the sediment was rich in sorbed bases. Above that level exchangeable cations were uniformly low.

This chemical, mineralogical, and sedimentary profile shows that prior to 12,000 years ago the lake was closed, and had been closed for a sufficiently long time to accumulate a considerable load of dissolved salts—enough to precipitate calcite, but apparently not enough to precipitate the more soluble salts of the common major ions. At one time 14,000 years ago, the water level at the coring site fell at least 26 m below the modern level. Later coring by Livingstone (75) showed that the level actually fell at least 75 m. This is only 13 m less than the total closure of the basin, and shows that Victoria must have been a very much smaller lake prior to 12,500 years ago.

Although Victoria lies within the humid tropics, it is also within the anomalously dry part of eastern Africa, affected by divergent flows of air from the Indian Ocean and probably the easterly tropical jet stream as well. About 90% of the water loss is evaporative, with only 10% by outflow through the Nile. Such a lake is relatively sensitive to any climatic change that increases the ratio of evaporation to precipitation.

A core taken in shallower water showed that, about 10,000 years ago, there was a brief drop in water level, sufficient to close the lake but too short-lived to permit appreciable accumulation of dissolved salts (66).

Pollen analysis revealed a conformable vegetational history; the most important change was about 12,500 years ago, from an assemblage rich in grass to one rich in the pollen of forest trees. Kendall attributed the grass zone to savanna, and the presence of a trace of *Acacia* pollen suggests that this was probably its origin. The increasing forest pollen received a temporary check about 10,000 yr ago, at the same time that the lake level fell temporarily below the Nile outlet.

The initial forest assemblage was rich in *Alchornea,* Moraceae, and Urticaceae, suggesting evergreen forest. About 6000 years ago, the pollen assemblage changed its composition slightly, with *Celtis* and *Holoptelea* increasing at the expense of evergreen forest taxa. This suggests a change to semideciduous forest, which dominated until the last few thousand years, when all forest elements declined in abundance. In an ordinary pollen plot of relative percentages this change is most conspicuous for a rise in the percentages of grass pollen, but precise radiocarbon control enabled Kendall to show that the absolute increase in grass grains was very slight. The main event was a decrease in forest pollen types, with nothing increasing to take their place.

The open vegetation of more than 12,000 years ago was probably due to lack of moisture. Sufficiently cold conditions could have had the same effect, but this site is at only 1134 m. Development of tundra on the equator at such an altitude requires a temperature lowering much in excess of that indicated by any other evidence. Deforestation by man or other mammals is almost equally unlikely; nothing in the archaeological or paleontological record suggests a drastic change in the potentialities of mammals to alter their environment 12,000 years ago. Thus moisture is the most likely control; this explanation is in accord with changes in lake level and composition of the water. The climate was dryer from at least 14,700 BP until 12,000 BP than it has been since. There was a temporary return to dry conditions about

10,000 years ago, after which evergreen forest expanded, reaching its zenith about 8000 BP. Since 6000 BP the climate has been somewhat dryer, more seasonal, or both.

The reduction in forest pollen during the past few millenia might result from decrease in moisture, from human activity, or both. For this epoch there is good archaeological evidence of the introduction of iron tools, domestic livestock, and agriculture, all of which may have reduced the growth of trees. The common African crops, such as bananas and millet, are poor pollen producers, so there are no positive pollen indicators of cultivation (64).

Kendall's (66) results show that the last dry period corresponded to at least the closing phase of the last ice age, and that the last pluvial period occurred during postglacial time, even overlapping with the postglacial thermal maximum of the temperate zone. They cannot be reconciled with at least one hypothesis of ice age causation, Simpson's cold-earth–hot-sun idea, according to which ice ages resulted from an increase in solar energy input, increasing the mean storminess and hence the cloudiness and snowfall of higher latitudes. Kendall's findings show that the destruction of rain forest is not an irreversible process. Artificial destruction of some existing rain forests may be irreversible, but it was clearly possible for rain forest to increase in parts of Uganda 10,000–12,000 years ago.

The paucity of forest trees on the East African plateau prior to 12,000 years ago poses a biogeographic puzzle. We have seen that forest trees disappeared from the mountains at that time. They did not simply move down to the plateau, for they were missing from the plateau as well. Presumably they were restricted to the best-watered parts of the lowland Zaire (Congo) basin and West Africa, but we must beware of facile explanations not supported by evidence.

Perhaps the most generally interesting aspect of Kendall's findings concerns the nature and antiquity of evergreen forest. Ecologists sometimes assume that the humid tropics have escaped the biogeographic disruptions of Pleistocene climatic change. At least in this area, however, evergreen tropical forest was considerably restricted during late-Pleistocene time. Far from being a serene untroubled ecosystem with its roots in early Cenozoic antiquity, it seems likely that the evergreen forest, like most major formations of the earth, is in a state of continual dynamic response to climatic change and that, given the considerable age of at least some species of evergreen forest trees, there may be too few generations between major changes of climate for the forest to achieve even a stable age distribution.

LAKE MOBUTU SESE SEKO (ALBERT)   Kendall's results also show that the upper part of the River Nile has flowed continuously for no more than 12,000, and probably for no more than 10,000 years. Hecky & Degens (61) and T. Harvey (personal communication) have endeavored to extend paleohydrological understanding to the other great lakes that form the headwaters of the White Nile. Coring in Lake Idi Amin Dada (Edward) has not yet raised a section with an age of 10,000–12,000 yr, but Harvey cored Lake Mobutu Sese Seko (Albert) in 46 m of water to a depth of 10.8 m. This core represents 28,000 yr, and Hecky & Degens have obtained shorter cores in somewhat deeper water. Both investigations yielded evidence that the lake level fell substantially during late-Pleistocene time. Harvey showed that the lake was at least 53 m lower 12,500 years ago when it would have begun to receive outflow from the Victoria Nile. Until cores of sufficient length have been taken at 58 m in the deepest part of the lake, we will not be able to tell if it dried up completely, but we know that at most it was a small part of what it is today. The lake basin is very old, for it contains some 2500 m of sediment (59), and the faults bounding the graben were active as early as the Miocene (5, 24). The modern episode of continuous existence of a body of standing fresh water may have been much shorter than the history of the basin.

At present the lower Nile receives water from two sources: the Ethiopian plateau via the Atbara and the Blue Nile, and equatorial East Africa via the White Nile. The Ethiopian contribution is by far the larger (63). It is, however, extremely seasonal, for the Blue Nile and Atbara are torrential mountain streams nourished by a single annual rainy season. The White Nile flow is much steadier, draining an area much of which has two rainy seasons a year, and the flow is buffered by passage through enormous lakes. As a consequence, although the White Nile contributes a small part of the annual flow, it provides most of the dry season flow. Prior to 12,500 yr ago this mighty river must have been reduced seasonally to a relative trickle.

This relative dryness may account for the poverty of the riverine fish fauna of the Nile as compared to other African rivers. Although the great Nilotic lakes support a rich and highly endemic fauna, the strictly riverine fauna includes only 115 species of fish (53), as compared with 639 species in the riverine fauna of the Zaire (Congo), a somewhat shorter river. The Zaire, however, is much less seasonal. Its total discharge is also considerably greater than that of the Nile. It is not clear what geometrical measure of a river provides the best index of its potential number of fish niches; the total wetted bottom surface might be a reasonable first approximation, and the equilibrium fish fauna may depend on the size of this parameter at the season of minimum discharge. Geographers do not tabulate the rivers of the world by wetted bottom area, for good practical reasons. Of the things they do measure, neither discharge nor length seems likely to be the best measure of fish ecospace, but rather some indeterminate quantity intermediate between them.

LAKE RUDOLF   Although it has no modern outlet, Rudolf has been joined to the Nile repeatedly during Quaternary time and is conveniently discussed with the other Nilotic lakes. Evidence for its past connection to the Nile is of two types: the fish species are a subset of Nilotic ones, and raised strand lines of several ages exist around the lake at an altitude permitting overflow first to the Lotigipi depression west of it, and then, at a slightly higher stage, to the Nile (23).

Lake Rudolf is now about 80 m below the sill of the depression in which it lies. Both strandlines around the lake and the stratigraphy of the deltas, particularly that of the Omo River, the lake's main tributary, provide a detailed record of times when the water level was higher than it is now. The main findings have been summarized by Butzer and co-workers (19, 23), who provide citations to the earlier literature.

For a long time after 35,000 BP the lake was low, but it rose prior to 9500 BP, by which date it reached +80 m. It fluctuated between +80 and +60 m for about

2000 yr, and then fell to the modern level. A further transgression began before 6600 BP, and reached +65 m by 6200 BP. This level was maintained until after 4400 BP, when the lake fell to an unknown depth only to recover to +70 m about 3000 BP. It has been relatively low since. The lake continues to oscillate, having occupied levels from +15 m to –5 m between 1897 and 1955, but it has not recently come close to establishing an outlet. The lake water is becoming more alkaline and saline by evaporative concentration of runoff, and now has a conductivity of 3300 μmhos per cm (121).

Bonnefille [(10), see also (7–9, 11–14)] has analyzed more than a hundred large samples of sediment of paleontological significance from the Omo Valley for pollen, and has found useful quantities of pollen in three of them. Under these circumstances, the risk of differential preservation is particularly severe, but her polliniferous samples contain a rich and internally consistent assemblage of well-preserved grains. One of the assemblages is a good match for that found in the silt load of the Omo River. The second suggests moister conditions, the third dryer conditions, than those of the present day. All three samples were collected between horizons potassium-argon dated at two and three million years. This tantalizing glimpse into the early Pleistocene suggests that climatic changes of the same order as those established for the past 15,000 years have been occurring in tropical Africa for a very long time.

Stoffers & Holdship (120) have examined a core from Lake Manyara in the Eastern Rift in northern Tanzania for diatoms and sedimentary minerals, particularly zeolites. The core is 55 m long and has been radiocarbon dated in 13 places. Extrapolation suggests that the base of the core, which is beyond the reach of radiocarbon dating, may have an age of about 60,000 yr, making it the longest continuous record from Africa.

The authors found a mutually exclusive distribution of diatoms, indicating moist conditions, and the zeolite analcime, indicating dry ones. The zeolites chabazite and erionite, which are likely to be formed in less concentrated water than analcime, overlap the vertical occurrence of diatoms, suggesting intermediate aridity. The last 5000 years are represented by analcime-rich sediments, the time from 5000 to 12,000 BP by diatomaceous sediments lacking zeolites. Conditions appear to have been dry from 16,000 to 12,000 BP, somewhat more humid again from about 22,000 to 16,000 BP, and then dry back to the limit of radiocarbon dating at 35,550 yr except for a brief moist episode at 27,000 BP. There is a final moderately moist episode that may have lasted for a few thousand years around 50,000 BP. All dates are approximate, based on interpolation and extrapolation from the dated levels.

A more precise paleoclimatic interpretation will be possible when the diatoms have been identified to species, but for the time being the Manyara record seems consistent with other tropical African localities, at least during the last 15,000 years. The late Pleistocene moist episodes seem to have been moister than the past 5000 years, which is somewhat surprising. Most indicators of conditions during the last glaciation suggest some climatic amelioration, especially prior to 22,000 BP, but they do not indicate that conditions were less glacial than they are today. For example,

Coetzee's pollen diagram from Sacred Lake contains broad *Podocarpus* and montane forest peaks several meters above the level of a 33,350 carbon date, but the vegetation was obviously much more open than that of the past few thousand years. Understanding of the nature of interstadial conditions, however, rests largely upon evidence from pollen analysis and glacial geology. Both these indicators have a slow response time, and when information comes in from faster-responding indicators, such as thermophilous beetles, which are able to extend their ranges much faster than forest trees when conditions improve, it frequently seems surprising. Manyara is a shallow lake with large inflows and large evaporative loss. No doubt the lake's chemical composition changes very rapidly in response to climatic change and it may have responded to climatic changes too short-lived to register on pollen diagrams or drift stratigraphy.

Other Rift Valley Lakes

Lakes Nakuru, Elmenteita, and Naivasha in the Eastern Rift in Kenya have provided some of the clearest evidence of tropical climatic change since the pioneering studies of Nilsson (93, 95, 97). The existing lakes are mere remnants of the bodies of water that once filled their common depression to overflowing, and at various stages of their expansion or retreat they have cut and built a series of raised strandlines. Modern leveling combined with radiocarbon dating (65, 133) has required a reassessment of the number of levels represented and their ages, but has done nothing to change the basic conclusion that there have been times in the past that were very much wetter than the present.

The Richardsons (101–104) have been coring these lakes for fifteen years, and have confirmed the main climatic features suggested by the strandline system and provided a detailed record of events while the lake level was low.

Recent cores represent almost 30,000 yr; the longest to be completely described in print comes from the Crescent Island Crater of Lake Naivasha and represents only the past 9200 years. At the beginning of that period the lake level was high, the water relatively dilute, and the productivity apparently high. Because the deep crater basin forms a low energy sediment trap with respect to the shallow main lake to which it has been intermittently connected, indications of past productivity are less firm than those of high lake level. One of the diatom species, *Nitzschia fonticola,* can be interpreted as an indicator of somewhat warmer temperatures than modern ones. Richardson has essayed a paleohydrologic budget for the high-water episode, and postulates precipitation 65% greater than that which now prevails.

Between 5650 and 3040 BP the lake shrank until it finally dried up briefly. Since 3000 BP, a small lake has existed in the basin, but it has frequently been smaller and sometimes contained much more concentrated water than the modern lake. For the past 5650 years, the lake apparently has not had an outlet.

At present the Naivasha area has two dry and two wet seasons each year, but prior to 2500 BP it seems to have had a single wet and a single dry season. This conclusion is based on thin light and dark laminae in the sediments that apparently were deposited at the rate of approximately one couplet per radiocarbon year.

Details of the longer sections are not yet available, but the general finding seems to be (J. L. Richardson 103 and personal communication) that high lake levels began about 12,500 BP and were preceded by a dry period of low lake levels that had persisted without essential interruption since at least 28,000 BP. This agrees with the lack of great changes in the pollen spectra of the same age from Cherangani and Mt. Kenya.

LAKES TANGANYIKA AND KIVU   These deep lakes of the western Rift differ conspicuously in the richness of their faunas. For example, Tanganyika contains 193 species of fishes, of which 173 are endemic, whereas Kivu has a total known fauna of about 18 species, only 8 of which are endemic (48). The difference has been attributed by biologists to a difference in age, and it has been maintained that Kivu is no more than 12,000 years old, because volcanism of this age is believed to have contributed to the volcanic dam closing the basin. Degens and his co-workers (36), however, using a sonic profiler, found that the basin contains about 1 km of fine-grained lake sediment. This thickness is consistent with an age of one million or more years. The poverty of the fish fauna is probably due to intermittent overturn, mixing the methane-rich anoxic deep water into the surface waters and poisoning the fish.

Tanganyika has been cored by Livingstone (72) and Degens (Degens, von Herzen & Wong 35) and Kivu by Degens et al (36). Because of the great depth of water (almost 500 m in Kivu, 1470 m in Tanganyika) coring with rods has not been attempted, and wire-operated piston samplers have provided a record of only 20,000–30,000 yr. Pollen analyses have not been completed for any of the cores, but mineralogical, chemical, and diatom analyses have provided an interesting record of changes in water level and water chemistry (61). The record of the two lakes is linked not just by common climatic changes affecting both, but also by the Ruzizi River, which forms the outlet of Lake Kivu and the principal inflow of Tanganyika. The response time of Kivu is much the shorter, and it has gone through more pronounced changes in water chemistry than the larger downstream Tanganyika, but the influence of each major change in Kivu, including times when its level has fallen sufficiently to cut off flow to the Ruzizi, is shown in muted form in the record of Tanganyika. Although both lakes have gone through long periods without an outlet, the mass of Tanganyika water is so large, and the total influx to be evaporatively concentrated is so low, especially during dry times when the Ruzizi delivers no Kivu water, that the lake never becomes very saline. This has been a critical feature in permitting the evolution of the large endemic flocks of fishes and other organisms in Tanganyika.

The major changes in levels of the lakes appear to have coincided with those of the lakes already considered. Kivu was 310 m lower than it is now, apparently about 13,700 BP. The level then rose, with a transitory fall between 11,000 and 10,000 PB, until it established its outlet. Tanganyika has gone through similar level changes, but the extent of lowering is less well established. Hecky & Degens (61) believe it may have been as much as 600 m in the north basin, but Livingstone, by extrapolating a sedimentation rate based on a date of 11,000 yr in gyttja, concluded that there has been continuous sedimentation for 22,000 yr at a station now covered by 400 m of water in the south basin. More radiocarbon dates are needed to resolve the difficulty. Perhaps late Quaternary tectonic activity is involved.

THE PLATEAUS OF CENTRAL AFRICA

The plateaus of Central Africa are less well supplied with basins of sedimentation than the highlands and rift valleys of East Africa, but are not without suitable coring prospects. The results to date, however, have not been encouraging.

Van Zinderen Bakker has analyzed samples from the archaeologically important river floodplain deposits at Kalambo Falls (127, 132), and drawn major climatic conclusions from them. Problems of differential preservation, differential sorting, local over-representation and long-distance stream transport are all present in a serious degree in this sort of depositional environment. There appear to be significant differences in the percentages of some elements, notably grasses, Myrtaceae, *Olea,* and Ericaceae, but van Zinderen Bakker's interpretation of these differences is purely intuitive. Until matching surface spectra from similar environments are available, the climatic interpretations should not be accepted. My intuition does support one of Bakker's conclusions: that the spectra he lumps as Zone Z suggest cooler conditions that the other spectra examined.

Lawton (68) cored the Bangweulu Swamps, a most promising paleoecological locality, but found only a poor and sparse pollen assemblage that was extremely difficult to interpret. Livingstone (74) raised a core representing 22,000 yr at Ishiba Ngandu that was very rich in pollen, but virtually devoid of compelling evidence for vegetational change.

This part of Africa is covered largely by a series of grassland, savanna, and woodland vegetations. The arboreal element, where present, tends to be leguminous, and the legumes are mostly entomophilous plants of such low pollen productivity that they do not register clearly in the fossil record.

SOUTHERN AFRICA

Although dated information is scarce, there is an abundance of undated geomorphological evidence for climatic change in the dry parts of southern Africa. Grove (55), for example, has shown that in the Kalahari and Ngamiland regions of Botswana there are extensive fields of inactive dunes that imply a rainfall reduction of at least 50% from the present half-meter per annum. Past pluvial conditions are indicated by river valleys that have never carried water in modern times, and by raised strandlines. One former lake, in the Makarikari basin, had an area of 34,000 km^2, a depth of 45 m, and would have required an inflowing river bigger than the Zambesi to sustain it.

The pollen stratigraphy of a saline hot spring deposit at Florisbad near Bloemfontein in the Orange Free State has been described by van Zinderen Bakker (126). Pollen preservation is intermittent in the deposit, and the only useful indicator taxa are Compositae, high near the base of the deposit, and grasses, high near the top.

The grass-dominated assemblage compares well with the modern pollen rain at Bloemfontein, and suggests a similar dry *Cymopogon-Themeda* veld nourished by 46–51 cm of annual precipitation. The composite-dominated assemblage is taken to represent a Karroo vegetation with 13–25 cm of rain, but no modern pollen spectra are presented from a matching Karroo locality. The climatic interpretation seems reasonable, though not conclusively demonstrated, but unfortunately the age of the deposit is not established. Two carbon dates on the upper part gave discordant dates of 6,700 and 19,600 yr, two in the middle discordant dates of 9,100 and 29,000 yr, and a single sample from the base of the deposit gave an age of greater than 41,000 yr. It is not clear whether the discrepancy is due to geochemical effects or churning of the deposit, which sometimes occurs during geyser-like eruptions of the shifting spring vents; until the chronology is settled no paleoclimatic use can be made of the results.

Coetzee (29) presented several short sections from another spring deposit at Aliwal North (30° S) and divided them into seven pollen zones of vegetational and climatic significance. The differences in pollen percentages on which most of these zones are based are too small to be significant, however, and are not consistent from one core to another. Possibly the basal part of her oldest section, about 12,000 or more years old, represents conditions that were cooler, moister, or both than the succeeding time. An abundance of sedge and *Typha* pollen together with a relative scarcity of pollen of the chenopod-amaranth type suggests abundant fresh water, and a peak of *Stoebe* pollen suggests that this composite moved down from the well-watered hills to the northwest where it is now an important part of the vegetation. Slightly above a carbon date of 11,250 yr, there is a slight increase in the pollen of miscellaneous Compositae, with spectra that are conformable with one from Middleburg in the dry False Upper Karroo vegetation. This is suggestive of further increase in dryness, possibly coming to an end somewhat before 9650 BP. Pollen analysis and interpretation of spring deposits are not easy, but one can safely conclude that at this latitude there was no sudden increase in effective humidity around 12,000 yr ago. Rather, the change seems to have been in the other direction.

Butzer et al (21) have radiocarbon dated a deposit of calcium carbonate 19 m above the floor of the closed Alexandersfontein depression near Kimberly at 16,010 ± 185 BP. The deposit does not seem to contain aquatic fossils, but if they are correct in identifying it as lake marl, and if there is no serious geochemical error in the date, this is evidence of a moist climate at a time when more equatorial parts of Africa were dry. Even making allowance for the reduced evaporation resulting from a probable 6°C reduction in temperature, Butzer et al concluded that the rainfall was more than twice its modern value at Alexandersfontein a little more than 16,000 years ago. Although this finding is difficult to reconcile with what is known about the climate of tropical Africa during that era, it accords with the history of the arid southwestern United States (47).

Many claims have been advanced for fossil frost phenomena in southern Africa. These claims have been examined by Butzer (20), who concluded that there is good evidence for a former colder climate in the Drakensberg, around 29° south latitude, with nivation niches down to 2100 m on the eastern windward slopes and to 2900 m on the western leeward ones. He also believes that the screes of valleys in the Cape Folded Ranges around 34° S would not form under the present conditions of rare and mild frosts, but would form under winter conditions that were colder by 10°C. Other workers, especially Linton (70), working near 25° in the Transvaal, and Harper (58), working near 18° S in Rhodesia, interpreted geomorphological evidence at lower latitudes in terms of a similar cooling. There is nothing improbable about so much low latitude cooling—it will be remembered that Osmaston's estimate of 5–6° for Kilimanjaro must be increased to some unknown amount to allow for decreases in humidity—but much of the evidence for periglacial activity is difficult to interpret and even more difficult to date.

The terraces of the Vaal River in South Africa have been interpreted paleoclimatologically for a long time. Butzer et al (22) provided a recent critical reappraisal. River terraces are notoriously difficult to interpret and not always easy to date. Those of the Vaal include a rich assemblage of stone industries that can be used for stratigraphic purposes, indicating that this major tributary of the Orange River, at 28° S, has been undergoing changes of regimen consistent with a changing climate since Pliocene time.

Pollen diagrams from coastal South Africa have been published by Martin (80) and Schalke (108). Martin made a painstaking analysis of sediments representing the last 8000 yr in Groenvlei, a small lake in the forested Knysna region, and concluded, with ample stratigraphic justification, that an increase in humidity 7000 years ago permitted forest to increase at the expense of macchia. A further increase in forest occurred about 2000 years ago, which Martin attributed with somewhat less assurance to still moister conditions. Martin was properly cautious about the climatic interpretation of pollen diagrams in an area where the effects of temperature and moisture on the vegetation are not easily untangled, even in modern times, but did point out interesting parallels between the record of vegetational change at Groenvlei and the standard sequence of temperate western Europe.

Schalke's cores penetrate interfingering continental and marine beds that underlie the Cape Flats near the Cape of Good Hope. The beds consist of sand and more or less sandy clay, with sandy peat, gravel, and shells in places. The material is not rich in pollen, and the percentages, especially for minor elements, do not have great statistical reliability. Schalke believes that the Holocene parts of his sections reflect climatic changes similar to those of temperate Europe at the same time.

One of Schalke's conclusions is very firmly established, and also of major paleoclimatic importance. In deposits of a minor marine transgression, when sea level was at least 18–24 m below that of modern times, there is excellent evidence, from pollen, stomata, and wood, of a former wide extension of forest of the Knysna type into an area now covered with macchia. The base of this deposit lies slightly below several carbon dates in excess of 40,000 yr. The top has an age of less than 28,500, and more than 10,000, yr. Clearly, forest was thriving at the southern tip of Africa; it expanded almost 300 km beyond its present range during the height of the last glaciation. This advance could be due to some combination of increased rainfall and lower temperature, or even, conceivably, to either acting alone. However, considering the climatic context of the southern tip of Africa, it is most likely that low temperatures and high rainfall were common results of a northward shift of the belt of westerly winds.

LOWLAND RAIN FOREST

The large areas of tropical evergreen forest in the Zairean cuvette and in West Africa are paleoclimatically poorly known. Most of the data are from their peripheries, and although these are suggestive of large and frequent changes in the climate of the humid equatorial part of Africa, the exact nature and timing of those changes have not yet been determined. This part of Africa is not well supplied with ancient lakes, and it will probably never be as well known paleoclimatically as the Rift Valley region to the east. There are excellent prospects of a long and informative record from Bosumtwi (119), an explosion crater in Ghana that appears to be of great age and is very well situated on the gradient of rainfall to provide a detailed and comprehensible record of vegetational change.

Kendall (66) showed that the evergreen forest was more extensive in Uganda during the early part of Holocene time than it has been during the past few thousand years, and also that forest essentially disappeared from the record prior to 12,500 BP. Langenheim (69) showed the common occurrence of amber coming apparently from *Copaifera mildbraedii,* in reworked sands of northeastern Angola, along the southern edge of the forested area. At present, this tree grows no closer to the fossil localities than the gallery forests lying 200 km to the north. One sample of the amber has been dated at 1650 $\pm$ 60 BP, a composite of two other samples at 2830 $\pm$ 80 BP, indicating that the forest retraction of the past few millenia was not restricted to the eastern boundary in Uganda, but occurred along the southern boundary as well. Evidence from the southern Sahara discussed below suggests that the climate of the past few millenia has been drying along the northern border also, and it seems reasonable to conclude that there has been a general retrenchment of the forested area during the past 2000–3000 years. This retraction has probably been caused by increasing aridity and human land-use.

A much more important matter is the state of the lowland forest during the dry time more than 12,000 years ago, and here relevant data are very scarce. Van Zinderen Bakker has carried out a few pollen analyses of material from archaeological sites. An initial set of conclusions, without the supporting evidence, may be found in Clark & van Zinderen Bakker (27). A later conclusion seems to be (130, p. 196), that conditions were dry during the late-Pleistocene hypothermal, with very open vegetation on the ridges, favoring reworking of the Kalahari sand. Van Zinderen Bakker (128) presented a number of pollen spectra, some radiocarbon dated, but without surface spectra. The environments of deposition of the samples are not clear; possibly some of the differences between spectra are functions of them rather than climatic change. The paleoclimatic significance of these isolated spectra from Angola is not clear, but they do suggest that changes have occurred.

Geomorphology provides some insight into the climatic history of the Zaire cuvette. It is well known that there was at some time or times extensive blowing of sand in western equatorial Africa. It is sometimes said that every tree in the Zaire rainforest has its roots in wind-blown sand, but this is an exaggeration, as can be seen by comparing maps of the areal distribution of the sand (32, 33) with a vegetational map (3). The inactive dunes and the evergreen forest have largely complementary distributions. It is nevertheless true that windblown sands do extend from the desert area of southwest Africa across the equator; in the Zaire basin, although they may not be covered with classic Guinean forest, they are covered with a woodland receiving 2000 mm of annual precipitation, a combination of circumstances so incompatible with active blowing as to suggest a very great change of climate indeed.

When that change occurred, or just what it was, is largely uncertain. The sands are an old feature of the geography of Africa. Some part of the material was reworked during the Pleistocene (91), but a closer dating is generally not possible, and even the total area of Pleistocene reworking has not been mapped. It may conceivably be the total area of the sands themselves.

An important attempt to characterize the climatic change by pedological, sedimentological, and geomorphological methods was made by de Ploey (37, 38) in the area around Stanley Pool and adjacent parts of the lower Zaire valley. In a number of localities there is a very well developed lateritic paleosol of a sort that could only be produced under long-sustained warm wet conditions. Overlying this paleosol is a body of colluvium and slope-wash deposits indicating steppe conditions. The base of the colluvium is somewhat beyond the reach of radiocarbon dating—from the thickness of the deposit, de Ploey suggests an age of 50,000 yr. The transition from colluvium to overlying alluvium and other material indicative of moister conditions is not well dated, but, from a combination of archaeological material and radiocarbon dates, de Ploey suggests an age of 10,000 yr. The moister Holocene conditions find their expression also in rejuvenation of small spring-fed valleys and of the Zaire River. Further data are required, but from present evidence it seems that the central part of the lower Zaire basin was substantially dryer during the late Pleistocene than it is today, and that the Holocene was, in general, a moist era. The magnitude of the climatic changes involved is not known, but was obviously large; one suspects that a major biogeographic problem will be locating and mapping the refugia in which moist forest species survived the dry late-Pleistocene time.

SAHARA

There is wealth of information concerning climatic change in the southern part of the Sahara, but some parts of this tremendous reach of country are better known than others. The greatest body of information comes from the interdisciplinary investigations of the French office de la Recherche Scientifique et Technique Outre-Mer (ORSTOM) team working in the closed depressions of the Chad region. A recent review (18) is the main source of the synthesis presented here. Those without access to the *West African Journal of Archaeology* may wish to consult the original papers (6, 31, 41–45, 79, 110–118).

Near the mouth of the Senegal River (18) reddened sand dunes plunge below present sea level, showing that the climate was dry enough for blowing to occur during the glacial eustatic lowering of sea level. This indicates, quite independently of radiometric dating, that the last ice age was dry.

The main evidence for climatic change consists of sand dunes, many of them large, in places too moist for such dunes to form today; the subsequent reddening of the dunes during a regime of seasonally moister climate; and former lake beds in depressions that are now dry, or in which the lake depth has been much reduced. The lacustrine evidence has been strengthened in places by detailed analysis of diatoms in the lake beds, and the first major pollen investigation in the Chad basin is beginning to bear fruit (77, 78). For the most part, however, Saharan evidence for climatic change consists of thin beds of marl laid down in evanescent lakes. The beds gleam in the desert sun and are easily located, even from the air, but their study and interpretation is complicated by the usual geochemical difficulties associated with radiocarbon dating of carbonate [see, however, the reassuring papers of Geyh & Jäkel (50–52)], the shortness of the sections, and the poor organic preservation. Under these circumstances some internal inconsistencies in the evidence are to be expected, and it is most gratifying that the broad picture seems to be the same all across northern Africa.

Prior to 30,000 BP the climate seems to have been moist. For a period of 10,000–25,000 yr centered on 20,000 BP, the climate was dry, and then, at a time that varies slightly from place to place but is commonly close to 12,000 BP, the climate became wet. It stayed wet, with a minor transitory dry phase between 5000 and 7000 BP, until about 3000 years ago. The agreement with conditions in tropical East Africa is excellent.

One can say with assurance that the climate has changed throughout the Sahara. In a sense, there are even written records of the changes, for the desert is dotted with rock engravings by Stone-Age men of elephant, rhinoceros, hippopotamus, and giraffe far from their modern ranges (30, 34, 81).

Many of these petroglyphs lie along trade routes and may represent the hopeful expectations of people who knew the animals in other places, rather than sketches from life; a former much wider range of large vertebrates is, however, attested by fossil remains. The presence of neolithic cultures in places too dry today to support neolithic people shows that the climate was once wetter. The changes of climate involved may not have been very great—see, for example, Sanford's remarks following a paper by Grove (54)—and do not seem to have generated large, deep, and permanent lakes.

Pollen analysis in the northern and central Sahara is rendered particularly difficult not only by an acute shortage of good polliniferous material, but also by the very low pollen productivity of the desert vegetation, even on relatively well-vegetated mountains such as Hoggar and Tibesti. Schulz (109) has shown that, even at Tibesti, the temperate pollen blown in on the wind from the Mediterranean or beyond constitutes an appreciable percentage of the modern pollen rain. Many interesting finds have been made of temperate tree pollen in Quaternary beds of the Sahara, and in some cases the percentages are so high as to suggest a local origin, but this finding of Schulz is very sobering, casting some doubt on the otherwise attractive idea that some substantial part of the flora of Europe may have spread during Pleistocene glaciations to the mountains of the Sahara.

The evidence is hard to obtain and to interpret (for introductions, see 6, 31), and there are some troubling inconsistencies in the radiocarbon dates [see for example, Alimen (1)], but a considerable body of evidence (2) suggests that conditions were moist in the northern Sahara while they were dry in more tropical parts of Africa prior to 12,000 years ago. For example, deposition of the Saourian sediments, which are taken to represent moist conditions, began between 20,000 and 40,000 BP and came to an end by 14,500. There are also indications of temperature change: Rognon (2, 106, 107) has provided evidence that the massif of Atakor (Hoggar) bears, between 2030 and 2370 m, névé niches, solifluction phenomena, and even moraines of small glaciers.

SUMMARY OF CLIMATIC CHANGE IN AFRICA

No part of the continent of Africa seems to have escaped serious climatic change during the past 20,000 years. Glaciers, lakes, aggrading and degrading rivers, and active dunes have all been more widespread during some part of that time than they are today. Forests, large vertebrates, molluscs, and human cultures have advanced and retreated.

The time of these changes has not been established everywhere, but a considerable number of carbon dates suggests a major change in climate 10,000 or 12,000 years ago. In tropical parts of the continent the climate increased in humidity at that time. In the extreme south temperate part, and apparently in the north temperate part as well, the climate became more arid. There is excellent evidence at 4000 m on the equator for an increase of mean annual temperature of the order of 6°C, and some evidence from lower altitudes and higher latitudes for a temperature change of the same magnitude, between 10,000 and 15,000 BP. Experience in other parts of the world suggests that the actual warming took place during a small part of that time, and was probably synchronous over the continent, but the strictly African data do not fix the time of warming precisely.

At least part of the continent was moister from early Holocene time until 7000 BP than it was during the next few thousand years. There was a tendency for conditions to be moist again in about 5000 BP, followed by gradual drying to conditions somewhat drier than today by 3000 BP. The past few thousand years have been dry over a wide area, but the effects of drought on the vegetation are compounded by the influence of agricultural and pastoral people. These Holocene changes may have been continent-wide, and the sporadic evidence for them may be due to no more than differences in the sensitivity of different sites, or there may have been genuine regional differences in paleoclimate. In any case, the changes involved are much smaller than the major change of 12,000 BP.

The earlier climatic history is less well known and much less well dated. There is good evidence that the climate of Africa has undergone repeated change since at least the beginning of Quaternary time, but in few places have those changes been dated with satisfactory precision. It is not yet possible to specify the time of onset of the last major tropical dry episode, but in a few places at least it seems to be at least 22,000 years ago. There is some tendency for conditions between 22,000 and 28,000 years ago to have been less dry and possibly warmer than the very last part of pre-Holocene time.

BIOLOGICAL CONSEQUENCES
OF CLIMATIC CHANGE IN AFRICA

The climatic history outlined above provides no neat explanations for outstanding biological puzzles. Rather, it generates new puzzles and casts doubt on some explanations that seemed adequate only a few years ago.

For example, the ornithological and entomological affinities of Mt. Cameroon and the highlands of East Africa were attributed quite reasonably by Moreau (86) to a cool time when the montane vegetation belts were depressed to the altitude of the broad plateaus connecting East and West Africa. This cool time Moreau took to be the last ice age. As we have seen, this theory must perish in the face of stratigraphic fact, for the last ice age was a time when forest trees, at the altitude of both the mountains and the plateau, were extremely scarce. At least in intramontane East Africa, the prevailing vegetation seems to have been some kind of savanna or grassland. It is still possible that—at some time too remote to have come to our stratigraphic attention—there was a montane forest corridor across Africa, but this is not likely. We have data now for a specimen ice age and a specimen interglacial, the Holocene. Neither is characterized by widespread montane forest at low altitudes, and it seems more reasonable to seek an explanation elsewhere for bird and butterfly distribution patterns.

Two recent developments, in island biogeographic theory and in the very pollen analysis that invalidates Moreau's hypothesis, may combine to form the answer. Pollen analysis indicates (73) that plant communities do not react to climatic change as integrated entities, but rather on an individualistic basis, with some species reacting one way, other species another. In fact, the pollen record of the past 15,000 years, not only in Africa but elsewhere, indicates that the plant-climate interaction is a very dynamic one. Climate changes and vegetational adjustments are not rare and isolated events, they are the norm.

During the early Holocene, there was a great expansion of evergreen forests in tropical Africa. Conditions were probably warm rather than cool, but they were certainly moist. Perhaps some of the species of birds and butterflies that are restricted, under the somewhat dryer conditions of today, to the montane forest at moderate altitudes were able to extend down into the lush evergreen forests of 10,000–7000 BP.

This is not to suggest that they formed continuous breeding populations through the plateau forest across Africa. As MacArthur & Wilson have shown (76), such continuous ranges are not essential for the colonization of small isolated habitats. The process of colonization, like the interaction of trees and climate, is a dynamic one. Immigration and extinction are both involved, and the dispersal powers of organisms have been considerably underestimated by classical biogeographers who failed to take extinction into account. Further, the effect of steppingstones on the spread of species between isolated habitats is very large. It is particularly significant that the long-distance affinities occur in two animal groups that include strong flyers and have much higher vagility than most of the montane fauna. A strong flyer is in a better position to bridge the gap through not completely hostile forest environments from one steppingstone to another. The failure of birds, though not butterflies, to cross the relatively short but very severe stretch of desert country between the mountains of East Africa and the Ethiopian highland shows that there are limits to the dispersal capacity of even strong fliers.

The African environment is capricious, not stable, and apparently has been so for at least several million years. As a corollary, one should expect to find the African biota changing ecologically, evolutionarily, and biogeographically, in response to continuing climatic change. In particular, one may expect to find indications of the dry tropical conditions of 13,000 years ago, or of the moist conditions of 8000 years ago, in the living biota of today. Haffer (56, 57) has shown that the biota of the Amazonian forest is a mosaic of differentiated populations, each derived from a forested refugium of the dry late-Pleistocene. It might be profitable to view the African biota from this point of view; J. L. Richardson (personal communication) suggests the alcephaline antelopes as a promising group with which to begin, and Booth (15) has already considered the West African primates from a similar point of view.

One longstanding geographical problem becomes even more difficult. The rich Cape flora, restricted to a tiny area at the southern tip of Africa, is possibly the most bizarre example of a relict biota in the world today. Its survival was puzzling when one could speculate that it inhabitated a climatically stable culdesac. Now that there is good fossil evidence (108) of profound climatic change in the very heart of the Cape region, persistence of the flora seems little short of marvelous. No doubt the answer is to be found in evolutionary and ecological adaptation, not in special features of the Cape environment.

The present equatorial biota represents a lag concentrate of species that were able to persist through much dryer and cooler conditions than those of today. From the accessible African terrestrial record, and especially by analogy with the marine stratigraphic record, it seems that present warm, moist conditions are unusual, at least by Quaternary standards, and that the dry, cool climate of the late-Pleistocene has prevailed for a much larger part of the past two million years. Against that temporal perspective it might be better to consider the surviving equatorial species as a selected set of the Tertiary biota, the organisms that were able to tolerate the cool, dry ice ages as well as the hot and humid conditions of the past 10,000 yr.

ACKNOWLEDGMENTS

This manuscript was prepared with the financial assistance of a National Science Foundation grant to Duke University. I am indebted to Ms. Susan Dickerson for secretarial assistance and to Ms. Marilyn Loveless for drafting the figures, and to

Dr. Robert Kendall, Ms. Marilyn Loveless, Mr. John Melack, and Dr. Charles Michener for critical commentary.

Literature Cited

1. Alimen, H. 1965. The Quaternary era in the Northwest Sahara. *Geol. Soc. Am. Spec. Pap.* No. 84, ed. H. E. Wright, D. G. Frey, 273–91
2. Alimen, H., Faure, H., Chavaillon, J., Taieb, M., Battisini, R., 1969. *Les Études Françaises sur le Quaternaire,* p. 201–14. Présentées à l'occasion du VIIIᵉ Congrès International de l'INQUA, Paris, 1969
3. Aubreville, A., Duvigneaud, P., Hoyle, A. C., Keay, R. W. J., Mendonça, A., Pichi-Sermolli, R. E. G. 1958. Vegetation map of Africa south of the Tropic of Cancer. *Assoc. pour l'Etude Taxonomique de la Flore d'Afrique Tropicale.* London: Oxford Univ. Press. 24 pp.
4. Baker, B. H. 1967. Geology of the Mount Kenya area. *Geol. Surv. Kenya Rep.* No. 79, 78 pp.
5. Bishop, W. W. 1967. The later Tertiary in East Africa—volcanics, sediments and faunal inventory. In *Background to Evolution in Africa,* ed. W. W. Bishop, J. D. Clark, 31–56. Chicago: Univ. Chicago Press
6. Beucher, F. 1971. Etude palynologique de formations néogènes et quaternaires au Sahara Nord-Occidental. Thèse, l'Université de Paris. 796 pp.
7. Bonnefille, R. 1968. Contribution à l'étude de la flore d'un niveau pléistocène de la haute vallée de l'Aouache (Ethiopie). *C. R. Acad. Sci.* (D)266(12)1229–32
8. Bonnefille, R. 1969. Analyse pollinique d'un sédiment recent: Vases actuelles de la Rivière Aouache (Ethiopie). *Pollen Spores,* 11(1):7–16
9. Bonnefille, R. 1969. Indication sur la paleoflore d'un niveau du Quaternaire moyen du site de Melka Kontouré (Ethiopie). *C. R. Som. Seances Soc. Geol. Fr.* 7(3):238 pp.
10. Bonnefille, R. 1970. Premiers resultats concernant l'analyse pollinique d'échantillons du Pléistocène inferieur de l'Omo (Ethiopie). *C. R. Acad. Sci.* Paris, (D) 270:2430–33
11. Bonnefille, R. 1971. Atlas des pollens d'Ethiopie. *Adansonia* 2(11):463–518
12. Bonnefille, R. 1971. Atlas des pollens d'Ethiopie; principales éspèces des forêts de montagne. *Pollen Spores* 13: 15–72
13. Bonnefille, R. 1972. Associations polliniques actuelles et quaternaires en Ethiopie (Vallés de l'Awash et de l'Omo). Thèse, l'Université de Paris. 513 pp.
14. Bonnefille, R. 1972. Aperçu sur les recherches palynologiques en Ethiopie. *Palaeoecology of Africa, the Surrounding Islands and Antarctica,* ed. E. N. van Zinderen Bakker, Sr., 7:19–28. Cape Town: Balkema
15. Booth, A. H. 1958. The zoogeography of West African primates. *Bull. Inst. Franc. Afr. Noire* 20:587–622
16. Brown, J. H. 1971. Mammals on mountaintops: nonequilibrium insular biogeography. *Am. Natur.* 105:467–78
17. Bryson, R. A., Kutzbach, J. E. 1974. On the analysis of pollen-climate canonical transfer functions. *Quat. Res.* 4:162–74
18. Burke, K., Durotoye, A. B., Whiteman, A. J. 1971. A dry phase south of the Sahara 20,000 years ago. *W. Afr. J. Archaeol.* 1:1–8
19. Butzer, K. W. 1971. *Recent History of an Ethiopian Delta.* Univ. Chicago Dept. Geogr. Res. Pap. No. 136. 184 pp.
20. Butzer, K. W. 1973. Pleistocene "periglacial" phenomena in southern Africa. *Boreas* 2(1):258–97 Oslo. Universitetsforlaget.
21. Butzer, K. W., Fock, G. J., Stuckenrath, R., Zilch, A. 1973. Palaeohydrology of Late Pleistocene Lake Alexandersfontein, Kimberley, South Africa. *Nature* 243:328–30
22. Butzer, K. W., Helgren, D. M., Fock, G. J., Stuckenrath, R. 1973. Alluvial terraces of the Lower Vaal River, South Africa: A reappraisal and reinvestigation. *J. Geol.* 81(3):341–62
23. Butzer, K. W., Isaac, G. L., Richardson, J. L., Washbourn-Kamau, C. 1972. Radiocarbon dating of East African lake levels. *Science* 175:1069–76
24. Cahen, L. 1954. *Géologie du Congo Belge.* Liège: Vaillant-Carmanne. 577 pp.
25. Carcasson, R. H. 1964. A preliminary survey of the zoogeography of African butterflies. *E. Afr. Wildlife J.* 2:122–57

26. Clark, J. D. 1967. *Atlas of African Prehistory*. Chicago: Univ. Chicago Press. 62 pp.
27. Clark, J. D., van Zinderen Bakker, E. M. 1964. Prehistoric culture and Pleistocene vegetation at the Kalambo Falls, Northern Rhodesia. *Nature* 201:971–75
28. Coetzee, J. A. 1964. Evidence for a considerable depression of the vegetation belts during the Upper Pleistocene on the East African mountains. *Nature* 204:564–66
29. Coetzee, J. A. 1967. Pollen analytical studies in East and Southern Africa. *Palaeoecol. Afr.* 3:1–146
30. Cole, S. 1963. *The Prehistory of East Africa*. London: Weidenfeld & Nicolson. 383 pp.
31. Conrad, G. 1969. L'evolution continentale post-Hercynienne du Sahara Algérien. Centre de Recherches sur les Zones Arides. *Sér. Géol. No. 10.* 527 pp.
32. Cooke, H. B. S. 1958. Observations relating to Quaternary environments in East and southern Africa. *Geol. Soc. S. Afr. Bull.* 20 (annexure). 73 pp.
33. Cooke, H. B. S. 1964. The Pleistocene environment in southern Africa. *Ecological Studies in Southern Africa. Monogr. Biol.* 14:1–23
34. Cooke, H. B. S. 1964. Pleistocene mammal faunas of Africa, with particular reference to southern Africa. *Anthropology*, ed. F. Howell, F. Bourlière 65–241, New York: Viking
35. Degens, E. T., von Herzen, R. P., Wong, H. K. 1971. Lake Tanganyika: Water chemistry, sediments, geological structure. *Naturwissenschaften* 58: 229–41
36. Degens, E. T., von Herzen, R. P., Wong, H. K., Deuser, W. G., Jannasch, H. W. 1973. Lake Kivu: structure, chemistry and biology of an East African Rift Lake. *Geol. Rundsch.* 62: 245–77
37. de Ploey, J. 1963. Quelques indices sur l'évolution morphologique et paléoclimatique des environs du Stanley-Pool (Congo). *Stud. Univ. Lovanium Fac. Sci.* 17:1–16
38. de Ploey, J. 1965. Position géomorphologique, génèse et chronologie de certains dépôts superficiels au Congo occidental. *Quaternaria* 7:131–54
39. Downie, C. 1964. Glaciations of Mount Kilimanjaro, northeast Tanganyika. *Bull. Geol. Soc. Am.* 75:1–16
40. Downie, C., Humphries, D. W., Wilcockson, W. H., Wilkinson, P. 1956.

Geology of Kilimanjaro. *Nature* 178: 828–30
41. Faure, H. 1966. Evolution des grands lacs sahariens à l'Holocene. *Quaternaria* 8:167–75
42. Faure, H. 1967. Le problème de l'origine et de l'age de l'eau des oasis sahariennes du Niger. *International Association of Hydrology* 7:277–78
43. Faure, H. 1969. Lacs quaternaires du Sahara. *Mitt. Int. Ver. Limnol.* 17: 131–46
44. Faure, H., Manguin, E., Nydal, R. 1963. Formations lacustres du Quaternaire supérieure du Niger oriental: diatomites et ages absolus. *Bull. Bur. Rech. Geol. Mineral.* 3:41–63
45. Faure, H., Servant, M. 1970. Évolution récente d'un bassin continental: le Tchad; programme d'étude français, *Office de la Recherche Scientifique et Technique Outre-Mer, Cah.*, Sér. Géol. 2(1):5–8
46. Flint, R. F. 1971. *Glacial and Quaternary Geology*. New York: Wiley 892 pp.
47. Flint, R. F., Gale, W. A. 1958. Stratigraphy and radiocarbon dates at Searles Lake, California. *Am. J. Sci.* 256:689–714
48. Fryer, G., Iles, T. D. 1972. *The Cichlid Fishes of the Great Lakes of Africa*. Edinburgh: Oliver & Boyd. 641 pp.
49. Garrels, R. M., Mackenzie, F. T. 1971. *Evolution of Sedimentary Rocks*. New York: Norton. 397 pp.
50. Geyh, M. A., Jäkel, D. 1974. *C-Altersbestimmungen in Rahmen der Forschungsarbeiten der Aussenstelle Bardai/Tibesti der Frien Universität Berlin*. Germany. 107–117
51. Geyh, M. A., Jäkel, D. 1974. Late Glacial and Holocene climatic history of the Sahara Desert derived from a statistical assay of C^{14} dates. *Palaeogeogr. Palaeoclimatol. Palaeoecol.* 15:205–8
52. Geyh, M. A., Jäkel, D. 1975. Spaetpleistozaene und holozaene Klimageschichte der Sahara aufgrund zugaenglicher C^{14} Daten. *Z. Geomorphol.* 18:82–98
53. Greenwood, P. H. 1975. Zoogeography and history. In *Biology of the Nile*, ed. J. Rzoska. The Hague: Junk. In press
54. Grove, A. T. 1960. Geomorphology of the Tibesti region. *Geogr. J.* 126(1): 18–31
55. Grove, A. T. 1969. Landforms and climatic change in the Kalahari and Ngamiland. *Geogr. J.* 135(2): 191–212
56. Haffer, J. 1969. Speciation in Am-

azonian forest birds. *Science* 165: 131–37

57. Haffer, J. 1974. Avian speciation in tropical South America. *Nuttall Ornithol. Club Publ.* No. 14, 390 p.

58. Harper, G. 1969. Periglacial evidence in southern Africa during the Pleistocene epoch. In *Palaeoecology of Africa and of the Surrounding Islands and Antarctica* 4:71–101. Cape Town: Balkema

59. Harris, N., Pallister, J. W., Brown, J. M. 1956. Oil in Uganda. *Mem. Geol. Surv. Uganda* No. 9. 33 pp.

60. Hamilton, A. C. 1972. The interpretation of pollen diagrams from highland Uganda. *Palaeoecol. Afr.* 7:45–149

61. Hecky, R. E., Degens, E. T. 1973. Late Pleistocene-Holocene chemical stratigraphy and paleolimnology of the Rift Valley lakes of Central Africa. *Woods Hole Oceanogr. Inst. Tech. Rep. May 1973.* WHOI 73–28. Unpublished manuscript

62. Hedberg, O. 1954. A pollen-analytical reconnaissance in tropical East Africa. *Oikos* 5:137–66

63. Hurst, H. E. 1952. *The Nile.* London: Constable. 331 pp.

64. Iversen, J. 1949. The influence of prehistoric man on vegetation. *Dan. Geol. Unders. Afh. Raekke* 3(6):1–25

65. Washbourn, C. K. 1967. Lake levels and Quaternary climates in the Eastern Rift Valley of Kenya. *Nature,* 216: 672–73

66. Kendall, R. L. 1969. An ecological history of the Lake Victoria basin. *Ecol. Monogr.* 39:121–76

67. Krebs, C. J. 1972. *Ecology.* New York: Harper & Row. 694 pp.

68. Lawton, R. M. 1963. Palaeoecological and ecological studies in the northern province of Northern Rhodesia. *Kirkia* 3:46–77

69. Langenheim, J. H. 1972. *Botanical Origin of Fossil Resin and its Relation to Forest History in Northeastern Angola.* Subsidios Para a Historia, Arqueologia e Etnografia dos Povos da Lunda, Museu do Dundo, Lisboa, 15–36

70. Linton, D. L. 1969. Evidences of Pleistocene cryonival phenomena in South Africa (with discussion). In *Conf. Quaternary Stud. Palaeoecol. Afr.* 5:71–89

71. Livingstone, D. A. 1962. Age of deglaciation in the Ruwenzori range, Uganda. *Nature* 194:859–60

72. Livingstone, D. A. 1965. Sedimentation and the history of water level change in Lake Tanganyika. *Limnol. Oceanogr.* 10:607–10

73. Livingstone, D. A. 1967. Postglacial vegetation of the Ruwenzori Mountains in Equatorial Africa. *Ecol. Monogr.* 37:25–52

74. Livingstone, D. A. 1971. A 22,000-year pollen record from the plateau of Zambia. *Limnol. Oceanogr.* 16:349–56

75. Livingstone, D. A. 1975. Paleolimnology of headwaters. In *Biology of the Nile,* ed. J. Rzoska, The Hague: Junk. In press

76. MacArthur, R. H., Wilson, E. O. 1967. *The Theory of Island Biogeography.* Princeton, NJ: Princeton Univ. Press. 203 pp.

77. Maley, J. 1970. Contributions à l'étude du Bassin tchadien. Atlas de pollens du Tchad. *Bull. Jardin Bot. Nat. Belg.* 40(1):29–48

78. Maley, J. 1972. La sédimentation pollinique actuelle dans la zone du lac Tchad (Afrique centrale). *Pollen Spores,* 14(3):263–307

79. Maley, J., Cohen, J., Faure, H., Rognon, P., Vincent, P. M. 1970. Quelques formations lacustres et fluviatiles associées à différentes phases du volcanisme au Tibesti (nord du Tchad.) *Cah. Office de la Recherche Scientifique et Technique Outre-Mer, Sér. Géol.* 2(1):127–52

80. Martin, A. R. H. 1968. Pollen analysis of Groenvlei Lake sediments Knysna (South Africa). *Rev. Palaeobot. Palynol.* 7:107–44

81. Mauny, R. A. 1957. Répartition de la grande faune ethiopienne du nordouest africain du Paléolithique à nos jours. *Pan-Afr. Congr. 3d.* Livingstone, 1955, 102–50 pp. London: Chatto & Windus

82. Moreau, R. E. 1933. Pleistocene climatic changes and the distribution of life in East Africa. *J. Ecol.* 21:415–35

83. Moreau, R. E. 1952. Africa since the Mesozoic: with particular reference to certain biological problems. *Proc. Zool. Soc. London* 121:869–913

84. Moreau, R. E. 1963. The distribution of tropical African birds as an indicator of climatic change. In *African Ecology and Human Evolution,* ed. F. Howell, F. Bourlière. Chicago: Aldine, 28–42

85. Moreau, R. E. 1963. Vicissitudes of the African biomes in the late Pleistocene. *Proc. Zool. Soc. London* 141:395–421

86. Moreau, R. E. 1966. *The Bird Faunas of Africa and its Islands.* New York: Academic. 424 pp.

87. Morrison, M. E. S. 1961. Pollen analysis in Uganda. *Nature* 190:483–86

88. Morrison, M. E. S. 1966. Low-latitude vegetation history with special reference to Africa. *R. Meteorol. Soc. Proc. Int. Symp. World Climate from 8000 to 0 B. C.,* 142–48 pp.

89. Morrison, M. E. S. 1968. Vegetation and climate in the uplands of Southwestern Uganda during the later Pleistocene period. I. Muchoya Swamp, Kigezi District, *J. Ecol.* 56:363–84

90. Morrison, M. E. S., Hamilton, A. C. 1974. Vegetation and climate in the uplands of south-western Uganda during the later Pleistocene Period. II. Forest clearance and other vegetational changes in the Rukiga Highlands during the past 8000 years. *J. Ecol.* 62:1–32

91. Mortelmans, G. 1957. Le Cénozoique du Congo Belge. *Pan-Afr. Cong. Prehist. 3d., Livingstone, 1955,* J. D. Clark, 23–50 pp.

92. Nilsson, E. 1929. Preliminary report on the Quaternary geology of Mount Elgon and some parts of the Rift Valley. *Geol. Foeren. Stockholm Foerh.* 51:253–61

93. Nilsson, E. 1931. Quaternary glaciations and pluvial lakes in British East Africa. *Geogr. Ann.* 13:249–348

94. Nilsson, E. 1935. Traces of ancient changes of climate in East Africa. *Geogr. Ann.* 17:1–21

95. Nilsson, E. 1938. Pluvial lakes in East Africa. *Geol. Foeren. Stockholm Foerh.* 60:423–33

96. Nilsson, E. 1940. Ancient changes of climate in British East Africa and Abyssinia. *Geogr. Ann.* 22:1–97

97. Nilsson, E. 1949. The pluvials of East Africa. *Geogr. Ann.* 31:204–11

98. Nilsson, E. 1953. Pleistocene climatic changes in East Africa. *Proc. Pan-Afr. Congr. Prehist. Nairobi, 1947,* 45–54. Oxford: Blackwell.

99. Osmaston, H. A. 1958. Pollen analysis in the study of the past vegetation and climate of Ruwenzori and its neighborhood. B.Sc. thesis. Oxford, Univ. Oxford, Engl. 44 pp.

100. Osmaston, H. A. 1965. The past and present climate and vegetation of Ruwenzori and its neighborhood. Ph.D. thesis. Oxford Univ., Oxford, Engl. n.p. Mimeo.

101. Richardson, J. L. 1964. Plankton and fossil plankton studies in certain East African lakes. *Mitt. Int. Ver. Theor. Angew. Limnol.* 15:993–99

102. Richardson, J. L. 1966. Changes in level of Lake Naivasha, Kenya, during postglacial times. *Nature* 209:290–91

103. Richardson, J. L. 1972. Palaeolimnological records from rift lakes in central Kenya. *Palaeoecol. Afr.* 6:131–36

104. Richardson, J. L., Richardson, A. E. 1972. History of an African rift lake and its climatic implications. *Ecol. Monogr.* 42:499–534

105. Roccati, A. 1970. Nell'Uganda e nella Catena del Ruwenzori. *Bull. Soc. Geol. Ital.* 26:127–58

106. Rognon, P. 1962. Observations nouvelles sur le Quaternaire du Hoggar. *Travaux de l'Institut de Recherches Sahariennes.* 21:57–80

107. Rognon, P. 1967. Le massif de l'Atakor et ses bordures Sahara central étude géomorphologique. *Cent. Rech. Zones Arides, Sér. Géol.* 9:559

108. Schalke, H. J. W. G. 1973. The Upper Quaternary of the Cape Flats Area (Cape Province, South Africa). *Scripta Geologica 15.* Leiden:Rijksmuseum van Geologie en Mineralogie. 57 pp.

109. Schulz, E. 1974. Pollenanalytische Untersuchungen quatärer Sedimente des Nordwest-Tibesti. In *Forschungsstation Bardai FU Geologen in der Zentralsahara.* 59–69. Pressedienst Wissenschaft FU Berlin 5.

110. Servant, M. 1967. Nouvelles données stratigraphiques sur le Quaternaire supérieur et récent au nord-est du Tchad. *Pan-Afr. Prehist. Congr. 6th, Dakar, 1967.* 20 pp. Mimeo.

111. Servant, M. 1970. Données stratigraphiques sur le quaternaire supérieur et récent au nord-est du lac Tchad. *Fr. Off. Rech. Sci. Tech. Outre-Mer, Cah., Sér. Géol.* 2(1):95–114

112. Servant, M. 1972. Les séries sédimentaires continentales du Plio-Quaternaire dans le bassin du Tchad. *Assoc. Sénégal. Etude Quat. Ouest Afr. Bull.* 35:36–84 (Abstr.)

113. Servant, M., Servant, S. 1970. Les formations lacustres et les diatomées du Quaternaire récent du fond de la cuvette Tchadienne. *Rev. Géogr. Phys. Géol. Dyn.* 12(2):63–76

114. Servant, M., Servant, S., Delibrias, G. 1969. Géologie du Quaternaire. Chronologie du Quaternaire récent des basses regions du Tchad. *C. R. Acad. Sci.* 269:1603–6

115. Servant, M., Servant-Vildary, S. 1972. Nouvelles données pour une interpretation paléoclimatique de séries continentales du Bassin Tchadien (Pléistocène récent, Holocène). *Palaeoecol. Afr.* 6:87–92

116. Servant, S. 1967. Répartition des diatomées dans les séquences lacustres Holocènes au Nord-Est du Lac Tchad. *Pan.-Afr. Prehist. Congr. 6th., Dakar, 1967.* 9 pp. Mimeo.

117. Servant, S. 1972. Le rôle des études sur les diatomées dans l'interpretation des séries continenties Plio-Quaternaires du Tchad. *Assoc. Sénégal. Etude Quat. Ouest Afr. Bull.* 35:36–85

118. Servant-Vildary, S. 1973. Néotectonique. Stratigraphie et néotectonique du Plio- Pléistocène ancien du Tchad d'après l'étude des diatomées. *C. R. Acad. Sci.* 276:2633–36

119. Smit, A. F. J. 1962. Origin of L. Bosumtwi and other problematic structures. *Ghana J. Sci.* 2(2):176–96

120. Stoffers, P., Holdship, S. 1975. Diagenesis of sediments in an alkaline lake: Lake Manyara, Tanzania. *Int. Congr. Sedimentol. 9th., Nice.* In press

121. Talling, J. F., Talling, I. B., 1965. The chemical composition of African lake waters. *Int. Rev. Gesamten. Hydrobiol.* 50(3):421–63

122. Temple, P. H. 1964. Evidence of lake-level changes from the northern shoreline of Lake Victoria, Uganda. In *Geographers and the Tropics,* ed. R. W. Steel, R. M. Prothero, 31–56. Liverpool

123. Temple, P. H. 1966. Lake Victoria levels. *Proc. E. Afr. Acad.* 2:50–58

124. Temple, P. H. 1967. Causes of intermittent decline of the level of Lake Victoria during the late Pleistocene and Holocene. In *Liverpool Essays in Geography.* 43–63. London: Longmans

125. Thompson, B. W. 1965. *The Climate of Africa.* London: Oxford Univ. Press. 132 pp.

126. van Zinderen Bakker, E. M. 1957. A pollen analytical investigation of the Florisbad deposit (South Africa). *Pan-Afr. Congr. Prehist. 3d, Livingstone, 1955,* 56–67

127. van Zinderen Bakker, E. M. 1962. A late-glacial and post-glacial climatic correlation between East Africa and Europe. *Nature* 194:201–3

128. van Zinderen Bakker, E. M. 1963. Analysis of pollen samples from northeast Angola. In *Prehistoric Cultures of Northeast Angola and Their Significance in Tropical Africa,* ed. J. D. Clark, Pt. 1, 213–22. Museu do Dondo, Subsidios para a História, Arqueologia e Etnografia dos Povos do Lunda. Lisboa: Campanhia de Diamantes de Angola

129. van Zinderen Bakker, E. M. 1964. A pollen diagram from equatorial Africa Cherangani, Kenya. *Geol. Mijnbouw,* 43(3):123–8

130. van Zinderen Bakker, E. M. 1966. *Paleoecology of Africa, Vol. I.* Capetown:Balkema. 270 pp.

131. van Zinderen Bakker, E. M. 1969. The Pleistocene vegetation and climate of the basin. In *Kalambo Falls Prehistoric Site I,* ed. J. D. Clark, 57–84. London: Cambridge Univ. Press

132. van Zinderen Bakker, E. M., Clark, J. D. 1962. Pleistocene climates and cultures in north-eastern Angola. *Nature* 196(4855)639–42

133. Washbourn-Kamau, C. K. 1971. Late Quaternary lakes in the Nakuru-Elementeita basin, Kenya. *Geogr. J.* 137:522–35

134. Webb, T. 1973. A comparison of modern and presettlement pollen from southern Michigan (U.S.A.) *Rev. Palaeobot. Palynol.* 16:137–56

135. Webb, T. 1974. Corresponding patterns of pollen and vegetation in lower Michigan: a comparison of quantitative data. *Ecology* 55:17–28

136. Webb, T., Bryson, R. A. 1972. Late- and postglacial climatic change in the northern Midwest, U.S.A.: quantitative estimates derived from fossil pollen spectra by multivariate statistical analysis. *Quat. Res.* 2:70–115

137. Western, D., van Praet, C., 1973. Cyclical changes in the habitat and climate of an East African ecosystem. *Nature* 241:104–6

EXPERIMENTAL STUDIES OF THE NICHE

❖4095

Robert K. Colwell

Department of Zoology, University of California, Berkeley, California 94720

Eduardo R. Fuentes

Instituto de Ciencias Biológicas, Universidad Católica de Chile, Santiago, Chile

INTRODUCTION

Ideally, experimental studies in ecology examine the response of genetically similar organisms to different environments or of genetically different organisms to similar environments, where *environment* is taken to include both biotic and abiotic components. Since environmental differences in nature may rapidly lead to subtle genetic differences among populations (e.g. 72, 177, 178), it is sometimes perilous to compare the response of a species among differing natural environments by observation alone. Likewise, simply observing or measuring the response of clearly different organisms in a single natural environment may fail to reveal the extent of their interaction. This means that experimental intervention or manipulation is usually required to establish controls for any rigorous test of hypotheses concerning biotic interactions or the differential response of organisms to environments.

The concept of the ecological niche is a useful device for describing concisely the patterns and limits of response of organisms and for abstracting certain generalities from the vast array of particulars comprising the literature of ecology. As Elton put it fifty years ago: "The importance of studying niches is partly that it enables us to see how very different . . . communities may resemble each other in the essentials of organization" (56).

In its broadest sense the title of this review could include most of experimental ecology. Besides the obvious impossibility of reviewing the literature of such a broad area, we believe that the niche concept is most useful in a much more restricted domain. It is almost entirely at the focus of communities (or at least interspecific relationships) that the concept of the niche has figured in ecological theory (31, 109, 116, 121, 123, 159–161, 190, 194) and in the large number of empirical studies of competition and coexistence not based on experiment. [T. W. Schoener (165) recently reviewed studies concerned with three or more species; see also (31).] Consequently we have generally not considered studies of an autoecological or demographic character, thereby excluding those portions of the literature of physiological ecology, behavioral ecology, ecological genetics, plant and animal breeding, and wildlife management, which could be considered "experimental studies of the niche" in a broad sense.

By far the most extensive experimental investigations of interactions between species have concerned organisms of economic or medical importance, and many of these studies can be analyzed in terms of niche theory. Included in this category is the vast literature of pollution biology, waste treatment, biological control, economic entomology, plant pathology, parasitology, microbiology, and several areas of human and veterinary medicine. We have not attempted to review these studies. This can be seen either as a practical necessity or as a restriction of our attention primarily to investigations of species that may reasonably be presumed to have evolved together long enough to reach evolutionary equilibrium, as defined by Wilson (201).

It is our goal here to bring together a representative collection of experimental studies that involve two or more species in or from the same co-evolved community, using the concept of the niche as an organizing principle. We have included laboratory experiments only when designed to elucidate more natural situations, and field experiments only when controlled or at least in principle subject to experimental control. These criteria thus exclude, on the one hand, most "population cage" experiments and laboratory microcosms and, on the other hand, most "natural experiments" (31, 39). Typically, in the latter, the responses of two or more populations of the same species (presumed genetically similar) are examined in two or more natural environments (believed to differ only in some known factor, such as the presence or absence of certain other species).

Even with these seemingly stringent restrictions, the remaining literature is large and widely scattered, and we have undoubtedly missed many relevant references. It is the purpose of this paper to point out the major ways in which biotic dimensions of the niche and the niche structure of communities have—and have not—been examined experimentally, and to attempt a synthesis of the major findings of existing studies.

THE NICHE

The origin and evolution of the niche concept have been reviewed at intervals during the past ten years (31, 39, 47, 191, 194), and are not recounted here. As we use the term, the *niche* is a phenotypic attribute of a population of conspecific individuals, a statistical entity that changes whenever the members of the population change in their response to the biotic and abiotic environment (34, 116, 193). (The gene pool of a population is a similarly abstract entity.) The biotic context of a population is the community, an association of coexisting populations bounded functionally by the domain of their interactions, or spatially by their co-occurrence in a habitat or biotope (156, 194). The biotic context of the niche is the set of niches belonging to the populations in the community, or the *columbarium*[1] of the community. The niche of a population is a hypervolume in a space defined by axes representing the biotic and abiotic factors to which populations in the community respond differentially. (Just as not all gene loci and allele frequencies in the gene pool need be known to permit meaningful studies of the evolution of a population, neither is it necessary to know all the axes of the niche hyperspace to permit meaningful studies of community structure.) A particular niche axis may represent the intensity or concentration of a physical or chemical factor; the density, biomass, or productivity of a population (or phenotype) in the community; the degree of patchiness in space or predictability in time (33) of a biotic or abiotic factor; or a simple or complex combination of the foregoing.

The response of organisms to different environments (different points in niche space) is an essential component of the niche (34, 56a, 109, 121, 194, 203). A "positive" response by a given population in an environment represented by some particular point in niche space is the criterion for inclusion of that point in the niche of the population. The level of response is taken to be a measure of population fitness in that environment. In practice, the response may be persistence, growth, abundance, biomass, metabolic or reproductive rate, numerical or functional dominance, or some other measure, depending upon the circumstances. The fitness of a population (or genotype) in niche space may be represented on an additional axis (56a, 121, 194), or it may be envisioned as the probability density of the niche hypervolume.

BIOTIC INTERACTIONS AND THE NICHE

The effect of one species on the niche of another within a community has been studied experimentally using a great variety of techniques. Almost without exception, these techniques involve either altering the density of one or both species or manipulating some aspect of their common environment. In the field, density alterations may be accomplished by installing enclosures or exclosures, by removal or addition of individuals, or by some combination or sequence of these techniques. Environmental manipulation takes many forms. The level of a physical or chemical factor may be raised or lowered. The experimenter may alter the relation between supply and demand for a resource (food or nutrients; refugia; germination, attachment or nesting sites) either by enriching or supplementing the resource, or else by removing it or making it inaccessible.

Whether a field experiment involves density alteration or environmental manipulation, subsequent changes in the response of species to one another or to aspects

---

[1] In general, ecologists seem to have several terms for any given concept, and not a few terms with no useful referent at all. However, there seems to be no word for the representation of a community in Hutchinsonian hyperspace. We hope that *columbarium* has a useful referent, and that, in any case, we will be forgiven this modest contribution to the proliferation of ecological jargon. (A columbarium is a vault filled with niches for cinerary urns.)

of their environment provide the basis for inferring the nature of their interaction under undisturbed conditions. Arranging for replication, appropriate controls, and sham operations is frequently no easy matter under typically heterogeneous field conditions, but that same heterogeneity makes replication and control all the more essential (39, 67).

Laboratory experiments designed to elucidate biotic interactions in nature can be categorized in roughly the same way as field experiments. Increased ease of control and replication in the laboratory is offset by a loss of realism, but careful comparison of the conditions of laboratory and field environments may permit reliable inferences.

In the following sections we discuss some representative studies of the effects of various kinds of interactions between species upon the breadth, overlap, and fitness pattern of their niches. Results are presented in qualitative terms as either the "expansion" or the "contraction" of the niche under experimental conditions, relative to the pattern observed under control or undisturbed conditions.

A number of authors (e.g. 99, 131, 152) have applied Hutchinson's (92, 93) distinction between the fundamental and realized niche to experimental studies of competition, referring to the niche under reduced competition as *fundamental,* and under control or undisturbed conditions as *realized.* While the latter case is a reasonably correct application of Hutchinson's original definition of the realized niche, the notion of the fundamental niche, strictly interpreted, under either Hutchinson's (92) or Vandermeer's (very different) (191) definition is difficult or impossible to define operationally except in the most artificial of circumstances. Because we have restricted our application of the niche concept to populations within a local community (rather than including aspects of the global ecology of each species), and since we have included niche changes under all kinds of biotic interactions (rather than competition only), it is tempting to take the conservative approach of simply describing the effects of experimental conditions on the niche of target species, avoiding entirely the use of the terms *fundamental* and *realized.* However, the usefulness of the concepts in theoretical contexts (e.g. 25, 39, 160) and the need for a concise way of relating diverse empirical patterns to one another and to theory have persuaded us to use the fundamental/realized dichotomy in a broad sense. The meaning should be clear in context.

Niche Expansion under Experimental Conditions

A great variety of experiments lead to the general conclusion that populations in nature often do not occupy as wide a variety of habitats or microhabitats as they are physiologically capable of occupying. In fact, in a surprising number of cases, a species[2] is rare or even absent in microhabitats for which it apparently is optimally adapted. Experimental evidence for the restriction of a species in nature to parts of the environment corresponding to a subset of the region of niche space it could

[2]In order to avoid the awkwardness of such phrases as "a population of a given species" and "two populations of different species," we frequently use the term *species* in this review to mean a local population of a species.

potentially occupy characteristically involves a demonstration of *niche expansion* under experimental conditions. Biotic interactions that have been shown to restrict the niche of species in nature include interspecific competition, predation, and parasitism. The degree of niche expansion under experimental conditions indicates the extent to which the fundamental niche of a population exceeds its realized niche, with respect to the factors altered by the experimenter.

COMPETITION   The restriction of the realized niche of one species by competition with another has been demonstrated experimentally for a wide range of organisms, and a great variety of mechanisms are known to be employed by organisms to gain and maintain access to resources potentially usable by their competitors. (A selection of case studies is detailed in succeeding sections.)

The distinction between "exploitation" and "interference"[3] mechanisms (25, 43, 117, 131) in competition is an important one. Exploitation competition between two species is mediated indirectly through a resource or resources, the supply of which, in extreme cases, is sufficiently lowered by the "winner" so that the "loser" is excluded from some region of niche space common to the fundamental niches of the two species. Thus exploitation competition is based entirely on differential efficiency of resource utilization. Interference between two species capable of exploiting the same resources requires a more direct interaction that in some way prevents the loser from realizing its maximum exploitation efficiency, while the winner is either unaffected or less affected by the interaction.

Once resource competition is experimentally demonstrated or strongly inferred, it is sometimes quite clear that interference (as defined above) is involved in the maintenance of distinct realized niches. Mortal combat between competing species (32, 202) is an indisputable case of interference. The problem arises in cases where competition can be demonstrated, but in which no interference mechanism is found. It is very difficult to prove that interference is not involved in the interaction, and thus that the outcome of competition is based on exploitation efficiency alone (99).

A hypothetical example may illustrate the predicament. Suppose that plant species *A*, a rapidly growing annual, is normally replaced in old-field succession by the taller plant species *B*, a slow-growing perennial. Greenhouse experiments with single-species plantings show that *A* cannot grow in reduced light, while *B* grows equally well in shade or sun. Allelochemic interactions are excluded, and the experimenter concludes that *B* replaces *A* by "shading it out." Is this pure exploitation competition, or is *A* eventually prevented from realizing its maximum exploitation efficiency by interference from *B*? The question may appear to be a word game, but it has an evolutionary answer. Of two genotypes of species *B* having equivalent fitness in the absence of species *A*, but differing in leaf arrangement such that one genotype produces a greater potential shading effect on species *A*, the shading genotype would have a higher relative fitness in the presence of species *A*. In fact,

---

[3]In this review we use the term *interference* in the general sense of Miller (131) and others, as formalized by Case & Gilpin (25). The relationship between this use of the term and that of plant ecologists [e.g. Harper (73)] has been clarified by Tinnin (185).

the shading genotype of species *B* might have a selective advantage in the presence of species *A* even if its fitness in the absence of *A* were less than that of the nonshading genotype. The frequency of the two genotypes in a natural community will of course depend upon a number of factors (patterns of vegetation and disturbance, dispersal mechanisms, etc), but it seems clear that fitness can in principle be increased by selection for interference by shading, with no increase whatsoever in photosynthetic efficiency per se. [Experimental studies by Gadgil & Solbrig (62a) illustrate most aspects of this hypothetical example, while Case & Gilpin (25) provide an explicit model separating exploitation and interference mechanisms in competition.] The foregoing argument is so easily adapted to a wide range of known examples of competition that one might well expect interference components of competitive interactions to be involved in many more cases than presently known.

From experimental studies of competitive interactions, some rather general patterns emerge. We illustrate each with an example, followed by discussion and briefer reference to other studies showing the same general pattern.

Niche inclusion: example   The gut of a vertebrate in nature is typically inhabited by a characteristic community of metazoan parasites, each species attaching itself or confining its movements to a particular region of the intestine. Schad (162) showed, for example, that up to ten congeneric species of helminths may coexist in the colon of the European tortoise. Those pairs of species not differing in location along the colon differ in radial distribution (lumen vs mucosa) or in the type of particles ingested. (See 68, 86, 179 for related examples.)

Holmes (84–86) performed an elegant and rigorous series of experiments demonstrating the effect of intra- and interspecific competition on the niches of two parasitic helminths, the tapeworm *Hymenolepis dimunuta* and the acanthocephalan (spiny-headed worm) *Moniliformis dubius,* both found in the intestine of wild rats. Using initially parasite-free laboratory rats as experimental hosts, Holmes produced single-species and mixed infections with known numbers of individual parasites. The results are shown in Figure 1. When both species are present (realized niches), there is practically no overlap in the distribution of individuals along the intestine. In single-species infections (fundamental niches), the tapeworm has a broader niche than the acanthocephalan, and there is a high degree of overlap. At low densities in single-species infections, the region of the gut selected by the tapeworm for attachment (the anterior half) coincides rather closely with the region used by the acanthocephalans in single-species infections at both high and low densities. At high densities in single-species infections, the tapeworm niche is broader, and includes the region occupied in mixed infections.

For each species, Holmes demonstrated significant reductions in both size and weight of parasites as a function of parasite population density in single-species infections. In mixed infections, size and weight in both species were reduced to an even greater degree, relative to total parasite density (84). In another series of experiments (85), newly introduced acanthocephalans successfully displaced established tapeworms from the anterior region of the intestine, while tapeworms invading established populations of acanthocephalans caused no significant shift in attachment sites of the latter species.

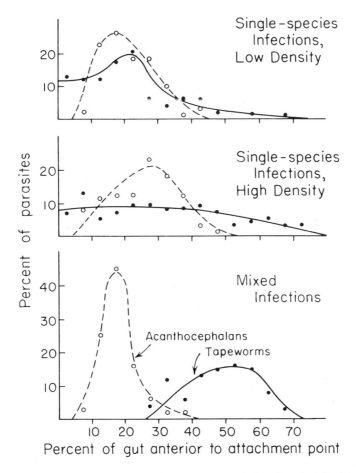

Figure 1 Effects of single-species and mixed infections on the intraintestinal distribution of the tapeworm *Hymenolepis diminuta* and the acanthocephalan *Moniliformis dubius.* Tapeworms are represented by solid dots and lines, and acanthocephalans by open dots and broken lines. The data points are from Holmes (84). The curves were drawn by eye.

The intestine represents a complex overlay of chemical, physical, and structural gradients, an extreme case of correlation between microhabitat and niche parameters. Thus we have freely taken differences in microhabitat (attachment site) as indicative of niche differences. Holmes (84, 86) was unable to discover the exact mechanism by which the acanthocephalan excludes the tapeworm from the anterior intestine, but suggests as most likely some modification of the microenvironment. Competition between these species (153) and among many other intestinal helminths (86) appears to be primarily for carbohydrates, which decrease in concentration with distance from the stomach. Thus the anterior region of the gut is richer in terms of the critical resource than the posterior region.

Niche inclusion: discussion and other examples   As a model of a two-species interaction, the experiments of Holmes, outlined in Figure 1, exemplify the expansion of the niche of one species (the tapeworm) under reduced competition with a second species (the acanthocephalan), while the latter shows relatively little expansion under reduced competition. Such asymmetry is by no means uncommon and is typical of cases in which the niche of one species is a subregion of the niche of another, when both are released from competition. This is our criterion for *niche inclusion* (92, 99, 131).

In two-species competitive interactions fitting this criterion, the fundamental niche of one species (in the context of the local community) is necessarily broader than that of the other. Relatively speaking, one species is a generalist and the other a specialist. The two can coexist indefinitely if either species can successfully enter a community in which the other species is already at equilibrium (117). [Holmes' (85) sequential introductions, discussed above, come close to showing this for the system outlined in Figure 1.] With the specialist already present, the generalist can become established simply by using resources corresponding to the region of its fundamental niche not shared with the fundamental niche of the specialist. However, when the generalist is already present, the invading specialist must somehow gain access to resources already being used.

The simplest means by which the specialist may gain and maintain exclusive access to resources is by efficient exploitation. If the specialist can reduce a critical resource below the maintenance threshold of the generalist and still maintain itself, the generalist will be excluded (117). Alternatively, the specialist may be equally or even less efficient than the generalist in terms of exploitation, but still be able to exclude the latter by some form of interference.

In the case of the two helminths studied by Holmes (Figure 1), a resource gradient clearly exists, from the carbohydrate-rich anterior region of the intestine to the carbohydrate-poor posterior region (86). The fact that the acanthocephalan does not use the posterior region, even at high density in the absence of the tapeworm, argues against its having a lower maintenance threshold, and thus against using such an advantage to exclude the tapeworm from the anterior region through pure exploitation competition. Indeed, Holmes (86) suggests that interference is involved, although the exact mechanism is as yet unknown. Case & Gilpin (25) predict on theoretical grounds that interference competitors should occupy "richer" regions of "resource density gradients" [see also (102a)] and should undergo little reduction in niche breadth under competition.

Jaeger's study of the terrestrial salamanders *Plethodon cinereus* and *P. richmondi shenandoah* (96–99) presents a different case. Using a combination of laboratory studies and field experiments, Jaeger showed that *shenandoah* is a physiological generalist (with regard to desiccation), whereas *cinereus* is a specialist requiring continuous expanses of moist soil. For all parameters studied, the fundamental niche of *cinereus* is included in that of *shenandoah,* and both show maximal survival in moist soil. In nature, *cinereus* occurs widely in moist soils, while *shenandoah* is restricted to small patches of drier soil in talus slopes. Jaeger was unable to find any evidence that interference by *cinereus* prevents *shenandoah* from entering areas of moist soil at the edge of talus slopes, and he tentatively concluded that pure exploitation competition for food was responsible.

Another example of an included niche on a well-defined physical gradient, studied by Castenholz and his associates, is illustrated by the interaction between two species of blue-green algae, *Oscillatoria terebriformis* and *Synechococcus lividus,* on temperature gradients in certain hot springs outflow channels (26, 27). The range of temperatures supporting growth in *Synechococcus* strains in isolation (73°C to about 30°C) includes the growth range of *Oscillatoria* (53°C to about 30°C), which grows at maximum rate up to 53°C, but is killed by sustained exposure to 54°C. In hot spring outflows, an *Oscillatoria* mat prevents *Synechococcus* from exploiting regions with water temperatures below 54°, resulting in an abrupt discontinuity in the flora of outflow channels at that temperature. *Oscillatoria* mats simply grow (or move) over *Synechococcus,* shading it out at lower temperatures. The relationship is illustrated in Figure 2. (The lower limit of *Oscillatoria* in outflow channels and other aspects of Figure 2 are discussed in subsequent sections.) Experimental evidence and the dynamics of seasonal shifts in the location of the *Synechococcus-Oscillatoria* interface implicate the shading effect of *Oscillatoria,* the physiological specialist, as the primary mechanism of competition—an example of interference.

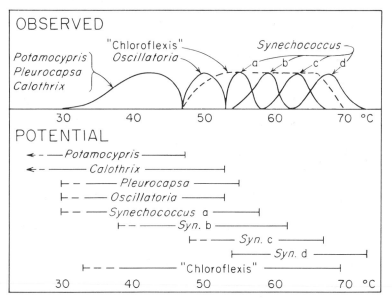

Figure 2   Observed and potential distributions of species inhabiting a thermal outflow stream at Hunter's Hot Springs, Oregon. The "observed" frequency distributions are intended to show approximate temperature limits only, and do not represent actual density measurements. [The "observed" ranges for individual *Synechococcus* strains are inferred from the description given by Castenholz (27).] The "potential" distributors show the approximate thermal limits for growth in laboratory culture. [Adapted from Figure 19.2 of Castenholz (29).] See text for discussion.

Although such clear-cut continua as hot spring outflow channels and intestinal microcosms present an appealing simplicity for experimental niche studies (other examples of niche inclusion on such gradients appear in 40, 143, 144, 167, 168, 187), resource states (34) in nature are frequently not neatly ordered in physical space, but appear as a mosaic. In some cases the states of a resource are nonetheless orderable on a specified axis of niche space. A good example is corolla length in animal-pollinated flowers.

Many observers have noted that bee species differing in tongue length visit flowers differing in mean corolla depth (e.g. 15, 83). A. D. Brian showed that the (significantly different) tongue length of four sympatric species of bumblebees (*Bombus*) was highly correlated with the depth of the flower tube in 15 native plant species visited differentially by the bees (15). In nature, longer-tongued bee species visit deeper-tubed flowers, and shorter-tongued bees visit shallower-tubed flowers, with only partial niche overlap between morphologically "adjacent" bee species.

However, when Brian provided paper flowers (partially filled with sugar-water) with a corresponding range of tube depths, visiting bumblebees of all species preferred shallow flowers. Thus we may infer that long-tongued bees are at least capable of feeding on shallow-tubed flowers in nature, while short-tongued species are prevented by simple morphological constraints from extracting nectar in the usual way from deep-tubed flowers. The fundamental niches of the four bee species therefore apparently form a series of progressively smaller subregions of niche space with respect to flower tube depth; the niche of each bee species is included in that of the next longer-tongued species (except for the longest-tongued species, whose fundamental niche includes all the others).

Why do the realized niches show only partial overlap? Brian found evidence of interspecific aggression among the bumblebees visiting her paper flowers, with the dominance hierarchy in precisely the reverse order formed by tongue-length. [See (32) for a parallel case involving bill length in hummingbirds.] That is, in any given interspecific encounter, the shorter-tongued bee typically displaced the longer-tongued bee from the experimental flowers. In nature, aggressive encounters between coexisting *Bombus* spp. are rarely seen, though well known among other assemblages of native bees (e.g. 102a, 113). But Brian's experiments also demonstrated a rapid conditioning of individual bumblebees to the color of paper flowers, and a survey of the local flora revealed a correlation of tube depth with color. The importance of using artificial flowers lies in the tendency of bumblebees to investigate any novel object in their environment, allowing perturbation of conditioned foraging patterns. Thus it is possible, although still far from proven, that in nature even infrequent aggressive encounters might account for the observed partitioning of resources among *Bombus* species. In any case, Brian's study (15) is an excellent example of the utility of the experimental approach in revealing unsuspected interference mechanisms in competition. Other workers have also found species interactions at baits or lures a useful means of inferring social dominance relations among uncaged animals (e.g. 21; 110; 202, pp. 447–48).

There exist several other clear cases of niche inclusion in mosaic environments demonstrated by experimental studies. A classic example for higher plants is detailed by Kruckeberg (105a), who showed that species and races restricted in nature to serpentine outcrops actually grow better (or at least as well) on nearby nonserpentine soils, but only in the absence of nonserpentine competitors. The latter are characteristically unable to grow on serpentine soils, even without competition. The fundamental niche of serpentine endemics thus includes and exceeds in breadth the fundamental niche of nonserpentine plants in the same community, with regard to soil chemistry. Studies by M. V. Brian (16, 17) and Pontin (146, 147) demonstrate niche inclusion in microhabitat segregation among ant species. In other studies yielding strong evidence for niche inclusion, Hunt (91) manipulated temporal (diel) activity patterns in competing ant species by altering the thermal environment of nests in the field. Using laboratory stream tanks to study microhabitat selection, behavioral interactions, and tolerance to physical factors, Bovbjerg (14) was able to account for patterns of realized niches in nature for two species of crayfish. Studies of intertidal organisms provide further examples of niche inclusion. Stimson (182, 183) investigated mosaic patterns in two species of grazing limpets and found that the larger, territorial species aggressively excluded the smaller, opportunistic species that would probably win in pure exploitation competition. B. A. Menge (127) and Menge & Menge (128) showed a similar relationship between two species of starfish, with niche expansion of the smaller species under reduced competition tied to morphological changes, rather than to spatial distribution alone. Connell (37, 38) has reviewed the literature of experimental intertidal ecology and discussed a number of other studies approximating the included niche model, including his own widely known study of barnacles in Scotland (35).

Experimental studies of interactions between vertebrate species that demonstrate niche expansion under reduced competition are few in number, and fewer yet approximate the criterion for niche inclusion. Jaeger's work with salamanders was discussed above. Among mammals, only rodents provide some possible examples. Removal experiments in the field include Sheppe's study with two species of *Peromyscus* (170) and Koplin and Hoffmann's work with two species of *Microtus* (104). Laboratory studies of behavioral interactions (in conjunction with field observations and distributions) by Cameron (23) with woodrats and Heller (80) with chipmunks reveal dominance relations supporting the existence of niche inclusion. Only one experimental study of birds known to us suggests niche inclusion: Davis (45) demonstrated a reversible niche expansion in juncos by removing and later releasing golden-crowned sparrows. The reciprocal experiment was not done, but it is unlikely that the sparrows would have moved into the microhabitats occupied by juncos upon removal of the latter. If this assumption is valid, then for the time, place, and variables studied, the fundamental niche of the sparrows is included in that of the juncos. Controlled experimental studies of vertebrate niche relationships are very difficult to arrange, and the studies mentioned here are almost all subject to a variety of valid methodological criticisms (see 39, 67), as are many of the studies with plants and invertebrates discussed in this section.

Nonetheless, a striking concordance emerges from an overview of the proven or strongly inferred cases of niche inclusion reviewed here. In every case in which interference mechanisms were shown to be involved in the competitive interaction between the generalist and the (included) specialist, it was the specialist that successfully interfered with the generalist. Moreover, in all but a few cases, interference was clearly demonstrated. Morse (134) recently came to roughly the same conclusion after reviewing and categorizing a great many studies (mostly nonexperimental) involving presumed or proven competition, for which there was evidence of interspecific social dominance.

Reciprocal niche overlap: example   The mussels *Mytilus edulis* and *M. californianus* are common inhabitants of the intertidal zone along the Pacific coast of North America. *M. edulis* predominates in quiet bays or other sheltered areas, while *M. californianus* is most common in areas exposed to heavy wave action. In an imaginative series of experiments, Harger revealed the mechanisms of competition and the adaptations accounting for this distribution [references and a summary for seven papers in *The Veliger* appear in (71)].

Combining results from quantitative observations, behavioral tests, and artificial substrates "seeded" with one or both species and exposed to various degrees of wave action, Harger concluded that young *M. californianus* are smothered by silt accumulation in habitats well protected from wave action, while in exposed situations, *M. edulis* is torn free from the substrate by water movement. (*M. edulis* escapes smothering by "crawling free" from accumulated sediment, while *M. californianus* survives the battering of waves by virtue of its stronger shell and tougher byssal threads.) In intermediate habitats the pattern of occurrence of the two species is determined by several factors, including the temporal pattern of wave action and silting, differential predation, and the crushing of *M. edulis* by *M. californianus*. In these habitats, the growth rate of *M. edulis* is decreased and mortality is increased in mixed patches.

Reciprocal niche overlap: discussion and other examples   Niche inclusion is of course a special form of niche overlap; the fundamental niche of the "specialist" is completely overlapped by that of the "generalist" for some critical parameter. The previous example of the mussels represents a formally different case, in which the fundamental niches of two competing species overlap, but each fundamental niche contains an *exclusive* region of niche space corresponding to environmental conditions under which the other species cannot persist even without interspecific competition—our criterion for *reciprocal niche overlap*.

Biologically, such a case differs fundamentally from niche inclusion in that coexistence is assured (in sufficiently heterogeneous environments, and barring other factors), regardless of which species is competitively superior under conditions corresponding to the zone of overlap in niche space. Expansion of the realized niche (or at least increased fitness) of one species upon reduction or elimination of the other species in the zone of overlap may be reciprocal, as suggested by the data of Harger (71), or may occur in only one of the two species.

For an example of "one-sided" expansion, we return to Figure 2. Castenholz and his associates [references in (27), see also (20)] isolated a series of four genetic strains (clones) of the blue-green alga *Synechococcus lividus* from a single hot spring outflow channel, each with a different thermal maximum and range for growth. If we take the temperature range in pure culture (see the lower part of Figure 2) as a measure of the fundamental niche for each strain, it is clear that each pair of strains fits the criterion for reciprocal overlap—for each member of the pair, there exists a range of temperatures at which the other member cannot reproduce. Considering all four strains together, however, strains *b* and *c* have no exclusive region, nor does *a*, when the tolerance range of *Oscillatoria* is taken into account (see section on niche inclusion).

In nature, no more than two strains of *Synechococcus* are found in abundance at any one temperature along the channel (27), as indicated in the upper part of Figure 2, each strain being most abundant at the upper end of its tolerance range. Thus, in the absence of strain *c*, for example, little or no expansion should occur in the realized niche of strain *b*, while the niche of strain *d* would be expected to expand considerably. Although, to our knowledge, this experiment has not been done it is difficult to imagine any other outcome. In passing, it is worth noting the potential of thermophilic algae for investigation of the theories of limiting similarity (118) and diffuse competition (117) under constant and variable thermal regimes.

Communities in which a single critical factor (or a set of highly correlated factors) determines the niche relationships of certain competing species are appealing for their clarity and deserve the special attention they receive from ecologists. However, it is clear that competition and coexistence in nature are commonly much more complex and, among other complications, involve strong factor interactions (106a). A relatively simple example of factor interaction is provided by the work of Shontz & Oosting (171) on two early colonists of old fields in North Carolina: yellow aster (*Haplopappus divaricatus*) and horseweed (*Conyza canadensis*). Greenhouse and controlled transplant experiments revealed that yellow aster cannot germinate in clay soils, while horseweed does very well, with or without disturbance (plowing). In sandy, nutrient-poor soils both reproduce, but horseweed is dominant only with the addition of the disturbance factor. Without disturbance, aster is dominant in sandy soils. Thus the case adequately meets the criterion for reciprocal niche overlap, but two roughly orthogonal axes (soil type and degree of disturbance) are required to represent the fundamental niches, since the factors are not correlated. (In the mussel example, wave action and degree of silting are not independent factors, since silting only occurs with low degrees of wave action.) We have simplified the results of Shontz & Oosting's study to illustrate this point, but a more complete analysis would only add more complicated factor interactions.

Coextensive niches   At the scale of the local community, the majority of reasonably well documented examples of competition, apart from those clearly demonstrating niche inclusion, lack evidence of exclusive regions of the fundamental niches of *either* species in a competing pair. In many cases, fundamental niches that appear to overlap completely may actually involve exclusive regions for one species (niche inclusion) or both species (reciprocal overlap) on the basis of unknown factors or factor interactions, demonstrable by further investigation at the focus of the local community. On the other hand, there are a few strong cases for the existence of *coextensive* or completely overlapping fundamental niches. Numerous other examples may eventually prove to fit this model; in spite of considerable uncertainty in most of these examples, we have listed them at the end of the present section.

All that is required for the coexistence of two competing species with coextensive fundamental niches is that each be a superior competitor in a different subregion of the fundamental niche (121), whether by virtue of greater exploitation efficiency or of more effective interference. An example elegant in its simplicity concerns two species of *Eucalyptus*, one (*E. rossii*) common in a dry sclerophyll forest and the other (*E. melliodora*) common in an adjacent savannah woodland in Australia. Moore (132) grew seedlings of each species in pure and mixed stands, using soil from each habitat. In pure stands, there was no significant difference between species in plant height or yield on either soil. In mixed stands, each species grew better on its own soil than did its competitor. The mechanism of competition is apparently not known, but the absence of differential response in pure stands suggests (but does not prove) some form of interference.

A second case of apparently complete overlap of fundamental niches in which coexistence demonstrably depends upon strong interference concerns a pair of species in the mite genus *Rhinoseius*, studied by Colwell (32) in a tropical highland forest. The mites feed on nectar and also breed within the corolla tube of certain hummingbird-pollinated flowers, and are carried between flowers on hummingbirds. The two mite species are found on distinct host-plant species in nature, but sometimes occur together on the same hummingbird, and two of the three bird species involved regularly visit host plants of both mite species. Reciprocal introductions of adult mites of one species into unoccupied flowers of a host plant of the other mite species show that each is capable of successful reproduction in alien flowers. However, laboratory experiments reveal the capacity of males of each species to maim and kill their congeners, and a limited series of reciprocal introductions into occupied flowers in nature suggests successful defense against invasion of the alien species. Coexistence is permitted by a combination of defensive behavior, active host-flower selection, and the relatively predictable foraging patterns of the birds.

The importance of population fitness in the presence of competing species is evident from the experiments of Ellenberg (55) on the effects of the water table on the growth of pasture grasses. He showed that several species grown in pure stands have the same optimum but when grown together in mixed stands, their response to the height of the water table changed greatly, each species having a distinct optimum.

Likewise, Fellows & Heed (59) found roughly coextensive fundamental niches with respect to the host plant for three species of "cactiphilic" *Drosophila* (*D. mojavensis, D. arizonensis,* and *D. nigrospiraculata*). There is little or no realized niche overlap; in nature, each fly species feeds on distinct host species. However, in population cages, each of the three *Drosophila* species (in single-species cultures) can reproduce using tissue from the entire range of host species occupied by the three in nature. Coexistence is made possible by some degree of active host selection, differential fitness in the presence of congeners (probably involving interference mechanisms in the case of *D. nigrospiraculata*), and perhaps interaction with ambient temperature and humidity.

A number of additional experimental studies of interspecific competition, or at least of putative competition, cover a wide variety of organisms with apparently coextensive or perhaps reciprocally overlapping fundamental niches. The organisms studied include protozoans (65), cellular slime molds (88), phytoplankton (155), higher plants (50, 62, 74, 103, 105, 125, 126, 130, 152, 155, 163, 176), trematodes (111), crustaceans (2, 43, 188), insect larvae (53, 57, 120, 122, 191a) and adults (3, 58), fish (95, 115, 148), frogs (9, 94, 186), rodents (67, 80, 133, 157, 158, 169, 174, 175), and birds (129, 180). This is by no means an exhaustive list, and several studies, rather incomplete for the present purposes, are included here primarily to suggest further work (9, 57, 80, 105, 111, 115, 120, 122, 129, 148, 163, 175, 180, 186). Numerous experimental studies provide fairly clear evidence of interference mechanisms (9, 43, 50, 62, 67, 80, 88, 103, 111, 115, 126, 133, 134, 155, 163, 169, 180, 186, 188, 191a), while others provide reasonable documentation of pure exploitation competition, or at least of differential fitness in distinct regions of the fundamental niches of competing species (53, 57, 88, 94, 120, 122, 125, 152, 157, 191a).

In theory, the final step in the progression from coextensive, to included, to reciprocally overlapping fundamental niches is a state of nonoverlapping fundamental niches. Although the last term implies no competition, a fondly held tenet of community ecology provides that nonoverlap of the fundamental niches of closely related sympatric species is often the evolutionary result of interspecific competition. An especially convincing example is provided by *Drosophila pachea*, a fourth species included in the study of Fellows & Heed (59) discussed earlier. *D. pachea* is apparently completely monophagic on the senita cactus (*Lophocereus schottii*), which contains a sterol required by the fly, as well as alkaloids toxic to all other sympatric *Drosophila* (77, 78). The first step toward forming this relationship may have been the development in *D. pachea* of some means of detoxifying or tolerating alkaloids already present in senita. Acting at this stage, competition from other *Drosophila* species may have restricted *D. pachea* to senita, with sterol dependence developing as a final step. Numerous examples of feeding specificity among insects provide good material for such speculations, and perhaps for experimentation (e.g. 52, 54, 64, 75, 135).

PREDATION AND PARASITISM   The historical fragmentation of biology along taxonomic and functional lines surely retarded the development of community ecology and continues to be honored by ecologists to a surprising degree. The influence of herbivores, carnivores, parasites, and pathogens on the ecological distribution of prey or hosts, and the importance of higher trophic levels in modifying biotic interactions between prey or host species have received until recently considerably less attention than either competition or the study of predation and parasitism at a more restricted focus. In this section we briefly summarize some representative studies of the influence of predators and parasites on the niches of prey and host species. We use the term *predation,* roughly following MacArthur (117), to mean the ingestion of live organisms, or parts of organisms, by free-living animals (or by "carnivorous" plants, although we have nothing to say about them).

Niche expansion in prey species under experimental reduction of predator density often follows a pattern quite similar to niche expansion of the "generalist" under reduced competition from the "specialist" in cases of niche inclusion. This pattern occurs in situations in which the fundamental niche of the prey species contains a region from which the predator is excluded by some extrinsic factor. Thus at high levels of predation, the realized niche of the prey consists only of this refuge, while at low levels of predation considerable niche expansion may occur.

For a simple example of this pattern we return to Figure 2. Although the alga *Oscillatoria terebriformis* is capable of substantial growth at temperatures as low as 35°C, in hot springs outflows in which an ostracod of the genus *Potamocypris* occurs, the alga is limited to temperatures above about 46–48°C. Wickstrom & Castenholz (197) have shown that algal glazing by the ostracod is responsible for this limitation, and that the ostracod cannot survive sustained temperatures above 48°C. Thus the alga has a refuge from predation between this temperature and its own thermal maximum of about 53–54°C. In hot springs where the ostracod is absent, the *Oscillatoria* mat extends to much lower temperature regions, but is grazed back to higher temperature after introduction of the ostracod (73). A similar pattern has been reported for algal-feeding ephydrid flies in other hot springs (198).

A surprising number of experimental studies of predator-prey (including herbivore-plant) coexistence demonstrate the limitation of the predator to a narrower fundamental niche than its prey, with respect to some critical factor. Rather than recounting any further studies here, we refer the reader to two recent reviews by Connell (37, 38), who treats this phenomenon in detail.

Although we know of no complete experimental studies extending the principle to three trophic levels, we suggest the following example. In habitats with a mosaic of dense shrubs and herbs, narrow regions with little or no herbaceous cover are sometimes found adjacent to the shrubs. That grazing by small mammals is partly responsible for these bare zones has been demonstrated by the growth of herbs in exclosures placed near the shrubs (8, 69, 70). The presumption is that avian (or other) predation prevents grazing at any great distance from the shrubs, which serve as a refuge for the grazers.

The "critical factor" in this case is the gradient from dense shrub cover to open ground some distance from shrubs. Along this gradient, the herbs have a refuge from grazing in the open ground, and their realized niche expands under decreased grazing. Likewise, the grazers have a refuge from avian predation at the shrub end of the gradient, and presumably would also show niche expansion (into more open areas) in the absence of raptors.

This example points out the importance of the biotic context of predator-prey interactions. The presence of avian predators of grazers is presumably an essential prerequisite for the refuge of the herbs, and, if so, must be included as a critical dimension of niche space in abstracting the grazer/herb interaction. Likewise, the presence of shrubs provides the refuge for grazers, whose activities may in turn benefit the shrubs by reducing competition with herbs for nutrients and water, and by actually importing nutrients to the shrubs in the form of excretory products. The relative importance of the many possible interactions in this semihypothetical example are not known, but it might well be possible to explore them with imaginative experimentation.

The niche representation of predator-prey interactions may not always permit the separation of a prey refuge from regions of greater vulnerability along niche axes presenting abiotic or biotic gradients (or ordered phases of a mosaic environment). If prey escape predation by chance alone, their refuge is a probabilistic one, which may be "available" only in environments of adequate complexity or unpredictability. In such cases, the critical dimension of niche space may be habitat complexity itself. The well-known experiments of Huffaker (89) provide an excellent example. To our knowledge, no field experiments in natural communities have demonstrated the role of habitat complexity, per se, in predator-prey interactions.

The role of parasites, parasitoids, and microbial pathogens in altering the potential ecological distribution of their hosts in natural communities has apparently received rather little experimental study. However, the literature of biological control provides a wealth of examples for introduced species (48, 49, 90, 189), and suggests a significant role for parasitoids and pathogens (as well as predators) in structuring natural communities (90a). Although we have not attempted to survey the literature of parasitology, two patterns of relevance to the topic of this review bear mention.

Some parasites may play a role similar to that of predators, restricting hosts to either an ecological or a probabilistic refuge. Barrow (7), for example, showed that at temperatures above 20°C the salamander *Triturus viridescens* is relatively unaffected by infections of *Trypanosoma diemyctyli,* while at lower temperatures the infection becomes increasingly pathogenic. The literature of parasitology abounds with studies demonstrating different levels of pathogenicity under differing physical and host-nutritional regimes that could potentially affect the relationship of hosts to environmental gradients and mosaics (e.g. 28), but little information is available relating these studies to the realized niche of host species in nature.

A number of cases are known in which parasites alter the realized niche of intermediate hosts by inducing behavioral modifications in infected individuals [see the review by Holmes & Bethel (87)]. For example, formicine ants infected with the metacercariae of liver flukes move to the tops of forage plants of their definitive ungulate hosts during the hours of the day when grazing activity is at its peak, facilitating transmission from ant to ungulate. Uninfected ants show no such behavior. The behavioral alteration in infected ants is thought to be associated with proximity of metacercariae to the suboesophageal ganglion of the ant (24). In a similar case, the amphipod *Gammarus lacustris* normally remains in relatively deep water, but individuals infected with cystacanths of the acanthocephalan *Polymorphus paradoxus* become positively phototactic, rising to the surface of the lake where they fall easy prey to ducks, a host of the adult worm (10, 87). In both examples, the parasites apparently cause their intermediate host to leave a refuge from predation, forcing the host into microhabitats where the host is more vulnerable to particular predators—a very special kind of "niche expansion."

Niche Contraction under Experimental Conditions

In previous sections of this review we have focused on patterns of interaction between species that restrict the breadth (and in some cases the overlap) of realized niches, such that a species in the role of competitor, prey, or host undergoes niche expansion under experimentally reduced competition, predation, or parasitism. We turn now to a brief consideration of interactions between species that involve either one-sided or reciprocal dependence, as demonstrated by contraction of the realized niche (or at least decreased fitness) of a dependent species, upon experimental dissociation from another species. Here we include only relatively direct or intimate dependencies, deferring indirect "positive" effects of one species on another to the section of this review on complex interactions.

MUTUALISM   As we use the term, the relationship between two species is *mutualistic* if the fitness of each is increased by some relatively direct interaction with the other in some specified environment. Gilbert (63) has pointed out that even though a great deal is known about mutualistic relationships in nature, only in recent years have ecologists begun to recognize the importance of mutualism in the origin and maintenance of community structure. Moreover, this phenomenon has yet to be satisfactorily integrated into theoretical community models, although J. Roughgarden (205) has taken the first steps in this direction.

The pollination of a plant by an animal visitor is a mutualistic relationship if the plant supplies sufficient rewards to the animal in the form of energy, nutrients, or sometimes non-nutritional chemical substances (79). In spite of the enormous number of observational studies on pollination ecology, surprisingly little experimental manipulation of pollination systems has been carried out in natural communities. The most common demonstration of dependence simply involves the exclusion of visitors from plants by "bagging," or covering the plants with an exclosure (e. g. 119). To the degree that seed set is reduced as compared to that of control plants, the fitness of the plant is shown to be dependent on animal visitors, assuming no effect of the exclosure on other potential means of pollination.

The degree of dependence of pollinators upon substances obtained from flowers, and the effect of the floral environment on the realized niche of pollinators with respect to extrinsic factors (especially temporal and spatial parameters), has been much studied in the honeybee *Apis mellifera,* but deserves more experimental attention from ecologists interested in pollination systems in natural communities. Manipulation of the floral environment by removal of flowers or by altering experimentally the rewards flowers offer could provide much information on both the plasticity of pollinator response and the effects of reduced or increased pollinator service on the fitness and ecological distribution of the plants. To the best of our knowledge, however, relatively little experimentation along these lines has been done.

Kuglar (106) enriched (with sugar-water) the nectar of one of two color morphs of *Lathyrus odorata* previously visited indiscriminately by bumblebees. As a result, the bees began to concentrate their visits on flowers of the enriched morph.[4] Dilution of nectar in the other morph had the same effect. By changing the number of

[4]A manuscript by B. Heinrich describes similar results using two co-occurring species in Maine.

honeybee hives in an orchard with several varieties of apple trees, Filmer (60) showed that at low bee density only a few "preferred" varieties were visited, while at high densities the bees visited all varieties. Levin and co-workers (among others) have exploited mixed plantings of morphs or species differing in flower characteristics to investigate the effects of the floral environment on pollinator behavior and seed set [e. g. (107, 108), see also (13, 113)].

Janzen's well-known studies of the mutualistic association between certain *Acacia* species and their ant inhabitants (*Pseudomyrmex* spp.) involved experimental demonstration of dependency of the plants on the activities of the ants [see (101) and references in (102)]. The ants protect the acacias from both competitors and herbivores. In a formally similar case, Gill (65) showed that the outcome of competition between species of *Paramecium* differed from expectation based on exploitation efficiency in single-species cultures, and suggested that intracellular symbionts ("kappa particles"), lethal to noncarriers, increased the fitness of their host species in competition.

A great variety of organisms have been shown experimentally to use nutrients produced by intimately associated symbionts, while in return providing a suitable environment for growth and reproduction of the symbiont. For a simple example we return to Figure 2. In the temperature zone of hot springs outflow channels kept free of *Oscillatoria* by grazing ostracods (*Potamocypris*), much of the substrate may be covered by an association of a blue-green alga of the genus *Pleurocapsa* and another of the genus *Calothrix*. The thick-walled cells of *Pleurocapsa* form crusts and nodules resistant to ostracod grazing, which also protect *Calothrix* filaments imbedded in the mass of *Pleurocapsa* cells. In turn, there is evidence that *Calothrix*, which can fix free nitrogen, is required for the persistence of *Pleurocapsa*, which cannot [experiments of Wickstrom, as reported by Castenholz (27)].

Another mutualism is also depicted in Figure 2. The algae *Synechococcus* and *Oscillatoria* require, as a mechanical substrate, the underlying dense mat of a filamentous microorganism that Castenholz (27) tentatively called "Chloroflexis" (of uncertain taxonomic affinities). In turn, the Chloroflexis organisms cannot live without organic matter derived from the algae (19).

An important generality that may be derived from the foregoing examples is that many mutualistic relationships between species are adaptive for both partners only in the context of the community in which they evolved. An ant-acacia derives no benefit from ants in the absence of herbivores and competing plants. The hot springs *Calothrix* grows well in culture without *Pleurocapsa*, in the absence of grazing ostracods. The reciprocal benefits exchanged by a particular plant species and a particular species of pollinator are functions of the local availability of other plants and pollinators. Thus in representing a mutualistic interaction in niche space, not only the densities of the partners, but the densities of other species must also be included as critical factors.

The study of classic examples of mutualism in the context of natural communities has received little attention, compared to physiological and behavioral aspects. However, a limited amount of information is available on the ecological context of mutualisms involving microorganisms (81, 100, 138), lichens (1, 42, 81), symbiotic blue-green algae (195), nitrogen-fixing organisms (181), and a variety of higher plants, invertebrates, and vertebrates (4, 12, 29, 30, 66, 81, 114, 124, 204).

ONE-SIDED DEPENDENCY   For the sake of completeness, it should be pointed out that niche contraction under experimental alteration of population densities or abiotic factors should also occur in a variety of interactions in which the fitness of only one member of a pair of species is decreased by the experimental alterations. Thus removal of the host of a commensal or a parasite, or the removal of the prey of a predator results in decreased niche breadth or lowered fitness (or in the extreme, local extinction) of the commensal, parasite, or predator species, while the niche breadth or fitness of the host or prey species is either unchanged (commensalism) or increased (parasitism or predation, as discussed in a previous section).

COMPLEX INTERACTIONS

Up to this point we have focused primarily on the various kinds of interactions that occur between two species—"simple" interactions. Nonetheless it has been necessary at several points to emphasize the importance of the community context of pairwise interactions, and in this section we discuss a few of the many possible ways in which simple interactions may be linked or compounded.

If we focus on one species in a community, the *target* species, we may inquire to what degree the realized niche of this species (including its fitness topography) is affected by interactions with other species in the community. Assuming that appropriate methods can be found to examine, one at a time, the effect of each significant pairwise interaction involving the target species, we may then have a set of planar projections (with fitness represented as topographic contours) of the niche hypervolume.

To take a very simple example, Press and co-workers (149) conducted a laboratory investigation of interactions between the granivorous larvae of a pyralid moth, a species of braconid wasp that parasitizes the moth larvae, and a species of anthocorid bug known to be a predator of the moth larvae. Replicated cultures with different combinations of the three insects were begun, and examined at a later date. The authors do not state how the time interval was chosen. With moths alone, a mean of 183 moth larvae per culture was present at the end of the experiment. With moths plus bugs, there were 142 moth larvae, and with moths plus wasps, only 48 moth larvae.

As a qualitative prediction for the cultures with moths plus wasps plus bugs (with the same initial numbers of each as in the pairwise cultures), one would expect at the very most 48 moth larvae, probably fewer, at the end of the experiment. However, there was in fact a mean of 87 moth larvae in the three-species cultures (all differences significant). The reason: the bugs eat adult wasps, as well as moth larvae. [Rosenzweig (156a) postulated a structurally similar situation among two weasel species and their prey, and explored its theoretical properties.]

The point of this simple example is that the niche hypervolume is not necessarily, and perhaps hardly ever, a simple intersection of the pairwise species interactions with the target species, extended into the niche hyperspace. Interactions between species other than the target species may greatly alter the shape and fitness topography of its niche. This is a special instance of nonindependent niche axes or, more broadly, factor interaction.

Even if *all* pairwise interaction patterns are known for the set of species affecting the realized niche of a target species (e.g. 141); the niche hypervolume may not be fully specified, as higher-order interactions may also be operating. In other words, the pattern of interaction of a given pair of species may depend upon the density or activities of a third (or third and fourth . . .) species. Although higher-order interactions seem to be found even in simple communities (61, 137, 150, 199, 200, but see 190) when looked for, most current mathematical models in community ecology are based on pairwise interactions only (109, 117, 123). While this simplification may be an adequate approximation in many cases, there are some kinds of interactions that intrinsically seem to involve higher-order effects. For example, taking a classical Batesian mimic as the target species, its fitness is a function of its own density, the density of at least one model species, and the density (or level of activity) of at least one predator species. The abiotic components of its niche (time and place of its own activity) are similarly constrained.

Batesian mimicry exemplifies another complication involved in certain complex species interactions, namely, frequency-dependent effects. Batesian mimics (see 5, 13, 154, 196), species exploiting other kinds of protective resemblance [e. g. "stick caterpillars" (51), see also (11)], and prey species of "switching" general predators (136), supposedly all have disproportionately higher fitness when rare, while Müllerian mimics and plants dependent on general pollinators that form search images (107, 108, 145) are thought to have disproportionately greater fitness when common. We do not pursue this topic in any greater depth, since Ayala & Campbell recently provided an able review of frequency-dependent selection (5).

At this stage of the development of community ecology, relatively few general patterns emerge from the existing experimental studies of complex interactions, as compared with the frequently recurring themes discernible at the focus of simple interactions. Perhaps the most important compounding of simple interactions involves predation and competition. In some communities, selective or at least differential predation reduces the effects of competition among prey species, permitting coexistence of species with largely coextensive fundamental niches, which would otherwise compete to the point of exclusion. Connell (39) has recently treated this pattern at length, thoroughly reviewing the evidence from experimental studies [see also (36, 44, 46, 139) and references therein].

For a simple example we return once again to Figure 2. The fundamental niches of the blue-green algae *Oscillatoria, Pleurocapsa,* and *Calothrix* are essentially coextensive. Because the (mutualistic) *Pleurocapsa-Calothrix* complex is attached to the substrate, the motile *Oscillatoria* could potentially exclude the complex by overgrowing it and shading it out completely (27). However, grazing by the ostracod *Potamocypris* eliminates both *Oscillatoria* and the underlying mat of Chloroflexis up to the ostracod's own thermal maximum, allowing coexistence of all three algal species. This example almost exactly parallels several examples discussed by Connell

(39), and points up the importance of differential predation on the dominant competitor.

Holmes (86) discusses an interesting variation on the same theme, involving parasite communities in vertebrate hosts. To the degree that the immune response of the host is specific to particular species of potentially competing parasite species, the immune system may play a role similar to that of a selective predator, limiting the population size of the more common (dominant) species of parasites, allowing others with similar site preferences to share the same host individual. [See (140) for an example.] This phenomenon, if general, might prove to be of central importance in accounting for the enormous diversity of parasites commonly found in vertebrates in nature.

Barbahenn (6) and Cornell (41) discuss the role that parasites (and pathogens) may play in interspecific competition between closely related host species, each relatively immune to its own parasites but sensitive to the parasites of its competitor. [See also (90a).] Evidence is as yet in short supply, but the hypothesis is intriguing. The parallel with such defensive mutualisms as Janzen's (102) ants and acacias is worth pointing out. (A formally similar situation among three species of Nicaraguan cichlid fishes is described in a manuscript by K. McKaye. Adults of one species often guard the fry of a second species, which is the principal predator of a third species, which is the main competitor of the first.) In each of these cases (see also 65), an indirect interference component is added to a simple competitive interaction by means of a mutualistic relationship, on the part of one competitor, with a species whose activities are detrimental to the other competitor.

Protection from predation can be gained through mutualism in the same way, as in the ant/acacia and *Pleuroscapsa/ Calothrix* examples, as well as in plants with ant-attracting extrafloral nectaries, in aphid/ant mutualism, and in the local enrichment of the soil by the excretory products of vertebrates habitually nesting beneath shrubs that protect them from predators. A striking example of the last category can be seen in the seaside vegetation of the Straits of Magellan. Stands of the composite shrub *Chiliotrichium diffusum* on otherwise nitrogen-poor strand soils are dense only where thousands of Magellanic penguins (*Spheniscus magellanicus*) nest annually in shallow burrows beneath the shrubs (M.-C. King and R. K. Colwell, unpublished observation).

Ontogenetic changes in the relationship between species is another kind of complex interaction. For example, in many fish and amphibians, adults of one species may prey on the young of another but compete with adults of the other species (e.g. 82, 199), either simultaneously or sequentially. Predation in such cases may be viewed as a form of interference competition, or the competition between the species as adults may be considered an additional complexity in a more general pattern in which prey escape predation by simply becoming too large for predators to capture or subdue. Connell (39) reviewed the literature supporting the widespread occurrence of this latter pattern (see also 46).

Species interactions may also change with population density. Broadening of the realized niche of a population with increased density, as in the tapeworms of Figure 1, or in Filmer's apple-orchard honeybees (60, see also 191), can alter the trophic structure or spatial pattern of communities, with secondary effects on the niches of other species. A parasite-host relationship may become mutualistic if the host's food supply becomes nutritionally unbalanced (a condition more likely at high host density), since some parasites supply vitamins and other substances to their hosts (112). (The implications of this phenomenon for the evolution of parasitism, and its possible role in amplifying cyclic population fluctuations in vertebrates, bears further investigation.)

The importance of priority and other stochastic effects in the development and structure of natural communities is gaining wider recognition (184). A variety of techniques involving artificial or defaunated natural substrates or microhabitats (22, 46, 110, 142, 164, 166, 173) lead to the conclusion that community structure may be relatively predictable in similar abiotic environments, while community composition (the species list) is much less so [see Heatwole & Levins (76), and the review by Simberloff (172)]. This discovery is of great significance, since it implies the existence of general patterns in community structure. Thus the findings of ecologists who have intensively studied complex biotic interactions and the niche structure of particular species assemblages (18, 27, 32, 35, 46, 103, 126, 139, 145, 151, 192, 199) may indeed yield generalities—as well as a wealth of intriguing differences to be explained.

ACKNOWLEDGMENTS

We thank Mary-Claire King for her assistance at all stages of the preparation of this review. We are also grateful to J. Filp, T. Asami, W. Glanz, M. Greenstone, and the copy editors of *ARES* for their help and criticism. Emily Reid prepared the figures. Preparation of this review was supported by NSF Grant GB-31195 to Colwell. Fuentes was supported by Ford Foundation funds through the University of California—Universidad de Chile Convenio.

Literature Cited

1. Ahmadjian, V. 1967. *The Lichen Symbiosis.* Waltham, Mass: Blaisdell. 152 pp.
2. Allan, J. D. 1973. Competition and the relative abundances of two cladocerans. *Ecology* 54:484–98
3. Anderson, N. H. 1962. Growth and fecundity of *Anthocorus* species reared on various prey. *Entomol. Exp. Appl.* 5:40–52
4. Aschner, M., Ries, E. 1933. Das Verhalten der Kleiderlaus bei Ausschaltung ihrer Symbionten. Eine Symbiosestudie. *Z. Morphol. Ökol. Tiere* 26: 529–90
5. Ayala, F. J., Campbell, C. A. 1974. Frequency-dependent selection. *Ann. Rev. Ecol. Syst.* 5:115–38
6. Barbehenn, K. R. 1969. Host-parasite relationships and species diversity in mammals: an hypothesis. *Biotropica* 1:29–35
7. Barrow, J. H. Jr. 1958. The biology of *Trypanosoma diemyctyli,* Tobey. III. Factors influencing the cycle of *Trypanosoma diemyctyli* in the vertebrate host *Triturus v. viridescens. J. Protozool.* 5:161–70
8. Bartholomew, B. 1970. Bare zone between California shrub and grassland communities: the role of animals. *Science* 170:1210–12
9. Berger, L. 1968. The effect of inhibitory agents in the development of green-frog tadpoles. *Zool. Pol.* 18:381–90
10. Bethel, W. M., Holmes, J. C. 1973. Altered evasive behavior and responses to light in amphipods harboring acanthocephalan cystacanths. *J. Parasitol.* 59:945–56

11. Blest, A. D. 1957. The function of eye-spot patterns in the Lepidoptera. *Behaviour* 11:209–56

12. Blewett, M., Fraenkel, G. 1944. Intracellular symbiosis and vitamin requirements of two insects, *Lasioderma serricorne* and *Sitodrepa panicea. Proc. R. Soc. London B* 132:212–21

13. Bohn, G. W., Davis, G. N. 1964. Insect pollination is necessary for the production of muskmelons (*Cucumis melo v. veticutatus*). *J. Apic. Res.* 3:61–63

14. Bovbjerg, R. V. 1970. Ecological isolation and competitive exclusion in two crayfish (*Orconectes virilis* and *Orconectes immunis*). *Ecology* 51:225–36

15. Brian, A. D. 1957. Differences in the flowers visited by four species of bumblebees and their causes. *J. Anim. Ecol.* 26:71–98

16. Brian, M. V. 1952. The structure of a dense natural ant population. *J. Anim. Ecol.* 21:12–24

17. Brian, M. V. 1956. Segregation of species of the ant genus *Myrmica. J. Anim. Ecol.* 25:319–37

18. Brinkhurst, R. O., Chua, K. E., Kaushik, N. K. 1972. Interspecific interactions and selective feeding by tubificid oligochaetes. *Limnol. Oceanogr.* 17:122–33

19. Brock, T. D. 1969. Vertical zonation in hot spring algal mats. *Phycologia* 8:201–5

20. Brock, T. D., Brock, M. L. 1968. Relationship between environmental temperature and optimum temperature of bacteria along a hot spring thermal gradient. *J. Appl. Bacteriol.* 31:54–58

21. Brown, J. H. 1971. Mechanisms of competitive exclusion between two species of chipmunks. *Ecology* 52:305–11

22. Cairns, J., Dahlberg, M. L., Dickson, K. L., Smith, N., Waller, W. T. 1969. The relationship of freshwater protozoan communities to the MacArthur-Wilson equilibrium model. *Am. Natur.* 103:439–54

23. Cameron, G. N. 1971. Niche overlap and competition in woodrats. *J. Mammal.* 52:288–96

24. Carney, W. P. 1969. Behavioral and morphological changes in carpenter ants harboring diroceliid metacercaria. *Am. Midl. Nat.* 82:605–11

25. Case, T. J., Gilpin, M. E. 1974. Interference competition and niche theory. *Proc. Nat. Acad. Sci. USA* 71:3073–77

26. Castenholz, R. W. 1968. The behavior of *Oscillatoria terebriformis* in hot springs. *J. Phycol.* 4:132–39

27. Castenholz, R. W. 1973. The ecology of blue-green algae in hot springs. *The Biology of Blue-Green Algae,* ed. N. G. Carr, B. H. Whitten, 379–414. London: Blackwell

28. Cheng, T. C. 1964. *The Biology of Animal Parasites.* Philadelphia: Saunders. 727 pp.

29. Cheng, T. C. 1967. Marine molluscs as hosts for symbioses. *Adv. Mar. Biol.* 5:1–424

30. Cheng, T. C., ed. 1971. *Aspects of the Biology of Symbiosis.* Baltimore: Univ. Park. 327 pp.

31. Cody, M. L. 1974. *Competition and the Structure of Bird Communities.* Princeton: Princeton Univ. Press. 318 pp.

32. Colwell, R. K. 1973. Competition and coexistence in a simple tropical community. *Am. Natur.* 107:737–60

33. Colwell, R. K. 1974. Predictability, constancy, and contingency of periodic phenomena. *Ecology* 55:1148–53

34. Colwell, R. K., Futuyma, D. J. 1971. On the measurement of niche breadth and overlap. *Ecology* 52:567–76

35. Connell, J. H. 1961. The influence of interspecific competition and other factors on the distribution of the barnacle *Chthamalus stellatus. Ecology* 42:710–23

36. Connell, J. H. 1971. On the role of natural enemies in preventing competitive exclusion in some marine animals and in rain forest trees. *Proc. Adv. Study Inst. Dynamics Numbers Pop.* (Oosterbeek, 1970) 298–312

37. Connell, J. H. 1972. Community interactions on marine rocky intertidal shores. *Ann. Rev. Ecol. Syst.* 3:169–92

38. Connell, J. H. 1974. Ecology: field experiments in marine ecology. In *Experimental Marine Biology,* ed. R. Mariscal, 21–54. New York: Academic

39. Connell, J. H. 1975. Some mechanisms producing structure in natural communities: a model and evidence from field experiments. *The Ecology and Evolution of Communities,* ed. M. Cody, J. Diamond. Cambridge: Harvard Univ. Press. In press

40. Cook, S. A. 1965. Population regulation of *Eschscholzia californica* by competition and edaphic conditions. *J. Ecol.* 53:759–69

41. Cornell, H. 1974. Parasitism and distributional gaps between allopatric species. *Am. Natur.* 108:880–88

42. Culberson, W. L. 1970. Chemosystematics and ecology of lichen-forming fungi. *Ann. Rev. Ecol. Syst.* 1:153–70

43. Culver, D. C. 1970. Analysis of simple cave communities: niche separation and species packing. *Ecology* 51:949–58

44. Davies, R. W. 1969. Predation as a factor in the ecology of triclads in a small weedy pond. *J. Anim. Ecol.* 38:577–84

45. Davis, J. 1973. Habitat preferences and competition of wintering Juncos and Golden-crowned Sparrows. *Ecology* 54:174–80

46. Dayton, P. K. 1971. Competition, disturbance, and community organization: the provision and subsequent utilization of space in a rocky intertidal community. *Ecol. Monogr.* 41:351–89

47. DeBach, P., Sundby, R. A. 1963. Competitive displacement between ecological homologues. *Hilgardia* 34:105–66

48. DeBach, P., ed. 1964. *Biological Control of Insect Pests and Weeds*. New York: Reinhold. 844 pp.

49. DeBach, P. 1974. *Biological Control by Natural Enemies*. Cambridge, Engl.: Cambridge Univ. Press. 323 pp.

50. Del Moral, R., Cates, R. G. 1971. Allelopathic potential of the dominant vegetation of western Washington. *Ecology* 52:1030–37

51. de Ruiter, L. 1952. Some experiments on the camouflage of stick caterpillars. *Behaviour* 4:222–32

52. Dethier, V. G. 1970. Chemical interactions between plants and insects. *Chemical Ecology,* ed. E. Sondheimer, J. B. Simeone, 83–102. New York: Academic

53. Edington, J. M. 1968. Habitat preferences in net-spinning caddis larvae with special reference to the influence of running water. *J. Anim. Ecol.* 37:675–92

54. Ehrlich, P. R., Raven, P. H. 1964. Butterflies and plants: A study in coevolution. *Evolution* 18:586–608

55. Ellenberg, H. 1956. Grundlagen der Vegetationsgliederung. 1 Tiel: Aufgaben und Methoden der Vegetationskunde. *Einfuhring in die Phytologie,* ed. H. Walter, 4:1–136. Stuttgart:Ulmer

56. Elton, C. 1927. *Animal Ecology.* London: Sidgewick & Jackson. 204 pp.

56a. Emlen, J. M. 1973. *Ecology: An Evolutionary Approach.* Reading, Mass.: Addison-Wesley. 493 pp.

57. Eriksen, C. H. 1968. Ecological significance of respiration and substrate for burrowing Ephemeroptera. *Can. J. Zool.* 46:93–103

58. Eyles, A. C. 1964. Feeding habits of some Rhyparochrominae (Heteroptera: Lygaeidae) with particular reference to the value of natural foods. *Trans. R. Entomol. Soc. London* 116:89–114

59. Fellows, D. P., Heed, W. B. 1972. Factors affecting host plant selection in desert-adapted cactiphilic *Drosophila. Ecology* 53:805–8

60. Filmer, R. S. 1941. Honeybee population and floral competition in New Jersey orchards. *J. Econ. Entomol.* 34: 198–99

61. Frank, P. W. 1957. Coactions in laboratory populations of two species of *Daphnia. Ecology* 38:510–19

62. Friedman, J. 1971. The effect of competition by adult *Zygophyllum dumosum* Boiss. on seedlings of *Artemesia herbaalba* Asso in the Negev Desert of Israel. *J. Ecol.* 59:775–82

62a. Gadgil, M., Solbrig, O. T. 1972. The concept of *r*- and *K*-selection: evidence from wild flowers and some theoretical considerations. *Am. Natur.* 106:14–31

63. Gilbert, L. E. 1975. Ecological consequences of a coevolved mutualism between butterflies and plants. *Coevolution of Animals and Plants,* ed. L. E. Gilbert, P. H. Raven, 210–40. Austin: Univ. Texas Press

64. Gilbert, L. E., Singer, M. C. 1975. Butterfly ecology. *Ann. Rev. Ecol. Syst.* 6:365–97

65. Gill, D. E. 1972. Intrinsic rates of increase, saturation densities, and competitive ability. I. An experiment with *Paramecium. Am. Natur.* 106:461–71

66. Graham, K. 1967. Fungal-insect mutualism in trees and timber. *Ann. Rev. Entomol.* 12:105–26

67. Grant, P. R. 1972. Interspecific competition among rodents. *Ann. Rev. Ecol. Syst.* 3:79–106

68. Hair, J. D., Holmes, J. C. 1975. The usefulness of measures of diversity, niche width and niche overlap in the analysis of helminth communities in waterfowl. *Acta Parasitol. Pol.* In press

69. Halligan, J. P. 1973. Bare areas associated with shrub stands in grasslands: the case of *Artemesia californica. Bioscience* 7:429–32

70. Halligan, J. P. 1974. Relationship between animal activity and bare areas associated with California sagebrush and annual grassland. *J. Range Manage.* 27:358–62

71. Harger, J. R. E. 1972. Competitive coexistence among intertidal invertebrates. *Am. Sci.* 60:600–7

72. Harlan, H. V., Martini, M. L. 1938. The effect of natural selection on a mixture of barley varieties. *J. Agric. Res.* 57: 189–99

73. Harper, J. L. 1964. The nature and consequences of interference amongst plants. *Proc. XI Int. Congr. Genet.* (*1963*), pp. 465–82

74. Harper, J. L., Clatworthy, J. N., McNaughton, I. H., Sagar, G. R. 1961. The evolution and ecology of closely related species living in the same area. *Evolution* 15:209–27

75. Heatwole, H., Davis, D. M. 1965. Ecology of three sympatric species of parasitic insects of the genus *Megarhyssa* (Hymenoptera: Ichneumonidae). *Ecology* 46:140–50

76. Heatwole, H., Levins, R. 1972. Trophic structure stability and faunal change during recolonization. *Ecology* 53: 531–34

77. Heed, W. B., Jensen, R. W. 1966. *Drosophila* ecology of the senita cactus, *Lophocereus schottii. Dros. Inf. Serv.* 43:94

78. Heed, W. B., Kircher, H. W. 1965. Unique sterol in the ecology and nutrition of *Drosophila pachea. Science* 149:758–61

79. Heinrich, B., Raven, P. H. 1972. Energetics and pollination. *Science* 176:597–602

80. Heller, H. C. 1971. Altitudinal zonation of chipmunks (*Eutamias*): Interspecific aggression. *Ecology* 52:312–19

81. Henry, S. M., ed. 1966, 1967. *Symbiosis,* Vols. I & II. New York: Academic. 921 pp.

82. Heusser, H. 1970. Spawn eating by tadpoles as possible cause of specific biotope preferences and short breeding times in European Anurans (Amphibia, Anura). *Oecologia* 4:83–88 (in German)

83. Hobbs, G. A., Nummi, W. O., Virostek, J. F. 1961. Food-gathering behavior of honey, bumble, and leaf-cutter bees (Hymenoptera: Apoidae) in Alberta. *Can. Entomol.* 93:409–19

84. Holmes, J. C. 1961. Effects of concurrent infections on *Hymenolepis diminuta* (Cestoda) and *Moniliformis dubius* (Acanthocephala) I. General effects and comparison with crowding. *J. Parasitol.* 47:209–16

85. Holmes, J. C. 1962. Effects of concurrent infections on *Hymenolepis diminuta* (Cestoda) and *Moniliformis dubius* (Acanthocephala). II. Effects on growth. *J. Parasitol.* 48:87–96

86. Holmes, J. C. 1973. Site selection by parasitic helminths: interspecific interactions, site segregation, and their importance to the development of hel-minth communities. *Can. J. Zool.* 51:333–47

87. Holmes, J. C., Bethel, W. M. 1972. Modification of intermediate host behaviour by parasites. *Zool. J. Linn. Soc.* 51: Suppl. 1, 123–49

88. Horn, E. G. 1971. Food competition among the cellular slime molds (Acrasieae). *Ecology* 52:475–84

89. Huffaker, C. B. 1958. Experimental studies on predation: dispersion factors and predator-prey oscillations. *Hilgardia* 27:343–83

90. Huffaker, C. B., ed. 1971. *Biological Control.* New York: Plenum. 511 pp.

90a. Huffaker, C. B. 1974. Some implications of plant-arthropod and higher level arthropod-arthropod food links. *Environ. Entomol.* 3:1–9

91. Hunt, J. H. 1974. Temporal activity patterns in two competing ant species (Hymenoptera: Formicidae). *Psyche* 81:237–42

92. Hutchinson, G. E. 1957. Concluding remarks. *Cold Spring Harbor Symp. Quant. Biol.* 22:415–27

93. Hutchinson, G. E. 1965. *The Ecological Theater and the Evolutionary Play.* New Haven, Conn.: Yale Univ. Press. 139 pp.

94. Inger, R. F., Greenberg, B. 1966. Ecological and competitive relations among three species of frogs (Genus *Rana*). *Ecology* 47:746–59

95. Ivlev, V. S. 1961. *Experimental Ecology of the Feeding of Fishes.* New Haven, Conn: Yale Univ. Press. 302 pp.

96. Jaeger, R. G. 1971. Competitive exclusion as a factor influencing the distribution of two species of terrestrial salamanders. *Ecology* 52:632–37

97. Jaeger, R. G. 1971. Moisture as a factor influencing the distributions of two species of terrestrial salamanders. *Oecologia* 6:191–207

98. Jaeger, R. G. 1972. Food as a limited resource in competition between two species of terrestrial salamanders. *Ecology* 53:535–46

99. Jaeger, R. G. 1974. Competitive exclusion: comments on survival and extinction of species. *Bioscience* 24:33–39

100. Jakob, H. 1961. Compatibilités, antagonismes et antibioses entre quelques algues du sol. *Rév. Gén. Bot.* 68:5–72

101. Janzen, D. H. 1966. Coevolution of mutualism between ants and acacias in Central America. *Evolution* 20:249–75

102. Janzen, D. H. 1974. Swollen-thorn acacias of Central America. *Smithson.*

Contrib. Bot. No. 13. Washington DC: Smithsonian Inst. Press. 131 pp.

102a. Johnson, L. K., Hubbell, S. P. 1974. Aggression and competition among stingless bees: Field studies. *Ecology* 55:120–27

103. Keever, C. 1950. Causes of succession on old fields of the Piedmont, North Carolina. *Ecol. Monogr.* 20:229–50

104. Koplin, J. R., Hoffmann, R. S. 1968. Habitat overlap and competitive exclusion in voles (*Microtus*). *Am. Midl. Nat.* 80:494–507

105. Kozlowski, T. T. 1949. Light and water in relation to growth and competition of Piedmont forest tree species. *Ecol. Monogr.* 19:207–31

105a. Kruckeberg, A. R. 1954. The ecology of serpentine soils. *Ecology* 35:267–74

106. Kuglar, H. 1955. *Einführung in die Blütenökologie.* Stuttgart: Fisher

106a. Levandowsky, M. 1972. Ecological niches of phytoplankton species. *Am. Natur.* 106:71–78

107. Levin, D. A. 1972. Low frequency disadvantage in the exploitation of pollinators by corolla variants in *Phlox. Am. Natur.* 106:453–60

108. Levin, D. A., Schaal, B. A. 1970. Corolla color as an inhibitor of interspecific hybridization in *Phlox. Am. Natur.* 104:273–83

109. Levins, R. 1968. *Evolution in Changing Environments.* Princeton, NJ: Princeton Univ. Press. 120 pp.

110. Levins, R., Pressick, M. L. Heatwole, H. 1973. Coexistence patterns in insular ants. *Am. Sci.* 61:463–72

111. Lie, K. J., Basch, P. F., Hoffman, M. A. 1967. Antagonism between *Paryphostomum segregatum* and *Echinostoma barbosai* in the snail *Biomphalaria straminea. J. Parasitol.* 53:1205–9

112. Lincicome, D. R. 1971. The goodness of parasitism: A new hypothesis. *Aspects of the Biology of Symbiosis,* ed. T. C. Cheng, 139–227. Baltimore: Univ. Park

113. Linsley, E. G., Cazier, M. A. 1970. Some competitive relationships among matinal and late afternoon foraging activities of caupolicanine bees in southeastern Arizona (Hymenoptera, Colletidae). *J. Kansas Entomol. Soc.* 43:251–61

114. Longhurst, W. M., Oh, H. K., Jones, M. B., Kepner, R. E. 1968–1969. A basis for the palatability of deer forage plants. *N. Am. Wildlife Natur. Res. Conf.* 33–34:181–89

115. Low, R. M. 1971. Interspecific territoriality in a pomacentrid reef fish, *Poma-*

centrus flavicauda Whitley. *Ecology* 52:648–54

116. MacArthur, R. H. 1968. The theory of the niche. *Population Biology and Evolution,* ed. R. C. Lewontin, 159–76. Syracuse: Syracuse Univ. Press. 205 pp.

117. MacArthur, R. H. 1972. *Geographical Ecology: Patterns in the Distribution of Species.* New York: Harper & Row. 269 pp.

118. MacArthur, R. H., Levins, R. 1967. The limiting similarity, convergence, and divergence of coexisting species. *Am. Natur.* 101:377–85

119. Macior, L. W. 1970. The pollination ecology of *Pedicularis* in Colorado. *Am. J. Bot.* 57:716–28

120. Madsen, B. L. 1968. A comparative ecological investigation of two related mayfly nymphs. *Hydrobiologia* 31: 337–49

121. Maguire, B. Jr. 1973. Niche response structure and the analytical potentials of its relationship to the habitat. *Am. Natur.* 107:213–46

122. Maitland, P. S., Penney, M. M. 1967. The ecology of the Simuliidae in a Scottish river. *J. Anim. Ecol.* 36:179–206

123. May, R. M. 1973. *Stability and Complexity in Model Ecosystems.* Princeton, NJ: Princeton Univ. Press. 244 pp.

124. McBee, R. H. 1971. Significance of intestinal microflora in herbivory. *Ann. Rev. Ecol. Syst.* 2:165–76

125. McKell, C. M., Duncan, C., Muller, C. H. 1969. Competitive relationships of annual ryegrass (*Lolium multiflorum* Lam.). *Ecology* 50:653–57

126. McLay, C. L. 1974. The distribution of duckweed *Lemna perpusilla* in a small southern California lake: an experimental approach. *Ecology* 55:262–76

127. Menge, B. A. 1972. Competition for food between two intertidal starfish species and its effect on body size and feeding. *Ecology* 53:635–44

128. Menge, J. L., Menge, B. A. 1974. Role of resource allocation, aggression, and spatial heterogeneity in coexistence of two competing intertidal starfish. *Ecol. Monogr.* 44:189–209

129. Mewaldt, L. R. 1964. Effects of bird removal on a winter population of sparrows. *Bird-Banding* 35:184–95

130. Miles, J. 1972. Experimental establishment of seedlings on a southern English heath. *J. Ecol.* 60:225–34

131. Miller, R. S. 1967. Pattern and process in competition. *Adv. Ecol. Res.* 4:1–74

132. Moore, C. W. E. 1959. Interaction of species and soil in relation to the distri-

bution of eucalypts. *Ecology* 40:734–35

133. Morris, R. D., Grant, P. R. 1972. Experimental studies of competitive interaction in a two-species system. IV. *Microtus* and *Clethrionomys* species in a single enclosure. *J. Anim. Ecol.* 41:275–90

134. Morse, D. H. 1974. Niche breadth as a function of social dominance. *Am. Natur.* 108:818–30

135. Mulkern, G. B. 1967. Food selection by grasshoppers. *Ann. Rev. Entomol.* 12:59–78

136. Murdoch, W. W. 1969. Switching in general predators: experiments on predator specificity and stability of prey populations. *Ecol. Monogr.* 39:335–54

137. Neill, W. E. 1974. The community matrix and interdependence of the competition coefficients. *Am. Natur.* 108:399–408

138. Orenski, S. W. 1966. Intermicrobial symbiosis. *Symbiosis,* ed. S. M. Henry, Vol. 1, 1–33. New York: Academic

139. Paine, R. T. 1974. Intertidal community structure. Experimental studies on the relationship between a dominant competitor and its principal predator. *Oecologia* 15:93–120

140. Paperna, I. 1964. Competitive exclusion of *Dactylogyrus extensus* by *Dactylogyrus vastator* (Trematoda: Monogenea) on the gills of reared carp. *J. Parasitol.* 50:94–98

141. Parker, B. C., Turner, B. L. 1961. Operational niches and community interaction values as determined from in vitro studies of some soil algae. *Evolution* 15:228–38

142. Patrick, R. 1967. The effect of invasion rate, species pool, and size of area on the structure of the diatom community. *Proc. Nat. Acad. Sci. USA* 58:1335–42

143. Phleger, C. F. 1971. Effect of salinity on growth of a salt marsh grass. *Ecology* 52:908–11

144. Platt, R. B. 1951. An ecological study of Mid-Appalachian shale barrens and of the plants endemic to them. *Ecol. Monogr.* 21:269–300

145. Platt, W. J., Hill, G. R., Clark, S. 1974. Seed production in a prairie legume (*Astragalus canadensis* L.). Interactions between pollination, predispersal seed predation, and plant density. *Oecologia* 17:55–63

146. Pontin, A. J. 1960. Field experiments on colony foundation by *Lasius niger* (L.) and *L. flavus* (F.) (Hym., Formicidae). *Insectes Soc.* 7:227–30

147. Pontin, A. J. 1963. Further considerations of competition and the ecology of the ants *Lasius flavus* (F.) and *L. niger* (L.). *J. Anim. Ecol.* 32:565–74

148. Powers, D. A. 1972. Hemoglobin adaptation for fast and slow water habitats in sympatric catostomid fishes. *Science* 177:360–62

149. Press, J. W., Flaherty, B. R., Arbogast, R. T. 1974. Interactions among *Plodia interpunctella, Bracon hebetor,* and *Xylocoris flavipes. Environ. Entomol.* 3:183–84

150. Preston, E. M. 1973. A computer simulation of competition among five sympatric congeneric species of xanthid crabs. *Ecology* 54:469–83

151. Price, P. W. 1971. Niche breadth and dominance of parasitic insects sharing the same host species. *Ecology* 52:587–96

152. Putwain, P. D., Harper, J. L. 1970. Studies in the dynamics of plant populations. III. The influence of associated species on populations of *Rumex acetosa* L. and *R. acetosella* L. in grassland. *J. Ecol.* 58:251–64

153. Read, C. P. 1959. The role of carbohydrates in the biology of cestodes. VIII. Some conclusions and hypotheses. *Exp. Parasitol.* 8:365–82

154. Rettenmeyer, C. W. 1970. Insect mimicry. *Ann. Rev. Entomol.* 15:43–74

155. Rice, E. L. 1974. *Allelopathy.* New York: Academic. 353 pp.

156. Ricklefs, R. E. 1973. *Ecology.* Newton, Mass.: Chiron. 861 pp.

156a. Rosenzweig, M. L. 1966. Community structure in sympatric carnivores. *J. Mammal.* 46:602–12

157. Rosenzweig, M. L. 1973. Habitat selection experiments with a pair of coexisting heteromyid rodent species. *Ecology* 54:111–17

158. Rosenzweig, M. L., Sterner, P. W. 1970. Population ecology of desert rodent communities: body size and seed-husking as bases for heteromyid coexistence. *Ecology* 51:217–24

159. Roughgarden, J. 1972. The evolution of niche width. *Am. Natur.* 106:683–718

160. Roughgarden, J. 1974. The fundamental and realized niche of a solitary population. *Am. Natur.* 108:232–35

161. Roughgarden, J. 1974. Species packing and the competition function with illustrations from coral reef fish. *Theor. Pop. Biol.* 5:163–86

162. Schad, G. A. 1963. Niche diversification in a parasite species flock. *Nature* 198:404–6

163. Schlatterer, E. F., Tisdale, E. W. 1969. Effects of litter of *Artemesia, Chrysothamnus,* and *Tortula* on germination and growth of three perennial grasses. *Ecology* 50:869–73

164. Schoener, A. 1974. Experimental zoogeography: colonization of marine mini-islands. *Am. Natur.* 108:715–38

165. Schoener, T. W. 1974. Resource partitioning in ecological communities. *Science* 185:27–39

166. Scholes, R. B., Shewan, J. M. 1964. The present status of some aspects of marine microbiology. *Adv. Mar. Biol.* 2:133–69

167. Seneca, E. D. 1972. Seedling response to salinity in four dune grasses from the outer banks of North Carolina. *Ecology* 53:465–71

168. Sharitz, R. R., McCormick, J. F. 1973. Population dynamics of two competing annual plant species. *Ecology* 54: 723–39

169. Sheppard, D. H. 1971. Competition between two chipmunk species (*Eutamias*). *Ecology* 52:320–29

170. Sheppe, W. 1967. Habitat restriction by competitive exclusion in the mice *Peromyscus* and *Mus. Can. Field Natur.* 81:81–98

171. Shontz, J. P., Oosting, H. J. 1970. Factors affecting interaction and distribution of *Haplopappus divaricatus* and *Conyza canadensis* in North Carolina old fields. *Ecology* 51:780–93

172. Simberloff, D. S. 1974. Equilibrium theory of island biogeography and ecology. *Ann. Rev. Ecol. Syst.* 5:161–82

173. Simberloff, D. S., Wilson, E. O. 1969. Experimental zoogeography of islands: the colonization of empty islands. *Ecology* 50:278–96

174. Smigel, B. W., Rosenzweig, M. L. 1974. Seed selection in *Dipodomys merriami* and *Perognathus penicillatus. Ecology* 55:329–39

175. Smith, C. C., Follmer, D. 1972. Food preferences of squirrels. *Ecology* 53: 82–91

176. Snaydon, R. W. 1962. The growth and competitive ability of contrasting populations of *Trifolium repens* on calcareous and acid soils. *J. Ecol.* 50:439–47

177. Snaydon, R. W. 1970. Rapid population differentiation in a mosaic environment. I. The response of *Anthoxanthum odoratum* populations to soils. *Evolution* 24:257–69

178. Snaydon, R. W., Davies, M. S. 1972. Rapid population differentiation in a mosaic environment. II. Morphological variation in *Anthoxanthum odoratum. Evolution* 26:390–405

179. Sommerville, R. I. 1963. Distribution of some parasitic nematodes in the alimentary tract of sheep, cattle, and rabbits. *J. Parasitol.* 49:593–99

180. Stewart, R. E., Aldrich, J. W. 1951. Removal and repopulation of breeding birds in a spruce-fir forest community. *Auk* 68:471–82

181. Stewart, W. D. P. 1966. *Nitrogen Fixation in Plants.* London: Athlone. 168 pp.

182. Stimson, J. 1970. Territorial behavior of the owl limpet, *Lottia gigantea. Ecology* 51:113–18

183. Stimson, J. 1973. The role of the territory in the ecology of the intertidal limpet *Lottia gigantea* (Gray). *Ecology* 54:1020–30

184. Sutherland, J. P. 1974. Multiple stable points in natural communities. *Am. Natur.* 108:859–73

185. Tinnin, R. O. 1972. Interference or competition? *Am. Natur.* 106:672–75

186. Toporkova, L. Ya. 1973. Growth and development of larvae of the genus *Rana* in metabolites of related species. *Sov. J. Ecol.* 4:542–45 (Transl. from *Ekologiya*)

187. Vaillant, F. 1960. Experiments in regard to the resistance of some madicolous insects to modification of their habitat. *Bull. Soc. Entomol. France* 65:7–16

188. Vance, R. R. 1972. Competition and mechanism of coexistence in three sympatric species of intertidal hermit crabs. *Ecology* 53:1062–74

189. Van den Bosch, R. 1971. Biological control of insects. *Ann. Rev. Ecol. Syst.* 2:45–66

190. Vandermeer, J. H. 1969. The competitive structure of communities: an experimental approach with Protozoa. *Ecology* 50:362–70

191. Vandermeer, J. H. 1972. Niche theory. *Ann. Rev. Ecol. Syst.* 3:107–32

191a. Viktorov, G. A. 1970. Interspecies competition and coexistence of ecological homologues in parasitic Hymenoptera. *Zh. Obshch. Biol.* 31:247–55

192. Wavre, M., Brinkhurst, R. O. 1971. Interactions between some tubificid oligochaetes and bacteria found in the sediments of Toronto Harbour, Ontario. *J. Fish. Res. Board Can.* 25:2365–85

193. Whittaker, R. H. 1969. Evolution of diversity in plant communities. *Brookhaven Symp. Biol.* 22:178–96

194. Whittaker, R. H., Levin, S. A., Root, R. B. 1973. Niche, habitat, and ecotope. *Am. Natur.* 107:321–38

195. Whitten, B. H. 1973. Interactions with other organisms. *The Biology of Blue-Green Algae,* ed. N. G. Carr, B. H. Whitten, 415–33. London: Blackwell

196. Wickler, W. 1968. *Mimicry.* New York: World Univ. Library. 249 pp.

197. Wickstrom, E. E., Castenholz, R. W. 1973. Thermophilic ostracod: highest temperature tolerant aquatic metazoan. *Science* 181:1063–64

198. Wiegert, R. G., Mitchell, R. 1973. Ecology of Yellowstone thermal effluent systems: Intersects of blue-green algae, grazing flies (*Paracoenia,* Ephydridae) and water mites (*Partnuniella,* Hydrachnellae). *Hydrobiologia* 41:251–71

199. Wilbur, H. M. 1972. Competition, predation, and the structure of the *Ambystoma-Rana sylvatica* community. *Ecology* 53:3–21

200. Wilson, D. S. 1973. Food size selection among copepods. *Ecology* 54:909–14

201. Wilson, E. O. 1969. The species equilibrium. *Brookhaven Symp. Biol.* 22:38–47

202. Wilson, E. O. 1971. *The Insect Societies.* Cambridge: Belknap. 548 pp.

203. Wuenscher, J. E. 1969. Niche specification and competition modeling. *J. Theor. Biol.* 25:436–43

204. Youngbluth, M. J. 1968. Aspects of the ecology and ethology of the cleaning fish *Labroides phthirophagus* Randall. *Z. Tierpsychol.* 25:915–32

205. Roughgarden, J. 1976. Marine symbiosis: A simple cost-benefit analysis. *Ecology* In press

SIMULATION MODELS OF ECOSYSTEMS

♦4096

Richard G. Wiegert

Department of Zoology, University of Georgia, Athens, Georgia 30602

> *Come forth into the light of things,*
> *Let Nature be your teacher.*
>
> Wordsworth

MODELS AND NATURE

Simulations of the kind considered in this review are models of nature, living nature. As such, they are subject not only to the physical constraints on flow of matter-energy, but must conform to the multitudinous constraints required by the many species present in even the simplest of ecosystems. Indeed, it pays the ecological modeler to heed the words quoted above and to use the natural world, if not as teacher, as his guide when resolving questions of model form and function. The results of ignoring this advice are at best useless and at worst positively dangerous, for we live in a world where, more and more, decisions are shaped by the products of science and technology. One of those products that has heralded an exciting new area of ecology research is computer simulation modeling of ecosystems. I propose in this review to outline the origins, structure, and possible future of such models.

My dictionary defines simulation as "the act of feigning to be what one is not; the assumption of a deceitful appearance or character." Despite the somewhat pejorative character of this definition, it does describe well enough the modeler's intent, to simplify or substitute a model for the real thing, the ecosystem. But this "simulation model" should reflect the true nature of one or more aspects of reality and should not be a "dissimulation" model, one which, by definition, conceals the truth. In scope, simulation models vary from the detailed but small models of simple thermal communities (108) or microcosms (96) to simulation models of the world (52).

A word is needed about the scope of this review. I have deliberately avoided discussion of single population models or of two populations or competition models when these primarily concern predator-prey or competition theory. In the first place, the literature of predation and competition models is too vast an assemblage

311

even to attempt to summarize here. Second, such models are easily formulated without much attention to questions of matter-energy flows between compartments, the major concern both of simulation models of ecosystems and of this review. Some of the results of population modeling are discussed, however, in the context of equations describing the interactions between components.[1]

Simulation modeling of a natural ecosystem is, to quote Steele (92),

> ... bound to involve extreme simplifications. It does not in any sense produce new facts, but merely permits the evaluation of laboratory experiments carried out on different components in isolation. By forcing one to produce formulas to define each process and put numbers to the coefficients, it reveals the lacunae in one's knowledge. Although the output of the model can be tested against existing field observations and experimental results, the main aim is to determine where the model breaks down and use it to suggest further field or experimental work.

It is my purpose in this review to essay a judgment about the success of simulation modeling in fulfilling and perhaps exceeding the functions outlined by Steele.

ORIGINS

Simulation modeling of ecosystems is a young subdiscipline even when compared to the relatively recent origins of ecology itself. Before the rapid development of analog and digital computers in the 1950s and early 1960s, the simulation modeling now taken for granted was impossible.

To my knowledge, the earliest actual applications of the idea of simulating the dynamics of ecosystems on computers were those of H. T. Odum (59) and J. S. Olson (124), on an analog device, and Garfinkel (30), using differential equations solved by numerical integration on a digital computer. By 1966, the field of systems ecology had been defined and incorporated in university level course work (72), and papers began to appear which discussed both the desirability and applicability of simulating ecosystems (24).

With the initiation of the International Biological Program (IBP) in 1966, simulation modeling received a great stimulus. The major theme of this worldwide cooperative program was the management of large biome-level ecological associations from the standpoint of long-term benefit to man. It was natural to approach most of these studies from a system-oriented standpoint, thus making simulation modeling and perturbation analysis a major and important part of each of the biome level studies. This emphasis has been particularly apparent in the US/IBP program (see for example, 10, 11, 35, 42, 69, 94).

Before leaving the topic of origins, it seems well to devote a few words to the distinction between simulation models, at least as they are defined and treated in

---

[1]The literature search was aided greatly by the computer search profile service provided by the University of Georgia. All modeling and simulation titles were listed from four catalogues covering the years 1968–1969 through the first half of 1974. These were 1. Biological Abstracts, 2. Cataloguing and Indexing Bibliography of Agriculture, 3. Bioresearch Index, and 4. United States Government Research and Development Reports.

this review, and model or systems analysis. The pioneering series edited by Patten (74, 76, 78, 119) quite correctly made this distinction in its title, *Systems Analysis and Simulation in Ecology*. Constructing, validating, and running the model and assimilating the information derived—all these constitute the simulation process. Mathematical manipulations of the model designed to produce generalizations about such characteristics as stability, sensitivity, and frequency response are part of the analysis process. Patten (73) gave a comprehensive introduction to simulation, including some elementary programming for both the analog and digital computers. A different introductory approach has been outlined by Caswell et al (18). The analysis process is not considered in this review except insofar as it intrudes into simulation procedures, mainly concerning questions of the equations of interaction: how they should be written and methods of solution.

Taxonomy of Models and Their Uses

Simulation models of ecosystems may be separated according to a variety of criteria. As in any classification, much depends upon the viewpoint of the classifier. A basic distinction hinges on whether the model is theoretical or empirical. Theoretical simulation models attempt to provide some insight into the real world organization and operation of the modeling system. Perhaps these models can be said to be hypotheses as postulated by Reddingius (83). Or maybe Levins (46) was correct when he argued that a model cannot be a hypothesis because it cannot be directly verified by experiment. But one can at least, often rather easily, categorize models according to the above criteria. Empirical models make no pretense of explaining the real operation of the system. At best, they may faithfully reproduce the behavior of the system under a variety of conditions. Walsh & Dugdale (97) distinguished between these categories of model on the basis that a theoretical model uses causal relationships between variables and measured rate constants as opposed to the curve-fitting multiple regression techniques used to construct more empirical models. This is also similar to the distinction made by Caswell (17) whose ideas are considered in more detail when validation procedures are discussed later in this review.

Theoretical models, in common with all theory, cannot be proved correct; one can only attempt invalidation. Sufficient unsuccessful attempts to invalidate strengthen confidence in the model, but predictions are still justified only within the range of conditions spanned by the parameters of the model. Empirical models, known from the beginning to be developed directly from measured data representing known behavior of the system, are confined, in the sense of valid prediction, to that data set. Such models are of little value in those cases, probably a majority, where simulation is needed because experimentation is not possible (86), such as perturbations of large scale or dangerous proportions.

The semantic difficulties of trying to decide on a classificatory scheme for models have been explored thoroughly by Skellam (88). He pointed out the difficulty of deciding what is model and what is mimic by noting that DaVinci spoke of nature as his model! Biometricians interchangeably speak of fitting data to the model and of fitting the model to the data. Speaking on theoretical models, Skellam cautioned that ". . . modelmakers dwell at rarefied conceptual levels and are prone to be carried away, particularly at their own constructions," or again, "Modelling is an area where that profound and disturbing remark of Lao Tse appears to hold: 'Those who know do not tell; those who tell do not know.' "—somewhat pessimistic perhaps, but nevertheless well to keep in mind when occupied in putting together a simulation model of a system as complex and obscure as an ecological community.

Models must have an explicit statement of purpose, else they lose their relevance (88). Besides as a statement of theory, models have been constructed for a variety of reasons. Some of these were alluded to in the earlier quote from Steele (92). A simple organization of thought and identification of gaps in the research plan may have great value. For this purpose, the actual simulation is secondary and it matters little whether the model is theoretical or empirical, correct or incorrect. To a great extent the world model put together by Forrester (27) like its sequel, *Limits to Growth* (53), was this kind of model; its purpose was to call attention to a little known problem and thus the numerous, often heated arguments about its validity are often superfluous (52).

Many present-day simulation models have been constructed for the express purpose of aiding in the management of a valuable resource. Through the use of such models the yield to man can be optimized (104) for a long-term stable system. Some currently popular areas of modeling emphasis range from vertebrate grazing systems (3, 56, 93), insect pest management (4), recreational game populations (99, 100), and fisheries management (86) to the control of oil refinery waste treatment facilities (48). A more detailed consideration is given some of these models in the latter half of this review. However, the management simulation models take us somewhat far afield of the central theme of ecosystems, for many of them consider only single populations or food chains. The principles needed in the development of such models are, however, identical with those of total ecosystem models.

Simulation models of ecosystems need not be all inclusive but they must be all encompassing, i.e. they must cover all the kinds of interactions present in the system without including all the interactions (42). This rather wide scope means that the modeling effort often requires a team approach; the team must be somewhat regimented yet simultaneously productive of theory. Regrettably, little information is available on how a good modeling team can be organized and maintained; for one short account of this type, see Innis (41).

MODEL STRUCTURE

Trophic Detail

Upon initiation of a simulation effort, an immediate decision must be made as to whether the model will reflect primarily a population or a trophic level organization. In other words, will the flow of matter-energy be directed through a set of variables of state representing the levels of a trophic flow diagram (113), or will the flow pass through a food web representing individual species? This decision is, for the most part, separate from the problem of model condensation (the lumping or combining of species into state variables) which is dealt with later. The decision at this stage is more one of emphasis. The primary interest may be in simulating the flows and learning, for example, whether flow to the first order biophage consumer level is much less in a given system than is flow from the autotrophic level to the first order saprophage consumer level. In this case, the trophic level model is appropriate. If, however, the primary stress is on the dynamics and fate of individual species or species groups, one must model the species food web as a flow network. The trophic level approach offers (usually) the advantage of a much smaller number of state variables, but the construction of realistic interaction equations is made more difficult if not impossible. The species population food web modeling approach requires a sometimes bewildering array of state variables and parameters, but the chore of writing an equation for each interaction is lessened materially, since at most two species or species groups are involved with each equation. The latter approach has the additional advantage that the trophic model can be reconstructed from the simulation flow data of the population-oriented model, but the reverse is not true.

A variant of the population approach, which perhaps should be called a hybrid technique, and which melds both approaches, has been proposed by H. T. Odum (61). He has argued for simplicity in structure coupled with a total system overview, while maintaining the capability of a population orientation. Because this approach also employs a language variant, it is discussed later.

Model Construction—General

The steps necessary to construct a simulation model of an ecological system follow a fixed, generally agreed upon, logical sequence (4, 10, 34, 44, 104). The first stage begins with the initial ecological survey, which describes the list of biotic and abiotic variables. Generally, initial knowledge of the ecological relationships is slight, but all available information must be used to select a list of state variables. Naturally this initial list reflects not only the relative importance attached to the various species and/or trophic levels, but is also responsive to the objectives or purposes of the simulation model itself. For example, a model whose objective is to simulate and predict the effects of short-term (several years) management of quail populations in south Georgia stands of longleaf pine could ignore certain aspects of the population dynamics of the trees themselves, an oversight that would be ludicrous if the model also had the purpose of simulating a program of forestry management for timber production.

With the initial list of state variables selected, some indication of the interaction or degree of connectedness of the state variables must be drawn up. This often takes the form of a trophic model or a population food web. Some divergence at this point is expected depending on the type of model structure to be employed, i.e. on what "model" of modeling is chosen. For example, the kind of state space model proposed by Caswell et al (18) requires modeling the input and output relationships of each state variable by itself, i.e. free of constraints imposed by any other state variable. In general, however, a food web static model or diagram is sufficient.

Model Condensation

At this point, some latitude exists for different approaches to decreasing the complexity of the model organization. Broadly one may elect: 1. to set up smaller submodels, each having few state variables and interactions, 2. to make some gross simplifications in the model, or 3. to utilize a two-stage modeling approach in which a first crude approximation is used to provide more refined parameter values for a second more detailed model (98).

The first approach is widely used when modeling large complex systems (e.g. see 75). With the gain in simplicity and the computer time economies gained by splitting the model into submodels come problems in properly interconnecting all the submodels. A major problem arises when the integration intervals needed for numerical solution of the differential equations differ between submodels. This is sometimes very difficult to implement in the program. Only if the submodels can be run or *exercised* alone is option 1 desirable.

The disadvantages of option 2, extreme condensation, are obvious. For a savings in computer time and programming difficulty, one sacrifices an (unfortunately unknown) amount of confidence in the predictions of the model. Although all simulation models, no matter how detailed they pretend to be, represent compromises with the complexity and diversity of the real world, the rule of thumb should be that only as much condensation is applied as is consistent with the needs and desires of the model objectives and the personnel involved. The degree of detail in simulation models is greatest when the time behavior of individuals must be simulated (38), and least when the trophic level approach is used or when complexity is deliberately suppressed to gain simplicity (61).

The third option for condensation, the two-stage approach, has the disadvantage of decreasing the objective a priori nature of the model by changing the parameter values in the second stage model on the basis of a previous simulation run. This first stage, a very much condensed version, may or may not be providing usable information. Thus the method seems to increase the chance of going wrong without materially helping with the problem at hand. Either methods 1 or 2, judiciously employed, should be better choices.

The majority of simulation models of ecosystems is of food webs, i.e. the dynamics of species populations are mimicked. Because of the vast assemblage of species in most communities, to say nothing of the large number of life history stages that should be accommodated as distinct state variables in the model, some condensation of almost all models is required at the outset.

When models are condensed there are certain obvious required constraints if the modeling of food web transfers is the objective. First, all species or life history stages that feed from the same resources must be lumped together in the same state variables or else considered as individual state variables. To be lumped together, these species should have similar life history characteristics; they may then be considered a single paraspecies (7). However, some caution must be exercised when combining species, and there exist few guidelines for this sort of procedure (36). A study having the specific objective of exploring the rules of model condensation is being conducted on the simple thermal algal-fly ecosystems of Yellowstone National Park (108, 112). To date, the best idea seems to be to lump together species that, as noted above, have similar food sources and life history characteristics, as long as they also possess similar intrinsic rates of increase. In cases where the different life history stages of a species are themselves ecologically distinct, but several species are very similar with respect to each life history stage, it is best to lump the species within the life history stage, but to keep the stage itself as a separate state variable in the model (see 111). The next few years should witness a great improvement in our procedures for this phase of model construction.

Having decided on the objectives and structure of the model, with a food web diagrammed and questions of model condensation settled, the modeler can proceed to that phase of construction which is simultaneously the most ecologically demanding and scientifically controversial, the verbal and mathematical description of compartment interaction. The verbal description is essential to avoid nonsense biology creeping into the model. The mathematical description is necessary to implement the model. (Mathematical is used here in the sense of any set of formal symbols denoting relationships and implementable on analog or digital computers.)

Equations of Interaction

The central problem of this section is the translation of ecology into mathematical notation, a step with many logical and philosophical pitfalls (9). But before coming to grips with this central problem, a number of peripheral issues needs attention: stochastic vs deterministic models, the role of curve-fitting in modeling (realism), linear vs nonlinear equations, differential vs difference equations, the question of thresholds, spatial-temporal heterogeneity, and feeding preferences. All of these have been discussed at length in modeling literature.

Most ecologists and modelers have definite views on each of these topics, and I am no exception. I have reviewed the pertinent evidence in each case, but express some definite personal viewpoints in several instances.

STOCHASTIC VS DETERMINISTIC MODELS   Because the individual organism is constantly faced with choices, and because the vagaries of the abiotic environment are largely unpredictable, particularly on a small scale, the argument for incorporating probabilistic elements into simulation models is a strong one. Garfinkel, MacArthur & Sack (31) described the circumstances under which either a Monte Carlo method or a Markov chain process could be used in place of a conventional set of equations to simulate the dynamics of simple ecosystems. To some extent, however, the need for stochastic elements in the model is tied to the level at which the modeling is to be done, i.e. upon which is to be simulated the behavior of an individual or the behavior of a large number of individuals (the population). If the latter, then many processes that are truly stochastic can be represented by a mean with very little effect on the resultant predictions. For example, birth and death must be regarded as probabilistic events at the individual level, but in a population of thousands a mean birth and death rate predicts quite well.

Some models dealing with the behavior of individual organisms have been presented in both deterministic (39) and stochastic forms (63). However, models as complex and detailed as those using the Holling approach have not yet been incorporated into simulations of ecosystems. In the early 1960s, the conclusion was that stochastic modeling would have only limited impact (103). Stochastic modeling is employed in ecosystem simulations but usually to model a specific process, for example, the change in condition of algal mats (112). It usually adds only a small amount of realism; therefore at this time I see no need to revise Watt's opinion. An interesting sidelight, however, is the recent finding (50, 67) that even completely deterministic models may exhibit uncertain behavior because of mathematical properties of the feedback terms.

REALISM   A vital question that must be decided before equations of interaction can be written concerns the degree of realism desired in the simulation model. Sufficient realism must be present to make the model suitable to achieve its stated objectives. Most disagreements originate over interpretation of the word *sufficient.* The issue is, to what extent should the model advance theory, i.e. how far should it be from a purely predictive model? This question is approached from a somewhat different viewpoint in the discussion of validation. Here I am concerned more with the degree to which system output should be used in the construction of the model.

Levins (46) suggested that a productive approach in model building is an emphasis on generality and realism, even if done at the expense of precision. A number of modelers have subsequently made a case for abandoning the idea that the model parameters should be arbitrarily adjusted or tuned until the model output resembles the time series behavior. Boling, Petersen & Cummins (7) emphasized the translation from the qualitative biological concepts to a mathematical model, i.e. from a verbal to a mathematical description. The older method of finding a mathematical function that can be fitted to a set of data is less useful. Ecologists regrettably have a common tendency to measure everything, to regard all patterns as meaningful, and to revere mathematics. Bunnell (11) found little use in accumulating elegant solutions of equations that make no sense biologically, and Lassiter & Hayne (45) argued that to admit mathematical functions to the model without a biological analog explains nothing. This kind of model would suffice only if the objective were to simulate a data set already at hand. Walsh & Dugdale (97) argued for a causal relationship to be established between variables and measured rate constants (parameter values). Goodall (34), in an excellent general article on construction and validation of ecosystem models, argued that only in the simplest of models can parameter values be estimated from observation of the whole system. Ad hoc studies of some kind are usually indicated. In many cases, data obtained separately from experiments or measurement of individual populations can be used to set the value of the model coefficients (57).

In general, a second prediction by Watt (103) is borne out, namely that the tendency is away from curve-fitting multiple regression techniques and toward models based on reasonably realistic equations.

The question now arises of the degree of nonlinearity required of the equation.

LINEAR VS NONLINEAR EQUATIONS   Because this review specifically excludes the area of systems analysis and concentrates on simulation, it cannot cover the full scope of the controversy over the use of linear vs nonlinear equations in systems ecology. A good flavor of the arguments pro and con is obtained from three recent papers (5, 66, 77). Most of the justification for the use of totally (or at least piecewise) linear models (79) is based on advantages gained in systems analysis, where a large body of theory is readily available to handle linear systems.

But I do want to consider the more limited question: Is there any advantage in restricting ecosystem simulation models to linear equations of interaction? In answering this question, one must keep in mind the purposes of simulation models are 1. to predict the effects of perturbations applied to the system, and 2. to explain some of the operation of the ecosystem itself, i.e. to contribute to theory.

Consider any two state variables x_i and x_j and F_{ij} a transfer of matter–energy from x_i to x_j, then if

$$F_{ij} = f(x_j, t), \hspace{4cm} 1.$$

F_{ij} may be described by a linear transfer equation, for example $\tau_{ij} x_j$. However, this is an uncontrolled relation, for x_j would grow without limit. Therefore, the more usual kind of linear transfer function encountered in ecosystem models is

$$F_{ij} = f(x_i, t). \hspace{4cm} 2.$$

An example might be $\tau_{ij} x_i$ where τ_{ij} is the specific rate of transfer with dimensions time $^{-1}$. This relationship is spoken of as a donor-controlled, or more properly as a donor-determined, donor-controlled, transfer because the state of the donor (resource) variable determines both the maximum rate and the realized rate of transfer. Common examples are respiratory energy loss and nonpredatory mortality. Indeed, any transfer of matter-energy to an abiotic compartment has a strong chance of fitting this model, assuming that the effects of other biotic and abiotic variables are negligible.

Transfers of matter energy to living or biotic components, however, often have the form

$$F_{ij} = f(x_i, x_j, t). \hspace{4cm} 3.$$

If an explicit form of this function involves (as it usually does) the product of the two variables, the function is nonlinear. A common form used in models of ecosystems is $\tau_{ij} x_i x_j$. In general, the linear formulations cannot represent some very common ecological phenomena. Specifically these are 1. the asymptotic approach of rates to a maximum or minimum value, 2. switching of behavior or rate of transfer when a threshold concentration is encountered, and 3. the multiplicative dependence exhibited by the example above, where a nonzero transfer rate depends upon a nonzero value of more than one state variable.

The consensus among ecological modelers is that many, if not most, processes operating to control the transfer of energy-matter between populations are nonlinear (5, 42–45, 47, 66). In contrast, a strong argument made for the consistent use of linear equations in simulation models (77) is that the nominal behavior of ecosystems, as their state variables fluctuate around some operating point, is linear, despite the prevalence of nonlinear mechanisms in the interactions. This can easily be shown to be false for a number of ecosystems where a nonlinear response to a change in state variable or flux is observed. The nonlinear relationship between solar radiation intensity and photosynthesis exhibited by most C_3 plants is one case in point.

However, there are cases where linear models do predict changes reasonably well. If a system is in steady state and a radionuclide tracer is introduced, linear equations may describe adequately the movement through the system and the eventual establishment of an equilibrium (82). But the use of such linear equations, even where they may predict an adequate result, i.e. show how the system operates, cannot explain why it operates, because the interactions are not realistic (117).

Linear models have been justified on the grounds that they are well behaved and nonlinear models are not. Although nonlinear models employing simple, somewhat unrealistic, cross-product control functions are notoriously unstable, even they can be made to behave well (116). Realistic nonlinear models have been developed (108, 112, 115) that are well behaved (stable) with a single steady state or limit cycle trajectory. In fact none of the above criticisms applies to nonlinear models constructed in a manner reflecting the underlying ecological reality of the system (see also 42). On balance then, for simulation models at least, there seems no good reason for restricting the interaction equations to a linear model. Further arguments showing the positive advantages of nonlinear forms are reviewed in the following sections.

DIFFERENTIAL VS DIFFERENCE EQUATIONS   The differential equation for exponential growth is

$$dx/dt = \rho{\cdot}x, \qquad\qquad 4.$$

where ρ equals the specific rate of growth or decline. This equation postulates a continuous growth process. The alternative consideration of growth as a discrete process demands a difference form of this equation:

$$x_{t+1} = x_t\exp(\rho) \qquad\qquad 5.$$

where t is the discrete interval of time.

At least one generalized ecosystem model relies on difference equations (45). There are some compelling arguments for the use of difference equations in such general models as well as in specific site models. Most ecological measurements are made during intervals that vary depending on the process studied (42). However, even though measurements are made over intervals, specific rates may still be calculated if the functional relationship is known. For example, in a population growing at an exponential rate, one may measure or count the population at two times t_0 and t_1, find the ratio of x_t/x_0, and take the logarithm to base e of this value. This specific (often called instantaneous) rate is in units t^{-1}. In equation 5, the value of $\exp \rho{\cdot}t$ equals the ratio of population sizes over the intervals, thus the measured rate can be used directly in the calculated solution.

Difference equations also may conform more closely to the way ecologists think of the dynamic processes in their systems than do differential equations (40, 42). That observation, however, is a value judgment greatly dependent on individual circumstances. Certainly today the training of most ecologists includes considerable exposure to calculus.

The use of differential equations can be justified on the basis that the growth and decline of large populations of organisms approximate continuous processes very closely. Although rates cannot be measured instantaneously, the specific rates can be calculated, as above, from the discrete ratios of change. Unfortunately, few modeling studies employing specific rates and differential equations seem to recognize the problem involved in computing specific rates from measurements over discrete intervals. For example, assume one is measuring net photosynthesis of an aquatic community in a light bottle for one hour, and the populations fixed 100 mg of C, ending with 27,100 mg of C in the bottle. The specific rate of photosynthesis would usually be simply reported (incorrectly) as 100/27,000 or 0.0037037 mg C $\times$ mgC^{-1} $\times$ hr^{-1}. In fact, the correct computation, assuming an underlying *exponential growth process,* is $\log_e$ (27,100/27,000) or 0.0036969. Not a large practical difference as long as the measured change is small compared to the standing stock, but, in any case, worth remembering.

Watt (106) also discussed the pros and cons of differential equations. Probabilistic alternatives to either equation, for example Monte Carlo models and Markov chain processes, were outlined by Garfinkel, MacArthur & Sack (31). Which form to use probably depends, in most cases, on personal preference as much as anything. In fact, when no analytical solutions are available, the solution of sets of differential equations involves approximating them with difference equations and solving the latter, choosing an appropriate step size (40). Major differences in overall system performance or prediction should seldom result with either.

THRESHOLDS   At the outset of writing the equations of interaction, the modeler of ecosystems must decide whether to incorporate threshold responses. To do so means introducing discontinuities into the equations. This not only ensures nonlinearity, but makes any single general analytical solution impossible. On the other hand, to rule out thresholds may render the equations so unrealistic or imprecise as to destroy their utility.

The thresholds considered here are not responses to such things as radiation and pollutants, which involve physiological responses, as some would argue thresholds do not exist in such instances, or at least can be ignored with little damage to the reality of the models (118). The thresholds of interest in ecosystem models involve the responses of consumers to: 1. changes in their own density, and 2. changes in the densities of their resources.

If, for example, a population shows distinct breaks in the relationship of ingestion to density, thresholds may be identified. Thus below a given threshold density one often observes a population growing, i.e. ingesting, at the maximum genetically determined rate, assuming food and other material resources are optimally abundant. As the density increases, the mean interorganism distance decreases to the point at which perception of the neighbor occurs and a negative interaction is possible. At this point, density-dependent feedback has begun to decrease the growth rate. Similarly, at some upper density, negative interactions may become so severe that all feeding activity is interrupted and ingestion drops to zero. This must be represented as an upper threshold with respect to the relationship between density and ingestion.

Alternatively, one may examine the relationship of ingestion of a consumer population to the density of a resource. Again, if breaks in the curve are noted, threshold responses may be identified. A common threshold is a "refuge" level below which the resource is unavailable to the consumer. Such refuge densities may occur in the case of a bacterium unable to utilize dissolved organic compounds present in less than a threshold concentration. Or the threshold might be determined by the lower limit of prey density available to a predator because of physical structure in the environment suitable for a prey refuge.

Upper threshold concentrations of a material resource may induce saturation of the consumer. The population, if space is still optimum, may be growing, i.e. ingesting, at the maximum possible physiological rate. Thus further increases of that particular material resource would have no discernible effect, a useful test for a true threshold.

Some will argue that abrupt breaks or discontinuities do not occur in biological functions, that they are always smooth curves (85), but the weight of observational evidence is against this interpretation. Such threshold phenomena have been observed in ecosystems and have been incorporated into simulation models (1, 37, 45, 92). A simple method of incorporation is to use logical *IF* statements in the computer program [(37), see also the grassland model (42)]. This technique, however, is notationally clumsy. Abrosov, Kovrov & Rerberg (1) made use of the function MIN (a,b) i.e. the smaller value, a or b, to set threshold constraints in their model of an algal-bacterial community. The most generally useful technique is the use of the DIM function, where DIM (a,b) specifies that argument b is subtracted from argument a, returning the least positive difference, i.e. a positive value or zero (108). Thus whenever $b \geq a$, the value returned is zero. In mathematical notation, this function is closely related to a unit step function and is written $(a-b)_+$, where the subscript plus takes the place of the FORTRAN DIM. Examples of the use of this function in threshold equations are given in the section on equations of interaction.

HETEROGENEITY   In addition to thresholds, the interactions between components of ecosystems operate within a mosaic of heterogeneity. Actions and reactions are delayed in time, and components are often clumped or otherwise nonrandomly arranged in space. In many cases, the importance of temporal and spatial heterogeneity is increased by greater diversity, and the relationship of these factors with stability (of various kinds) has been the subject of considerable discussion (49, 91). There is general agreement that this heterogeneity is vital to the validity or realism of an ecosystem simulation model (36, 47).

The incorporation of a time delay (temporal heterogeneity) in the interaction equations can be accomplished by making change a function of past states of the system (16), or by using vectors of population age (112). The use of difference equations (42, 45) automatically introduces the capability for time delays; instead of X_{t+1} being a function of X_t, it can be a function of X_{t-n}, where n is any chosen number of time intervals.

Handling spatial heterogeneity is a more difficult problem. The simplest solution is effectively to run as many concurrent models as there are spatially distinct areas. This is a solution if the areas themselves do not change in size, but the problem becomes more complicated when the spatial heterogeneity itself changes with time (112). At present this difficulty remains one of the major unsolved problems of ecological modeling.

FEEDING PREFERENCES   Whenever any compartment in a model obtains resources from more than one source, some method of proportional allocation of these resources must be used to avoid biological unrealities in the equations when all resources are optimally abundant. Some simple and straightforward methods of accomplishing this have been discussed in the literature (62, 66, 108).

FUNCTIONAL FORMS   The number of mathematical functions available with which to represent a verbal description of an ecological interaction is literally infinite. The modeler is faced initially with the choice of "tailor-made" versus "off-the-shelf," i.e. he may use standard mathematical notation to build an equation that faithfully represents the ecological situation, or he may opt for one of the standard mathematical representations used to describe material transfers in an ecological context. Unfortunately for the realism, accuracy, and generality of most models, the latter course is usually adopted. Of the many (2, 105) functions available "off-the-shelf," the logistic-based equations of Lotka and Volterra and adaptations of the Michaelis-Menten equations of enzyme kinetics are the most commonly used.

The Lotka-Volterra Equations

Two forms of the Lotka-Volterra equations of population growth may be identified. The first is a model of a population growing alone or, with the addition of a competition term, with another population (Equation 6a). It uses an asymptotic constant K to set a maximum density or carrying capacity. The second covers the case of a prey limited by being eaten by a predator (Equation 6b).

$$\mathrm{d}x_j/\mathrm{d}t \;=\; x_j(\rho_j - \rho_j x_j/K_j), \qquad\qquad 6a.$$

$$\mathrm{d}x_j/\mathrm{d}t \;=\; x_j(\rho_j - \tau_i x_i). \qquad\qquad 6b.$$

Perhaps the greatest single disadvantage of the first of these equations is that depletion of the environment is measured by the single value K. This means that the species utilizes the different resources of the environment (space, food of all kinds, etc) in ways that cannot be separately represented (87, 109). Another major disadvantage is the linear relationship between the realized rate of increase and density. Often this is not observed. The second of the equations is notoriously unstable, for it permits growth at what may be biologically unreasonable rates. And finally, none of the Lotka-Volterra equations incorporates thresholds.

With these serious shortcomings, it is surprising that equations so devoid of biological reality (103) continue to be employed. Nevertheless, theoretical population biologists still make extensive use of the Lotka-Volterra equations, and they are found in simulation models as well, e.g. (48). Usually, however, some attention is paid to the shortcomings of these equations and variations are proposed. Thus O'Neill chose to represent the flux x_i to x_j as:

$$F_{ij} = kx_ix_j/(x_i + x_j).$$

Still, no thresholds are possible, and there is no incorporation of a density-dependent feedback from increases in x_j; indeed, increasing x_j can only increase the ingestion of x_j, except insofar as very high values of F_{ij} may tend to reduce the resource x_i.

Lassiter & Hayne (45) discussed the pros and cons of the Lotka-Volterra equations at length, but ended up using similar equations, modified only insofar as separate feedback controls of birth and death rates were included.

Michaelis-Menten Equations

A second model utilizes the analogy between the feeding, assimilating living population and the equations describing the enzyme-catalyzed reactions taking place in the cell, the Michaelis-Menten formula. This is a two-step process involving the formation of an enzyme-substrate complex and the subsequent separation of the enzyme from the new product. Two rate constants are employed: one for the formation of the enzyme-substrate complex, one for the formation of product.

$$F + x_t \rightleftharpoons Fx \rightleftharpoons_{\tau^1} x_{t+1}, \qquad\qquad 7.$$

where F = food (substrate), x = population (enzyme), τ = rate of assimilation, and τ^1 = rate of production.

In enzyme chemistry these processes are reversible, but because reversibility is not a characteristic of biological growth, it is eliminated in ecological adaptations of this model. Smith (90) used this model to represent population growth, letting food and the feeding portion of the population combine at one rate to produce the biological equivalent of the enzyme-substrate complex. The latter is turned into new enzyme (population) at a second rate, then the total new production is computed. Because of the autocatalytic nature of growth, this new production is equivalent to new enzyme.

Where it applies, this formulation of the feeding equation is considerably more realistic than the Lotka-Volterra equations. Unfortunately, the ingestion of few populations can be considered even formally analogous to enzyme reactions in cells.

The assumptions of Michaelis-Menten-type models do not hold even for protozoa unless the cell size is constant (23). The Michaelis-Menten formula is best applied to the single-celled populations present in open water phytoplankton systems. Thus it is not surprising that this type of feeding interaction equation dominates the aquatic population model literature and is commonly used in simulation models of such ecosystems.

Choosing mathematical relationships that represent biological and ecological phenomena accurately is not simple. As Innis (40) noted in a refutation of the idea (20) that higher order derivatives are needed:

> The established quantitative techniques were not developed for biological applications, but for physical applications. It results therefore, that these techniques may not be ideally suited to the biological and ecological realm and that, until some segment of the ecological community takes it upon itself to investigate these fundamentals more carefully, we may continue trying to fit square pegs into round holes.

A Generalized Realistic Model

Better interaction equations than the logistic Lotka-Volterra and Michaelis-Menten can be built if a suitable general, realistic foundation is used. One such model, with the capacity for incorporating techniques for building temporal and spatial heterogeneity, thresholds, and feeding preferences has been proposed (108–112, 115). An outline of the main arguments for this model follows.

For a model to be both realistic and general, the structure of the equations must: 1. reflect the important basic ecological parameters and 2. be flexible enough to accommodate additional special parameters whenever they are desired or deemed necessary.

Consider the transfer of matter-energy from a source (x_i) to a living component of the system (x_j). Without specifying the specific functional form of change in ingestion, five necessary parameters can be identified. The rate τ_{ij} is the maximum physiological rate of ingestion possible by x_j under a specified physical environment. If the latter does not vary, and if evolutionary change in physiological characteristics is negligible, then τ_{ij} is a constant. Otherwise it may be represented as a variable (28).

The flux of matter-energy into a population is represented by

$$F_{ij} = \tau_{ij}x_j \cdot f(x_i,x_j) \quad 0 \leq f(x_i,x_j) \leq 1. \qquad 8.$$

When the feedback term $f(x_i,x_j)$ is unity, the population is ingesting at the maximum rate τ_{ij}; when $f(x_i,x_j)$ is zero, ingestion is also zero. Between these two limits, x_j is ingesting at a rate between 0 and τ_{ij}.

If the limitation is imposed wholly by the condition of the food resource x_i, three basic pieces of information are needed in order to construct the proper feedback function $f(x_i)$. 1. The value of α_{ij}, the level of food density at which limitation first begins (the saturation threshold) must be ascertained. 2. The value of γ_{ij}, the lower limit of food availability (the refuge threshold) must be measured, and 3. the relationship between the change in density of the food resource and the change in the function $f(x_i)$ must be chosen.

As an example, assume a linear relationship for 3, for which the feedback function is written

$$f(x_i) = [1 - \{(\gamma_{ij} - x_i) / (\gamma_{ij} - \alpha_{ij})\}_+]_+, \qquad\qquad 9.$$

where the subscript $+$ notation places the desired constraints through the convention that

$$(\cdot)_+ = \begin{cases} 0 \text{ if } (\cdot) \leq 0 \\ (\cdot) \text{ if } (\cdot) > 0 \end{cases}.$$

Note that $f(x_i)$ is zero whenever $x_i \leq \gamma_{ij}$ and unity whenever $x_i \geq \alpha_{ij}$, and that the change of $f(x_i)$ with change in x_i between α_{ij} and γ_{ij} is linear.

If the limitation on F_{ij} is wholly imposed by intrapopulation strife in x_j, that is, assume $x_i \geq \alpha_{ij}$, then evolution of the feedback function $f(x_j)$ depends upon knowing at least 1. the lower limit of density of x_j, where the effects of intrapopulation interaction first interfere with ingestion, i.e. α_{jj} or the threshold response density, 2. the upper asymptotic level of population density where interference has risen to the point at which ingestion is just equal to the maintenance costs of the population, i.e. the equilibrium threshold or carrying capacity, and 3. the relationship between the change in x_j and change in $f(x_j)$.

Again as an example, assume 3 to be linear, whence

$$f(x_j) = [1 - \{1 - \lambda_j/\tau_{ij}(1 - \epsilon_j)\} \{(x_j - \alpha_{jj})/(\gamma_{jj} - \alpha_{jj})\}_+]_+. \qquad 10.$$

The additional term involving λ_j (the specific maintenance losses) and ϵ_j (the proportion of ingestion that is egested) is necessary because the upper density level γ_{jj} is defined as the equilibrium point, not the point at which ingestion becomes zero. Thus when $x_j \leq \alpha_{jj}$, the flux F_{ij} is maximum, and when $x_j = \gamma_{jj}$, the flux F_{ij} equals $x_j [\lambda_j/(1 - \epsilon_j)]$, the amount of maintenance losses.

When both scarcity of resources (x_i) and intrapopulation (x_j) interaction are factors in limiting F_{ij}, equation 8 can be rewritten

$$F_{ij} = [1 - f(x_i) - \{1 - \lambda_j/\tau_{ij}(1 - \epsilon_i)\}f(x_j)]_+. \qquad\qquad 11.$$

Equation 11 assumes that the repressive effects of each feedback term are additive. This is the simplest assumption, but in the face of experimental evidence to the contrary, other arrangements can be made.

The use of interaction equations constructed as outlined above produces models that combine aspects of generality, reality, and precision. Additional modifications in the mathematical functions can simulate the effects of time delays in development, spatial and temporal heterogeneity, variable parameters, and stochastic terms (111).

The equations are nonlinear and discontinuous, yet the models constructed from them are stable in the sense that no compartment can increase to infinity nor can a compartment go to extinction without a biological reason. There is only one positive steady state or limit cycle possible for a given set of parameter values. Models of this kind can be used to generate hypotheses concerning perturbations (109). Most important, every state variable and parameter in such models can be defined, and a procedure outlined for its observational or experimental measurement.

SIMULATION

Digital vs Analog

The mechanics of translating from a set of mathematical equations to a computer program are highly machine dependent. The initial choice facing many ecological modelers is whether to use an analog or a digital computer, assuming of course that both are available.

The analog computer is inherently less accurate than the large digital computers. But for solving simultaneous sets of differential equations, the analog machine has the advantage of not requiring a numerical approximation, i.e. its solutions are continuous and thus free of inherent error. In this application, the usual four-place accuracy of the analog machine may more than equal that of the high speed digital computer.

The analog computer also offers the advantages of low running cost and rapid operator-machine interaction. Once programmed, many simulations can be done rapidly. Because the output commonly shows up as a graph on an oscilloscope tube, or XY plotter, the modeler can quickly make decisions about the effects of parameter changes and perturbations.

Unfortunately, the process of programming and debugging the analog computer is relatively time consuming, and scaling restrictions reduce the dynamic range of model behavior. A shortcut method of programming the analog computer for models of ecosystems solves some of the worst of the scaling problems (61). The greatest single drawback of analog computers for ecological modeling is simply their unavailability, at least in the larger sizes, to most modelers. Models incorporating nonlinearities, spatial and temporal heterogeneities, and thresholds use components, particularly amplifiers, at a very great rate, and can be handled only by the largest analog machines, usually found only in schools of engineering. Digital computers capable of handling such models are commonly accessible, often via a terminal in the investigator's office.

Digital computers are ideal for handling spatial and temporal heterogeneities, threshold limits, and other nonlinearities (38). Their disadvantages are the cost of the simulations, a definite and sometimes great delay between making the parameter changes or perturbations and seeing the result, and the necessity of numerical methods for solving sets of differential equations.

The first two disadvantages are being eased by the rapid technological breakthroughs now being made in the computer industry. Unit costs are going down and speed is rising, causing a reduction in running expenses and making the computer more accessible. Most ecological modelers can now arrange to spend time at an interactive terminal or console, where turnaround time is measured in seconds or, at most, minutes.

With regard to the third disadvantage, numerous numerical methods are available for the solution of simultaneous sets of differential equations with the digital computer (15). Probably because of the nature of ecological systems simulations (relatively imprecise parameters values and oscillating state variables) inaccuracies in the integration methods will remain relatively unimportant for some time. Distinct differences in results of simulations using four different integration techniques were noted in one unpublished study (R. V. O'Neill, An introduction to the numerical solution of differential equations in ecosystem models. ORNL-IBP-70-4. IBP UC-48-Biology and Medicine, 1970). Magnitude of the error with any of the techniques was relatively low. Usually, the simpler and more economical of time the method, the more inaccurate its results. However, it is sometimes possible to choose a shorter interval of integration and obtain equal results with a simpler technique at a much lower cost (114). Another way around the problem of numerical integration is simply to use difference equations in the model.

Simulation Language

Actually to simulate the dynamics of an ecosystem on a computer, the modeler is faced with the problem of translating a set of mathematical statements of the rules governing the operation of the system into something the computer can understand. For this purpose he has at his disposal a number of standard programming languages for the digital computer.

The most flexibility and often the most efficiency (from the standpoint of computer time) is frequently achieved by translating the mathematical model into a standard general purpose language such as FORTRAN. Many useful special on-line functions are available. Common examples are the logical *IF* statements and the replacement for the subscript plus notation [*DIM* (*a*, *b*)] discussed earlier. In addition, a number of other kinds of simulation languages have been developed to aid either in visualizing the system or in running it on the computer.

Both the system of diagrams developed by Forrester (26) and the energy circuit language developed by Odum (60, 61) are intermediary languages of the first type. As visual spans across the gap between a concept of the real ecosystem and its mathematical or computer language representation, both of these languages are extremely useful. The Odum circuit language, because of its greater numbers of symbols, is more explicit about relationships, but also more difficult to learn to use well. However, neither comes close to the richness of mathematical notation, and thus these languages are not suitable for direct implementation of the model from conception to simulation on the digital computer. The energy circuit diagram, with appropriate mathematical notation as addenda, can be implemented directly on the analog computer by someone experienced with both the analog and the energy circuit diagrams (61).

The second type of special category comprises those simulation languages developed to avoid the sometimes lengthy programs arising from a complete *FORTRAN* translation of the mathematical model. In essence, these special languages act as partial translators and enable simulation modeling to be accomplished with a minimum of knowledge of and concentration on the programming itself.

Radford (81) presented an interesting and thorough discussion of the relative merits of three of these simulation languages. DYNAMO was discussed in the context of modeling the upstream migration of salmon. An application of the simulation language DSL 1130 to a lake ecosystem was illustrated, as was the use of the CSMP 360 language applied to management of a grassland cut for hay. Radford conceded somewhat more versatility to the CSMP language in terms of changing parameters between runs. The utility of CSMP 360 was also stressed by Brennan et al (9) who discussed its use in simulation models of four different kinds of system. However, Martin (48), comparing CSMP with the Dynamic Simulation Program (DYSIMP), was in favor of the latter on the basis of lower cost and the presence of optimization capabilities. SIMCOMP, a special simulation language, was developed specifically to ease the problems of communication of model changes between a large group of ecologists and modelers working on a common system, the short grass prairie (42).

The choice of programming approach and computer language depends on a combination of preference, availability, and training of the modeler, although some ecologists (97) see a trend from the special simulation languages and toward common use of FORTRAN by modelers of large biological systems.

Validation

The procedure necessary to validate a simulation model of an ecosystem depends on the type of model and the purpose for which it was designed. In an excellent general treatment of the difficult question of validation, Caswell (17) distinguished between predictive models, designed solely to provide information on the future behavior of the state variables, and theoretical models, those that purport to provide insight into how the real system operates internally. The validation of the latter must involve sufficient attempts at disproof so that confidence in the theory is established. Caswell noted the impossibility of proving the truth of a universal statement by inductive logic. Thus the validation (Caswell suggests the term *corroboration*) procedure for theoretical models must include hypothesis testing, the hypotheses being derived from the universal statements in the model plus certain initial conditions of the variables and parameters. A single instance of disagreement can be considered sufficient evidence for refutation of the model as a theoretical statement. This distinction between the two kinds of models was also implied by Morales (58) when exhorting modelers to criticize models not as true or false, but as useful/insightful vs ineffectual.

One point not made by Caswell concerns the role of measurement. The successful refutation of any scientific theory requires that the accuracy of measurement used to distinguish between alternative hypotheses be equal to the task. Thus the Newtonian view of the world could be refuted only when our standards of physical measurement became sufficiently precise to detect the minute discrepancies (under ordinary conditions and velocities) between theory and the real world. Relativity now forms the theoretical model of our concept of the relation of mass and velocity, yet the predictive model used in everyday engineering retains the Newtonian ideas because they are simpler, and adequate under ordinary conditions.

In the ecological context, we are now at the stage where the refutation of theoretical models is possible only at the grossest level because of the relative crudity of our measurements. For example, Wiegert (108), in an ecosystem model simulating the growth of brine flies on thermal algal mats, employed a feedback control relationship in which the reduction of ingestion by flies upon an increase of the density was a nonlinear function. Further consideration of the biological factors involved prompted a revision of this theoretical model to incorporate a linear relationship into a feedback control of the flies (112) but to retain the experimentally corroborated inverse (nonlinear) feedback for the algae (28). Nevertheless, despite their different theoretical basis, the two models produced pictures of the successional dynamics of the thermal system that were virtually indistinguishable from each other, given the variance in the field data.

The opposite result was obtained in the case of a preliminary salt marsh model (115), where the initial linear feedback function used to represent the effects of resource scarcity on growth of anaerobic sediment bacteria produced such grossly aberrant estimates of the carbon dynamics of the salt marsh as to prompt reevaluation and subsequent substitution of a new theoretical feedback relationship.

Predictive models, according to Caswell, are an entirely different matter. Because the reality or universal truth of the model is not an issue, the validation procedure is simply one of establishing the degree of accuracy with which the model predicts the system behavior, and the range of conditions and parameter values over which this degree of accuracy holds. In this sense, I suppose, one cannot really invalidate a predictive model, but only accept or reject its range of error in predictive ability. Thus, because Caswell argued for corroboration instead of validation for testing theoretical models, and because acceptance might be preferable to validation for predictive models, it may be wisest simply to drop the term *validation* from the simulation model literature. Perhaps the term can be used in the very general sense of suggesting the degree of confidence one has in models that combine both theoretical and predictive functions.

Simulated Ecosystems

At present, there are relatively few simulation models of ecosystems and fewer still attempt to include in detail more than a few of the component species or abiotic compartments. Nevertheless, in this review, space permits no more than a cursory description of several such models.

The simulations were divided into those whose object was to deal with the management of economically important systems and those primarily directed toward an understanding of a hypothetical or a natural system or process. In the former category are 1. terrestrial grazing systems, 2. marine fisheries, and 3. pest management. Although there is a vast literature on all three of these areas, relatively few ecosystem simulations model practical management problems.

In the latter category of simulation models are those dealing with 1. general (i.e. theoretical) ecosystems, 2. succession, 3. aquatic ecosystems, 4. forest, 5. grassland, 6. tundra, 7. desert, and 8. decomposers.

This is not an exhaustive list, yet it represents the total of relevant papers gleaned from a rather extensive search of the recent literature. Space does not permit more than a brief comment about each paper, but the following annotated list should serve to give the reader some appreciation for the kinds of model being developed and the future needs and possibilities of simulation modeling.

MANAGEMENT MODELS   Five papers addressed the general topic of management of grazing systems. Goodall (32, 33) demonstrated the use of computers in the grazing management of semi-arid lands and simulated the grazing by a nonreproductive population of sheep in Australia with rainfall, the major abiotic variable, treated stochastically. Milner (56) also simulated the dynamics of vegetation and sheep on the St. Kilda Nature Reserve heath. The model defined harvest of vegetation by sheep and the resultant effects on sheep and vegetation, but lacked any feedback effect of vegetation changes in the present on vegetation and thus sheep in the future. Swartzman & Singh (93) discussed the value of the computer model in assessing the worth of alternative management schemes in a successional tropical grassland.

Many innovative ideas and techniques have been pioneered in the management of fisheries, an applied ecological area. Indeed, Volterra's original stimulus for his population work was an interest in an applied fisheries problem. Thus it is no surprise to find some early applications of simulation models in this field. Steele (92) provided a stimulating introduction to all phases of model building, with an example drawn from the phytoplankton-zooplankton food chain in the North Sea. An energy flow simulation model of a pelagic ecosystem in the tropical ocean (95) dealt with the problem of modeling vertical zones in an upwelling system. The state of the system was determined by intensity and nutrient dynamics. Models of both upwelling and marine outfalls (97) illustrate the use of submodels of nutrients. Saila (86), in a model of the Georges Bank haddock fishery, was concerned with economic factors as they interact with the biological model to influence catch. A model of a lobster fishery was also discussed briefly. Paulik (80) published a useful review of the economic simulation models used for biological management and for bioeconomic purposes relevant to fisheries. It contained a selected bibliography of references to computer modeling and simulation.

The third group of management models, those dealing with insect pests, was difficult to condense into a small list because of the vast literature on models of insect pest population (see 105). Berryman & Pienaar (4) developed an argument for simulation as a tool for investigating the dynamics and management of insects, and Walters & Peterman (101) provided an example applied to the spruce budworm in New Brunswick. A useful exposition of methods is found in Watt (106).

GENERAL MODELS   Generalized models of ecosystem (no direct application to any existing community) were proposed for a resource population system (29), a predator-prey food chain (45), and simple three-species predator-prey and competition ecosystems (109–111). Menshutkin (54) proposed a general algorithm for developing mathematical models of aquatic biological systems as did Walters, Park & Koonce (102). Park et al (122) proposed a generalized lake ecosystem model (CLEAN) for which some documentation and output were obtained for several lakes studied in IBP programs (123). But only Bledsoe & Van Dyne (6) developed a compartment model (linear) simulating successional changes on an old field.

AQUATIC MODELS   A large number of simulation models describe aquatic ecosystems. Chen & Orlob (19) discussed a number of models ranging from the pollution-recovery system in Lake Washington to a hydrodynamic model, interfaced with an ecologic model of San Francisco Bay. Parker (70, 71) reviewed models of Kootenay Lake in British Columbia. He first concluded (70) that more knowledge of the nutrient cycling system was necessary, but later found (71) that phosphorus apparently did not limit the algae and that the length of the integration interval may have caused problems with the accuracy of the simulation predictions. Kremer & Nixon (121) developed a realistic model of the phytoplankton component of Narragansett Bay that employs the concept of maximum rates under optimum conditions. Because of stability problems, which necessitate a small integration interval, it is best used only for short-term simulations. Williams (116) based an aquatic model on the Cedar Bog Lake, Minnesota data as a class exercise to compare linear and nonlinear models, and Fagerstrom & Asell (25) used a model of methyl mercury movement to focus research effort on the ecology of aquatic food chains. This use was also one objective of the salt marsh model of Wiegert et al (115), which incidentally predicted that the *Spartina* marsh was a net producer, not importer, of organic carbon.

One of the most extensive aquatic models persuasively arguing the linear point of view was that of a cove in Lake Texoma (79). The opposite point of view, i.e. a philosophy of nonlinear, discontinuous yet realistic and stable models, was pursued in the detailed simulation model of a simple algal-fly-predator thermal spring community (108, 112).

FOREST MODELS   Perhaps because of their complexity and the relative difficulties of sampling, ecosystem simulation models of forests are scarce. The IBP deciduous and coniferous biome modeling projects will help remedy this deficiency. O'Neill (65) provided a general discussion of the work on the eastern deciduous forest with a summary table showing type of model, statistics, and sites. A general description of the modeling process with specific reference to a tulip poplar forest was published (84), as well as a detailed nonlinear model of a forested ecosystem and its response to temperature and light (120).

Overton (68, 69) discussed the modeling problems encountered in the coniferous biome and presented a general model for forested ecosystems. Botkin has been very active in developing simulation models of forests that deal with the dynamics of individual trees (8).

GRASSLAND MODELS   The grassland biome is another that has been the focus of much of the attention of IBP studies. Unfortunately, few of the models of this type of ecosystem have been published. A general description of the grassland research program, with modeling as a major objective, was described by Van Dyne (94). Connor, Brown & Trlica (21) published a description of a photosynthesis submodel of the shortgrass prairie. The latest detailed nonlinear version of the grassland simulation model (ELM) was discussed. Also unpublished was the earlier nonlinear version (PWNEE), developed by Bledsoe. Both the ELM and PWNEE models are described in *US IBP Grassland Biome Technical Reports,* numbers 64 and 156. (United States International Biological Program, Ecosystem Analysis Studies, Grassland Biome, Natural Resource Ecology Lab., Colorado State University, Fort Collins, Colorado 80521.) A linearized version of the PWNEE model has been published (75).

TUNDRA, DESERTLAND, SOIL-LITTER MODELS   Models of the tundra biomes are under development, but the only published works at the time of writing are those describing the effects of wind and radiation on primary production of the arctic tundra (55) and a 14-compartment carbon-flux simulation model (13). Unfortunately, the preprint copy of the latter, which I examined, did not contain the interaction equations, so I do not know what type of model was constructed. A history and description of the tundra modeling project can be obtained from the US IBP Tundra Biome Center *Tundra Biome Report No. 73–2.*

A general description of the simulation modeling efforts in the desert biome was given by Goodall (35, 37), but the extensive models themselves were unpublished. They are described in an extensive series of *US IBP Desert Biome Reports, Modelling Series 1–15.*

A number of specialized simulation models of the important soil-litter decomposition process have been constructed. Some of these are of course integral parts of larger overall ecosystem models. Others have been published by themselves. In the latter category are a generalized model of soil-litter decomposers (44), models of tundra (12, 14), a model of the soil-litter system in an eastern deciduous forest (120), a nitrogen flow model (96), and a bat guano ecosystem in a cave (51).

Cooper et al (22) presented a number of simulation models showing the effect of climatic change on a number of different biome models. Watt (107) has offered a critique of the entire biome modeling program and pointed out some of the major problems facing the reviewer when much of the material is in the "not for citation" literature.

CONCLUSIONS

One fact emerges clearly from the simulation models now in the literature. That is: the science (or art?) of simulation modeling, so concerned with the fluxes of matter-energy in ecosystems, is itself in a state of rapid flux. There seems to be a steady, if sometimes confused, search for new ways of approaching the modeling process and new sets of objectives for the models themselves. Simulation models could have advantages for prediction, for organizing research efforts, and as generators of new ecological hypotheses; all of these are being tried and discussed.

Modern high speed computers have made available to us a tool perhaps as important to ecology as the electron microscope was to cell biology. As with any new tool, much misuse and many distorted creations have followed in the wake of its first employment. The mistakes should wane and the important advances increase in number as our first clumsy efforts improve and as we employ ecological insights in the planning and construction of models, in other words, as we begin to let nature be our teacher. I believe that simulation models have great promise and that their use will advance ecology as a science. I hope this review will in some manner encourage this progress, thereby aiding the development of ecological theory and its application to pressing environmental problems.

ACKNOWLEDGMENTS

I am deeply indebted to Ms. Susan Bellinger for her capable handling of the long and tiresome job of finding and collating the numerous reprints and journals consulted and of checking the citations. Without her assistance this review could not have been written. I would also like to thank Hal Caswell and Jack Waide for critically reading an earlier draft of the manuscript, and for pointing out several errors.

Support for the preparation of this paper was received from NSF Grants GB 21255 and DES 72–01605 A02.

Literature Cited

1. Abrosov, N. S., Kovrov, B. G., Rerberg, M. S. 1972. Dynamics of the bacterial components of algo-bacterial coenosis (mathematical model). *Biofizika* 17(5):910–18
2. Allen, K. R. 1971. Relation between procedure and biomass. *J. Fish. Res. Board Can.* 28:1573–81
3. Armstrong, J. S. 1971. Modeling a grazing system. *Proc. Ecol. Soc. Aust.,* 6:194–202
4. Berryman, A. A., Pienaar, L. V. 1974. Simulation: a powerful method of investigating the dynamics and management of insect populations. *Environ. Entomol.* 3(2):199–207
5. Bledsoe, L. J. 1975. *Linear and Nonlinear Approaches for Ecosystem Dynamic Modeling. Systems Analysis and Simulation in Ecology,* ed. B. C. Patten, Vol. 4. New York: Academic. In press
6. Bledsoe, L. J., Van Dyne, G. M. 1971. A compartment model simulation of secondary succession. See Ref. 5, 1:479–511
7. Boling, R. H. Jr., Petersen, R. C., Cummins, K. W. 1975. Ecosystem modeling for small woodland streams. See Ref. 5, 3:183–204
8. Botkin, D. B., Miller, R. S. 1974. Complex ecosystems: models and predictions. *Am. Sci.* 62(4):448–53
9. Brennan, R. D., De Wit, C. T., Williams, W. A., Quattrin, E. V. 1970. The utility of a digital simulation language for ecological modeling. *Oecologia* 4:113–32
10. Bunnell, F. 1973. Decomposition: models and the real world. *Bull. Ecol. Res. Commun.* 17:401–15
11. Bunnell, F. 1973. Theological ecology or models and the real world. *Forest. Chron.* 49(4):1–5
12. Bunnell, F. L., Dowding, P. 1974. ABISKO—A Generalized Decomposition Model for Comparisons Between Tundra Sites. *Soil Organisms and Decomposition in Tundra,* ed. A. J. Holding et al. 227–47. Stockholm: Tundra Biome Steering Committee
13. Bunnell, F. L., Scoullar, K. A. 1975. ABISKO II—a computer simulation model of carbon flux in tundra ecosystems. *Structure and Functioning of Tundra Ecosystems,* ed. O. W. Heal, T. Rosswall. *Bull. Ecol. Res. Comm.* Swedish Nat. Sci. Res. Counc. In press
14. Bunnell, F. L., Tait, D. E. N. 1974. Mathematical simulation models of de-

composition processes. See Ref. 12, 207–25

15. Carnahan, B., Luther, H. A., Wilkes, J. O. 1969. *Applied Numerical Methods.* New York: John Wiley. 604 pp.

16. Caswell, H. 1972. A simulation study of a time lag population model. *J. Theor. Biol.* 34(3):419–39

17. Caswell, H. 1975. The validation prob lem. See Ref. 5, Vol. 4:In press

18. Caswell, H. Koenig, H. E., Resh, J. A., Ross, Q. E. 1972. An introduction to systems science for ecologists. See Ref. 5, 2:3–78

19. Chen, C. W., Orlob, G. T. 1975. Ecologic simulation for aquatic environments. See Ref. 5, 3:475–588

20. Clark, J. P. 1971. The second derivative and population modeling. *Ecology* 52(4):606–13

21. Connor, D. J., Brown, L. F., Trlica, M. J. 1974. Plant cover, light interception, and photosynthesis of short grass prairie. A functional model. *Photosynthetica* 8(1):18–27

22. Cooper, C. F. et al 1975. Simulation models of the effects of climatic change on natural ecosystems. *Proc. Clim. Impact Assessment Prog. Conf., 3rd,* Cambridge, Mass. In press

23. Curds, C. R. 1971. A computer-simulation study of predator-prey relationships in a single-stage continuous-culture system. *Water Res.* 5(10): 793–812

24. Davidson, R. S., Clymer, A. B. 1966. The desirability and applicability of simulating ecosystems. *Ann. NY Acad. Sci.* 182(3):790–94

25. Fagerstrom, T., Asell, B. 1973. Methyl mercury accumulation in an aquatic food chain: a model and some implications for research planning. *Ambio* 2(5): 164–71

26. Forrester, J. W. 1961. *Industrial Dynamics.* Cambridge, Mass.: MIT Press. 464 pp.

27. Forrester, J. W. 1971. *World Dynamics.* Cambridge, Mass.: Wright-Allen 142 pp.

28. Fraleigh, P. C., Wiegert, R. G. 1975. A model explaining successional change in standing crop of thermal blue-green algae. *Ecology* In press

29. Gallopin, G. C. 1971. A generalized model of a resource-population system. *Oecologia* 7:382–413

30. Garfinkel, D. 1962. Digital computer simulation of ecological systems. *Nature* 194(4381):856–57

31. Garfinkel, D., MacArthur, R. H., Sack, R. 1964. Computer simulation and analysis of simple ecological systems. *Ann. NY Acad. Sci.* (115)2:943–51

32. Goodall, D. W. 1969. Simulating the grazing situation. *Biomathematics,* Vol. 1: *Concepts and Models of Biomathematics,* ed. F. Heinmets, 211–36. New York: Dekker

33. Goodall, D. W. 1970. Use of computers in the grazing management of semi arid lands. *Proc. Int. Grassland Congr. XI,* 917–21. St. Lucia: Univ. Queensland

34. Goodall, D. W. 1972. Building and testing ecosystem models. *Mathematical Models in Ecology,* ed. J. N. R. Jeffers, 173–94. Oxford: Blackwell

35. Goodall, D. W. 1973. *Ecosystem Simulation in the US/IBP Desert Biome.* Presented at Summer Computer Simulation Conference, Montreal

36. Goodall, D. W. 1974. Problems of scale and detail in ecological modelling. *J. Environ. Manage.* 2:149–57

37. Goodall, D. W. 1975. Ecosystem modeling in the desert biome. See Ref. 5, 3:73–94

38. Holling, C. S. 1966. The strategy of building models of complex ecological systems. *Systems Analysis in Ecology,* ed. K. E. F. Watt, 195–214. New York: Academic

39. Holling, C. S., Ewing, S. 1971. Blind man's bluff: exploring the response space generated by realistic ecological simulation models. *Statistical Ecology,* ed. G. P. Patil, E. C. Pielou, W. F. Waters, Vol. 2, *Sampling and Modelling Biological Populations and Population Dynamics.* 207–29. Univ. Park, Pa.: Pa. State Univ. Press

40. Innis, G. S. 1972. The second derivative and population modeling: another view. *Ecology* 53(4):720–23

41. Innis, G. S. 1972. Simulation of ill-defined systems: some problems and progress. *Simulation* 19(6)(Suppl.) *Simulation Today* 33–36

42. Innis. G. S. 1975. Role of total systems models in the grassland biome study. See Ref. 5, 3:13–47

43. Jeffers, J. N. R. 1972. The challenge of modern mathematics to the ecologist. *Mathematical Models in Ecology,* ed. J. N. R. Jeffers, 1–11. Oxford: Blackwell 56(3):656–64

44. Kowal, N. E. 1971. A rationale for modeling dynamic ecological systems. See Ref. 5, 1:123–94

45. Lassiter, R. R., Hayne, D. W. 1971. A finite difference model for simulation of dynamic processes in ecosystems. See Ref. 5, 1:367–440

46. Levins, R. 1966. The strategy of model building in population biology. *Am. Sci.* 54(4):421–31

47. Margalef, R. 1973. Some critical remarks on the usual approaches to ecological modelling. *Invest. Pesq.* 37(3): 621–40

48. Martin, G. D. 1972. Optimal Control of an oil refinery waste treatment facility: a total ecosystem approach. *Cen. Syst. Sci., Okla. State Univ. Res. Rep. 73–10*

49. May, R. M. 1971. Stability in multispecies community models. *Math. Biosci.* 12(1–2):59–79

50. May, R. M. 1973. *Stability and Complexity in Model Ecosystems.* Princeton, NJ: Princeton Univ. Press. 235 pp.

51. Mazanov, A. 1973. A model for a class of biological systems. *J. Environ. Manage.* 1:229–38

52. McLeod, J. 1975. Simulating the world ecosystem. See Ref. 5, Vol. 4: In press

53. Meadows, D. H., Meadows, P. L., Randers, J., Behrens, W. W. II. 1972. *The Limits to Growth.* New York: Universe. 205 pp.

54. Menshutkin, V. V. 1974. Theoretical basis for mathematical modelling of aquatic ecological systems. *Zh. Obshch. Biol.* 35(1):34–42

55. Miller, P. C., Tieszen, L. 1972. A preliminary model of processes affecting primary production in the arctic tundra. *Arctic. Alp. Res.* 4(1):1–18

56. Milner, C. 1972. The use of computer simulation in conservation management. See Ref. 43, 249–75

57. Mobley, C. D. 1973. A systematic approach to ecosystems analysis. *J. Theor. Biol.* 41:119–36

58. Morales, R. 1975. A philosophical approach to mathematical approaches to ecology. *Ecosystem Analysis and Prediction, Proc. SIAM-SIMS Conf., Alta, Utah,* ed. S. A. Levin. 334–37. *Philadelphia: Soc. Ind. Appl. Math.*

59. Odum, H. T. 1960. Ecological potential and analogue circuits for the ecosystem. *Am. Sci.* 48(1):1–8

60. Odum, H. T. 1972. An energy circuit language for ecological and social systems: its physical basis. See Ref. 5, 2:139–211

61. Odum, H. T. 1975. Macroscopic minimodels of man and nature. See Ref. 5, Vol: 4: In press

62. O'Neill, R. V. 1969. Indirect estimation of energy fluxes in animal food webs. *J. Theor. Biol.* 22:284–90

63. O'Neill, R. V. 1971. A stochastic model of energy flow in predator compartments of an ecosystem. *Statistical Ecology,* ed. G. P. Patil, E. C. Pielou, W. E. Waters, Vol. 3, *Populations, Ecosystems, and Systems Analysis,* 3:107–21. Univ. Park, Pa.: Pa. State Univ. Press

64. O'Neill, R. V. 1971. Systems approaches to the study of forest floor arthropods. See Ref. 5, 1:441–77

65. O'Neill, R. V. 1975. Modeling in the eastern deciduous forest biome. See Ref. 5, 3:49–72

66. O'Neill, R. V. 1975. Dynamic ecosystem models: progress and challenges. See Ref. 58, 280–96

67. Oster, G. 1975. Stochastic behavior of deterministic models. See Ref. 58, 24–37

68. Overton, W. S. 1972. Toward a general model structure for a forest ecosystem. *Proc. Res. Conif. For. Ecosysts. Symp.,* ed. J. Franklin, L. J. Dempster, R. A. Waring. Portand, Ore.: Pac. NW For. Range Exp. Sta. pp. 37–47

69. Overton, W. S. 1975. The ecosystem modeling approach in the coniferous forest biome. See Ref. 5, 3:117–38

70. Parker, R. A. 1968. Simulation of an aquatic ecosystem. *Biometrics* 24: 803–21

71. Parker, R. A. 1973. Some problems associated with computer simulation of an ecological system. *Mathematical Theory of the Dynamics of Biological Populations,* ed. M. S. Bartlett, R. W. Hiorns, 269–88. London: Academic

72. Patten, B. C. 1966. Systems ecology: a course sequence in mathematical ecology. *Bioscience* 9:593–98

73. Patten, B. C. 1971. A primer for ecological modeling and simulation with analog and digital computers. See Ref. 5, 1:3–121

74. Patten, B. C., ed. 1971. See Ref. 5, Vol. 1. 607 pp.

75. Patten, B. C. 1972. A simulation of the shortgrass prairie ecosystem. *Simulation* 19:177–86

76. Patten, B. C., ed. 1972. See Ref. 5, Vol. 2. 592 pp.

77. Patten, B. C. 1975. Ecosystem linearization: an evolutionary design problem. See Ref. 58, 182–201

78. Patten, B. C., ed. 1975. See Ref. 5, Vol. 3. 601 pp.

79. Patten, B. C. et al 1975. Total ecosystem model for a cove in Lake Texoma. See Ref. 5, 3:205–421

80. Paulik, G. J. 1972. Digital simulation modeling in resource management and

the training of applied ecologists. See Ref. 5, 2:373–418

81. Radford, P. J. 1972. The simulation language as an aid to ecological modelling. See Ref. 43, 277–95

82. Raines, G. E., Bloom, S. G., Levin, A. A. 1969. Ecological models applied to radionuclide transfer in tropical ecosystems. *Bioscience* 19(12):1086–91

83. Reddingius, J. 1970. Models as research tools. *Proc. Adv. Study Inst. Dyn. Numbers Pop.*, 64–76. Oosterbeek

84. Reichle, D. E., O'Neill, R. V., Kaye, S. V., Sollins, P., Booth, R. S. 1973. Systems analysis as applied to modeling ecological processes. *Oikos* 24(3): 337–43

85. Ross, G. G. 1973. A population model for limited food competition *J. Theor. Biol.* 42(2):333–47

86. Saila, S. B. 1972. Systems analysis applied to some fisheries problems. See Ref. 5, 2:331–72

87. Scudo, F. M. 1971 Vito Volterra and theoretical ecology. *Theor. Pop. Biol.* 2(1):1–23

88. Skellam, J. G. 1972. Some philosophical aspects of mathematical modelling in empirical science with special reference to ecology. See Ref. 43, 13–28

89. Slobodkin, L. B. 1975 Comments from a biologist to a mathematician. See Ref. 58, 318–29

90. Smith, F. E. 1969. Effects of enrichment in mathematical models. *Eutrophication: Causes, Consequences, Correctives. Proc. Int. Symp. Eutrophication,* 631–45. Washington DC: Nat. Acad. Sci.

91. Smith, F. E. 1972. Spatial heterogeneity, stability, and diversity in ecosystems. *Trans. Conn. Acad. Arts Sci.* 44:307–35

92. Steele, J. H. 1974. *The Structure of Marine Ecosystems.* Oxford: Blackwell, 128 pp.

93. Swartzman, G. L., Singh, J. S. 1974. A dynamic programming approach to optimal grazing strategies using a succession model for a tropical grassland. *J. Appl. Ecol.* 11:537–48

94. Van Dyne, G. M. 1972. Organization and management of an integrated ecological research program—with special emphasis on systems analysis, universities, and scientific cooperation. See Ref. 43, 111–72

95. Vinogradov, M. E., Menshutkin, V. V., Shushkina, E. A. 1972. On mathematical simulation of a pelagic ecosystem in tropical waters of the ocean. *Mar. Biol.* 16(4):261–68

96. Visser, S. A., Witkamp, M., Dahlman, R. C. 1973. Flow of microbially fixed nitrogen in a model ecosystem. *Plant Soil* 38(1):1–8

97. Walsh, J. J., Dugdale, R. C. 1972. Nutrient submodels and simulation models of phytoplankton production in the sea. *Nutrients in Natural Waters,* ed. H. E. Allen, J. R. Kramer, 171–91. New York: Wiley

98. Walters, C. J. 1975. Dynamic models and evolutionary strategies. See Ref. 58, 68–82

99. Walters, C. J., Bunnell, F. 1971. A computer management game of land use in British Columbia. *J. Wildl. Manage.* 35(4):644–57

100. Walters, C. J., Gross, J. E. 1972. Development of big game management plans through simulation modeling. *J. Wildl. Manage.* 36(1):119–28

101. Walters, C. J., Peterman, R. M. 1974. A systems approach to the dynamics of spruce budworm in New Brunswick. *Quaest. Entomol.* 10:177–86

102. Waters, C. J., Park, R. A., Koonce, J. F. 1975. Dynamic models of aquatic systems. *Productivity of Freshwater Ecosystems,* ed. E. D. Le Cren, Chap. 10. Oxford: Blackwell. In press

103. Watt, K. E. F. 1962. Use of mathematics in population ecology. *Ann. Rev. Entomol.* 7:243–60

104. Watt, K. E. F. 1966. The nature of systems analysis. *Systems Analysis in Ecology,* ed. K. E. F. Watt, 1–14. New York: Academic

105. Watt, K. E. F. 1968. *Ecology and Resource Management.* New York: McGraw-Hill. 450 pp.

106. Watt, K. E. F. 1969. Methods of developing large-scale systems models. *Forest Insect Population Dynamics, USDA For. Serv. Res. Pap. NE–125,* 35–51

107. Watt, K. E. F. 1975. Critique and comparison of biome ecosystem modelling. See Ref. 5, 3:139–52

108. Wiegert, R. G. 1973. A general ecological model and its use in simulating algal-fly energetics in a thermal spring community. *Insects: Studies in Population Management,* ed. P. W. Geier, L. R. Clark, D. J. Anderson, H. A. Nix. (Mem. 1), 85–102. Canberra: Ecol. Soc. Aust.

109. Wiegert, R. G. 1974. Competition: a theory based on realistic, general equations of population growth. *Science* 185:539–41

110. Wiegert, R. G. 1974. A general mathematical representation of ecological flux

processes: description and use in ecosystem models. *Proc. Ann. SE Symp. Syst. Theory, 6th,* ed. R. Kinney, Session TP-2. Baton Rouge, La. 15 pp.

111. Wiegert, R. G. 1975. Mathematical representation of ecological interactions. See Ref. 58, 43–54

112. Wiegert, R. G. 1975. Simulation modeling of the algal-fly components of a thermal ecosystem: effects of spatial heterogeneity, time delays and model condensation. See Ref. 5, 3:157–81

113. Wiegert, R. G., Owen, D. F. 1971. Trophic structure, available resources and population density in terrestrial vs. aquatic ecosystems. *J. Theor. Biol.* 30:69–81

114. Wiegert, R. G., Wetzel, R. L. 1974. The effects of numerical integration technique on the simulation of carbon flow in a Georgia salt marsh. *Proc. Summer Computer Simulation Conf., Houston* 2:575–77

115. Wiegert, R. G. et al 1975. A preliminary ecosystem model of coastal Georgia *Spartina* marsh. *Recent Advances in Estuarine Research,* ed. R. Reimold. In press

116. Williams, R. B. 1971. Computer simulation of energy flow in Cedar Bog Lake, Minnesota based on the classical studies of Lindeman. See Ref. 5, 1:543–82

117. Williams, R. B. 1972. Steady-state equilibriums in simple nonlinear food webs. See Ref. 5, 2:213–40

118. Woodwell, G. M. 1975. The threshold problem in ecosystems. See Ref. 58, 9–23

119. Patten, B. C. 1975. See Ref. 5, Vol. 4: In press

120. Shugart, H. H., Goldstein, R. A., O'Neill, R. V., Mankin, J. B. 1974. TEEM: a terrestrial ecosystem energy model for forests. *Oecol. Plant.* 9(3): 231–64

121. Kremer, J. N., Nixon, S. W. 1975. An ecological simulation model of Narragansett Bay—the plankton community. *Recent Advances in Estuarine Research.* ed. R. Reimold. In press

122. Park, R. A., et al. 1974. A generalized model for simulating lake ecosystems. *Simulation* 23(2):33–50

123. Scavia, D., Bloomfield, J. A., Nagy, J., Park, R. A. 1974. Documentation of CLEAN: a generalized model for simulating the open-water ecosystems of lakes. *Simulation* 23(2):51–156

124. Olson, J. S. 1963. Analog computer model for movement of nuclides through ecosystems *Radioecology.* ed. V. Schultz and A. Klement, 121–25. New York: Reinhold

MODES OF ANIMAL
SPECIATION

♦4097

Guy L. Bush

Department of Zoology, University of Texas, Austin, Texas 78712

INTRODUCTION

The study of speciation is an ad hoc science. No one has yet observed the development from beginning to end of a new plant or animal species in nature. The fossil record, so useful in painting a broad picture of long-term evolutionary trends, lacks the temporal and biological sensitivity to dissect the multitude of processes involved in speciation. Population genetics, the extreme opposite of paleontology in its ability to measure and model gene frequency changes and stability, has also failed to provide much insight into the genetics of speciation (110).

At the molecular level, proteins used in the current flush of allozyme studies (110, 158, 174) have almost no direct bearing on speciation. Estimates of genetic differentiation based on allozyme data alone (i.e. 147, 189), therefore, provide only crude indexes of genetic divergence between species and so-called semispecies, most of which could easily be regarded as species (178). Moreover, animals used in these studies have often long since passed through the critical "point of no return," at which hybridization will no longer result in fusion of two races (24, 29, 110). They tell us little about how many or what kind of genes are directly implicated in speciation itself. Thus we are still forced to reconstruct events surrounding speciation from observations on extant species, a task that has fallen almost exclusively to the naturalist (see 44, 69, 72, 124 for historical aspects).

Speciation is ultimately an adaptive process that involves establishment of intrinsic barriers to gene flow between closely related populations by development of reproductive isolating mechanisms. A study of speciation is, to a considerable extent, a study of the genetics and evolution of reproductive isolating mechanisms. Most evolutionists therefore generally accept as a working definition the *biological species concept*: species represent groups of interbreeding natural populations reproductively isolated from other such groups (125). Although it has its weak points (44), alternatives (e.g. 65, 164) impose more problems than they solve.

Selection for reproductive isolation between closely related populations is fundamentally different from the process involved in local adaptation (28, 109, 181). The latter entails only minor genetic adjustments, whereas speciation frequently involves a reorganization of some crucial component in the genetic system that results in a

339

quantum step toward the origin of interspecific differences (109, 181). Although the genetic system embraces a constellation of biological properties, those most pertinent to modes of speciation to be discussed are presented in Table 1.

In this review, I show that the ways in which various groups of animals differ in these properties determine, to a great extent, the mode of speciation they are most likely to follow. I also reexamine the conventional wisdom that new species of sexually reproducing animals arise only after a period of complete geographic isolation and gradual genetic change, a viewpoint long held by most evolutionary biologists. Major advances in our understanding of the relationship between the structure and function of genetic systems and mechanisms of speciation in different animal and plant groups now make it almost impossible to accept the universality of allopatric speciation.

MODES OF SPECIATION AND THE GENETIC SYSTEM

Almost all evolutionary biologists agree that, in order to speciate, gene flow between diverging populations must be reduced to a level where "foreign" genes entering the population as a result of hybridization can be eliminated by natural selection. Disagreement arises, however, on whether this is accomplished by strictly extrinsic barriers (i.e. geographic isolation) or by intrinsic barriers arising within a population as certain members bearing unique genotypes shift into a new environment while still in contact with the parental population.

Three broad patterns of speciation involving either allopatric, parapatric, or sympatric development of reproductive isolation have been invoked to explain the origin of new species of sexually reproducing animals (44, 69, 124, 179, 181, 182). Others have been suggested—saltational or catastrophic speciation (66, 109); centrifugal speciation (17, 20, 40); stasipatric speciation (179); cascading speciation, etc (162); allelic contracomplementation (141)—but they all appear to represent only modifications of the three basic types.

Properties of the genetic system appear to restrict each group of animals to predominantly one mode of speciation. Although I recognize that some attributes may be relatively unimportant for certain animals, the combinations provided in Table 1 are those most frequently encountered in animal speciation by one of the modes discussed below.

ALLOPATRIC SPECIATION (TYPES Ia and Ib)

Without a doubt, allopatric (geographic) speciation is very common in almost all groups of sexually reproducing animals. Examples of ecogeographic races existing in various states of geographic speciation are so common that it is hardly necessary to discuss them here. Dobzhansky (44, 45), Grant (69, 72), Mayr (124), and others treat the subject thoroughly; here I touch only on features relevant to later topics.

Allopatric speciation, regarded by many evolutionary biologists as the major means of generating new species in sexually reproducing animals, can occur in basically two different ways (Type Ia and Type Ib, Table 1).

Type Ia: Speciation by Subdivision

In this classic model, a widely distributed species becomes subdivided into two or more relatively large populations (Figure 1). After gene flow is interrupted by some extrinsic barrier, genetic differences begin to accumulate between isolates as each population responds to its own array of selective forces and tracks its ever-changing environment. Any barriers to gene exchange that develop are usually the result of these fortuitous adaptive genetic changes (44, 72, 124, 128, 178). Populations will not fuse once they reestablish contact if enough genetic differences have accumulated during isolation to ensure that hybrids between them are of such low fitness that they are strongly selected against or eliminated. Reproductive isolation is therefore frequently incomplete between closely related allopatric populations, although genetical analyses and ecological studies necessary to demonstrate this quantitatively have not been made (110, but see 50, 51).

Premating reproductive isolating mechanisms are usually perfected first in the zone of contact so that hybrids eventually become rare or absent altogether (13, 44, 115, 128). This final stage in the speciation process is called the *Wallace effect* (72, 128). Although examples in nature are difficult to substantiate (68, 177), it has been covincingly demonstrated in frogs (111) and in lizards (81, 82). The process has been reproduced experimentally using various mating and selection schemes on sibling species (or so-called semispecies) of *Drosophila* (48, 70, 71, 93, 99, 101). It can occur very rapidly (e.g. 90), or occasionally may establish a more or less permanent narrow hybrid belt (124). Because this last stage of speciation requires accumulation of many genetic changes during the period of complete geographic isolation, allopatric speciation by subdivision is necessarily a relatively long-term process.

Speciation through subdivision is commonly regarded as widespread in animals and certain large animal species of high vagility that are typically K-selected (4), as defined by several authors (62, 115, 137, 138), are probably limited to this mode of speciation. This is apparent in vertebrates, such as the Carnivora, whose reproductive strategies are better known and hence more easily correlated with speciation.

Chromosome evolution, a good indicator of rapid speciation (see below), is minimal in such wide-ranging groups (181). All cats (*Felis*) in the northern hemisphere are $2n = 38$ and true dogs (*Canis*) all have $2n = 78$. In both genera, speciation appears to have been slow relative to that of foxes, which have a wide range of chromosome numbers ($2n = 38$, 40, 64, and 78 in 4 species). The reason for karyotype stability in dogs and not in foxes is related to their ecology and social structure. Most large carnivores, for instance, must space themselves out to ensure an adequate food supply; as a result, selection favors dispersal accomplished by defense of territory, by formation of small closed social groups with annual juvenile dispersal, or by pronounced natural avoidance. Diurnality and utilization of an open habitat in many species also favor group formation (*Canis, Panthera*, etc) and is accompanied by high mobility with extensive home ranges. Foxes are an exception in that they do not form cohesive units beyond a permanent pair association. Their home range is thus relatively much more restricted than canids, which often form

Table 1 Some correlates in the properties of the genetic systems and other attributes of species undergoing allopatric, parapatric, and sympatric speciation. These relationships pertain only to individuals within related taxonomic groups or ecological guilds (i.e. large carnivores or a guild of parasitic insects)

	Allopatric		Parapatric	Sympatric
	Type Ia	Type Ib	Type II	Type III
Reproductive strategy	Low reproductive rate; late sexual maturity, few offspring, long life span, high competitive ability (K-strategist)	High reproductive rate; early sexual maturity, large number of offspring, short life span, low competitive ability (r-strategist)	Same as Type Ib (r-strategist)	Same as Type Ib (r-strategist, but see text)
Vagility	High	High	Low	High (but niche limited) to low
Initial population size	Large	Small	Small	Variable
Ecological amplitude	Mostly broad. Environment utilization fine grained, i.e. generalized feeding habits	Type Ia or Type II	Mostly narrow. Environment utilization coarse grained i.e. specialized feeding habits	As in Type II
Change in niche	Speciation involves no radical shift to new niche.	Type Ia or Type II	Speciation accompanied by shift to new niche.	As in Type II
Mate selection	Not closely linked with niche selection	Same as Type Ia	As in Type Ia or III	Closely linked to niche selection
Breeding system	Normally outbreeding	Normally inbreeding or facultatively inbreeding	As in Type Ib	As in Type Ib
Selection	Heteroselection level high	Homoselection level high	As in Type Ib	As in Type Ib
Chromosome rearrangements	Little or no chromosome evolution; chromosome rearrangements if present not associated with speciation	Chromosome rearrangements may or may not be associated with speciation	Chromosome rearrangements frequently associated with speciation ("negative heterosis")	As in Type Ib

Table 1 (*Continued*)

	Allopatric		Parapatric	Sympatric
	Type Ia	Type Ib	Type II	Type II
Genetic changes	Speciation results from cumulative adaptive changes at structural and regulatory genes	Speciation may or may not be associated with radical alterations in structural and regulatory genes	Speciation frequently associated with radical alterations of regulatory systems, but minor structural gene mutation	Speciation frequently results from only minor changes in both regulatory and structural genes
Nongenetic DNA	Relatively large amounts of nongenetic DNA; genome size large	Comparatively small amounts of nongenetic DNA; genome size small	Same as Type Ib	Same as Type Ib
Evolution of reproductive isolating mechanisms (RIM)	Postmating RIM *fortuitous* result of long term adaptive genetic changes during isolation. Premating RIM perfected in and limited to hybrid zone	Same as Type I but genetic changes occur rapidly; RIM widespread	Premating and/or post-mating RIM directly selected for during or soon after shift to new niche and widespread	Premating RIM directly selected for and *precedes* shift to new niche
Gene flow	None	None	Some initially	Some initially
Distribution of semispecies and sibling species	Usually allopatric or parapatric; niche requirements similar; interspecific competition high in zone of contact	Allopatric or broadly sympatric; niche requirements vary	Usually parapatric Niche requirements usually somewhat different; interspecific competition moderate	Usually sympatric Niche requirements usually very different; no interspecific competition
Speciation rates	Slow	Rapid, sometimes passing through critical stages which guarantee speciation within a few generations	As in Type Ib	As in Type Ib

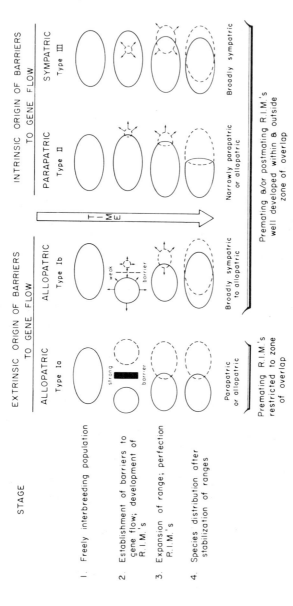

Figure 1 Diagrammatic representation of three basic modes of speciation. Post-mating reproductive isolation (RIM's) arise after geographic isolation in Types Ia and Ib situations; pre-mating RIM's are perfected once contact is reestablished. In Type Ib situations, populations of daughter species may still be small at time of contact. Pre- and postmating RIM's are therefore fully developed over the entire range of the daughter species. Niche preferences may also be somewhat different between recently evolved Type Ib species that may permit broad sympatry. Type Ia species are likely to share very similar ecologies and are rarely broadly sympatric; premating reproductive isolating mechanisms are usually restricted to or just beyond the zone of overlap. In parapatric and sympatric speciation pre- and postmating RIM's are developed *during* or *before* a shift to a new niche. In both modes, daughter species are never geographically isolated from parental species during speciation. Consequently, reproductive isolating mechanisms are fully developed throughout the range of daughter species. Sympatric speciation normally involves radical shifts in niche preferences. Competition is greatly reduced between the old and new species, which inevitably remain broadly sympatric. The ecologies of species that have undergone parapatric speciation are generally more similar. This fact, coupled with their low vagility, usually results in very narrow zones of overlap. See text for details.

wide-ranging packs (53). Other small carnivores have similar reproductive strategies and dispersal characteristics (118). The chance of a chromosome rearrangement becoming fixed in a small, isolated population founded by a single pair of foxes is considerably increased over wide-ranging carnivores, such as cats and dogs and constitutes a different kind of allopatric speciation (Type Ib).

Felids and canids represent the extreme opposite of the genetic system encountered in small, short-lived fossorial species of rodents, many of which have more typically r-selected strategies of reproduction. As a group they are the most diverse and dominant of all vertebrates (161) and are represented by many cryptic or sibling species and races distinguishable only on the basis of their karyotypes (181).

Many birds (157), certain reptiles (138) and amphibians (18), and most large marine (119), lacustrine, and riverine fish (113) would also share similar biological attributes with the large mammals. Most birds are highly vagile, have extensive ranges, and show remarkably little variation in chromosome number, form, or DNA content (8, 181). Speciation in birds is almost invariably by geographic speciation, with many species resulting from populations isolated in refugia established by climatic events of the Pleistocene (157), although founder events may also be important in some groups (124, 125). One case of sympatric speciation has recently been suggested in Darwin's finches (59), but the evidence is inconclusive.

From an ecological standpoint, many Type Ia species treat the environment in a fine-grained manner (107, 159). Their high vagility and relatively broad niche requirements ensure that suitable resources will be encountered frequently. When the distributions of such wide-ranging species are broken up into two or more populations, environmental conditions within each refugium remain similar (124, 125). Unless population size is greatly restricted, relatively high heteroselection levels (selection for heterozygosity) will be maintained.

Genetic changes in such species will occur slowly as such tightly integrated coadapted polymorphic systems exert considerable speciational inertia (27, 124). These populations rarely tolerate the levels of inbreeding necessary to go through a genetic bottleneck typically experienced in Type Ib speciation events. Many of the same coadapted gene complexes will be retained in large isolated populations. Most genetic changes therefore probably occur in structural genes rather than through any major reorganization of the regulatory system.

This view is supported by the fact that Type Ia mammalian species have undergone much less chromosomal evolution and also have retained relatively larger amounts of nongenetic DNA than have rapidly evolving rodents (127). Recent investigations have found that large amounts of nongenetic DNA prolong cell generation and life, minimize chromosomal errors during cell division, protect genes from mutagenic agents, and are essential for structural organization, functional relationships, and regulation of the cell nucleus (8, 56, 176, 181, 184, 185, 187, 188). DNA also plays a special inhibitory role on speciation. Small genome size is associated with rapidly evolving, specialized, short-lived groups (168). Accumulation of nongenetic DNA may actually be a form of evolutionary senescence that, unless reversed, might eventually contribute significantly to extinction.

Type Ib: Speciation by the Founder Effect

Probably a much more common form of geographic speciation occurs by way of the establishment of a new colony by a small number of founders (Figure 1) (121, 123, 124). What kind of species are most likely to provide founders that can survive the rigors of colonization and go on to evolve into new species? Carson (29, 30), who has studied the pattern of rapid speciation in Hawaiian *Drosophila* (600+ spp.), provided rather convincing evidence that small propagules—sometimes represented by only a single fertile female—coming from small semi-isolated peripheral populations, are most likely to be preadapted as colonists. These populations are usually adapted to some degree of inbreeding and are often products of a population "flush," a period of rapid population increase wherein natural selection is temporarily relaxed (26, 30). Such populations usually exist in peripheral habitats and are under relatively intense r-selection regimes (115). By existing at the periphery, they also are most likely to become permanently isolated from the parent population or to be close to an unexploited area suitable for invasion. Homoselection (selection for homozygosity) may also be more common in marginal environments, as less of the genome is tied up in elaborate coadapted complexes characteristic of large, stable central populations with high heterozygosity (29, 30, 44, 115, 124). Speciation under these conditions may occur rapidly (27, 45, 103, 123, 124, 157).

In endemic Hawaiian *Drosophila,* there has been little chromosome evolution and no evidence for a major genetic revolution during speciation by founder effects. A number of related species are homosequential (i.e. banding patterns of polytene chromosomes are identical) (29), and what little variation does exist involves paracentric inversions that are normally not involved in speciation. Only four of the many species examined cytologically have undergone chromosome fusions, although many have different levels and distributions of heterochromatin (181). Regulatory rather than structural gene changes may therefore be important in speciation in these animals. Allozyme variation suggests that species differ in gene frequency rather than allele form (88, 151). What little change has occurred involved enzymes like esterases and phosphotases, which are affected by changes in substrates and probably have no direct bearing on reproductive isolation separating closely related species (R. H. Richardson, personal communication). There has also been very little ecological divergence (29).

The major differences are in premating isolating mechanisms (28, 29, 37, 38). Carson proposed a plausible explanation for these differences, involving isolation of a fertile female on an unexploited kipuka, or island, followed by random genetic drift and abrupt nonadaptive changes in the gene pool following the founder event; Richardson (146) and Craddock (38) have suggested alternative means for some species. As the population size increases, normal interdemic processes of mutation, recombination, and selection take over.

In other invertebrate and vertebrate groups, chromosome evolution may be an important factor in speciation by the founder effect. Species of mammals of moderately low vagility that are subsocial or social frequently form small cohesive closed bands or small family groups (primates), harems, herds (some ungulates), or family groups with permanent pair bonds (e.g. foxes). Others are solitary, flightless, or cave-dwelling with homing behavior similar to that of bats (40). Most have fairly restricted home ranges, and a single pair or small group existing near the species' border may frequently become isolated from the parental population for a time. Many of these temporarily isolated populations either become extinct or are reabsorbed into the parental population. If the isolated population, however, includes an individual (possibly an α-male in a small herd, or one member of a pair) bearing a major adaptive chromosome rearrangement, conditions are ideal for its rapid fixation in homozygous condition within the small isolated group (181).

Major chromosome rearrangements (i.e. fusions, fissions, whole arm translocations, and pericentric inversions) provide one way whereby rapid reorganization of regulatory mechanisms can take place; their importance in speciation has been demonstrated experimentally in animals (102) and plants (72). They may cause enormous changes in the developmental process without a genic change (8, 19, 91, 92, 185), and may permit a population homozygous for the new karyotype to penetrate and exploit a new and previously unsuitable habitat at the periphery of a species' range. These rearrangements therefore appear to play a special role in speciation, as reorganization of control systems provides, more than genic changes, major innovation and novelty. Such rearrangements are essentially the systematic mutations, or macromutations, of Goldschmidt (66).

Some groups with highly variable karyotypes, such as certain carnivores (foxes, $2n = 38-78$), artiodactyles (horses, $2n = 44-64$; pigs, $2n = 30-38$; deer, $2n = 6-68$), and certain primate groups, including man ($2n = 20-80$) appear to fit this Type Ib mode of allopatric speciation. McFee et al (116), for instance, reported a case of Robertsonian fusion in a semi-isolated herd of wild pigs in Tennessee, introduced from Europe in 1912. The Tennessee populations have $2n = 36-37$, and the wild and domesticated pig have $2n = 38$. These three orders also represent the most widespread and successful groups of large mammals (161, 169).

In Type Ib speciation, incipient reproductive isolation arises fortuitously, just as in the Type Ia case, after the founders have established themselves in a new territory. If chromosome rearrangements are involved, as they frequently appear to be, then some postmating reproductive isolation will occur with the fixation of the new arrangement. There is, however, no direct selection for intrinsic barriers to gene flow, as hybrids are never produced between the parent and daughter populations.

If, as in the case of the homosequential Hawaiian *Drosophila*, no chromosomal changes are involved, rapid development of reproductive isolation is more difficult to explain. Carson (29, 30) suggested that this process is accomplished as a by-product of a "genetic revolution," although in this case it appears to be more of a local disturbance than a revolution. In an unsaturated island environment, minor genetic changes may permit new founder populations to exploit unoccupied niches and, in turn, promote some form of reproductive isolation. Perfection of postmating reproductive isolating mechanisms and selection for premating mechanisms in Type Ib speciation, as in Type Ia, will generally occur only if and when the sister populations reestablish contact (the Wallace effect). In Type Ib speciation, this may occur rapidly, as the new species is likely to reestablish contact with the parent population within a short time. The budded-off species will therefore carry fully developed reproductive isolating mechanisms with it as it expands its range into unoccupied territory (178), thus possibly accounting for the frequent lack of evidence for the Wallace effect [i.e. character displacement of Brown & Wilson (21)] in natural populations.

PARAPATRIC SPECIATION (TYPE II)

Parapatric speciation occurs whenever species evolve as contiguous populations in a continuous cline. White (179, 181, 182), who called this *stasipatric speciation,* and Murray (128) have offered interesting evaluations of the criteria necessary for parapatric speciation. Other authors have recently analyzed specific examples (34, 40, 87, 130, 131, 135, 156, 170, 175, 179, 180, 182) or presented theoretical models and experimental evidence (33, 54, 75, 140, 149, 150). Parapatric speciation superficially resembles speciation by the founder effect, but differs in three major ways: 1. no spatial isolation is required during speciation, 2. the level of vagility in the animals undergoing parapatric speciation is exceptionally low, and 3. reproductive isolating mechanisms arise by selection simultaneously with the penetration and exploitation of a new habitat by genetically unique individuals.

In general, all species implicated in parapatric speciation possess the combination of properties listed in Table 1. They are typically r-selected species existing in many small to medium sized semi-isolated peripheral populations, and thus adapted to some inbreeding. Although chromosome rearrangements frequently are responsible for initiation of speciation by altering major regulatory pathways, little or no genic differentiation occurs even after speciation has been completed (131, 132, 160). Populations also have relatively low levels of heterozygosity and are normally under regimes of homoselection (131).

Parapatric divergence and speciation are always associated with animals and plants of low vagility (72, 109, 128, 179, 181, 182). Adjacent populations in new habitats are established as an animal acquires the genetic tools to enter and exploit a new environment [Ludwig effect (114)]. No long-range dispersal takes place, and diverging populations are in constant contact. They are therefore never strictly geographically isolated from one another. Although some gene flow occurs initially only at the borders between adjacent populations, introgressing genes may penetrate no more than a few feet as observed in some natural populations of morabine grasshoppers (179, 181), snails (33, 34), and plants (3). Furthermore, a growing body of evidence suggests that effects of this gene flow on differentiation and parapatric speciation may be small (49, 54).

Note that some authors have regarded this as a case of disruptive selection (173) when in fact the majority of individuals in the two populations are under strong directed selection in different habitats (126). Only a few individuals at the borders may experience some disruptive selection, but even at this point gene flow from allopatric members of their population would tip the balance of selection in favor of one or the other (43). Disruptive selection may rarely be associated with speciation in nature.

Sessile animals share many features with annual plants (cf 72, 109). Dispersal powers of adults and young in snails, fossorial rodents, and some flightless insects may frequently be no greater than seed and pollen dispersal in annual plants. They also have similar reproductive strategies and genetic systems (76, 139). Many models proposed for the genetic differentiation of parapatric plant populations (3, 39, 43, 83), therefore, are applicable to these animals with low vagility.

Another attribute that differentiates parapatric speciation from allopatric modes is that both postmating and, more importantly, premating reproductive isolating mechanisms arise by selection as individuals penetrate a new habitat. A typical case that has been thoroughly studied in plants is the rapid evolution of heavy metal (copper, lead, zinc)-tolerant races of plants growing on contaminated mine tip soils (3, 83). Within a relatively short span of time, strong barriers to gene flow between the races have been erected, involving a shift in flowering time and increased selfing, even though individuals have always existed within only a few feet of one another at the racial border. Extensive studies on *Clarkia* by Lewis and his co-workers (109) and Gottlieb on *Stephanomeria* (67) are also pertinent here.

A very similar case of parapatric differentiation of adjacent populations in snails of the genus *Partula* has been investigated by Clarke (33, 34). Certain features of the *mooreana-olympia-tohiveana* series on the island of Moorea in the Pacific suggest that reproductive isolation has evolved parapatrically around vertical overhanging cliffs. The observed morph-ratio clines have probably resulted from sharp discontinuity between two incompatible coadapted gene complexes.

Even though these sharp genotypic changes do not necessarily coincide with an easily defined environmental change, morph-ratio clines ultimately depend on a change in some important environmental gradient. If the gradient is smooth, then an animal of low vagility will retain a coadapted gene complex until selective coefficients become so great that a complete reordering of the genome is necessary in order to cope with any further environmental changes. Because hybrids occurring at the morph-ratio cline interfaces are of reduced fitness, selection favors the evolution of reproductive isolation. More detailed accounts of this process can be found in Clarke (33), Murray (128), and Cook (35), and the subject is treated in a somewhat different way, using fitness sets and environmental grains, by Levins (107).

Other investigations into the genetic structure of natural animal populations reveal similar patterns of rapid differentiation over very short distances. Borisov (15) has found that there are urban and rural ecological races of *Drosophila funebris* clearly demonstrating ecogeographic divergence in inversion frequencies. Evidence of strong local parapatric racial differentiation for alcohol tolerance was found by McKenzie & Parsons (117) in a wine cellar population of *D. melanogaster*. However, the relatively high vagility of these flies probably precludes speciation unless a major chromosome rearrangement could become established, an unlikely possibility given the apparent low rate of incorporation of major chromosome rearrangements in this group of flies (181).

If sharp discontinuities in sedentary species can evolve by genic reorganization alone, any intrinsic factor that would reduce gene flow between parapatric races, such as a chromosome rearrangement, would be strongly selected for (186). The two most thoroughly studied examples of parapatric speciation involving chromosome rearrangement are a group of flightless morabine grasshoppers (179, 181, 182) and a fossorial group of mole rats of the genus *Spalex* (129, 133, 175).

The great majority of Australian morabine grasshopper species and races (300+) have unique karyotypes. The most conspicuous chromosome rearrangements involve fusions, dissociations, and pericentric inversions. The latter exist in several species as floating polymorphisms. Fusions or dissociations, on the other hand, are never found in the polymorphic state except in narrow hybrid zones existing between two chromosome races. It is unlikely any of these fusions and dissociations ever existed as balanced polymorphisms. They are therefore divisive rather than cohesive factors in phylogeny, and play important roles in speciation of these insects (182).

White (179, 181, 182) outlined step-by-step the way in which parapatric speciation occurred in these grasshoppers. Formation and fixation of these divisive rearrangements, which give adaptively superior homozygotes but inferior heterozygotes (*negative heterosis*), comprise the essential first step. Such rearrangements simultaneously permit the population to shift into a new niche and erect a strong postmating barrier to gene exchange. White considered it possible for this to happen most frequently well within the population range, rather than at the periphery where populations are smaller and ecologically more restricted. Yet, as pointed out by Lewis (109), rearrangements arising at the center of the range are the most likely to be lost, as there is little chance for the heterozygotes to mate before the rearrangement is eliminated by strong negative selection and chance. Only on the periphery, where inbreeding is high, do such rearrangements stand a chance to survive through the negative heterosis barrier. On the other hand, species existing in a patchy environment may have many populations that are *internally parapatric* or stasipatric (182). The possibility that an individual could shift to a new niche well within the range of the animal is still an open possibility warranting further study.

Key (94, 95) argued that all such rearrangements arise first in a propagule well beyond the general range of the parental species. He asserted, without presenting evidence, that all hybrid zones between chromosome races in morabines are the result of secondary range expansions of previously isolated populations, that the rearrangements act only as gene filters, and that introgression of some genes will occur.

His argument seems to pivot on whether small populations existing at the periphery of species range are allopatric or parapatric. To a morabine grasshopper, a snail, or a fossorial rodent, 100–1000 m is a formidable distance. Such distances may exist temporarily at the extreme limits of a species range when population densities become low during adverse years. Small, highly localized, and inbred family groups would be common under these circumstances. It seems realistic to envision a situation in which fixation of the new arrangement in only a few generations could occur in a small colony established by a single fertile female bearing a major chromosome rearrangement located only a few hundred meters from other members of her population. As most animals undergo similar density cycles without speciating, it seems rather arbitrary to single out these grasshoppers as unique. In the light of the genetic evidence from morph-ratio clines, even this degree of isolation is probably not essential.

Fixation of the new karyotype brings about major changes in regulatory function and a release from the ecogeographic restraints that the genetic architecture placed on the parental population (181). The new chromosome races are then free to expand into the unoccupied area to which they were previously poorly adapted. Hybridization might occur at the border between the old and new populations, but if a large enough reservoir of homozygotes with the new rearrangement existed beyond the hybrid zone, gene flow would be eliminated by selection. Furthermore, it is unlikely that much, if any, introgression would occur as proposed by Key (95), because of the high selective coefficients that act in the hybrid zones, and because invading genes between chromosome races are likely to be poorly coadapted.

The case of parapatric speciation in the mole rats *Spalex* is somewhat similar. Four chromosome races of mole rats in Israel have arisen by a series of dissociations (129, 132, 175). Races are distributed in a more or less north-south cline, with higher chromosome races clearly adapted to more arid zones. As in the case of *Partula,* the environmental gradient is a smooth one and no specific attribute of the habitat is correlated with the shift from one chromosome race to another (133). The same sort of negative heterosis found in the morabines accompanies heterokaryotypes that occur at low frequency at the boundaries of the parapatrically distributed races of species.

Many other animal species with low vagility appear to have speciated by the same parapatric process. Parapatrically distributed chromosomal forms of species or so-called semispecies have been reported in a wide range of fossorial mammals and insects reviewed by White (181). More recently, other examples have appeared involving pocket gophers (131, 135, 170), *Peromyscus* (156), shrews (64), *Sceloporus* lizards (74, 112), mole crickets (130), and stick insects (36), for which parapatric speciation has been invoked or is likely because of their population structure and other features of their genetic system.

One group in particular, the cichlid fish of the great lakes of East Africa, is of special interest. These animals have formed many huge endemic species flocks in short periods of time. In Lake Victoria more than 150 endemic species of *Haplochromis* evolved in less than 750,000 years (73). Lake Nabugabo, cut off from Victoria about 3500 years ago, contains five endemic species (61, 73). Without a doubt, all species have arisen by intralacustrine speciation. Both Greenwood (73) and Fryer & Iles (61) argue with good evidence that allopatric speciation by the founder principle augmented by strong predation can probably account for most species, although some forms may have speciated parapatrically. No chromosome studies have been published thus far, so little more can be said at this point.

SYMPATRIC SPECIATION (TYPE III)

Mayr (124) has defined sympatric speciation as the "origin of isolating mechanisms within the dispersal area of the offspring of a single cline." Superficially, there appears to be a fuzzy line between parapatric and sympatric speciation and, indeed, at least one author (128) regards the former to be a special case of the latter. On closer examination, the two are qualitatively different in several important features; probably the most important one is the way in which reproductive isolating mechanisms are established. In cases of parapatric speciation, strong pre- and postmating reproductive isolation develops as the population penetrates a new niche. In sympatric speciation, premating reproductive isolation arises before a population shifts to a new niche. Chromosome rearrangements appear to be rarely involved in the examples thus far studied. Furthermore, speciation frequently occurs at the center of the species range in a patchy environment rather than at the periphery, which results in the rapid generation of many sympatric sibling species.

Several authors have developed realistic mathematical models for sympatric speciation based on selection in a heterogenous environment (10, 114, 120). The first stage in sympatric speciation, development of a stable polymorphism, is easily attained in populations adapting to different niches in the absence of heterozygotic advantage. Maynard Smith (120) found that if such a polymorphism were accompanied by an assortative mating gene (i.e. individuals best adapted to a particular niche tend to mate with one another) and if there were some degree of habitat selection, then two reproductively isolated populations would evolve.

Attempts have been made to simulate sympatric speciation in the laboratory under different levels of artificial selection and isolation. Of those carried out with *Drosophila melanogaster* under disruptive selection regimes (32, 146–148, 152–155, 171, 172), only the one by Thoday & Gibson (171) was successful. Others were able to demonstrate that populations can undergo rapid divergence in the face of massive gene flow, but found little evidence for the development of strong reproductive isolation between lines. These results are not surprising. Most *Drosophila* possess a constellation of biological traits, particularly in female reproductive behavior (148), which make them unlikely to speciate sympatrically, especially under laboratory conditions and with the small population sizes used in the experiments.

Sympatric speciation appears to be limited to special kinds of animals, namely phytophagous and zoophagous parasites and parasitoids. However, this group encompasses a huge number of species (well over 500,000 described insects alone). A few other animals with special biological features may also speciate sympatrically, but they represent a small number of the world's fauna and are discussed later. The basic question is how can reproductive isolation arise before some barrier to gene flow is erected? Because parasites and parasitoids constitute the vast majority of animals speciating sympatrically, I examine them first.

Many parasites use their hosts as a rendezvous for courtship and mating (5, 22–24, 144). Mate selection in many of these groups therefore depends upon host selection (5, 22–24, 97, 134, 190). The principal isolating mechanism is ecological, not ethological, and postmating isolating mechanisms do not seem to be involved. Closely related sympatric sibling species of Tephritidae, for instance, can frequently be hybridized, and fertile F_1, F_2, and backcross progeny produced in the laboratory (14, 80; G. L. Bush unpublished). Under certain conditions, a shift to a new host can therefore have a profound effect on mate selection and provide a strong barrier to gene exchange between the parental and daughter populations.

The appearance of new host races of insects on introduced and native plants provides the best examples of this process. Some shifts such as the hawthorn fly, *Rhagoletis pomonella*, to apple in 1864, and later to cherries in Wisconsin in about 1960, have been studied in considerable detail (see 23, 24 for summaries). Similar cases of rapid sympatric host race formation have been reported for diprionid sawflies (97) and the codling moth, *Laspeyresia pomonella* (16, 136). Even some plant-infesting *Drosophila*, such as those associated with cacti (96) and certain endemic Hawaiian trees (38, 146), may have speciated sympatrically, but probably only in a few instances (78).

Genetics of host race formation and speciation has been reviewed by Bush (22–24) and Huettel & Bush (80). Two genetic components appear to play a primary role in many parasitic host shifts: 1. gene(s) controlling host recognition and selection and 2. genes involved with survival. Two other genetic factors may be of considerable importance: 3. those genes that determine to what degree insects can be induced to distinguish one host plant from another, and 4. in insects of temperate climates, those gene(s) that regulate diapause. They are not essential for sympatric speciation to occur, but serve to reinforce ecological isolation.

The number of major genes controlling host selection and survival may be relatively few (24, 106) and involved for the most part with chemoreception, as host discrimination is often determined by chemical rather than physical cues (42, 183).

No one has reported a case of genetic polymorphism for host selection within a natural population of parasites. Evidence that allelic variation is involved in host shifts comes from a few studies made on interspecific hybrids. The most extensive are those of Huettel & Bush (80). In the gall-forming fruitfly *Procecidochares*, one locus controls host plant recognition in males and females (80). Minor changes resulting from single mutations in receptor proteins of a chemoreceptor or decoding pathways in the central nervous system of the type demonstrated by Ferkovich & Norris (55) in roaches and Bentley (11) in crickets, respectively might well be involved.

The ability of a parasite to survive on a new host may also require the alteration of alleles at only a few minor loci (23, 24, 41, 63, 77, 106). A gene-for-gene relationship between host-resistant genes and parasite survival genes usually exists, and has been extensively studied in a wide range of coevolutionary associations between domesticated crops and their pest species (41). When the formal genetics of the relationship has been established, alleles for resistance are usually found to be dominant over those for susceptibility, and survival genes in the animal are frequently recessive (41). Genetic modifiers and epistatic interactions are also involved (106), but their relationship to speciation is not understood.

A shift to a new host, if it is to be permanent, must be preceded by the establishment of a new host recognition allele (or alleles), which permits the parasite to recognize and, preferentially, move to the new host. If the mortality level induced by the new host is not too high, further genetic changes are not initially required to establish a new host race. Mayr (124) has argued that it is "typological" to assume that a single gene could preadapt an individual to a new niche and lead to speciation. Such a gene, he contended, "would require a veritable 'systemic' mutation that simultaneously results in 1) a change in host preference; 2) a special adaptation to a new niche; and 3) a preference for mates with similar niche preferences." Yet, contrary to Mayr, there is strong evidence that a single allele substitution either in structural or regulatory genes can indeed fulfill all of these criteria.

If the new host has a radically different chemistry, new survival genes must also be incorporated into the genome of those parasites switching to the new host. A genetic model based on host recognition and survival genes has been proposed by Bush (22–24). It closely resembles the model of sympatric speciation rejected by Mayr (122, 124) as biologically untenable for reasons that no longer seem objectionable. There is now no question that host and mate selection are often closely linked and genetically controlled. It is also likely that the genetic variation needed to establish a new host race is present in the parent population even before the new host appears on the scene. This is evident in the speed at which phytophagous parasites can adapt to new insecticides and new plants to which they have never been previously exposed (23, 106).

Bush (23, 24) also pointed out that host shifts are most likely to occur when the original and new hosts occur together within the dispersal range of the parasite. In this situation, new genotypes can constantly be tested until the right combination results in a successful shift. New recombinants would statistically stand a much better chance of survival in an area where both host and parasite occur together in high densities along with the new host. Thus peripheral populations may be of less importance than in the other modes of speciation. It is improbable that random long-range dispersal of a few individuals beyond the normal range of the species would enable them to find a new host with which they are genetically compatible.

Another objection expressed by Mayr (124) against sympatric speciation by way of host race formation is that parasites are seldom truly monophagous, but often have secondary hosts. He contends that specialization occurs over a period of time in different areas as one host or the other becomes extinct. Specialization in isolation eventually results in the development of sufficient genetic divergence to guarantee speciation, even if the population becomes sympatric. Such a pattern of geographic isolation has undoubtedly occurred. Ample opportunity for such events have arisen during the Pleistocene.

However, the majority of parasites are, in fact, host specific (6, 57, 58, 144). In one well-studied dipteran family, the Agromyzidae, 73% are monophagous, 12% oligophagous, and 5% polyphagous (166). In light of the fact that parasites are probably the most abundant of all eukaryotes, sympatric divergence seems an equally probable, and possibly even the normal, mode of speciation in many groups. The number of zoophagous and phytophagous parasites is staggering. Price (144 and personal communication) estimated that about 72.1% of the British insects (among the best known in the world) are parasitic on plants or animals. Two species of oaks (*Quercus robae* and *Q. petraea*) alone sustain 284 species of parasites (herbivores) (165). If we consider that there are already 750,000 described species of insects worldwide (125), over 525,000 of these are parasites, a conservative figure as at least three times this number remains undescribed. This amounts to more than all other plant and animal species combined. It is difficult to believe that all of these speciated allopatrically.

When comparing the reproductive strategies of parasites and parasitoids with vertebrates and some large flying insects, it is clear that all possess many of the biological correlates generally attributed to r-strategists (6, 57, 58, 142, 143). Frequently they have very high reproductive rates, small body size, and short life spans; their mortality is likely to be density independent. However, when comparisons are made within a complex of parasitic or parasitoid species, sequential series of r- and K-strategists relative to one another within the same guild have obviously evolved (6, 57, 58, 142–144). This is the case with certain insect-host-parasitoid communities, such as the one consisting of the gall-forming midge (*Rhopalomyia californica*) and some 10 species or more of its hymenopterous parasitoids. Even more elaborate complements involving 30–40 parasitoids per host species are known (6).

Studies by Force (57, 58) and Askew (6) show that within such groups the r-strategists are the opportunists best able to penetrate and adapt to new environments. They have: 1. greater physiological tolerances to withstand physical perterbations, 2. greater dispersal capabilities so that new host populations can be located in case of local disaster, 3. high reproductive capacity, 4. adaptiveness to a variety of conditions, but 5. low competitive ability, and are 6. characterized as being monophagous. Parasites with these credentials are obviously the most likely to speciate sympatrically.

In sawflies and parasitoids such as the Chalcidoidea, host specificity is high and mating frequently occurs on the host. Sib-mating and inbreeding are common, and females usually mate only once. Furthermore, the haplodiploid genetic system exposes unfavorable genes directly to selection in the male and unfertilized females produce male progeny that will tend to increase the frequency of new host selection genes. Thus adaptation to a new host would occur rapidly (5, 97). Most parasitic species also appear to belong to groups that have small genome size (12), a feature associated with rapid speciation (168, 187).

High reproductive rates and other r-selected traits are also common to many other phytophagous insect parasites and to most other parasitic groups of animals. In many, a sequential series of r- to K-strategists exists similar to the ones found in hymenopterous parasitoids, but no relevant studies seem to have been published. Reproductive strategies, however, are probably less important to speciation in parasitic forms than other aspects of the genetic system, such as the amount of inbreeding, level of heterozygosity, the mode of host and mate selection, host specificity, and the distribution of potential host plants. K-selected species, which are also frequently host specific, might be capable of shifting just as readily to a new host under certain stable environmental conditions such as those encountered in the tropics (84–86).

Other Patterns of Sympatric Speciation

At least five other patterns of divergence involving temporal, chemical (pheromone), symbiotic factors, hybridization, and polyploidy suggest a sympatric origin of species. Examples of allochronic divergence in breeding time have been discussed for diprionid sawflies (97, 98), field crickets (1, 2), *Rhagoletis* fruit flies (22–24), and charr (60). In some cases, allochronic isolation has apparently led to speciation (1, 2, 24, 60), whereas in others it appears to have occurred along with a shift to a new host, thus serving to reinforce reproductive isolation (22–24, 98). All of these examples are as feasible as any allopatric explanation thus far offered (124).

Laven (104, 105) outlined a mechanism for sympatric speciation based on a cytoplasmic incompatibility system caused by an independent factor he called a *plasmon.* A similar case has been observed in Australian *Aedes* (163). A series of hybridization studies has revealed 20 "crossing types" representing different geographic races of *Culex pipiens.* Crosses between races are fertile in one direction, but sterile in the other. Caspari & Watson (31) pointed out that such a system would not necessarily lead to speciation because only the cytoplasm of one race would replace that of another, whereas genotypes would be mixed. A similar pattern of incipient speciation has been reported in semispecies of *Drosophila paulistorum,* caused by a mycoplasma-like intracellular symbiont (46–48, 52), and in *D. equinoxialis* [see (7)]. Lewis (108) has suggested that symbiosis of this type might permit exploitation of a marginal habitat.

Dobzhansky & Pavlovsky (47, 48) found that one laboratory strain of *D. paulistorum* had undergone speciation while in laboratory culture over a period of a few years. This phenomenon of symbiotic sterility interactions may be much more widespread in insects than is now recognized and deserves more attention. Its role in speciation, however, is still poorly understood, but may be more important in parapatric than in sympatric divergence.

Another possible mechanism for sympatric speciation is through a change in mating pheromones (145). Two pheromone "races" are known to exist in the European corn borer, *Ostrinia nubilalis.* The males of the eastern population are attracted to the natural female lure consisting of *trans*-11-tetradecenyl acetate (96%) and *cis*-11-tetradecenyl acetate (4%). Males of the western strain are attracted to the same isomeres but in the reverse concentrations (3 and 97%). Kochansky et al (100) correctly regarded these two forms, which occur sympatrically in Pennsylvania without interbreeding (25), as distinct species. Both species have slightly different biological traits and may represent separate introductions into North America from Europe. Irrespective of the origin of the two species in North America, a change in responsiveness could have arisen here or in Europe by only two independent mutations, one in the male altering a *trans* receptor protein to a *cis* receptor, and the second occurring in an enzyme used by the female to produce the *cis* rather than the *trans* form of the pheromone. Such a system could only arise successfully at the periphery of a large population. A more detailed study of these two populations and the forms from which they came is obviously warranted. The host plant may also alter sex pheromones and reinforce incipient isolation (79).

Interspecific hybridization and polyploidy, of major importance in plant speciation (72), appear to have played an insignificant role in animal speciation. The most clear-cut example of interspecific hybridization leading to speciation is that described by Johnston (89). He has presented convincing evidence that the Italian sparrow arose as the result of the hybridization of the house and willow sparrows some 3500 years ago. The final development of complete reproductive isolation appears to have occurred 300–400 years ago during the "Little Ice Age," and may well have involved a period of complete geographic isolation.

The only legitimate examples of polyploidy in bisexual species are in the cerato-phyrid Amphibia and two species of hylid frogs, a few teleost fishes, and possibly a few hermaphroditic turbellaria and annelids [(181), but see (9)]. Cases of poly-ploidy in asexual forms are more common, but because of the nature of sex chromo-some mechanisms, the high frequency of cross fertilization, and the overall cellular complexity of animals, the phenomenon is rare in sexual forms.

CONCLUDING REMARKS

The mode of speciation adopted by a species or group of related species of animals (or plants) is clearly determined to a considerable extent by the architecture of their genetic systems. We still have only fragmentary information on many aspects of the biological properties of most animals. Yet the emerging picture in recent years, gathered from independent molecular, organismal, and population studies, clearly indicates that a reconsideration of some long held attitudes toward the way animals speciate is warranted. A multidisciplinary approach to speciation problems is obvi-ously needed if we are ever to construct realistic models of speciation.

Genetic revolution and geographic isolation may not be required for speciation in many animals. Single gene substitution or a chromosome rearrangement can initiate speciation, if they have a drastic and permanent effect on gene flow between diverging populations. Reproductive isolation can arise during (parapatric) or even before (sympatric) new populations become adapted to new niches; such populations can speciate without loss of contact with parent populations. Based on new evidence, old models of sympatric and parapatric speciation are now being dusted off and reexamined.

Even the view of Goldschmidt (66) that a 'hopeful monster," a mutation that, in a single genetic step, simultaneously permits the occupation of a new niche and the development of reproductive isolation, no longer seems entirely unacceptable. In discussing the population size necessary for fixation of chromosomal rearrange-ments in parapatrically distributed European newts, Spurway (167) remarked: "The population size N of Wright should be so small that it may be profitable to think of some species originating from a single pair in a new Eden. Being an Adam and Eve gives a monster a chance to hope."

Because of the possibility that the number of animals that may be speciating sympatrically or parapatrically (i.e. rodents, parasites, flightless insects, etc) might exceed or at least equal the number of those speciating allopatrically, more emphasis should be placed on biological studies involving Types II and III species in the future for both academic and practical reasons. Included in this group of animals are almost all of the major pests of plants, animals, and man. We need, therefore, an emphasis on applied evolutionary biology. From an agricultural and medical stand-point, an understanding of speciation mechanisms is essential to the development of realistic pest control programs. It is time that more applied evolution be incorpo-rated into this essential field of biology.

ACKNOWLEDGMENTS

I would like to thank Tony Joern, Craig Jordan, Dan Otte, Eric Pianka, and Richard Williams, who read early drafts of this review, for their helpful comments. This work was supported in part by a grant from the National Science Foundation GB 40137.

Literature Cited

1. Alexander, R. D., Bigelow, R. S. 1960. Allochronic speciation in field crickets, and a new species, *Acheta veletis. Evolution* 24:334–46
2. Alexander, R. D. 1968. Life cycle origins, speciation, and related phenomena in crickets. *Quart. Rev. Biol.* 43:1–41
3. Antonovics, J. 1971. The effects of a heterogeneous environment on the genetics of natural populations. *Am. Sci.* 59:593–99
4. Árnason, Ú. 1972. The role of chromosomal rearrangement in mammalian speciation with special reference to Cetacea and Pinnipedia. *Hereditas* 70:113–18
5. Askew, R. R. 1968. Considerations on speciation in Chalcidoidea (Hymenoptera). *Evolution* 22:642–45
6. Askew, R. R. 1975. The organisation of chalcid-dominated parasitoid communities centred upon endophytic hosts. In *Evolutionary Strategies of Parasitic Insects*, ed. P. W. Price, 130–53. London: Plenum. 224 pp.
7. Ayala, F. J., Tracey, M. L., Barr, L. G., Ehrenfeld, J. G. 1974. Genetic and reproductive differentiation of the subspecies, *Drosophila equinoxialis caribbensis. Evolution* 28:24–41
8. Bachmann, K., Harrington, B. A., Craig, J. P. 1972. Genome size in birds. *Chromosoma* 37:405–16
9. Balsano, J. S., Darnell, R. M., Abramoff, P. 1972. Electrophoretic evidence of triploidy, associated with populations of the gynogenetic teleost *Poecilia formosa. Copeia* 1972:292–97
10. Basykin, A. D. 1965. On the possibility of sympatric species formation. *Byull. Mosk. Obshchest. Ispyt. Prir. Otd. Biol.* 70:161–65
11. Bentley, D. R. 1971. Genetic control of an insect neuronal network. *Science* 174:1139–41
12. Bier, K. Von, Müller, W. 1969. DNS-Messungen bei insekten und eine hypothese über retardierte evolution und besonderen DNS – Reichtum in Tierreich. *Biol. Zentralbl.* 88:425–49
13. Bigelow, R. S. 1965. Hybrid zones and reproductive isolation. *Evolution* 19:449–58
14. Boller, E. F., Bush, G. L. 1973. Evidence for genetic variation in populations of the European cherry fruit fly, *Rhagoletis cerasi* (Diptera: Tephritidae) based on physiological parameters and hybridization experiments. *Entomol. Exp. Appl.* 17:279–93
15. Borisov, A. I. 1970. Interaction of *Drosophila funebris* chromosomes from urban and rural races in experimental populations. *Genetika* 6:81–90
16. Bovey, P. 1966. Tortricoidea. In *Entomologie Appliqueé à L'agriculture*, ed. A. J. Balachowsky, 2:456–895. Paris: Masson
17. Bowers, J. H., Baker, R. J., Smith, M. H. 1973. Chromosomal, electrophoretic, and breeding studies of selected populations of deer mice (*Peromyscus maniculatus*) and black-eared mice (*P. melanotis*). *Evolution* 27:378–86
18. Blair, W. F. 1965. Amphibian speciation. VII Congr. Int. Assoc. Quaternary Res., Boulder, Colo. 543–56
19. Britten, R. J., Davidson, E. H. 1971. Repetitive and non-repetitive DNA sequences and a speculation on the origins of evolutionary novelty. *Quart. Rev. Biol.* 46:111–33
20. Brown, W. L. Jr. 1957. Centrifugal speciation. *Quart. Rev. Biol.* 32:247–77
21. Brown, W. L. Jr., Wilson, E. O. 1956. Character displacement. *Syst. Zool.* 5:49–64
22. Bush, G. L. 1969. Sympatric host race formation and speciation in frugivorous flies of the genus *Rhagoletis* (Diptera, Tephritidae). *Evolution* 23:237–51
23. Bush, G. L. 1974. The mechanism of sympatric host race formation in the true fruit flies (Tephritidae). In *Genetic Mechanisms of Speciation in Insects*, ed. M. J. D. White, 3–23. Sydney: Australia & New Zealand Book Co. 170 pp.

24. Bush, G. L. 1975. Sympatric speciation in phytophagous parasitic insects. In *Evolutionary Strategies of Parasitic Insects*, ed. P. W. Price, 187–206. London: Plenum. 224 pp.

25. Cardé, R. T., Kochansky, J., Stimmel, J. F., Wheeler, A. G. Jr., Roelofs, W. L. 1975. Sex pheromones of the European corn borer *Ostrinia nubilalis: cis* and *trans* responding males in Pennsylvania. *Environ. Entomol.* In press

26. Carson, H. L. 1968. The population flush and its consequences. In *Population Biology and Evolution*, ed. R. C. Lewontin, 123–37. Syracuse: Syracuse Univ. Press

27. Carson, H. L. 1970. Chromosome tracers of the origin of species. *Science* 168:1414–18

28. Carson, H. L. 1971. Speciation and the founder principle. *Stadler Genet. Symp.* 3:51–70

29. Carson, H. L. 1973. Reorganization of the gene pool during speciation. In *Genetic Structure of Populations*, ed. N. E. Morton, Popul. Genet. Monogr. 3:274–80. Honolulu: Univ. Press of Hawaii. 205 pp.

30. Carson, H. L. 1975. The genetics of speciation at the diploid level. *Am. Natur.* 109:83–92

31. Caspari, E., Watson, G. S. 1959. On the evolutionary importance of cytoplasmic sterility in mosquitoes. *Evolution* 13:568–70

32. Chabora, A. J. 1968. Disruptive selection for sternopleural chaeta number in various strains of *Drosophila melanogaster*. *Am. Natur.* 102:525–32

33. Clarke, B. 1966. The evolution of morph-ratio clines. *Am. Natur.* 100: 389–402

34. Clarke, B., Murray, J. 1969. Ecological genetics and speciation in land snails of the genus *Partula*. *Biol. J. Linn. Soc.* 1:31–42

35. Cook, L. M. 1961. The edge effect in population genetics. *Am. Natur.* 95: 295–307

36. Craddock, E. M. 1974. Chromosomal evolution and speciation in *Didymuria*. See Ref. 37, 24–42

37. Craddock, E. M. 1974. Degrees of reproductive isolation between closely related species of Hawaiian *Drosophila*. In *Genetic Mechanisms of Speciation in Insects*, ed. M. J. D. White, 111–39. Sydney: Australia & New Zealand Book Co. 170 pp.

38. Craddock, E. M. 1974. Reproductive relationships between homosequential

39. species of Hawaiian *Drosophila*. *Evolution* 28:593–606

39. Crosby, J. L. 1969. The evolution of genetic discontinuity: compter models of the selection of barriers to interbreeding between subspecies. *Heredity* 25: 253–97

40. Davis, B. L., Baker, R. J. 1974. Morphometrics, evolution, and cytotaxonomy of mainland bats of genus *Macrotus* (Chiroptera: Phyllostomatidae) *Syst. Zool.* 23:26–39

41. Day, P. R. 1974. *Genetics of Host-Parasite Interaction*. San Francisco: Freeman. 238 pp.

42. Dethier, V. G. 1952. Evolution of feeding preferences in phytophagous insects. *Evolution* 8:33–54

43. Dickinson, H., Antonovics, J. 1973. Theoretical considerations of sympatric divergence. *Am. Natur.* 107:256–74

44. Dobzhansky, T. 1970. *Genetics of the Evolutionary Process*. New York & London: Columbia Univ. Press. 505 pp.

45. Dobzhansky, T. 1972. Species of *Drosophila*. New excitement in an old field. *Science* 177:664–69

46. Dobzhansky, T., Pavlovsky, O. 1966. Spontaneous origin of an incipient species in the *Drosophila paulistorum* complex. *Proc. Nat. Acad. Sci. USA* 55:727–33

47. Dobzhansky, T., Pavlovsky, O. 1967. Experiments on the incipient species of the *Drosophila paulistorum* complex. *Genetics* 55:141–56

48. Dobzhansky, T., Pavlovsky, O. 1971. Experimentally created incipient species of *Drosophila. Nature* 230:289–92

49. Ehrlich, P. R., Raven, P. H. 1969. Differentiation of populations. *Science* 165:1228–32

50. Ehrman, L. 1965. Direct observation of sexual isolation between allopatric and between sympatric strains of the different *Drosophila paulistorum* races. *Evolution* 19:459–64

51. Ehrman, L. 1969. Genetic divergence in M. Vetukhiv's experimental populations of *Drosophila pseudoobscura*. 5. A further study of rudiments of sexual isolation. *Am. Midl. Natur.* 82:272–76

52. Ehrman, L., Kernaghan, R. P. 1971. Microorganismal basis of infectious hybrid male sterility in *Drosophila paulistorum*. *J. Hered.* 62:76–72

53. Eisenberg, J. F. 1966. The social organization of mammals. *Handb. Zool.* 8(10):1–92

54. Endler, J. A. 1973. Gene flow and popu-

lation differentiation. *Science* 179: 243–50

55. Ferkovich, S. M., Norris, D. M. 1971. Antennal proteins involved in the neural mechanism of quinone inhibition of insect feeding. *Experientia* 28:978–79

56. Flamm, W. G. 1972. Highly repetitive sequences of DNA in chromosomes. *Int. Rev. Cytol.* 32:1–51

57. Force, D. C. 1972. *r*-and *K*-strategists in endemic host-parasitoid communities. *Bull. Entomol. Soc. Am.* 18:135–37

58. Force, D. C. 1975. Succession of *r*- and *K*-stragetists in parasitoids. In *Evolutionary Strategies of Parasitic Insects,* ed. P. W. Price, 112–29. London: Plenum. 224 pp.

59. Ford, H. A., Parkin, D. T., Ewing, A. W. 1973. Divergence and evolution in Darwin's finches. *Biol. J. Linn. Soc.* 5:289–95

60. Frost, W. E. 1965. Breeding habits of Windermere charr, *Salvelinus willughbii* (Günther), and their bearing on speciation of these fish. *Proc. R. Soc. Ser. B. Biol. Sci.* 163:232–84

61. Fryer, G., Iles, T. D. 1972. *The Cichlid Fishes of the Great Lakes of Africa. Their Biology and Evolution.* Hong Kong: TFH. 641 pp.

62. Gadgil, M., Solbrig, O. T. 1972. The concept of *r*- and *K*-selection: Evidence from wild flowers and some theoretical considerations. *Am. Natur.* 106:14–31

63. Gallun, R. L. 1972. Genetic inter-relationships between host plants and insects. *J. Eviron. Qual.* 1:259–65

64. Genoways, H. H., Choate, J. R. 1972. A multivariate analysis of systematic relationships among populations of the short-tailed shrew (genus *Blarina*) in Nebraska. *Syst. Zool.* 21:10–16

65. Ghiselin, M. T. 1974. A radical solution to the species problem. *Syst. Zool.* 23:536–44

66. Goldschmidt, R. 1940. *The Material Basis of Evolution.* New Haven: Yale Univ. Press. 436 pp.

67. Gottlieb, L. D. 1973. Genetic differentiation, sympatric speciation and the origin of a diploid species of *Stephanomeria. Am. J. Bot.* 60:545–53

68. Grant, P. R. 1972. Covergent and divergent character displacement. *Biol. J. Linn. Soc.* 4:39–68

69. Grant, V. 1963. *The Origin of Adaptations.* New York & London: Columbia Univ. Press. 606 pp.

70. Grant, V. 1966. Selection for vigor and fertility in the progeny of a highly sterile species hybrid in *Gilia. Genetics* 53:757–75

71. Grant, V. 1966. The selective origin of incompatibility barriers in the plant genus *Gilia. Am. Natur.* 100:99–118

72. Grant, V. 1971. *Plant Speciation.* New York & London: Columbia Univ. Press. 435 pp.

73. Greenwood, P. H. 1974. The cichlid fishes of Lake Victoria, East Africa: The biology and evolution of a species flock. *Bull. Brit. Mus. (Natur. Hist.) Zool.* 6:134 pp.

74. Hall, W. P., Selander, R. K. 1973. Hybridization of karyotypically differentiated populations in the *Sceloporus grammicus* complex (Iguanidae). *Evolution* 27:226–42

75. Hanson, W. D. 1966. Effects of partial isolation (distance), migration, and different fitness requirements among environmental pockets upon steady state gene frequencies. *Biometrics* 22:453–68

76. Harper, J. L., White, J. 1974. The demography of plants. *Ann. Rev. Ecol. Syst.* 5:419–63

77. Hatchett, J. H., Gallun, R. L. 1970. Genetics of the ability of the Hessian fly, *Mayetiola destructor,* to survive on wheats having different genes for resistance. *Ann. Entomol. Soc. Am.* 63: 1400–1407

78. Heed, W. B. 1971. Host plant specificity and speciation in Hawaiian *Drosophila. Taxon* 20:115–21

79. Hendry, L. B. 1975. Evidence for the origin of insect sex pheromones present in food plants. *Science.* In press

80. Huettel, M. D., Bush, G. L. 1972. The genetics of host selection and its bearing on sympatric speciation in *Procecidochares* (Diptera: Tephritidae). *Entomol. Exp. Appl.* 15:465–80

81. Huey, R. B., Pianka, E. R. 1974. Ecological character displacement in a lizard. *Am. Zool.* 14:1127–36

82. Huey, R. B., Pianka, E. R., Egan, M. E., Coons, L. W. 1974. Ecological shifts in sympatry: Kalahari fossorial lizards (*Typhlosauris*). *Ecology* 55:304–16

83. Jain, S. K., Bradshaw, A. D. 1966. Evolutionary divergence among adjacent plant populations. 1. The evidence and its theoretical analysis. *Heredity* 22: 407–41

84. Janzen, D. H. 1970. Herbivores and number of tree species in tropical forests. *Am. Natur.* 104:501–28

85. Janzen, D. H. 1973. Comments on host-specificity of tropical herbivores and its relevance to species richness. In *Tax-*

onomy and Ecology, ed. V. H. Heywood, 201–12. New York: Academic. 370 pp.

86. Janzen, D. H. 1973. Tropical agroecosystems. *Science* 182:1212–19

87. John, B., Lewis, K. R. 1965. Genetic speciation in the grasshopper *Eyprepocnemis plorans, Chromosoma* 16: 308–44

88. Johnson, G. B. 1973. Relationship of enzyme polymorphism to species diversity. *Nature* 242:193–94

89. Johnston, R. F. 1969. Taxonomy of house sparrows and their allies in the Mediterranean basin. *Condor* 71: 129–39

90. Jones, J. M. 1973. Effects of thirty years hybridization on the toads *Bufo americanus* and *Bufo woodhousii fowleri* at Bloomington, Indiana. *Evolution* 27:435–48

91. Jones, K. W. 1973. Satellite DNA. *J. Med. Genet.* 10:273–81

92. Judd, B. H. 1975. Genes and Chromomeres of *Drosophila.* In *The Eukaryote Chromosome,* ed. J. W. Peacock, R. Brock. Canberra: Aust. Nat. Univ. Press. In press

93. Kessler, S. 1966. Selection for and against ethological isolation between *Drosophila pseudoobscura* and *Drosophila persimilis. Evolution* 20:634–45

94. Key, K. H. L. 1968. The concept of stasipatric speciation. *Syst. Zool.* 17: 14–22

95. Key, K. H. L. 1974. Speciation in the Australian Morabine grasshoppers—taxonomy and ecology. In *Genetic Mechanisms of Speciation in Insects,* ed. M. J. D. White, 43–56. Sydney: Australia & New Zealand Book Co. 170 pp.

96. Kircher, H. W., Heed, W. B. 1970. Phytochemistry and host plant specificity in *Drosophila. Recent Adv. Phytochem.* 3:191–209

97. Knerer, G., Atwood, C. E. 1973. Diprionid sawflies: Polymorphism and speciation. *Science* 179:1090–99

98. Knerer, G., Marchant, R. 1973. Diapause induction in the sawfly *Neodiprion rugifrons* Middleton (Hymenoptera: Diprionidae). *Can. J. Zool.* 51:105–08

99. Knight, G. R., Robertson, A., Waddington, C. H. 1956. Selection for sexual isolation within a species. *Evolution* 10:14–22

100. Kochansky, J., Cardé, R. T., Liebherr, J., Roelofs, W. L. 1975. Sex pheromones of the European corn borer (*Os-trinia nubilalis*) in New York *J. Chem. Ecol.* In press

101. Koopman, K. F. 1950. Natural selection for reproductive isolation between *Drosophila pseudoobscura* and *Drosophila persimilis. Evolution* 4:135–48

102. Kozhevnikov, B. T. 1936. Experimentally produced karyotypical isolation. *Biol. Zh.* 5:121–32

103. La Bounty, J. F., Deacon, J. E. 1972. *Cyprinodon milleri,* a new species of pup-fish (family Cyprinodontidae) from Death Valley, Calif. *Copeia* 1972: 769–80

104. Laven, H. 1959. Speciation by cytoplasmic isolation in the *Culex pipiens* complex. *Cold Spring Harbor Symp. Quant. Biol.* 24:166–73

105. Laven, H. 1967. A possible model for speciation by cytoplasmic isolation in the *Culex pipiens* complex. *Bull. WHO* 37:263–66

106. Levin, D. A. 1975. Pest pressure and recombination systems in plants. *Am. Natur.* In press

107. Levins, R. 1968. *Evolution in Changing Environments.* Princeton, NJ: Princeton Univ. Press. 120 pp.

108. Lewis, D. H. 1973. The relevance of symbiosis to taxonomy and ecology, with particular reference to mutualistic symbiosis and the exploitation of marginal habitats. In *Taxonomy and Ecology,* ed. V. H. Heywood, 151–72. New York & London: Academic. 370 pp.

109. Lewis, H. 1966. Speciation in flowering plants. *Science* 152:167–72

110. Lewontin, R. C. 1974. *The Genetic Basis of Evolutionary Change.* New York & London: Columbia Univ. Press. 346 pp.

111. Littlejohn, M. J., Loftus-Hills, J. J. 1968. An experimental evaluation of premating isolation in the *Hyla ewingi* complex (Anura: Hylidae). *Evolution* 22:659–63

112. Lowe, C. H., Cole, C. J., Patton, J. L. 1967. Karyotype evolution and speciation in lizards (Genus *Sceloparus*) during evolution of the North American desert. *Syst. Zool.* 16:296–300

113. Lowe-McConnell, R. H. 1969. Speciation in tropical freshwater fishes. *Biol. J. Linn. Soc.* 1:51–75

114. Ludwig, W. 1950. Zur Theorie der Konkurrenz Die Annidation (Einnischung) als fünfter Evolutions–faktor. *Neue Ergeb. Probleme Zool., Klatt-Festschrift* 1950:516–37

115. MacArthur, R. H., Wilson, E. O. 1967. *The Theory of Island Biogeography.* Princeton, NJ: Princeton Univ. Press. 203 pp.

116. McFee, A. F., Banner, M. W., Rary, J. M. 1966. Variation in chromosome number among European wild pigs. *Cytogenetics* 5:75

117. McKenzie, J. A., Parsons, P. A. 1974. Microdifferentiation in a natural population of *Drosophila melanogaster* to alcohol in the environment. *Genetics* 77:385–94

118. McNab, B. K. 1963. Bioenergetics and the determination of home range size. *Am. Natur.* 97:133–40

119. Marshall, N. B. 1963. Diversity, distribution and speciation of deep-sea fishes. In *Speciation in the Sea,* ed. J. P. Harding, N. Tebble, 181–95. London: Systematics. 199 pp.

120. Maynard-Smith, J. 1966. Sympatric Speciation. *Am. Natur.* 100:637–50

121. Mayr, E. 1942. *Systematics and the Origin of Species.* New York: Columbia Univ. Press. 334 pp.

122. Mayr, E. 1947. Ecological factors in speciation. *Evolution* 1:263–88

123. Mayr, E. 1954. Change of genetic environment and evolution. In *Evolution as a Process,* ed. J. Huxley, A. C. Hardy, E. B. Ford, 157–80. London: Allen & Unwin. 367 pp.

124. Mayr, E. 1963. *Animal Species and Evolution.* Cambridge, Mass.: Harvard Univ. Press. 797 pp.

125. Mayr, E. 1969. *Principles of Systematic Zoology.* New York: McGraw-Hill. 428 pp.

126. Mayr, E. 1974. The definition of the term disruptive selection. *Heredity* 32:404–6

127. Mazrimas, J. A., Hatch, F. T. 1972. A possible relationship between satellite DNA and the evolution of kangaroo rat species (Genus *Dipodomys*). *Nature New Biol.* 240:102–5

128. Murray, J. 1972. *Genetic Diversity and Natural Selection.* New York: Hafner. 128 pp.

129. Nevo, E. 1969. Mole rat *Spalax ehrenbergi:* Mating behavior and its evolutionary significance. *Science* 163:484–86

130. Nevo, E., Blondheim, S. A. 1972. Acoustic isolation in the speciation of mole crickets. *Ann. Entomol. Soc. Am.* 65:980–81

131. Nevo, E., Kim, Y. J., Shaw, C. R., Thaeler, C. S. Jr. 1974. Genetic variation, selection and speciation in *Thomomys talpoides* pocket gophers. *Evolution* 28:1–23

132. Nevo, E., Shaw, C. R. 1972. Genetic variation in a subterranean mammal, *Spalax ehrenbergi. Biochem. Genet.* 7:235–41

133. Nevo, E., Shkolnik, A. 1974. Adaptive metabolic variation of chromosome forms in mole rats, *Spalax. Experientia* 30:724–26

134. Nowakowski, J. T. 1962. Introduction to a systematic revision of the family Agromyzidae (Diptera) with some remarks on host plant selection by these flies. *Ann. Zool.* 20:68–183

135. Patton, J. L. 1972. Patterns of geographic variation in karyotype in the pocket gopher, *Thomomys bottae* (Eydoux and Gervais). *Evolution* 26:574–86

136. Phillips, P. A., Barnes, M. M. 1975. Host race formation among sympatric apple, walnut, and plum populations of the codling moth. *Ann. Entomol. Soc. Am.* In press

137. Pianka, E. R. 1970. On r- and K-selection. *Am. Natur.* 104:592–97

138. Pianka, E. R. 1972. Zoogeography and speciation of Australian desert lizards: An ecological perspective. *Copeia* 1972:127–45

139. Pianka, E. R. 1974. *Evolutionary Ecology.* New York: Harper & Row. 356 pp.

140. Pimentel, D., Smith, G. J. C., Soans, J. 1967. A population model of sympatric speciation. *Am. Natur.* 101:493–504

141. Portin, P. Allelic contracomplementation possible mechanism for initiating sympatric speciation. *Nature* 247:216–17

142. Price, P. W. 1973. Parasitoid strategies and community organization. *Environ. Entomol.* 2:623–26

143. Price, P. W. 1974. Strategies for egg production. *Evolution* 28:76–84

144. Price, P. W. 1975. Introduction: The parasitic way of life and its consequences. In *Evolutionary Strategies of Parasitic Insects,* ed. P. W. Price, 1–19. London: Plenum. 224 pp.

145. Priesner, E. 1970. Über die spezifität der Lepidopteren–Sexuallockstoffe und ihre rolle dei der Artbildung. *Verh. Dtsch. Zool. Ges.* 64:337–43

146. Richardson, R. H. 1974. Effects of dispersal, habitat selection and competition on a speciation pattern of *Drosophila* endemic to Hawaii. In *Genetic Mechanisms of Speciation in Insects,* ed. M. J. D. White, 140–64. Sydney: Aus-

tralia & New Zealand Book Co. 170 pp.

147. Richmond, R. C. 1972. Genetic similarities and evolutionary relationships among the semispecies of *Drosophila paulistorum*. *Evolution* 26:536–44

148. Robertson, A. 1970. A note on disruptive selection experiments in *Drosophila*. *Am. Natur.* 104:561–69

149. Robertson, F. W. 1966. A test of sexual isolation in *Drosophila*. *Genet. Res.* 8:181–87

150. Robertson, F. W. 1966. The ecological genetics of growth in *Drosophila*. 8. Adaptation to a new diet. *Genet. Res.* 165–79

151. Rockwood, E., Kanapi, C., Wheeler, M., Stone, W. 1971. Allozyme changes during the evolution of Hawaiian *Drosophila*. *Univ. Texas Publ.* 7103: 193–212

152. Scharloo, W. 1964. The effect of disruptive and stabilizing selection on the expression of a cubitus interruptus mutant in *Drosophila*. *Genetics* 50:553–62

153. Scharloo, W. 1971. Reproductive isolation by disruptive selection: Did it occur? *Am. Natur.* 105:83–86

154. Scharloo, W., Den Boer, M., Hoogmoed, M. S. 1967. Disruptive selection on sternopleural chaeta number in *Drosophila melanogaster*. *Genet. Res.* 9:115–18

155. Scharloo, W., Hoogmoed, M. S., Kuile, A. T. 1967. Stabilizing and disruptive selection on a mutant character in *Drosophila*. I. The phenotypic variance and its components. *Genetics* 56:709–26

156. Schmidly, D. J., Schroeter, G. L. 1974. Karyotypic variation in *Peromyscus boylii* (Rodentia: Cricetidae) from Mexico and corresponding taxonomic implications. *Syst. Zool.* 23:333–42

157. Selander, R. K. 1971. Systematics and speciation in birds. In *Avian Biology,* ed. D. S. Farner, J. R. King, K. C. Parkes, 57–147. New York: Academic. 586 pp.

158. Selander, R. K., Johnson, W. E. 1973. Genetic variation among vertebrate species. *Ann. Rev. Ecol. Syst.* 4:75–116

159. Selander, R. K., Kaufman, D. W. 1973. Genic variability and strategies of adaptation in animals. *Proc. Nat. Acad. Sci.* 70:1875–77

160. Selander, R. K., Kaufman, D. W., Baker, R. J., Williams, S. L. 1974. Genic and chromosomal differentiation in pocket gophers of the genus *Bursarius* group. *Evolution* 28:557–64

161. Simpson, E. E. 1959. The nature and origin of supraspecific taxa. *Genetics*

and the Twentieth Century Darwinism. *Cold Spring Harbor Symp. Quart. Biol.* 24:255–71

162. Slatkin, M. 1974. Cascading speciation. *Nature* 252:701–2

163. Smith-White, S., Woodhill, A. R. 1954. The nature and significance of nonreciprocal fertility in *Aedes scutellaris* and other mosquitoes. *Proc. Linn. Soc.* 79:163–76

164. Sokal, R. R. 1974. The species problem reconsidered. *Syst. Zool.* 22:360–74

165. Southwood, T. R. E. 1961. The number of species of insects associated with various trees. *J. Anim. Ecol.* 30:1–8

166. Spencer, K. A. 1964. The species-host relationship in the Agromyzidae (Diptera) as an aid to taxonomy. *XII Int. Congr. Entomol. Proc.* 1:101–2

167. Spurway, H. 1953. Genetics of specific and subspecific differences in European newts. *Symp. Soc. Exp. Biol.* 7:200–37

168. Stebbins, G. L. 1966. Chromosomal variation and evolution. *Science* 152:1463–69

169. Taylor, K. M., Hungerford, D. A., Snyder, R. L. 1969. Artiodactyl mammals: their chromosome cytology in relation to patterns of evolution. In *Comparative Mammalian Cytogenetics,* ed. K. Benirschke, 346–56. New York: Springer-Verlag. 473 pp.

170. Thaeler, C. S. Jr. 1974. Four contacts between ranges of different chromosome forms of the *Thomomys talpoides* complex. (Rodentia: Geomyidae). *Syst. Zool.* 23:343–54

171. Thoday, J. M., Gibson, J. B. 1962. Isolation by disruptive selection. *Nature* 193:1164–66

172. Thoday, J. M., Gibson, J. B. 1970. Reply to Scharloo. *Am. Natur.* 105:86–88

173. Thoday, J. M. 1972. Disruptive selection. *Proc. R. Soc. London Ser. B.* 182:109–43

174. Wagner, R. P., Selander, R. K. 1974. Isozymes in insects and their significance. *Ann. Rev. Entomol.* 19:117–38

175. Wahrman, J., Goitein, R., Nevo, E. 1969. Mole rat *Spalax:* Evolutionary significance of chromosome variation. *Science* 164:82–84

176. Walker, P. M. B. 1971. "Repetitive" DNA in higher organisms. *Progr. Biophys. Mol. Biol.* 23:145–90

177. Walker, T. J. 1974. Character displacement and acoustic insects. *Am. Zool.* 14:1137–50

178. Wallace, B. 1968. *Topics in Population Genetics.* New York: Norton. 481 pp.

179. White, M. J. D. 1968. Models of speciation. *Science* 159:1065–70
180. White, M. J. D. 1970. Cytogenetics of speciation. *J. Aust. Entomol. Soc.* 9:1–6
181. White, M. J. D. 1973. *Animal Cytology and Evolution.* Cambridge, Engl.: Cambridge Univ. Press. 961 pp. 3rd ed.
182. White, M. J. D. 1974. Speciation in Australian Morabine grasshoppers—the cytogenetic evidence. In *Genetic Mechanisms of Speciation in Insects,* ed. M. J. D. White, 57–68. Sydney: Australia & New Zealand Book Co. 170 pp.
183. Whittaker, R. H., Feeny, P. P. 1971. Allelochemics: Chemical interactions between species. *Science* 171:757–70
184. Wilson, A. C., Maxson, L. R., Sarich, V. M. 1974. Two types of molecular evolution. Evidence from studies of interspecific hybridization. *Proc. Nat. Acad. Sci. USA* 71:2843–47
185. Wilson, A. C., Maxson, L. R., Sarich, V. M. 1974. The importance of gene rearrangement in evolution: Evidence from studies on rates of chromosomal, protein, and anatomical evolution. *Proc. Nat. Acad. Sci. USA* 71:3028–30
186. Wright, S. 1956. Modes of selection. *Am. Natur.* 90:5–24
187. Yunis, J. J. 1974. Structure and molecular organization of chromosomes. In *Human Chromosome Methodology,* ed. J. J. Yunis, 1–15. New York: Academic. 2d ed. 377 pp.
188. Yunis, J. J., Yasmineh, W. G. 1971. Heterochromatin, satellite DNA, and cell function. *Science* 174:1200–1209
189. Zouros, E. 1973. Genic differentiation associated with early stages of speciation in the Mulleri subgroup of *Drosophila. Evolution* 27:601–21
190. Zwölfer, H., Harris, P. 1971. Host specificity determination of insects for biological control of weeds. *Ann. Rev. Entomol.* 16:159–78

BUTTERFLY ECOLOGY ❖4098

Lawrence E. Gilbert
Department of Zoology, University of Texas, Austin, Texas 78712

Michael C. Singer
Department of Agriculture and Horticulture, University of Bristol, Research Station, Long Ashton, England

> *"A collector who is a careful observer is often able to examine a terrain and to decide, intuitively as it were, whether a given butterfly will be found there . . . but this is a work of art rather than of science, and we would gladly know the components which make such predictions possible."*
>
> E. B. Ford (1945)

INTRODUCTION

Ecology concerns itself with explaining such things as why species live where they do, why they exist at certain densities, why they do or do not persist in a given habitat, why certain combinations of species coexist whereas others do not, and why communities are structured as they are. Although studies of other taxonomically well-known groups such as birds have contributed greatly to advances in general ecological theory, the contributions of ecological studies on butterflies have been minor. This could mean that general theory is not general, that butterflies are highly exceptional in their life histories, that better minds have worked with birds, and so forth. Although there may be some truth in each of these possibilities, it seems to us that the early interest in butterfly mimicry as a test of evolutionary theory and the relative ease of combining field observations with laboratory breeding studies on butterflies greatly influenced the interests of the most talented early students of butterflies. Moreover, as only a few butterflies are agricultural pests (*Pieris, Colias*), we find few contributions to insect population ecology coming from butterfly research.

Thus during the first half of this century, the main contributions of butterfly workers were in the areas of genetics (103, 220), evolution (104), and ethology (see 224). Ecological work over the same period is largely descriptive and nonsynthetic work [often an afterthought of systematic (e.g. 27, 163, 171) or genetic studies

365

(106)]. It suffers from a lack of general hypotheses and associated experiments. Much of the information of potential interest to ecologists is published in entomological rather than ecological journals by workers familiar with the taxonomic, but not the conceptual, context of their observations. Because this is the first review of butterfly ecology, we have attempted to pull together a fair sample of the worldwide literature as far back in time as appropriate work exists. We emphasize in our discussion, however, that research (mostly of the last decade) which has been most useful in forming our own view of the subject, and have ignored some ecologically relevant aspects of butterfly reproductive biology, including mate-locating behavior (12, 214, 222).

POPULATION ECOLOGY

Because butterfly pests, with few exceptions (6, 43, 185), utilize annual crops, there has been little economic incentive for long-term population dynamic studies equivilent to those carried out on tree-feeding moths (see examples in 251). Few life-tables have been compiled, even for single generations (22, 62, 80, 118, 199, 243), although demographic models have been developed that can be applied to insects, such as butterflies, which exist as several distinct stages from the standpoint of pathogens, parasites, and predators (143–145). Richards (199) has emphasized the difficulty of studying mobile species. He found that changes in numbers of *Pieris rapae* in any small area from year to year were attributable as much to changes in attractiveness of the area to *Pieris* as to changes in overall abundance. In view of the paucity of data, statements in the literature about factors that limit or regulate butterfly populations usually are guesses based on circumstantial evidence.

Limiting Factors

LARVAL RESOURCES   Dethier (65) observed: "It is generally agreed that insects whose larvae feed on plants do not increase to the larval food limit except in sporadic and unusual cases." However, the extent to which Dethier's generalization holds for butterflies is not yet clear. Most adults of *Melitaea harrisii* emigrated from a field in which their density had been artificially increased (67). The carrying capacity of the habitat for adults was well below that indicated by larval food supply. In a closely related species, *Euphydryas editha,* larval food supply was limiting in several populations and starvation of larvae at one stage or another occurred in almost all populations on several different host plant genera (263). Starvation was not necessarily associated with competition, however, and in those *E. editha* populations where starvation mortality was highest (99%), intraspecific competition was absent and population size was not regulated by larval food supply (227, 263). It is likely that other aspects of larval host plants, such as spatial pattern, size, and growth form or shape are equally important as absolute amount of host material in any attempt to estimate carrying capacities for particular butterfly species (see 66, 168).

ADULT RESOURCES   Clench (54) used the degree of overlap in flight season of several sympatric skippers, and the correspondence between flight seasons and nectar availability, to deduce that nectar supplies were limiting to the population densities of these species.They may indeed be so, but the different flight seasons imply very different conditions for the early stages also. Clench assumed that, because larval food was apparently abundant and shared among all species, the ecology of early stages was unimportant in this context. Scott (214) has used similar arguments to avoid considering larval ecology in detail. Studies of a long-lived tropical species, *Heliconius ethilla* (88), suggest that adult resources are one of several important factors that interact to limit both size and fluctuation of populations.

PREDATION AND PARASITISM   The extent to which parasitoids regulate butterfly populations is not adequately understood. A theoretical background for the types of possible interactions is given by Varley, Gradwell & Hassell (251). The relationships between *Pieris rapae* and its parasites vary geographically (62, 175, 188, 199); the same applies to *Euphydryas editha* (262) and *Plebejus icarioides* (74). In most populations of *E. editha* and *Lycaena dispar* (22, 80), the proportion of parasitism was not correlated with population trends. Harcourt (118), in what appears to be the most rigorous life table analysis of *P. rapae,* found that parasites accounted for about 4% of the total mortality per generation averaged over 18 generations.

Eggs and young larvae of *P. rapae* may suffer heavy predation from invertebrates (62) or birds (11), depending on the type of habitat (63). In agricultural situations, birds become major predators of the later larval stages (62). Dempster (62) used an immunological test to ascertain whether predators gathered in the field had recently eaten the abundant *Pieris rapae* larvae, but this method failed to identify predators of two scarcer Lycaenid species (243).

PATHOGENS   A viral, bacterial, or fungal disease is sometimes either the key mortality factor or a major one (22, 118, 173), often acting in a density-dependent fashion. Pathogens have been used to control pest species (5).

CLIMATE   Climatic factors may influence butterfly populations through effects on host quality (123, 227). Cold weather slows development of insects while often increasing the food requirements of their warm-blooded predators. Thus cold conditions may increase rates of predation, especially where the attainment of a relatively well-protected winter diapause stage (23) is delayed (193, 243). Bink (22) carried out a key-factor analysis on populations of *Lycaena dispar* and found that deaths of diapausing larvae (probably caused by fungi) during dull, humid weather constituted a major mortality factor.

Population extinction events have been attributed to unseasonal cold and snow in Colorado (89) and to poor host quality caused partly by low rainfall in California (227). The discovery that summer temperatures are correlated with changes of range in some British butterflies (see section on distributions) carries the implication that climatic factors may determine large-scale population extinctions and recolonizations. British workers tend to find that cold or dull weather reduces butterfly fecundity, estimated from egg counts. This is easy to demonstrate in species with well-synchronized flight seasons or diapausing eggs (243), when total generation egg counts can be made. In other species, the extent to which reduced daily oviposition may be counterbalanced by increased adult longevity is not clear. The converse of Dempster's (62) finding for *P. rapae* in England has been obtained in Japan, where Takata (241) found that fecundity in the field was depressed by hot weather.

Population Dynamics

Ford & Ford (105) have described changes in numbers of a population of *Euphydryas aurinia* over the period 1881–1930. After an initial increase from high to very high density, the insects reached a peak in 1896, defoliated their normal host, and turned to an unrelated secondary host. At this time, about 75% mortality was caused by parasites. This rose to 90–95% in subsequent years, and the population declined to near extinction in 1912. The extent to which this decline was due to parasitism or to long-term reduction of the perennial host, *Succissa pratensis* (Dipsacaceae), is not clear. From 1912 to 1920 the insect was very scarce; from 1920 to 1924 it increased rapidly again. The rate of increase then slowed, but was continuing in 1930 at the end of the study. This account suggests that *E. aurinia* may have semistable equilibria at two very different densities. The relationship between *E. aurinia* and its parasite(s) may be such that, at equilibrium, host densities are very low and extinction of either host or parasite is likely. If the parasite became locally extinct, then the butterfly would increase to a limit set by the host plant.

This hypothesis is supported by observations of *E. aurinia* in Europe (157). The insect occurs in isolated colonies, usually at low density, but occasionally reaching outbreak proportions. Local extinction is quite frequent (119), but introduction of a few parasite-free individuals to one site where extinction had occurred resulted in a rapid increase of density to the point at which the host plants were defoliated (200). This contrasts sharply with the failure or difficulty of other attempts to reintroduce British butterflies to habitats in which they had become extinct, even when such extinction was temporary and natural recolonization eventually occurred (80, 102). There is some evidence that the North American *Euphydryas editha* also undergoes crash/outbreak cycles in the northern parts of its range (226). However, three adjacent populations in the central portion of the species range showed independent fluctuations in size (41, 85), with one extinction and recolonization event. Changes in density from year to year were ascribed to changes in quality and quantity of larval resources and climate-caused differences in the relative timing of peak oviposition and host plant senescence (227). Density-dependent factors were not found and probably operate infrequently in these three populations. Parasites were present but did not seem to respond to changes in host density within wide limits (90).

A long-term study of *Heliconius ethilla* in Trinidad has shown remarkable constancy of adult numbers in two adjacent populations through wet and dry seasons (88). Several interacting factors, including egg parasites, the long (6 month) reproductive life of adults, and the resistance to dry seasons by the resources used by the adults account for these patterns. Larval resources seem in no way limited. This study points to the difficulty of applying common generalities about life history strategies to butterflies. First, Labine's (149) assumption that *Euphydryas editha* has a much higher reproductive effort than *Heliconius* turned out not to be true when *Heliconius* egg production was measured in nature. One big difference between these species has to do with which life history stage is responsible for most of the reproductive effort (see Gilbert 114). Second, another heliconiine, *Eueides aliphera,* shares the same host with *H. ethilla,* feeds on a more abundant and predictable part of the plant (older leaves), but lays much smaller eggs and is a smaller animal. Much of the *Eueides* reproductive strategy may have to do with its interactions with host chemistry and texture rather than environmental predictability. Ehrlich & Gilbert (88) pointed to the difficulty of drawing the "*r* vs *k*" selection dichotomy within closely related groups, and Wilbur, Tinkle & Collins (267) gave some general insight into this problem. Unfortunately, attempts to use butterfly data to support theories about life history strategies are likely to be overly simplistic, misguided, or *wrong* (e.g. 234) if they ignore the complex selective forces that shape egg size, body size, and voltinism (130, 230) (see below).

In another tropical work, Benson & Emmel (21) studied a roosting assemblage of *Marpesia berania* over a 6-month period. Their estimates of daily mortality and survivorship match similar statistics for *Heliconius* (88, 246). On the other hand, much lower adult survivorship indicated for *Parides* in Trinidad (55) is probably an artifact of the small study area and high vagility of *Parides.*

Acraea encedon in Uganda shows a greater fluctuation in population size than has been observed in neotropical rainforest species (185). Although this difference may be due in part to different climatic regimes, our observations of New World acraeines (*Actinote*) indicate that they also flux strongly in areas where other species are relatively constant.

In tropical wet forest areas, it is possible to observe as many generations in 2 yr (88) as would take 25–30 yr in a univoltine temperate species. It may thus be possible to make rapid headway in understanding tropical butterfly population dynamics.

Butterfly Movements and Population Structure

Many butterflies typically move very short distances (70, 84, 88, 245, 246, 274). Others are more wide-ranging (38, 111, 138, 166, 201–205, 215, 217) or undergo regular migrations (136, 180, 247, 257, 268). There is evidence that at least two species consist of populations that differ in their dispersal characteristics (64, 116), and this phenomenon may well be widespread. Various workers have been interested in quantifying butterfly movements, in assessing the effects of natural selection on distance and direction of movement, and in the evolutionary and genetic consequences of patterns of movement. We begin with a discussion of the movements of more sedentary species.

For some species, narrow bands of unsuitable habitat act as effective barriers to movement (71). This situation is not always easy to identify. For example, Ehrlich (84) selected a "population" of *Euphydryas editha* for study, and found that it was in fact a colony comprising three populations which exchanged very few individuals

($\sim$ 0.5%) per generation. Ehrlich felt that these populations were separated by suitable habitat and that the insects were, by their behavior, creating "intrinsic" barriers to dispersal. Subsequent findings have shown that resource requirements of *E. editha* larvae are so complex (227) that ovipositing females have no way of selecting suitable habitat for oviposition. Consequently it has been suggested (116) that the presence of other individuals is the best measure of suitable habitat and that the resulting tendency to stay with other individuals is the basis for the intrinsic barriers to dispersal in these populations. The fact that emigration is an inverse function of density (116) in this colony supports this idea. A study of movements within one of the three populations (subdivided into 30 m squares) showed that many individuals ranged through virtually the whole area, whereas others were recaptured many times in the same or adjacent squares. The population was considered to be effectively panmictic (41).

Many butterflies that are locally distributed and characteristic of specific habitat types probably maintain the integrity of their populations in a manner similar to that of *E. editha* (205). One alternative method involves the restriction of individuals to learned "home ranges," possibly centered on each emergence site. This is the tentative interpretation of Keller, Mattoni & Sieger (141) on the results of reciprocal transfers of adult lycaenids (*Philotes sonorensis*) between centers of abundance about 50 m apart. Half the insects caught in each area were released at a central site in the area of capture; the remainder were transferred to the other area. Displaced butterflies showed a significant tendency to return to the area of first capture, possibly because some of them were able to locate familiar landmarks from a distance (see 174, 240) or were displaced within their home ranges. A similar experiment with *E. editha* (briefly referred to, 90) failed to show a homing tendency by displaced butterflies.

A very different picture is emerging from studies of *Heliconius* populations, in which many individuals may share a home range that does not overlap with other home ranges, and in which each butterfly may show a strong tendency to visit the same place at the same time each day (114). Both *H. erato* (246) and *H. ethilla* (88) have proved just as sedentary as *E. editha* in the populations studied, although *H. ethilla* and other *Heliconius* have been recorded elsewhere as taking part in large-scale butterfly migrations (15). Two populations of *H. ethilla* (88) were contiguous, at least at some points, and hence not separated by even a narrow band of unsuitable habitat. Individuals from these adjacent populations may meet at the population boundary, but, with rare exceptions, each then returns to its population of origin. Neither the mechanism nor the selective advantage of this behavior is understood. Ehrlich & Gilbert (88) suggested that it results from the distribution of adult (and perhaps larval) resources as the adults learn optimum foraging routes. If new recruits learn by following other individuals (114), then there may be a tendency for population structures to be maintained by tradition. In addition, most *Heliconius* have gregarious roosting sites (137, 195) to which individuals are faithful (246) and which are thought to aid individuals in remembering foraging routes (114). Although other tropical butterflies roost gregariously (21, 178, 198), in no other cases are ecological details available to explain such behavior plausibly. How such roosting groups relate to local deme structure and to what extent they represent social groups are important areas for future study.

A corollary of *Heliconius* behavior and population structure is that kin selection may operate. Benson (19) inferred an inverse correlation between unpalatability and mobility in heliconiines, concluding that low dispersal rates have favored evolution of unpalatability, possibly by facilitating kin selection. Such a mechanism would avoid the necessity of assuming that individuals carrying genes for unpalatability must survive to reproduce after being attacked by the predators they are "educating." Such individuals could, by their demise, protect their close relatives, some of which would carry the same genes.

Among more mobile species, recaptures are fewer and patterns of movement more difficult to study by mark-release methods. Long-distance migrations of the Monarch, *Danaus plexippus,* in North America have, however, been documented by this method (247), as have shorter movements of *Pieris brassicae* and *Aglais urticae* in Germany (201, 202). Johnson (136) concluded that the two generations of *A. urticae* have differential tendencies to migrate, whereas Baker (12) argued that any such conclusion is unwarranted. Johnson regarded the prereproductive adult stage of many insects as specifically adapted for transit between habitats, and so may have been predisposed to the view that the sexually immature autumn generation of *A. urticae* is the more mobile. Baker, on the other hand, has stressed that *Pieris* spp. oviposit along their flight paths as they migrate. Evidently, in some species there is no correlation between sexual immaturity and migration.

Brussard & Ehrlich (38, 39), again using mark-release methods, have attempted a quantitative description of movements of *Erebia epipsodea* in Colorado. They concluded that insects marked at all their study sites were members of the same large panmictic population, which probably covered a very wide area. Individuals recognized suitable habitat and accumulated in it, but various types of unsuitable habitat did not act as barriers. These authors did not attach significance to the direction of movement. They contrasted population structures of ubiquitous butterflies, such as *E. epipsodea,* with those of more colonial species (40, 41), with a view to formulating and testing hypotheses about relationships between natural selection, gene flow, and genetic characteristics of populations (90). It was already evident (102, 104) that populations that exchange individuals may nonetheless become genetically distinct. Gel electrophoresis has confirmed this and facilitated an estimation of genetic similarity between two *Euphydryas editha* populations for which the rate of gene exchange has also been estimated (90).

Hard data on the nature and evolution of butterfly migration are elusive. Although many observations of butterfly migrations exist [summarized by Williams (268) and Johnson (136)], the sources of immigrants and destination of emigrants can rarely be ascertained. For example, 50 years ago, *Vanessa atalanta* and *V. cardui* were thought to be resident British butterflies subject to wide fluctuations in numbers. Then, largely as a result of Williams's work, they became accepted as nonresident migrants, arriving in spring in variable numbers, probably from the shores of the Mediterranean where swarms of *V. cardui* had been seen emerging from pupae and flying north as soon as their wings were dry. More recently, Baker (13) gave evidence that these species can survive the British winter as adults, and he deduced from phenotypic evidence that most individuals seen in spring in Britain are either resident or have traveled relatively short distances from northern Europe.

Another controversy stems from assessment of the influence of meteorological factors (wind direction and weather fronts) on the direction and distance of travel. Work on locusts and aphids has shown the paramount importance of wind direction, and students of insect migration are inclined to extrapolate these results to butterflies (136). Observers of butterflies, on the other hand, tend to the opinion that the insects normally retain control of their flight paths and cease flight when winds are too strong to achieve such control. Such observations can be very misleading, especially where characteristics of long-distance migrations are inferred from observations at one or a few geographical points. Johnson (136) interpreted Roer's (202) data as showing that *Aglais urticae* travels downwind, and suggested that even regular migrations, such as those of *Danaus plexippus,* are aided by wind. It is indeed possible that regular migration can be achieved in this way if winds are regular or if the insects are capable of detecting and utilizing favorable winds. It seems to us equally likely that butterflies are capable of compensating for wind-caused deviations from their flight paths, although this would be difficult to prove. Johnson's bias against this possibility is one of the factors leading him (and others) to be extremely critical of Baker's (8–10) theory of evolution of butterfly migration, because it implies that the directions of movement of *Pieris rapae* and several other species are determined by natural selection, independent of wind direction.

Baker has defined *voluntary displacement* as the act of leaving a patch of suitable habitat; he has followed Southwood (233) in arguing that such behavior should be more frequent in insects characteristic of temporary habitats. He has gathered some evidence (10) that the probability of observing an individual of any British butterfly species away from its habitat and flying in a straight line is positively correlated with the rate of formation of new habitat patches and with the extent of spatial separation of larval and adult resources in the species concerned. Larval habitat is defined solely in terms of presence of the host plant, an approximation that may frequently be wide of the mark. Some butterflies adapt to nonoverlapping larval and adult resources by commuting between them, presumably by memory, while maintaining the integrity of their populations (116, 164, 213).

Baker (8, 9) argued that insects undergoing voluntary displacement require a simple mechanism to avoid searching the same habitat patch twice, and that orienting at a constant angle to the sun's azimuth when traveling between habitat patches would provide such a mechanism. He showed that individuals of several species actually do this when followed over short distances, but that different angles are taken by different individuals in any set of observations. The average direction of movement can be calculated. Baker termed this the peak flight direction and showed that it can change seasonally. *P. rapae* changed peak direction abruptly in late August in both 1965 and 1966 at Bristol, England. Evidence that this is a response to photoperiod acting directly on the adults came from an experiment in which field-caught individuals kept at different photoperiods for a few days took the predicted directions on release. Differences in flight direction of individual *P. rapae* traveling between habitats were not traceable to genetic differences between them; laboratory-raised offspring of butterflies caught flying in opposite directions showed no differences in peak flight direction on release (10).

Johnson (136) considered Baker's report of a change in peak direction of *P. rapae* in two consecutive years to be inconsistent with the data. Although this criticism seems to us unjustified, Baker's calculation of selection pressures acting on direction of movement involves many explicit and some implicit assumptions. He rightly pointed out that first generation *P. rapae* that move to warmer areas stand a higher chance of giving rise to a third generation late in the summer. (This consideration seems applicable to second generation individuals also.) He then calculated from the relationship between temperature and pupal weight that individuals that move to cooler areas (within certain limits) give rise to more fecund offspring. The interaction of selection pressures on generation time and on fecundity will favor moving towards different temperatures at different times of year. Using climatological data, developmental times, and several assumptions, Baker calculated that *P. rapae* in southern England behave in the way that natural selection associated with movement towards optimum temperatures would dictate.

Baker observed that northward movement of *P. rapae* continues each year for a much longer period than southward movement, and concluded that the two are neither equal nor opposite. He concluded that a "return flight" is not a necessary component of migration, and that movement patterns of *P. rapae* involve continuous loss of individuals at the edge of the range (10). He included this loss in a model for a species reminiscent of *P. rapae,* calculating that, in a situation where selection favors moving north in the spring and summer and south in the autumn, individuals (termed type X) with a 45% probability of flying north, 15% probability of flying south, and 20% probability of flying east or west will come to outnumber type Y with random flight direction or any type with fixed flight direction, provided that the respective probabilities of flying north and south are reversed in the third generation. This does not mean that if the gentoype were to influence flight direction of individuals in a deterministic rather than stochastic manner (i.e. a genotype for flying north, one for east, etc), equilibrium gene frequencies would be such that 45% of individuals would fly north in the spring and south in the autumn, etc. Baker's model sets up a geographical range, initially containing types X and Y distributed equally and evenly with no interbreeding between them. The number of butterflies moving in a particular direction is multiplied by the selective advantage or disadvantage of doing so, and an even geographical distribution is then restored by a density-dependent mortality factor. This allows the fecundity advantage of moving northwards to act a generation too soon. Insects moving north, whose larvae have developed at higher temperatures, are accorded an immediate advantage over those already there, when the reverse should be the case from the effect of temperature on fecundity. We think that specific fecundities should be allocated to larvae developing at each latitude and that mortality factors should act in a specific sequence. If we do this, with reference to the situation described by Baker for *P. rapae,* in which northward flights always increase fecundity of offspring, whereas southward flight in the third generation increases *survival* of the resulting larvae to pupation, then we find that type X can indeed displace randomly moving individuals from the whole species population. Whether or not it does so depends on the manner in which the probability of larval survival varies with latitude. Baker has himself (8) noted that some *P. rapae* larvae in each year fail to reach pupation owing to the onset of cold weather.

Another *Pieris* species, *P. protodice,* occupies ephemeral successional habitats, and its populations usually last for no more than four years. Dispersal in this species has been described by Shapiro (217). Repeated interaction with males causes mated females to emigrate from high density areas with the result that such high density populations acquire heavily male-biased sex ratios. Individuals caught in peripheral areas of the species' range are almost all mated females, indicating that this density-related migration results in wide dispersal of females. There has been no suggestion of similar behavior in *P. rapae.*

Evidence that selection pressures on dispersal rates can produce quite rapid effects has been given by Dempster (64). Study of museum specimens has shown a decrease in wing size and thoracic width of *Papilio machaon* adults at Wicken Fen, England, after this habitat had become isolated (by >50 mi) from the remainder of the species' British range. This is interpreted as morphological evidence of reduced dispersal tendency. The Wicken Fen population has been extinct for about twenty years; attempts to recolonize it by releasing individuals from Norfolk have failed. Dempster found that mortality of early stages (introduced) at Wicken Fen is consistently lower than in the remaining natural *P. machaon* populations. He concluded that the reason for the failure of introduction attempts is that adults leave the habitat because of their (genetically) high dispersal rate. A possible alternative explanation of these very interesting results is that Norfolk individuals simply do not recognize Wicken Fen as suitable habitat, and leave for this reason.

A further example of intraspecific differences in dispersal tendency is given by Gilbert & Singer (116). They found that *Euphydryas editha* were predictably mobile in successional habitats, but a comparison of populations in two climax habitats also showed great differences in mobility. Insects in a chaparral habitat had to commute between larval host plants and nectar sources, while those in a grassland community found these resources superimposed. In addition, emigrant grassland butterflies were less capable of responding to the correct cues for selecting oviposition sites than were sedentary individuals.

Distributions of "Nonmigratory" Species

We have already described studies on microdistributions of single species. On a larger scale, most zoogeographical works (e.g. 1, 3, 34, 42, 44, 79, 99, 107, 128, 129, 131, 139, 140, 152, 172, 212, 228) have little ecological content, but some authors have related spatial and temporal distributions of the butterflies present in small areas to soil and habitat types and to floral associations. Such data are available for mountain ranges in Colombia (1) and California (91–94) and for Contra Costa County, California (182). Some taxonomic generalities have emerged. For example, Adams (1) found that 71% of lycaenid species were habitat specific, i.e. occurred in no more than one of the eight habitat types which he recognized. Papilionidae also contained a high proportion of "endemic" species, but other groups, particularly heliconiines, nymphalines, pierids, and danaines, were much less habitat specific. Williams (270) has analyzed Emmel & Emmel's data and also concluded that lycaenids are more habitat specific than other groups. One necessary limitation of this kind of study is that we cannot conclude that the habitats are suitable for all the species sampled from them. We know nothing of the distribution of early stages, which may occasionally be very different from that of the adults, and, in the absence of life-tables, the extent to which each species may be transient or dependent upon immigration for periodic replenishment is intangible.

Studies of butterfly distribution have also revealed the restriction of some species or subspecies to specific geological formations. Examples are the lycaenids, *Lysandra coridon,* and *L. bellargus,* on chalk and limestone (102); *Boloria napaea* and *Papilio machaon* on basic sedimentary rock (109); *Oeneis chryxus* along a junction between Ordovician sediment and granite (109); and *Euphydryas editha bayensis* on serpentine (116, 135). The two *Lysandra* spp. are host specific, and their host is confined to chalk and limestone. The same cannot be said for *E. editha,* which is not restricted significantly by the distributions of its various host plants (225, 264).

British authors have discussed the extent to which present geographical distributions of British butterflies are determined by their current ecological requirements and by historical factors such as glaciation. Beirne (17, 18) stressed the effects of glaciation on modern butterfly and moth distributions, with the assumption that most nonmigratory species have not yet recolonized areas that became favorable to them as the ice retreated northwards. He used this argument to account for the restriction of many species to southern or southeastern England. Ford (102) categorized all British butterflies in terms of their patterns of geographical distribution and used this evidence to deduce the relative antiquity of species in each category as inhabitants of the British Isles. He further noted which species have evolved distinctive British races. He argued that differentiation in these species, many of which are at the edges of their ranges, should be more rapid than average [see Raven (196) for a more detailed argument], and concluded that, in some species, differentiation to the subspecific level has occurred in less than 8700 years.

Ford is undoubtedly correct in his assumptions that most resident Irish butterflies arrived before the land bridge to Scotland was broken and that most resident British butterflies arrived via the land bridge to Continental Europe. Thus some European insects are absent from Britain and some British insects from Ireland for historical reasons. There is evidence that at least one French butterfly is well able to maintain itself in England if introduced (102). In general, effects of climatic changes are also apparent in that some northern species which possessed more southerly distributions during glaciations are now found in isolated populations at high altitudes far to the south of the bulk of their ranges (131).

However, the effects of past glaciation on the present distributions of butterflies within mainland Britain have been the subject of controversy, the outcome of which seems to be that these distributions are not amenable to analysis by the methods of "classical zoogeography" (72). Ford (102) criticized Beirne's (17) strictly historical interpretation on the grounds that several British species, notably the (dismorphiine) pierid *Leptidea sinapis* and the nymphalines *Ladoga camilla* and *Polygonia c-album,* had undergone dramatic changes of range in the preceding hundred years.

These species were widespread through most of the nineteenth century, became restricted to very small areas in the early twentieth century, and, from about 1920, have increased again. Since Ford wrote, other species, notably the satyrine *Pararge aegeria,* have extended their ranges (121, 156) and several species have declined considerably (121).

In many parts of the world, especially in the tropics, butterflies adapted to both early and late successional stages have increased in distribution and abundance as a result of human activities. However, man is unlikely to be responsible for documented increases in ranges of British butterflies, especially because *L. sinapis* and *L. camilla* are strictly woodland insects for which the amount of apparently suitable habitat has declined and fragmented. If not caused by man, such changes are probably not confined to recent history, and we can conclude that the distributions of these species are sensitive to and determined by current rather than past events. Scott (212) has reached similar conclusions for antillean butterflies.

Other British species have not shown recent changes in distribution, and Ford interpreted some aspects of their mainland distributions as effects of glaciation (72, 102). Downes's (72) criticism of this view shows that the butterflies of England and Scotland seem much more efficient than most authors have supposed in occupying almost the full area of currently suitable habitat. It is particularly impressive that three species typical of woodland, *Ladoga camilla, Pararge aegeria,* and *Leptidea sinapis,* have undergone extensions of range that must have involved them in crossing tracts of unsuitable habitat, because most of the woodland areas of England occur as small isolated pockets. In contrast, the woodland hairstreak *Strymonidia pruni* seems to oviposit on the same few host plant individuals year after year and to colonize very slowly or not at all other areas that have become suitable and that will support introduced populations (243). The distribution records of all British butterflies are now gathered in systematic fashion, and current distribution maps on a 10 km grid (some of which are very different from Ford's 1945 maps) have been published by Heath (119–121). Where changes in distribution are accurately recorded we have more hope of interpreting them in terms of the ecology of the species concerned, although it does seem that the data already available are susceptible to more detailed analysis than has been reported. Although Ford (102) was unable to account for changes of range in terms of climatic trends, Pollard (193) stated that the increase of *Ladoga camilla* (and presumably several other species) in this century has been closely related to changes in summer temperature. Population studies in England have shown great variation in fecundity dependent on temperature and isolation (62, 80, 243), and solar energy is very important to adult flight activity (53, 122, 131, 142, 253, 258, 272), so that effects of short-term variation in climate on species' range seem quite probable.

COMMUNITY ECOLOGY

Niche Components

The evolution of multispecies communities has resulted from the fact that no one genotype has high enough fitness over all the different ecological circumstances to prevent more specialized genotypes (and species) from evolving in or invading the community. Although we are a long way from being able to explain why local butterfly communities are structured as they are, we can, on the basis of current knowledge, point to those niche dimensions most important in understanding the ecological segregation of butterfly species.

Owen (184), using information provided by Ford (102), provided the first synthesis of this subject from the standpoint of ecological theory. In his analysis of ecological segregation of British butterflies, he noted the following niche differences between butterflies: 1. larval food plant, 2. part of host used, 3. time of appearance (phenology) and number of generations (voltinism), 4. habitat, and 5. flowers visited by adults. We would add: 6. parasite and predator escape (including mutualistic relations with ants); and change 5. to read "adult resources." In the following paragraphs, we discuss the state of our knowledge concerning each of these niche components. In real life, these categories interact in complex ways but they are kept separate here for purposes of discussion; complications are noted where necessary.

LARVAL RESOURCES: TAXONOMIC PARTITIONING   Butterfly species are generally confined to taxonomically and chemically related larval host plants and sometimes to a single species (163). Moreover, the guild found on a taxon of plants is almost always composed of groups of closely related species. This pattern led Brues (36, 37) to suggest strong evolutionary interaction between insects and plants. Kusnezov (148), looking at host relations in the Pieridae, drew similar conclusions but focused on the role of host plant restriction in limiting geographical distributions of butterflies.

Ehrlich & Raven (87) summarized a vast amount of worldwide host plant information for butterflies and developed a model for the coevolution of butterflies and plants. The basic steps of coevolution involve adaptive radiation of plants that have evolved relative chemical protection from herbivores, followed by adaptive radiation on these plant groups of herbivores able to circumvent their defenses. (See also 100, 112).

A major question deserving much further work involves the idea (87) that such coevolution has generated much of the observed organismic diversity. Unfortunately, much of the published literature on butterfly host plants (e.g. 61) is not useful for investigation of the contemporary ecological and microevolutionary dynamics of butterfly-plant interaction. Answers to questions as to what limits the number of species within a local host plant guild or what factors determine the microevolution of host plant preference depend upon the collection of locality-specific host plant data for interacting sets of species. For example, for many years, the reported host plant of *Euphydryas editha* was simply *Plantago erecta*. Recent studies have shown, however, that different local populations having similar or identical arrays of potential host plants available are likely to show differences both in which plants are preferred and in the rank of preference shown (225, 264).

Similar detailed studies of the geographical ecology of host usage, although few, are invaluable for understanding the factors involved in the partitioning of host plants among members of a guild. Two components of larval host plant choice must be kept in mind in discussing these factors: that of the ovipositing female and that of the larval stage. Because the larva has the burden of eating and assimilating the host plant, its physiology (shaped in part by coevolution with the host) must determine the fundamental niche of the species with respect to host plants. Consequently we can generally view adult oviposition preference (in the field) as an indication of the realized niche of the species at any given locality.

In theory then, analysis of the nature and degree of differences between adult and larval host preference among species and among populations should be a powerful tool for understanding why and how host plants are partitioned within local guilds. For instance, one population of *E. editha* does not oviposit on *Pedicularis,* yet larvae can feed on the plant. Proximate cues for this preference had nothing to do with plant chemistry but rather were related to avoidance of oviposition in the shade (225). The presence of ants on *Pedicularis* in that locality may be the basis for the evolution of this behavior (116). In contrast, among certain lycaenids that are mutualistic with ants, the presence of ants on a plant is required for oviposition (14, p. 34).

Competition with closely related butterflies may help to shape the evolution of oviposition preference. For example, Downey & Dunn (78) have shown that local populations of the blue butterfly *Plebejus icarioides* are almost always monophagous, in spite of using 28 species of lupines over their range and in spite of the ability of larvae to develop on most of the lupines not used locally. In one of the few areas where *P. icarioides* used two species of lupine, ants were found to be transporting young larvae to the lupine species ignored by ovipositing females. Geographical comparison of *P. icarioides* hosts with those of *Glaucopsyche lygdamus* (76) in northwestern America indicate that *P. icarioides* populations are always on Lupine *A* except in areas of overlap with *G. lygdamus.* In such areas, *P. icarioides* uses Lupine *B* whereas its presumed competitor occupies *A.*

In a more recent study, Goodpasture (117) has found similar patterns of host plant usage in the *Plebejus acmon* group. Again, larvae have a greater range of host acceptance than oviposition preference indicates. Butterfly species in this group have mutually exclusive diets where they overlap. For example, *P. acmon* does not utilize the shrubby *Eriogonum* hosts of *P. lupini* if in sympatry with *P. lupini,* but occurs on a wide variety of these plants where allopatric to this species.

In a similar study, Straatman (237) observed two species of *Ornithoptera* feeding on two *Aristolochia* spp. on six Pacific islands (Malaita, Nggela, Guadalcanal, San Cristobal, Santa Ysabel, and Rendova). His results indicate that in areas where plant II is absent, it is found that either both butterfly species use plant I or one of the butterfly species is missing. Where plant II is present but rare, it is preferred by *O. victoriae,* but plant I is in fact used as second choice. Where both plants are common, the host plants are cleanly divided between the butterfly species, and larvae are not able to survive on the nonhost species. Straatman took *O. victoriae* larvae from Guadalcanal where they feed on I and tested them on *Aristolochia* II used by *O. victoriae* in Malaita. That they cannot survive indicates differentiation among the islands with respect to host tolerance (see also 236). Superimposed on this interaction with host plants is possible competition with *O. urvilleanus* indicated by the restriction of *O. victoriae* to II in the presence of *O. urvilleanus* on three islands, even though it is evolutionarily capable of exploiting plant I. In this case, the relationship between competition and coevolutionary interactions with hosts is not clear.

Another way to gain insight into host plant partitioning is to observe the same complex through time. Emmel & Emmel (95) observed that two desert swallowtail butterflies, *Papilio indra* and *P. rudkini*, normally restricted to different host plants during most years, both switch as larvae to the other's food plant under conditions of high density and host defoliation. Likewise, Brower (28) provided evidence that oviposition preference in *Danaus gilippus* females changes through a season depending on the presence or absence of its migratory congener *D. plexippus*. Such behavior indicates possible recognition of eggs and larvae on host plants and represents a more flexible (phenotypic) response to competition than those previously discussed. The ability to recognize and avoid the eggs of competitors is strongly suspected for *Heliconius* butterflies (114) where larvae are often cannibalistic. However, such behavior would not account for partitioning of host plants; rather, it is an alternative strategy.

Heliconius butterflies, like *Euphydryas* and *Plebejus,* show a much broader range of host acceptance as larvae than as adults. Gilbert (114) has suggested that the highly visual searching behavior of *Heliconius* has been an evolutionary force in determining leaf shape and stipular morphology of their host plants. Indeed, in each locality, *Passiflora* species are remarkably distinct in appearance. Such interaction between the butterflies and plants would enhance the host partitioning, as would, in general, any factor increasing the distinctness of the different host species.

The excellent work of Chew (47) on a complex of *Pieris* species and their crucifer host plants shows some striking departures from the patterns discussed to this point. In Rocky Mountain populations of *P. napi* and *P. occidentalis,* females oviposit on four crucifer species in proportion to their abundance, in spite of the fact that, for each butterfly, there is at least one crucifer species best for larval development and one which kills young larvae. Adult oviposition preference is thus broader than the feeding tolerance of larvae (see also 239). Chew suggested that as long as the unsuitable plants are often in low frequency relative to more suitable species, selection on oviposition-level discrimination will not occur. She points out that within- and between-season fluctuation in the relative abundances of the crucifers (caused by climatic factors) retards refinement of female oviposition choice; at the same time the low growth form and high density of the host plants permit larval mobility, giving these populations the evolutionary opportunity to develop greater selectivity in the larval stage. Not unexpectedly, therefore, *Pieris* larvae select the host species that provides the most rapid development.

Neck (179) found a similar pattern in composite-feeding *Chlosyne lacinia* populations in Texas. During the growing season, the host plant community changes in composition, requiring the butterfly to use several different hosts to produce the five generations observed. These successional habitats are more unstable between seasons than those observed for the other species described in this discussion. The resulting dispersiveness of *Chlosyne* may account for the lack of geographical differences in host use and the lack of multispecies assemblages of *Chlosyne* dividing up Texas sunflowers.

In none of the cases presented so far can we be sure of the ultimate basis for within-guild partitioning of host plants. *E. editha,* for example, is often highly restricted in the absence of both competition and chemical barriers (98) in unexploited but suitable hosts. Although the development of general niche theory has been aided by easily quantifiable resource spectra (prey size) and the related beak or jaw sizes of birds and lizards, the analogous ranking of possible resource plants within a guild of butterflies on the basis of chemical distance is currently not possible and may never be feasible. For one thing, plants vary both within and between species in many classes of compounds (see 69 for discussion), nutritional content (97, 229), water content (100, 230), and pubescence (25, 76). One approach to developing more rigorous theory in the absence of full understanding of chemical differences between resources is that of Colwell & Futuyma (57). They were faced with the analogous problem of assigning values to the various chemical states of a continuously decaying resource. The linear distance between two resource states is determined by the total community of arthropods occupying them. In like fashion, we suggest that a study of the total herbivore assemblage of each host plant might indicate the degree of similarity of these plants from the standpoint of an average phytophagous insect's digestive physiology.

The observed breadth of larval feeding is not determined by beak or jaw size, but by a complex system of sensory coding (68) in adults and larvae, and by the particular mode of biochemical processing of host plant chemicals [including, in some cases, bacterial symbionts (165)]. We know that polyphagous species differ from monophagous species in the nature of their digestive enzymes (147) and in the efficiency with which they assimilate the host plant (183). MacArthur & Levins (158) have discussed a theoretical model relevant to the evolution of monophagy and polyphagy as strategies.

LARVAL RESOURCES: PARTITIONING THE PARTS OF HOST PLANTS   To this point we have stressed that coevolutionary interactions with hosts and competitors have tended to segregate butterfly species onto different plant species. However, a plant species is not a single resource but several: roots, stems, old leaves, buds, flowers, fruits, etc, each of which has its own chemical, nutritional, and textural qualities. Butterflies are primarily restricted to flowers, buds, young leaves, and old leaves, although a few root feeders and fruit feeders are known. Moths, on the other hand, although they tend as a group to be less specific to host taxonomy, have partitioned the parts of plants much more finely. A comparison of moth and butterfly host plant segregation would be of great interest in understanding the basis for these major adaptive radiations of Lepidoptera.

There is no synthesis available concerning the segregation of butterflies on different plant parts. However, within some guilds such as the *Plassiflora*-feeding heliconiines, we find that feeding on the growing point of these vines seems to be a recent adaptation characteristic of the genus *Heliconius* (K. Brown and W. Benson, unpublished), whereas other genera, such as *Eueides,* feed generally on old leaves (2). If host plants are often scattered and ephemeral (as in successional habitats or highly seasonal areas), the new growth of plants will be even more restricted in time and space. Moreover, new growth supports an insect community high in honeydew-producing homopterans along with their ant attendants. Not surprisingly, the butterfly group (Lycaenidae) that has adaptively radiated on new growth is characterized by various levels of association with ants and Homoptera, ranging from feeding the ants and receiving protection from parasites in some species, to specialized feeding on homopterans or ant larvae (14). Lycaenids are, on average, the smallest of adult butterflies, possibly because new growth often comes in small isolated patches insufficient for the support of large larvae. Parasitic mistletoes, which may share the general features of new growth judging from the number of independent shifts to these plants among the Lycaenidae (e.g. 52), represent larger, less temporary host patches, and, consistent with our idea, support some of the larger lycaenids. Consistent also is the fact that ant-feeding species (e.g. *Liphyra brassolis, Maculinea arior*) tend to be much larger than most.

ADULT RESOURCES   The broad outlines of adult feeding habits in butterflies as provided by early naturalists (181) have been expanded only slightly by recent studies (4, 16, 73, 108, 113, 151, 153, 223). Adults are known to feed on floral nectar, pollen, honeydew, froghopper secretions, rotting fruit, urine, perspiration, dung, and carrion. All indications are that adults are, as species, less specific and more opportunistic in feeding than are their larval counterparts. However, definite patterns do exist, which are discussed below.

Surprisingly little is known about the relative importance of adult feeding compared to larval feeding in the overall matter and energy budget of butterflies. Not only are the various "foods" poorly characterized chemically, but also the physiological roles of certain known constituents of adult diets are not understood (4). Most studies of adult diet have emphasized the contribution of nectar carbohydrate to longevity, fecundity (235), and flight energetics (35). Recently, P. A. Labine and T. R. Turk (unpublished) have worked out a complete energy budget for *Pieris rapae,* showing that 56% of the adult budget is due to adult feeding.

But for herbivores, energy is not so often limiting as are various elements such as nitrogen. Gilbert (113) has recently shown that *Heliconius* butterflies collect pollen and extract free amino acids that are used in egg production. The fact that reducing pollen meals while keeping nectar available drastically reduces egg production in *Heliconius* (113) suggests that the amino acid content of butterfly nectars, now known to be ubiquitous (7), cannot be neglected in studies relating nectar or honey consumption to fecundity. In field studies, Watt, Hoch & Mills (259) have shown that flowers preferred in the field by *Colias* are those that supply high monosaccharide nectars of high water content containing polar, nitrogen-rich amino acids.

Of the non-floral foods of butterflies, little is known. Some, such as urine concentrations and dung, mysteriously attract almost entirely the male sex of butterflies and also of moths that visit the same resources by night (73). It was long ago suggested (194), and recently confirmed experimentally (4), that sodium salt is the primary attractant to "puddle"-visiting males. Is sodium used as a signal for the presence of nitrogenous products, or do males throughout the Lepidoptera differ drastically in physiology from females? Because partitioning of resources by sex is a well-known phenomenon in other animals, we suggest that male and female butterflies may partition nitrogen sources so that females obtain a larger share of nectar amino acids than males. This is certainly the case in those *Heliconius* species that temporally partition adult resources between the sexes. Earlier-flying females gather most of the pollen and thus the best nitrogen supply (114). In any event, studies of floral visits by the different sexes of puddle-visiting species would be useful. Other male-specific feeding behavior, such as that of male daniine and ithomiine butterflies on decaying borage plants (16, 115), can relate to the collection of sex pheromones or their precursors (82, 192) and has resulted in a male-specific pollination relationship (254).

Although the chemical and physiological details of adult feeding need much more attention, the studies summarized here leave little doubt of their importance. Moreover, several studies of adult behavior and population structure as they relate to adult resources (86, 88, 116, 164) support this view. We can now turn to the problem of ecological segregation on such resources.

As with larval feeding, it is possible to distinguish feeding guilds that, as a rule, characterize higher taxa. For example, within the family Nymphalidae, we find that the subfamily Danainae is primarily nectar feeding, the Ithomiinae feed on nectar and bird droppings, and the Charaxinae feed entirely on dung and rotting plant material (249). Another nymphalid subfamily, the Nymphalinae, cannot be characterized, but its component tribes can. The tribes Heliconiini and Argynnini rely heavily on nectar, whereas the tribes Apaturini and Nymphalini, except for a few genera, are not flower visitors.

Segregation occurring at the level of subfamily and tribe probably reflects noncompetitive kinds of restraints on adult feeding habits. In tropical regions, for example, most diurnal flower visitors tend to be 1. small and/or hard to catch (Lycaenidae, many Papilionidae and Pieridae), 2. distasteful and warningly colored, or 3. mimics of distasteful species. Dung, fruit, and sap feeders are cryptic and edible. Papageorgis (187) has made similar observations. One of the largest lycaenids, *Liphyra brassolis,* is crepuscular in flight habits (14).

Because larval host plant chemistry often (29), but not always (170), determines whether imagos have the evolutionary option to become unpalatable and warningly colored, larval host plants may thereby help determine whether evolution of flower feeding is feasible in the face of the presumed higher predator pressure associated with that resource. Whether, as we have just indicated, particular larval host plant specializations preadapt butterflies for invading certain adult food categories, or whether the reverse is true (187), is a question deserving study.

Although coevolution with nectar plants may tend to associate butterflies seeking particular kinds of nectar with particular floral patterns (259), relations with adult resources do not show nearly so high a degree of chemical specificity as is generated by larval-host plant coevolution [except in a few highly exceptional cases (192)].

Thus, adult resources are more likely to be partitioned among members of a guild because of contemporary competitive interaction than are larval resources. For example, the flowers of *Anguria* are almost entirely dominated by *Heliconius* in the field. In insectary conditions, however, *Anguria* flowers are routinely visited by species (*Danaus, Dryas, Itaballia*) that have never been seen to visit them in the field (L. E. Gilbert, unpublished observations). In nature, where these flowers are at low density, *Heliconius* exploit them so efficiently that they are never economically exploitable to nontraplining species. All *Heliconius* species so far studied feed on pollen (114), and as pollen is a limited and valuable resource (113), it might be expected that competitive partitioning would have occurred. So far, however, the only indications of segregation within this guild are the facts that different *Heliconius* species in insectaries differ in their efficiency in harvesting pollens of particular sizes under differing competitive situations, and that in the field, some species collect only large pollen whereas others collect size mixtures (Gilbert, in preparation).

Some of the most interesting problems in butterfly community biology are to be found in the area of adult resource ecology. For instance, dung- and urine-feeding species have definite microspatial segregation patterns on the resources that are determined in part by the color, and in part by the size and aggressiveness, of the species involved (56, 181, 185). If, in such species, size evolution is greatly influenced by competition on adult resources, compromises will likely be required with respect to larval biology, adult-predator interaction, and overall life history strategies that further complicate interpretation of community patterns.

Aside from brief morphological descriptions by Scudder (211) and a restricted discussion of functional morphology (113), the adult mouthparts of butterflies have been essentially ignored from every possible point of view. Within larval host plant guilds, species often differ in proboscis length (e.g. *E. editha* vs *E. chalcedona; Heliconius* species vs *Eueides* species) and therefore in the range of floral corollas exploitable. Some tropical skippers (Hesperioidea) and riodinids (e.g. *Eurybia patrona*) have very long tongues and visit flowers of a wide range of color and morphology, including "moth" and "bird" flowers (Gilbert, unpublished observation).

HABITAT PARTITIONING   It is often observed that butterfly species are much more restricted in their distributions than are their potential host plants (123, 238) and this can be true locally (219). Where the absence of one species is correlated with the presence of other ecologically similar species, it is often asserted that past competition has caused the contemporary segregation.

Petersen (190) suggested, for example, that in Italy sympatric *Pieris* species have responded to competition by evolving different habitat preferences with regard to whether host plants grow in sunny (*P. rapae*) or shaded (*P. napi*) situations. However, because shaded host plants have thin leaves compared to those in the sun, and because *P. napi* larvae have a lower temperature preference (28°C) than those of open habitat *Pieris* such as *brassicae* (33°C), thermal and host plant differences between sunny and shaded habitats, rather than competition, could account for such observations. Recall that at least one *E. editha* population restricts its oviposition to sunny situations in the absence of competition.

Shapiro & Cardé (218) also suggested that interspecific competition accounts for the habitat differences of *Lethe eurydice* (sunny, open areas) and *L. appalachia* (dense, shaded woods and brush), both sedge-feeding satyrines. As in the case of *Pieris napi* vs *rapae,* direct evidence for competitive displacement is lacking. The case for competition in another satyrine complex is more convincing. Lorković (162) has shown that two species of *Erebia, E. cassioides* and *E. tyndarus,* with similar altitude preferences, show strict geographical replacement with zones of contact but no coexistence. Similarly, *E. cassioides* and *E. nivalis* abruptly replace one another vertically when (geographically) sympatric, but in allopatry each occupies the full vertical range. A similar case is noted for *Argynnis* spp. in Britain (102).

It is our bias that only when more detailed and experimental population studies are conducted on complexes of ecologically similar sibling species (such as those mentioned above) will we be able to properly judge the role of competition (past or present) on habitat selection in butterflies.

TIME: PATTERNS OF PHENOLOGY AND VOLTINISM   In any seasonal habitat where resources are unavailable to butterflies for some part of the year, there is great variation between species in the timing of their first and last appearances of the season (phenology) and in number of generations during the "growing season" (voltinism). For nonmigrants, the physiological mechanism controlling phenology is diapause. Various environmental factors are known to govern the initiation and breaking of diapause (155), and genetic differentiation has occurred within species in response to these factors (60).

In theory, time of appearance could thus be accurately controlled to minimize overlap between competing species in seasonal habitats. Clench (53) suggested that the phenological displacement of adult hesperiines within flower-feeding guilds represents competitive displacement in time. Owen (185) made similar assertions for African *Precis* and *Acraea.* Most variation of phenology and voltinism within butterfly communities, however, can be attributed to interactions between the life histories of larval resource plants and the butterflies that exploit them (e.g. 24). Thus Slansky (230) showed that multivoltine species tend to be associated with shrubs and trees, whereas univoltine species are more likely to be associated with more ephemeral herbs. Moreover, small butterflies, which develop faster, can have more generations per season than large species utilizing similar types of host plant. Multivoltine species in a Mediterranean climate are generally associated with plants that remain green during the hot dry summer in habitats such as stream banks (182). Weissman (260) has studied in detail the population biology of a bivoltine satyrine (*Coenonympha tullia*) at Jasper Ridge, California, and suggested that the necessity for a second generation that undergoes reproductive diapause (see also 83, 210) stems from the fact that larval diapause occurs in the first instar, which would be unable to survive both a long dry summer and part of a winter. The second adult generation then minimizes the time spent as diapausing larvae. However, other resident species on

Jasper Ridge, such as *Euphydryas editha*, persist with univoltine strategies, presumably because of host plant and other life history differences (227). In Egypt, a lycaenid, *Virachola livia,* is able to breed all year without diapause (in spite of the Mediterranean climate) by alternating among three cultivated host plants (6).

Sherman & Watt (221) have argued that some butterfly larvae have sacrificed maximum efficiency in the use of solar energy for their cryptic green color patterns evolved in response to visually hunting predators. The resulting metabolic organization means slower larval development and would, like tougher, more toxic, or more ephemeral host plants, alter the feasible set of phenologic and voltinism patterns in a given environment (and thus determine the diapause strategy evolved). That thermal differences between early and late parts of the growing season are important to some adult butterflies is evidenced by the photoperiodically controlled pigment changes between generations of *Colias* (126, 258). Thermal heterogeneity within seasons could, in the absence of such changes in development and in host plants, allow temporal separation of otherwise identical species (but we know of no unambiguous case of this in butterflies).

Other potential factors that would select for temporal separation (or, in some cases, synchrony) in activity periods are shared predators and parasitoids. Waldbauer & Sheldon (255) have shown that dipteran mimics of wasps appear after their models, presumably to allow time for the education of naive birds. Rothschild (170) has suggested similar reasons for the phenology of certain British Lepidoptera.

Shared parasites can represent a similar kind of selective pressure organizing species in time. The European pierids *Aporia crataegi* and *Pieris brassicae* share the braconid parasitoid, *Apanteles glomeratus* (266). Differences in their voltinism and phenologic patterns may in part be a parasite escape response, but studies on the life histories of these and other sets of butterflies overlapping in parasitoids are needed.

Although dormancy is a commonly used strategy for persisting through unsuitable periods (146, 178), emigration is a common alternative that increases in importance as one approaches Mediterranean deserts, the subtropics, and tropics where unfavorable seasons are dry and warm and where montane or riparian refugia are nearby. Janzen (134) has considered the costs of dormancy in tropical dry seasons, and his generalities apply to butterflies. Several long-term studies of tropical wet forest butterflies indicate relatively constant populations throughout the year (21, 88); in some areas the seasonal population changes suggested by samples (96, 186) may be due to activity differences from season to season (185, p. 72).

Any interpretation of phenological patterns in local butterfly communities must be based on thorough knowledge of the life histories of the species involved. For instance, in the deserts of the Mid-East, a moth, *Celerio,* and a butterfly *Melitaea,* both appear as adults and larvae each spring and fall. However, these similar phenologies are achieved in ecologically distinct ways, *Celerio* emigrating to higher elevations, the *Melitaea* diapausing as pupae through the dry summer (271). From the standpoint of understanding community evolution, resident (diapausing) species are perhaps of greatest interest, but it should always be kept in mind that they may occasionally interact, as larvae or adults, with migrants possessing a more global strategy. In particular, if migrations are predictable, resident competitors or mimics would be expected to adjust their phenologies accordingly.

PREDATOR ESCAPE   Clarke (51) was the first to suggest the possibility that apostatic selection, the frequency-dependent visual selection exerted by predators that develop searching images for prey, might not only affect allelic frequencies within species, but also allow the coexistence of two otherwise ecologically identical species that share such predators. The different color patterns adopted by various prey can thus be regarded as analogous to resources, and all species sharing a color pattern constitute a guild. The realized number of different prey color patterns depends upon the visual structure of the environment (substrate and other prey species) and the discriminatory ability of the predator, the latter a psychological dimension to ecology in need of much more study. The extent of mimetic polymorphism in butterflies (103, 104) and the observed correlations between mimetic morph frequency and model frequency (30, 33, 231) indicate that Clarke's model is applicable to butterflies. In this case, however, each model pattern is a resource to an edible mimic species. Only so many individuals, and thus species, can share such a resource.

Papageorgis's (187) most interesting and important study on neotropical mimicry complexes revealed vertical stratification of different color pattern groups within rain forests. Her data suggest that several different mimicry complexes can coexist (contrary to theory) in an area due to the different visual backgrounds at each level in the forest, each of which specifies different optimal wing colors for signals or crypsis. Different thermal conditions at each level seemed to have little significance, and vertical stratification of different predators was not required to generate the observed pattern. Presumably this means that at least some closely related species are segregated largely on the basis of predator escape modes.

Unfortunately, studies of mimicry complexes have not been accompanied by studies of host plant relations and life-table or key-factor analyses (251). So, although we know that birds attack butterflies (45, 46, 171, 240) and that these attacks are somewhat frequency dependent (58, 59, 127) and exert selective effects on color pattern (20), there is still no evidence to back the common assumption that the observed lower relative densities of batesian mimics (vs models) are maintained by bird predation on adults. Indeed, Rothschild (207) regarded low density as a preadaptation for mimicry. Thus although we understand the significance of color pattern within single mimicry associations, explaining why several mimicry systems coexist (187) or why some groups such as riodinids (see 216) have high color or pattern diversity whereas others, such as satyrines, have little (the reverse may hold for larvae) is a complex ecological problem that can be attacked only at a community level.

The larval stages of butterflies are, in theory, competing for the same kinds of predator escape niches. Brower (26) has suggested that because the host plant is often the background that the caterpillar must match to escape visually hunting predators, there is a limit imposed on the number of species that can use a plant species (just as there is a limit to the number of mimic species per model). Brower suggested that either the different species are forced onto different plants (or different plant parts?) or they become polymorphic. Rothschild (206) extended these arguments to predict the clumping of aposematic species on the same host plant.

Larval escape from invertebrate predators and parasites takes a different form. Some parasitoids locate prey by odors of the feces, and we find butterfly species that explosively eject frass up to 3 ft from the feeding position (110). Many nymphalines build chains of frass on the tip of the midrib of host leaves, and thus presumably avoid patrolling ants (154, 177). Many of the structures and behaviors of heliconiine larvae relate to defense against ants (2), and it appears that specialization on different parts of the host plant requires very different kinds of defensive strategies.

Another kind of predator escape is represented by the lycaenid-ant association that we have mentioned in previous sections. This is a fascinating subject, with a large descriptive literature that has been summarized several times (14, 77, 124, 150, 167, 169, 205). By feeding ants, lycaenids are not only immune to predation from their hosts, but also are better protected from other ant species and parasitoids than would otherwise be the case. The ants also transfer eggs and larvae to more suitable host plants and protect them in their pupal stage. Although some excellent ecological observations have been made on single species (75, 167, 205), the possibility of ecological segregation via choosing different ant species has not been considered. Because the Lycaenidae account for about half of all butterfly species and genera, it is tempting to suggest that mutualism with ants represents an important adaptive zone for butteflies.

Community Patterns

The number of important niche dimensions for butterflies reflects the diversity of strong trophic links to other groups of organisms: adult and larval host plants (usually different), parasites and predators of eggs and immature stages, and predators of adults. Although some data are available on the diversity and relative abundances of species in local butterfly communities (56, 81, 91–94, 96, 185, 186, 252, 273), there is little hope of explaining such patterns without a better understanding of the important ecological parameters that define each guild. Gilbert's (114) attempt to explain species packing in local heliconiine communities illustrates the complexities involved.

Butterflies like *Heliconius* are very often major herbivores in simple food webs consisting of coevolved, chemically distinct larval host plants and (quite likely) host-specific parasitoids. The fact that many butterflies store plant poisons (29, 32, 170, 208) and pass them on to the next trophic level (31) has many evolutionary and ecological ramifications at the food web level, some of which have been discussed above in other contexts. We believe that studying butterflies as elements in broader systems can help us to understand the evolution and ecology of this group and to render butterfly research more relevant to general problems of community biology.

ACKNOWLEDGMENTS

We should like to thank P. R. Ehrlich, T. R. E. Southwood, R. I. Vane-Wright, and the staff of Monks Wood Experimental Station for the use of their private and/or institutional library facilities; R. R. Baker, P. R. Ehrlich, D. A. Kendall, C. Norris, and M. E. Solomon for reading parts of the manuscript; K. Cooper, C. G. Gilbert, C. Powell, and P. A. Singer for their technical help; W. W. Benson, L. P. Brower, K. S. Brown, P. F. Brussard, F. Chew, J. A. Edgar, J. L. Gressitt, D. Høegh-Guldberg, C. Jordan, P. A. Labine, A. M. Lucas, A. K. Mecci, D. F. Owen, C. Papageorgis, A. P. Platt, T. E. Pliske, V. B. Polácek, J. A. Powell, M. Rothschild, F. H. Rindge, J. A. Scott, R. Silberglied, A. M. Shapiro, A. W. Skalski, J. Smiley, V. G. L. van Someren, J. H. Szent-Ivany, J. A. Thomas, J. R. G. Turner, L. Vári, K. Weissman, R. R. White, and A. M. Young for helpful discussions, correspondence, reprints, or loans of manuscripts and theses.

Literature Cited

1. Adams, M. 1973. Ecological zonation and the butterflies of the Sierra Nevada de Santa Marta, Columbia (Lep.) *J. Natur. Hist.* 7:699–718
2. Alexander, A. J. 1961. A study of the biology and behavior of the caterpillars and emerging butterflies of the subfamily Heliconiinae in Trinidad, West Indies. Part I. Some aspects of larval behavior. *Zoologica* 46:1–24
3. Amsel, H. G. 1933. Distribution and ecology of Lepidoptera of Palestine. *Zoogeographica* 2:1–146
4. Arms, K., Feeny, P., Lederhouse, R. C. 1974. Sodium: Stimulus for puddling behavior by tiger swallowtail butterflies, *Papilio glaucus*. *Science* 185:372–74
5. Arthur, A. P., Angus, T. A. 1965. Control of a field population of the introduced European skipper *Thymelicus lineola* (Ochsenheimer) (Lep. Hesperidae) with *Bacillus thuringiensis* Berlinen. *J. Invertebr. Pathol.* 7:180–83
6. Awadallah, A. M., Azab, A. K., El-Nahal, A. K. M. 1970. Studies on the promegranate butterfly, *Virachola livia* (Klug). *Bull. Soc. Entomol. Egypte* 54: 545–67
7. Baker, H. G., Baker, I. 1975. Studies of nectar-constitution and pollinator-plant coevolution. In *Coevolution of Animals and Plants*, ed. L. E. Gilbert, P. R. Raven, 100–40. Austin & London: Univ. Texas Press. 246 pp.
8. Baker, R. R. 1968. A possible method of evolution of the migratory habit in butterflies. *Phil. Trans. R. Soc. London* 253:309–41

9. Baker, R. R. 1968. Sun orientation during migration in some British butterflies. *Proc. R. Entomol. Soc. London* (A) 43:89–95
10. Baker, R. R. 1969. The evolution of the migratory habit in butterflies (Lepidoptera). *J. Anim. Ecol.* 38:703–46
11. Baker, R. R. 1970. Bird predation as a selective pressure on the immature stages of the cabbage butterflies, *Pieris rapae* and *P. brassicae. J. Zool.* 162: 43–59
12. Baker, R. R. 1972. Territorial behaviour of the nymphalid butterflies, *Aglais urticae* (L.) and *Inachis io* (L.). *J. Anim. Ecol.* 41:453–69
13. Baker, R. R. 1972. The geographical origin of the British spring individuals of the butterfly *Vanessa atalanta* and *V. Cardui. J. Entomol.* (A) 46:185–96
14. Balduf, W. V. 1939. *The Bionomics of Entomophagous Insects: Part II.* St. Louis: Swift. 384 pp.
15. Beebe, W. 1951. Migration of Nymphalidae (Nymphalinae), Brassolidae, Morphidae, Libytheidae, Satyridae, Riodinidae, Lycaenidae and Hesperiidae (butterflies) through Portachuelo Pass, Rancho Grande, North-Central Venezuela. *Zoologica* 36:1–16
16. Beebe, W. 1955. Two little known selective insect attractants. *Zoologica* 40: 27–32
17. Beirne, B. P. 1943. The distribution and origin of British Lepidoptera. *Proc. R. Irish Acad. Ser. B* 49:27–59
18. Beirne, B. P. 1947. The origin and history of the British macrolepidoptera.

Trans. R. Entomol. Soc. London 98: 273–372

19. Benson, W. W. 1971. Evidence for the evolution of unpalatability through kin selection in the Heliconiinae (Lepidoptera). *Am. Natur.* 105:213–26
20. Benson, W. W. 1972. Natural selection for Müllerian mimicry in *Heliconius erato* in Costa Rica. *Science* 176: 936–39
21. Benson, W. W., Emmel, T. C. 1973. Demography of gregariously roosting populations of the nymphaline butterfly *Marpesia berania* in Costa Rica. *Ecology* 54:326–35
22. Bink, F. A. 1972. Het onderzoek naar de grote vuurvlinder (*Lycaena dispar batava* Oberthür) in Nederland (Lep. Lycaenidae). *Entomol. Berichten* 32: 225–39
23. Borowski, S. 1961. Climatic conditions of wintering of the butterlfy *Gonepteryx rhamni* L. in the Bialowieza National Park. *Przegl. Zool.* 4:344–47
24. Breedlove, D. E., Ehrlich, P. R. 1968. Plant herbivore coevolution: Lupines and lycaenids. *Science* 162:672–73
25. Breedlove, D. E., Ehrlich, P. R. 1972. Coevolution: patterns of legume predation by a lycaenid butterfly. *Oecologica* (Berl.) 10:99–104
26. Brower, L. P. 1958. Bird predation and food plant specificity in closely related procryptic insects. *Am. Natur.* 92: 183–87
27. Brower, L. P. 1959. Speciation in butterflies of the *Papilio glaucus* group. II. Ecological relationship and interspecific sexual behavior. *Evolution* 13:212–28
28. Brower, L. P. 1962. Evidence for interspecific competition in natural populations of the monarch and queen butterflies, *Danaus plexipus* and *D. gilippus berenice* in South Central Florida. *Ecology* 43:549–52
29. Brower, L. P., Brower, J. V. Z. 1964. Birds, butterflies and plant poisons: a study in ecological chemistry. *Zoologica* 49:137–59
30. Brower, L. P., Brower, J. V. Z., Corvino, J. M. 1967. Plant poisons in a terrestrial food chain. *Proc. Nat. Acad. Sci. USA* 57:893–98
31. Brower, L. P., Brower, J. V. Z. 1962. The relative abundance of model and mimic butterflies in natural populations of the *Battus philenor* mimicry complex. *Ecology* 43:154–58
32. Brower, L. P., McEvoy, P. B., Williamson, K. L., Flannery, M. A. 1972. Variation in cardiac glycoside content of

monarch butterflies from natural populations in Eastern North America. *Science* 177:426–29
33. Brown, K. S., Benson, W. W. 1974. Adaptive polymorphism associated with multiple Müllerian mimicry in *Heliconius numata*. *Biotropica* 6: 205–28
34. Brown, K. S., Sheppard, P. M., Turner, J. R. G. 1974. Quaternary refugia in tropical America: evidence from race formation in *Heliconius* butterflies. *Proc. R. Soc. London* 187:369–78
35. Brown, J. J., Chippendale, G. M. 1974. Migration of the monarch butterfly, *Danaus plexippus:* energy sources. *J. Insect Physiol.* 20:1117–30
36. Brues, C. T. 1920. The selection of foodplants with special reference to lepidopterous larvae. *Am. Natur.* 54: 313–32
37. Brues, C. T. 1924. The specificity of food-plants in the evolution of phytophagous insects. *Am. Natur.* 58:127–44
38. Brussard, P. F., Ehrlich, P. R. 1970. The population structure of *Erebia epipsodea* (Lepidoptera: Satyrinae). *Ecology* 51:119–29
39. Brussard, P. F., Ehrlich, P. R. 1970. Adult behavior and population structure in *Erebia epipsodea* (Lepidoptera: Satyrinae). *Ecology* 51:880–85
40. Brussard, P. F., Ehrlich, P. R. 1970. Contrasting population biology of two species of butterfly. *Nature* 227:91–92
41. Brussard, P. F., Ehrlich, P. R., Singer, M. C. 1974. Adult movements and population structure in *Euphydryas editha*. *Evolution* 261:1–19
42. Buresch, I., Tuleschkow, K. 1930. Die horizontale Verbreitung der Schmetterlinge (Lepidoptera) in Bulgarien II. *Mitt. Naturwissen. Inst. Sofia* 3:145–248
43. Burns, J. M. 1966. Expanding distribution and evolutionary potential of *Thymelicus lineola* (Lepidoptera: Hesperiidae), an introduced skipper, with special reference to its appearance in British Colombia. *Can. Entomol.* 98: 859–66
44. Carcasson, R. H. 1964. A preliminary survey of the zoogeography of African butterflies. *E. Afr. Wildl. J.* 2:122–57
45. Carpenter, G. D. H. 1941. The relative frequency of beak-marks on butterflies of different edibility to birds. *Proc. Zool. Soc.* (London) 3:223–30
46. Carpenter, G. D. H. 1942. Observations and experiments in Africa by the late C. F. M. Swynnerton on wild birds

eating butterflies and the preference shown. *Proc. Linn. Soc. London* 154: 10–46

47. Chew, F. S. 1974. Strategies of food plant exploitation in a complex of oligophagous butterflies (Lepidoptera). PhD thesis. Yale University, New Haven, Conn.

48. Chlodny, J. 1967. The energetics of the development of cabbage white, *Pieris brassicae. Ekol. Pol.* (A) 15:553–61

49. Clark, A. H. 1932. The butterflies of the district of Colombia and vicinity. *Bull. U.S. Nat. Mus.* Vol. 157. 337 pp.

50. Clark, A. H. 1935. The butterflies of Virginia. *Smithson. Inst. Ann. Rep.* 1934:267–96

51. Clarke, B. 1962. Balanced polymorphism and the diversity of sympatric species. In Syst. Assoc. Publ. No. 4. *Taxonomy and Geography,* 47–70

52. Clark, G. C., Dickson, C. G. C. 1971. *Life histories of the South African lycaenid butterflies.* Cape Town: Purnell. 272 pp.

53. Clench, H. K. 1966. Behavioral thermoregulation in butterflies. *Ecology* 47:1021–34

54. Clench, H. K. 1967. Temporal dissociation and population regulation in certain hesperiine butterflies. *Ecology* 48:1000–1006

55. Cook, L. M., Frank, K., Brower, L. P. 1971. Experiments on the demography of tropical butterflies. I. Survival rate and density in two species of *Parides. Biotropica* 3:17–20

56. Collenette, C. L., Talbot, G. 1928. Observations on the bionomics of the Lepidoptera of Matto Grosso, Brazil. *Trans. Entomol. Soc. London* 76:391–416

57. Colwell, R. K., Futuyma, D. J. 1971. On the measurement of niche breadth and overlap. *Ecology* 52:567–76

58. Coppinger, R. P. 1969. The effect of experience and novelty on avian feeding behavior with reference to the evolution of warning coloration in butterflies. Part I. Reactions of wild-caught adult blue jays to novel insects. *Behaviour* 35:45–60

59. Coppinger, R. P. 1970. The effect of experience and novelty on avian feeding behavior with reference to the evolution of warning coloration in butterflies. Part II. Reactions of naive birds to novel insects. *Am. Natur.* 104:323–35

60. Danilevski, A. S. 1965. Photoperiodism and seasonal development of insects. Edinburgh: Oliver & Boyd. 283 pp.

61. Davenport, D., Dethier, V. G. 1937. Bibliography of the described life-histories of the Rhopalocera of America north of Mexico. *Entomol. Am.* 17: 155–94

62. Dempster, J. P. 1967. The control of *Pieris rapae* with DDT. I. The natural mortality of the young stages of *Pieris. J. Appl. Ecol.* 4:485–500

63. Dempster, J. P. 1969. Some effects of weed control on the numbers of the small cabbage white (*Pieris rapae* L.) on brussels sprouts. *J. Appl. Ecol* 6:339–45

64. Dempster, J. P. 1974. The swallowtail butterfly. *Rep. Monks Wood Exp. Sta. 1972–1973*

65. Dethier, V. G. 1959. Food-plant distribution and larval dispersal as factors affecting insect populations. *Can. Entomol.* 91:581–96

66. Dethier, V. G. 1959. Egg-laying habits of lepidoptera in relation to available food. *Can. Entomol.* 91:581–96

67. Dethier, V. G., MacArthur, R. H. 1964. A field's capacity to support a butterfly population. *Nature* 201:728–29

68. Dethier, V. G., Schoonhoven, L. M. 1969. Olfactory coding by Lepidopterous larvae. *Entomol. Exp. Appl.* 12:535–43

69. Dolinger, P. M., Ehrlich, P. R., Fitch, W. L., Breedlove, D. E. 1973. Alkaloid and predation patterns in Colorado lupine populations. *Oecologia* 13:191–204

70. Dowdeswell, W. H., Fisher, R. A., Ford, E. B. 1949. The quantitative study of populations in the Lepidoptera II. *Maniola jurtina* L. *Heredity* 3: 67–84

71. Dowdeswell, W. H., Fisher, R. A., Ford, E. R. S., Ford, E. B. 1940. The quantitative study of populations in the lepidoptera. I. *Polyommatus icarus* Rott. *Ann. Eugen.* 10:123–36

72. Downes, J. A. 1948. The history of the speckled wood butterfly (*Pararge aegeria*) in Scotland, with a discussion of the recent changes of range of other British butterflies. *J. Anim. Ecol.* 17:131–38

73. Downes, J. A. 1973. Lepidoptera feeding at puddle-margins, dung, and carrion. *J. Lepid. Soc.* 27:89–99

74. Downey, J. C. 1962. Variation in *Plebejus icarioides* (Lep. Lycaenidae). II. Parasites of the immature stages. *Ann. Entomol. Soc. Am.* 55:367–73

75. Downey, J. C. 1962. Myrmecophily in *Plebejus icarioides* (Lep. Lycaenidae). *Entomol. News* 73:57–66

76. Downey, J. C. 1962. Host-plant relations as data for butterfly classification. *Systematic Zool.* 11:150–59
77. Downey, J. C. 1966. Sound production in pupae of Lycaenidae. *J. Lepid. Soc.* 20:129–55
78. Downey, J. C., Dunn, D. B. 1964. Variation in the lycaenid butterfly *Plebejus icarioides*. III. Additional data on food-plant specificity. *Ecology* 45:172–78
79. Drenowski, A. K. 1931. Uebt die Meeresnachbarschaft einen Einfluss auf die Höhenverteilung der Begirgslepidoptera in Bulgerien aus? *Dtsch. Entomol. Z.* 1930:179–92
80. Duffey, E. 1968. Ecological studies on the large copper butterfly *Lycaena dispar* Haw. *batavus* Obth. at Woodwalton Fen National Nature Reserve. Huntingdonshire. *J. Appl. Ecol.* 5:69–96
81. Ebert, H. 1969. On the frequency of butterflies in eastern Brazil with a list of the butterfly fauna of Pecos de Caldas, Minas Gerais. *J. Lepid. Soc. Suppl.* 3:1–48
82. Edgar, J. A., Culvenor, C. C. J. 1974. Pyrrolizidine ester alkaloids in danaid butterflies. *Nature* 248:614–16
83. Edwards, E. D. 1973. Delayed ovarian development and aestivation in adult females of *Heteronympha merope merope* (Lepidoptera: Satyrinae). *J. Aust. Entomol. Soc.* 12:92–98
84. Ehrlich, P. R. 1961. Intrinsic barriers to dispersal in checkerspot butterfly. *Science* 134:108–9
85. Ehrlich, P. R. 1965. The population biology of the butterfly, *Euphydryas editha*. II. The structure of the Jasper Ridge Colony. *Evolution* 19:327–36
86. Ehrlich, P. R., Ehrlich, A. H. 1972. Wing shape and adult resources in lycaenids. *J. Lepid. Soc.* 26:196–97
87. Ehrlich, P. R., Raven, P. H. 1965. Butterflies and plants: a study in coevolution. *Evolution* 18:586–608
88. Ehrlich, P. R., Gilbert, L. E. 1973. Population structure and dynamics of the tropical butterfly *Heliconius ethilla*. *Biotropica* 5(2):69–82
89. Ehrlich, P. R., Breedlove, D. E., Brussard, P. F., Sharp, M. A. 1972. Weather and the "regulation" of subalpine populations. *Ecology* 53:243–47
90. Ehrlich, P. R., White, R. R., Singer, M. C., McKechnie, S. W., Gilbert, L. E. 1975. Checkerspot butterflies: A historical perspective. *Science* 188:221–28
91. Emmel, T. C. 1964. The ecology and distribution of butterflies in a montane community near Florissant, Colorado. *Am. Midl. Nat.* 72:358–73
92. Emmel, T. C., Emmel, J. F. 1962. Ecological studies of Rhopalocera in a high Sierran community—Donner Pass, California. I. Butterfly associations and distributional factors. *J. Lepid. Soc.* 16:23–44
93. Emmel, T. C., Emmel, J. F. 1963. Ecological studies of Rhopalocera in a high Sierran community—Donner Pass, California. II. Meteorologic influence on flight activity. *J. Lepid. Soc.* 17:7–20
94. Emmel, T. C., Emmel, J. F. 1963. Composition and relative abundance in a temperate zone butterfly fauna. *J. Res. Lepid.* 1(2):97–108
95. Emmel, T. C., Emmel, J. F. 1969. Selection and host plant overlap in two desert *Papilio* butterflies. *Ecology* 50:158–59
96. Emmel, T. C., Leck, C. F. 1969. Seasonal changes in organization of tropical rain forest butterfly populations in Panama. *J. Res. Lepid.* 8:133–52
97. Erickson, J. M. 1973. The utilization of various *Asclepias* species by larvae of the monarch butterfly, *Danaus plexippus*. *Psyche* 230–44
98. Erickson, J. M., Feeny, P. 1974. Sinigrin: A chemical barrier to the black swallowtail butterfly, *Papilio polyxenes*. *Ecology* 55:103–11
99. Esaki, T. 1929. Zoogeographical relationship between the islands of Yakushima and Kynshu according to the distribution of butterflies. *Bull. Biogeogr. Soc. Jpn.* 1:47–56
100. Feeny, P. 1975. Biochemical coevolution between plants and their insect herbivores. See Ref. 7, 3–19
101. Fisher, R. A., Ford, E. B. 1929. The variability of species in the Lepidoptera, with reference to abundance and sex. *Trans. Entomol. Soc. London* 1929:367–84
102. Ford, E. B. 1945. *Butterflies*. London: Collins. 368 pp.
103. Ford, E. B. 1953. The genetics of polymorphism in the Lepidoptera. *Advan. Genet.* 5:43–87
104. Ford, E. B. 1964. *Ecological Genetics*. London: Methuen 335 pp.
105. Ford, H. D., Ford, E. B. 1930. Fluctuation in numbers and its influence on variation in *Melitaea aurinia*, Rott (Lepidoptera). *Trans. Entomol. Soc. London* 78:345–51
106. Forman, B., Ford, E. B., McWhirter, K. 1959. An evolutionary study of the butterfly *Maniola jurtina* in the north of Scotland. *Heredity* 13:353–61

107. Forster, W. 1958. Die Tiergeographischen Verhaltinisse Bolviens. *Proc. Int. Congr. Entomol., 10th,* 843–46

108. Freeman, H. A. 1950. The distribution and flower preferences of the Theclinae of Texas (Lepidoptera, Rhopalocera, Lycaenidae). *Field Lab.* 18:65–72

109. Freeman, T. N. 1972. A correlation of some butterfly distributions with geological formations. *Can. Entomol.* 104:443–44

110. Frohawk, F. W. 1913. Fecal ejection in hesperids. *Entomologist* 49:201–2

111. Gilbert, L. E. 1969. On the ecology of natural dispersal: *Dione moneta* in Texas (Nymphalidae). *J. Lepid. Soc.* 23:177–85

112. Gilbert, L. E. 1971. Butterfly-plant coevolution: has *Passiflora adenopoda* won the selectional race with heliconiine butterflies? *Science* 172:585–86

113. Gilbert, L. E. 1972. Pollen feeding and reproductive biology of *Heliconius* butterflies. *Proc. Nat. Acad. Sci. USA* 69:1403–7

114. Gilbert, L. E. 1975. Ecological consequences of coevolved mutualism between butterflies and plants. See Ref. 7, 210–39

115. Gilbert, L. E., Ehrlich, P. R. 1970. The affinities of the Ithomiinae and the Satyrinae (Nymphalidae). *J. Lepid. Soc.* 24:297–300

116. Gilbert, L. E., Singer, M. C. 1973. Dispersal and gene flow in a butterfly species. *Am. Nat.* 107:58–72

117. Goodpasture, C. 1974. Food-plant specificity in the *Plebejus* (*Icarcia*) *acmon* group (Lycaenidae). *J. Lepid. Soc.* 23:53–63

118. Harcourt, D. G. 1966. Major factors in the survival of the immature stages of *Pieris rapae* (L.). *Can. Entomol.* 98:653–62

119. Heath, J., ed. 1970. Provisional atlas of the insects of the British Isles. Part I. Lepidoptera, Rhopalocera, Butterflies. Nature Conservancy, Biological Records Centre. Monks Wood Exp. Sta. Maps 1–57

120. Heath, J. 1973. Maps in Howarth, T. G. *South's British Butterflies.* London: Warne

121. Heath, J. 1974. A century of change in the Lepidoptera. *Syst. Assoc. Spec. Vol.* No. 6, 275–92

122. Heinrich, B. 1972. Thoracic temperatures of butterflies in the field near the equator. *Comp. Biochem. Physiol.* 43A:459–67

123. Heslop, I. R. P., Hyde, G. E., Stockley, R. E. 1964. *Notes and Views of the Purple Emperor.* Brighton: Southern. 248 pp.

124. Hinton, H. E. 1951. *Myrmecophilous Lycaenidae* and other Lepidoptera—a summary. *Trans. S. Lond. Entomol. Natur. Hist. Soc.* 1949–1950:111–75

125. Høegh-Guldberg, O. 1968. Evolutionary trends in the genus *Aricia* (Lep.) *Natura Jutl.* 14:7–77

126. Hoffman, R. J. 1973. Environmental control of seasonal variation in the butterfly *Colias eurytheme.* I. Adaptive aspects of a photoperiodic response. *Evolution* 27:387–97

127. Holling, C. S. 1965. The functional response of predators to prey density and its role in mimicry and population regulation. *Mem. Entomol. Soc. Can.* 45:1–60

128. Holloway, J. D. 1969. A numerical investigation of the biogeography of the butterfly fauna of India, and its relation to continental drift. *Biol. J. Linn. Soc.* 2:259–86

129. Holloway, J. D., Jardine, N. 1968. Two approaches to zoogeography: a study based on the distribution of butterflies, birds and bats in the Indo-Australian area. *Proc. Linn. Soc. London* 179:153–88

130. Hovanitz, W. 1942. Genetic and ecologic analyses of wild populations in Lepidoptera. I. Pupal size and weight variation in some California populations of *Melitaea chalcedona. Ecology* 33:175–88

131. Hovanitz, W. 1958. Distribution of butterflies in the New World. *Zoogeography,* 321–68. Washington DC: AAAS

132. Hovanitz, W. 1969. Inherited and/or conditioned changes in host plant preference in *Pieris. Entomol. Exp. Appl.* 12:729–35

133. Jackson, T. H. E. 1961. Entomological studies from a high tower in Mpanga Forest, Uganda. Part IX: Observation on Lepidoptera. *Trans. Entomol. Soc. London* 113:249–362

134. Janzen, D. H. 1973. Sweep samples of tropical foliage insects: effects of seasons, vegetation types, elevation, and insularity. *Ecology* 54:687–708

135. Johnson, M. P., Keith, A. D., Ehrlich, P. R. 1968. Population biology of the butterfly *Euphydryas editha.* VII. Has *E. editha* evolved a serpentine race? *Evolution* 22:422–23

136. Johnson, C. G. 1969. *Migration and Dis-*

persal of Insects by Flight. London: Methuen. 763 pp.
137. Jones, F. M. 1930. The sleeping *Heliconius* of Florida *Nat. Hist.* 30:635–44
138. Jumalon, J. N. 1970. Additional notes on the Palawan birdwing, *Trogonoptera trojana,* with comments on tagging. *Philipp. Entomol.* 1:473–78
139. Kaisila, J. 1962. Immigration und Expansion des Lepidopteren in Finland in den Jahren 1869 bis 1960. *Acta Entomol. Fenn.* Vol. 18. 452 pp.
140. Kano, T. 1931. Notes on the zoogeographical correlation between Kotosho and adjacent territories under the distribution of butterflies. *Bull. Biogeogr. Soc. Jpn.* 2:221–36
141. Keller, E. C., Mattoni, H. T., Sieger, M. S. B. 1966. Preferential return of artificially displaced butterflies. *Anim. Behav.* 14:197–200
142. Kevan, P. G., Shorthouse, J. D. 1970. Behavioral thermoregulation by high arctic butterflies. *Arctic* 23:268–78
143. Kiritani, K., Nakasuji, F. 1967. Estimation of the stage-specific survival rate in the insect population with overlapping stages. *Res. Popul. Ecol., Kyoto* 10:40–44
144. Kobayashi, S. 1966. Process generating the distribution pattern of eggs of the common cabbage butterfly *Pieris rapae crucivora. Res. Popul. Ecol. Kyoto* 8:51–61
145. Kobayashi, S. 1968. Estimation of the individual number entering each developmental stage in an insect population. *Res. Popul. Ecol., Kyoto* 10:40–44
146. Kowalski, W. 1965. Ethological and ecological observations on Lepidoptera in their subterranean hibernating places in the vicinity of Cracow. *Zesz. Nauk. Uniw. Jagiellon. Pr. Zool.* 103:97–161
147. Krieger, R. I., Feeny, P. P., Wilkinson, C. F. 1971. Detoxication enzymes in the guts of caterpillars: an evolutionary answer to plant defenses? *Science* 172:579–81
148. Kusnezov, N. J. 1930. Abhängigkeit der geographischen Verbreitung der Weisslling, Asciidae, von der Verbreitung ihre Futterpflanzen un den chemischen zuzammensetzung der letzteren. *Z. Morphol. Oekol. Tiere* 17:778–94
149. Labine, P. A. 1968. The population biology of the butterfly, *Euphydas editha.* VIII. Oviposition and its relation to patterns of oviposition in other butterflies. *Evolution* 22:799–805

150. Lamborn, W. A. 1914. On the relationship between certain West African insects, especially ants, Lycaenidae and Homoptera. *Trans. Entomol. Soc. London* 1914:438–524
151. Lane, C. 1961. A butterfly feeding on frog hopper larva (Hem. Cercopidae) secretion. *Entomol. Mon. Mag.* 96:130
152. De Lattin, G. 1967. *Grundriss der Zoogeographie.* Stuttgart: Verlag
153. Lederer, G. 1951. Biologie der Nahrungsaufnahme der Imagines von *Apatura* und *Limenitis,* sowie Versuche zur Featstellung der Gustorezeption durch die Mittel- und Hinter fusstarsen dieser Lepidopteren. *Z. Tierpsychol.* 8:41–61
154. Lederer, G. 1960. Verhaltsensweisen der Imagines und der Entwicklungsstadien von *Limenitis camilla camilla* L. (Lep. Nymphalidae). *Z. Tierpsychol.* 17:521–46
155. Lees, A. D. 1962. Phenological aspects of diapause. *Ann. Appl. Biol.* 50:596–99
156. Lees, E. 1962. Factors determining the distribution of the speckled wood butterfly (*Pararge aegeria* L.) in Great Britain. *Entomol. Gaz.* 13:101–3
157. Lenz, F. 1929. Massenauftneten von *Melitaea aurelia* vor *britomartis. Int. Entomol. Z.* 23:149–50
158. Levins, R., MacArthur, R. H. 1969. An hypothesis to explain the incidence of monophagy. *Ecology* 50:910–11
159. Long, D. B. 1953. Effects of population density on larvae of Lepidoptera. *Trans. R. Entomol. Soc. London* 104:543–84
160. Long, D. B. 1959. Observations on adult weight and wing area in *Plusia gamma* L. and *Pieris brassicae* L. in relation to larval population density. *Ent. Exp. Appl.* 2:241–48
161. Longstaff, G. B. 1909. Bionomic notes on butterflies. *Trans. Entomol. Soc. London* 1909:607–73
162. Lorkovic, Z. 1958. Some peculiarities of spatially and sexually restricted gene exchange in the *Erebia tyndarus* group. *Cold Spring Harbor Symp. Quant. Biol.* 23:319–25
163. Lorkovic, Z. 1968. Systematischgenetisch und Ökologische besonderheiten von *Pieris ergane* Hbn. (Lep., Pieridae). *Mitt. Schweiz. Entomol. Ges.* 41:233–44
164. MacNeill, C. D. 1964. The skippers of the genus *Hesperia* in western North America. *Univ. Calif. Publ. Entomol.* 35:20–38

165. McWhirter, K., Scali, V. 1966. Ecological bacteriology of the meadow brown butterfly. *Heredity* 21:517–21

166. Magnus, D. 1954. Methodik und Ergebnisse einer Populations markierung des Kaisermantels. *Verh. Dtsch. Entomol.* Hamburg 1953:187–97

167. Malicky, H. 1961. Uber die Okologie von *Lycaeides idas* L., insbesondere uber seine Symbiose mit Ameisen. *Z.*

167. *Arbeitsgem. Oesterr. Entomol.* 13: 33–52

168. Malicky, H. 1968. Uber einen Mortalitatsfaktor beim Schlupfen der Larve von *Maculinea alcon* (Schiff) (Lep. Lycaenidae). *Zool. Anz. Suppl.* 32: 575–79

169. Malicky, H. 1969. Versuch einer Analyse der öklogischen Beziehungen zwischer Lycaeniden (Lepidoptera) und Formiciden (Hymenoptera). *Tijdschr. Entomol.* 122:213–98

170. Marsh, N., Rothschild, M. 1974. Aposematic and cryptic Lepidoptera tested on the mouse. *J. Zool.* 174:89–122

171. Marshall, G. A. K. 1902. Five years' observations and experiments (1896–1901) on the bionomics of South African insects chiefly directed to the investigation of mimicry and warning colours. *Trans. Entomol. Soc. London* 1902:287–584

172. Mell, R. 1935. Grundzüge einer Oekologie der chinesischen Lepidopteren. I. Die "Bioklimatische Regel" und die Erscheinungszeiten von Lepidopteren. *Biol. Zentralbl.* 55:2–16

173. Michelbacher, A. E., Smith, R. F. 1943. Some natural factors limiting the abundance of the alfalfa butterfly. *Hilgardia* 15:369–97

174. Michael, O. 1894. Ueber den Fang und die Lebensweise der wichtigsten Tagfalter der Amazonasebene. *Dtsch. Entomol. Z. "Iris"* 217–23

175. Moss, J. E. 1933. The natural control of the cabbage caterpillars, *Pieris* spp. *J. Anim. Ecol.* 2:210–31

177. Mülier, W. 1886. Sudamerikanische Nymphalidenraupen. *Zool. Jahrb. Syst. Oekol. Geogr. Tiere* 1:1–255

178. Muyshondt, A., Muyshondt, A. Jr. 1974. Gregarious seasonal roosting of *Smyrna karwinskii* adults in El Salvador (Nymphalidae). *J. Lepid. Soc.* 28:224–29

179. Neck, R. W. 1973. Food-plant ecology of the butterfly *Chlosyne lacinia* (Geyer) (Nymphalidae). I. Larval foodplants. *J. Lepid. Soc.* 27:22–33

180. Nielsen, E. T. 1961. On the habits of the migratory butterfly *Ascia monuste* L. *Biol. Medd. Dansk. Vid Selskab.* 23:1–81

181. Norris, M. J. 1936. The feeding habits of the adult Lepidoptera Heteroneura. *Trans. R. Entomol. Soc. London* 85:61–90

182. Opler, P. A., Langston, R. L. 1968. A distributional analysis of the butterflies of Contra Costa County, California. *J. Lepid. Soc.* 22:89–107

183. Orians, G. H., Cates, R. G., Roades, D. F., Schultz, J. C. 1974. Producer-consumer interactions. In Structure, function and management of ecosystems. 213–16. *Proc. Int. Congr. Ecol., 1st.* Wageningen: Cent. Agric. Publ. Doc.

184. Owen, D. F. 1959. Ecological segregation in butterflies in Britain. *Entomol. Gaz.* 10:27–38

185. Owen, D. F. 1971. Tropical butterflies. Oxford: Clarendon. 214 pp.

186. Owen, D. F., Chanter, D. O. 1972. Species diversity and seasonal abundance in *Charaxes* butterflies (Nymphalidae). *J. Entomol.* (A) 46:135–43

187. Papageorgis, C. 1974. The adaptive significance of wing coloration of mimetic neotropical butterflies. PhD thesis. Princeton Univ., Princeton, NJ. 117 pp.

188. Parker, F. D. 1970. Seasonal mortality and survival of *Pieris rapae* (Lepidoptera: Pieridae) in Missouri and the effect of introducing an egg parasite, *Trichogramma evanescens*. *Ann. Entomol. Soc. Am.* 63:985–94

189. Perez-Ruiz, H. 1971. Algunas consideraciones sobre la poblacion de *Baronia brevicornis* (Ablu.) (Lepidoptera, Papilionidae, Baroniinae) en la region de Mezcala, Guerrero. *An. Inst. Biol. Univ. Nac. Auton. Mex. (Zool.)* 42: 63–72

190. Petersen, B. 1954. Egg-laying and habitat selection in some *Pieris* species. *Entomol. Tidskr.* 74:194–203

191. Pictet, A. 1935. Ecologie et genecologie de *Maniola nerine* Frr. (*alecto* Hb.) au Parc national Suisse et dans la Vallée du Munster. *Mitt. Schweiz. Entomol. Ges.* 16:378–94

192. Pliske, T. E. 1975. Attraction of Lepidoptera to plants containing pyrrolizidine alkaloids. *Environ. Entomol.* In press

193. Pollard, E. 1974. The white admiral butterfly *Ladoga camilla*. *Rep. Monks Wood Exp. Sta. 1972–1973*

194. Poulton, E. B. 1917. Salt (chloride of sodium) probably sought by the Hesperiidae. *Proc. Entomol. Soc. London* 1917:77

195. Poulton, E. B. 1931. The gregarious sleeping habits of *Heliconius charitonia* L. *Proc. R. Entomol. Soc. London* 6: 4–10

196. Raven, P. H. 1964. Catastrophic selection and edaphic endemism. *Evolution* 18:336–38

197. Rawson, G. W. 1945. Interesting problems associated with the checkered white butterfly *Pieris protodice. Bull. Brooklyn Entomol. Soc.* 40:49–54

198. Reichhoff, J. 1973. Communal roosting in the neotropical hesperiid *Sarbia demippe* Mabille et Bullet (Lep.). *Dtsch. Entomol. Z.* 20:355–56

199. Richards, O. W. 1940. The biology of the small white butterfly (*Pieris rapae*), with special reference to the factors controlling its abundance. *J. Anim. Ecol.* 9:243–88

200. Robins, E. J. 1972. Population dynamics of the marsh fritillary butterfly *Euphydryas aurinia.* M.Sc. thesis. Imperial College, London

201. Roer, H. 1959. Uber Flug und Wandergewohnheiten von *Pieris brassicae.* L. *Z. Angew. Entomol.* 44:272–309

202. Roer, H. 1968. Weitere Untersuchungen über die Auswirkungen der Witterung auf Richtung un Distanz der Flüge des Kleinen Fuchses (*Aglais urticae* L.) (Lep. Nymphalidae) im Rheinland. *Decheniana* 120:313–34

203. Roer, H. 1969. Zur Biologie des Tagpfauenauges, *Inachis io* L. (Lep. Nymphalidae), unter besonderer Beruchsichtigung der Wanderung im mitteleuropaishen Raum. *Zool. Anz.* 183: 177–94

204. Roer, H. 1970. Untersuchungen zum Migrationsverhalten des Trauermantels (*Nymphalis antiopa* L.) (Lep. Numphalidae). *Z. Angew. Entomol.* 65: 388–96

205. Ross, G. N. 1966. Life history studies on Mexican butterflies. IV. The ecology and ethology of *Anatole rossi,* a myr mecophilous metal mark (Lepidoptera: Riodinidae). *Ann. Entomol. Soc. Am.* 59:985–1004

206. Rothschild, M. 1964. An extension of Dr. Lincoln Brower's theory on bird predation and food specificity; together with some observations on bird memory in relation to aposematic colour patterns. *Entomologist* 1964:73–78

207. Rothschild, M. 1971. Speculations about mimicry with Henry Ford. In *Ecological Genetics and Evolution,* ed. R. Creed, 202–23. Oxford: Blackwells

208. Rothschild, M. 1972. Some observations on the relationship between plants, toxic insects and birds. In *Phytochemical Ecology,* ed. J. B. Harborne, 1–12. London & New York: Academic

209. Sachlová, R., Sachl, J. 1969. Beitrag zur Kenntnis von Tagfaltern der Umgegung von Podebrady (Mittlebōhman). *Práce Kraj. Mus. Hradci Králové (A)* 10:63–76

210. Scali, V. 1971. Imaginal diapause and gonadal maturation of *Maniola jurtina* (Lepidoptera: Satyridae) from Tuscany. *J. Anim. Ecol.* 40:467–72

211. Scudder, S. H. 1889. *The Butterflies of the Eastern United States and Canada.* Cambridge, Mass. 1774 pp.

212. Scott, J. A. 1972. Biogeography of antillean butterflies. *Biotropica* 4:32–45

213. Scott, J. A. 1973. Down-valley flight of adult Theclini (Lycaenidae) in search of nourishment. *J. Lepid. Soc.* 27:283–87

214. Scott, J. A. 1973. Convergence of population biology and adult behavior in two sympatric butterflies. *J. Anim. Ecol.* 42:663–72

215. Scott, J. A. 1973. Population biology and adult behavior of the circumpolar butterfly *Parnassius phoebus* F. Papilionidae Lep. *Entomol. Scand.* 4:161–68

216. Seitz, A. 1910–1924. *Macrolepidoptera of the World.* Vols. 1–5. *Butterflies.* Stuttgart: Kernen

217. Shapiro, A. M. 1970. The role of sexual behavior in density-related dispersal of pierid butterflies. *Am. Nat.* 104:367–72

218. Shapiro, A. M., Cardé, R. T. 1970. Habitat selection and competition among sibling species of satyrid butterflies. *Evolution* 24:48–54

219. Sharp, M. A., Parks, D. R., Ehrlich, P. R. 1974. Plant resources and butterfly habitat selection. *Ecology* 55:870–75

220. Sheppard, P. M. 1961. Some contributions to population genetics resulting from the study of Lepidoptera *Adv. Genet.* 10:165–215

221. Sherman, P. W., Watt, W. B. 1975. The thermal ecology of some *Colias* butterfly larvae. *J. Comp. Physiol.* 83:25–40

222. Shields, O. 1967 (1968). Hilltopping. *J. Res. Lepid.* 6:69–78

223. Shields, O. 1972. Flower visitation records for butterflies (Lepidoptera). *Pan-Pac. Entomol.* 48:189–203

224. Silberglied, R. E. 1975. Communication in the Lepidoptera. In *Animal Communication*, ed. T. A. Sebeok. Indiana Univ. Press. 2nd ed. In press

225. Singer, M. C. 1971. Evolution of foodplant preferences in the butterfly *Euphydryas editha*. *Evolution* 35:383–89

226. Singer, M. C. 1971. Ecological studies on the butterfly *Euphydryas editha*. PhD thesis. Stanford Univ., Stanford, California. 72 pp.

227. Singer, M. C. 1972. Complex components of habitat suitability within a butterfly colony. *Science* 173:75–77

228. Slansky, F. Jr. 1972. Latitudinal gradients in species diversity of the New World swallowtail butterflies. *J. Res. Lepid.* 11:201–7

229. Slansky, F. Jr. 1973. Energetic and nutritional interactions between larvae of the imported cabbage butterfly *Pieris rapae* L. and cruciferous foodplants. PhD thesis. Cornell Univ., Ithaca, NY. 303 pp.

230. Slansky, F. Jr. 1974. Relationship of larval food-plants and voltinism patterns in temperate butterflies. *Psyche* 81:243–53

231. Smith, D. A. S. 1973. Batesian mimicry between *Danaus chrysippus* (Danaidae) and *Hypolimnas missippus* (Nymphalidae) (Lepidoptera) in Tanzania. *Nature* 242:129–31

232. Smith, R. F., Bryan, D. E., Allen, W. W. 1940. The relation of flights of *Colias* to larval population density. *Ecology* 30:288–97

233. Southwood, T. R. E. 1962. Migration of terrestrial arthropods in relation to habitat. *Biol. Rev.* 37:171–214

234. Southwood, T. R. E., May, R. M., Hassell, M. P., Conway, G. R. 1974. Ecological strategies and population parameters. *Am. Nat.* 108:791–804

235. Stern, V. M., Smith, R. F. 1960. Factors affecting egg production and oviposition in populations of *Colias eurytheme* Boisduval (Lepid, Pieridae). *Hilgardia* 29:411–54

236. Straatman, R. 1962. Notes on certain Lepidoptera ovipositing on plants which are toxic to their larvae. *J. Lepid. Soc.* 16:99–103

237. Straatman, R. 1969. Notes on the biology and host-plant associations of *Ornithoptera priamus urvilleanus* and *O. victoriae* (Papilionidae). *J. Lepid. Soc.* 23:69–76

238. Szent-Ivany, J. J. H. 1971. Notes on the distribution and host plant of *Ornithoptera meridionalis* (Rothschild) (Lepi-

doptera: Papilionidae). *Papua New Guinea Sci. Soc. Proc.* 22:35–37

239. Swiecimskz, Z. 1958. Training butterflies to scents as a method to improve the pollination of flowers. *Zesz. Nauk. Uniw. Jagiellon. Zool.* 19:127–77

240. Swynnerton, C. F. M. 1926. An investigation into the defenses of butterflies of the genus *Charaxes*. III. *Int. Entomol. Kongr. Bd.* 2:478–506

241. Takata, N. 1961. Studies on the host preference of common cabbage butterfly, *Pieris rapae crucivora* (Boisduval) XII. Successive rearing of the cabbage butterfly larva with certain host plants and its effect on the oviposition preference of the adult. *Jpn. J. Ecol.* 11:147–54

242. Terofal, Fritz. 1965. Zum Problem der Wirtsspezifitat bei Pieriden (Lep.) *Münch. Entomol. Ges. Mitt.* 55:1–76

243. Thomas, J. A. 1974. Autecological studies of hairstreak butterflies. PhD thesis. Univ. Leicester, Leicester, England. 372 pp.

244. Tsujita, M., Nawa, S., Sakaguchi, B. 1959. Studies on a silkworm poison emanating from tobacco plants. *Proc. Jpn. Acad. Tokyo* 35:180–85

245. Turner, J. R. G. 1963. A quantitative study of a Welsh Colony of the large heath butterfly, *Coenonympha tullia* Muller (Lepidoptera). *Proc. R. Entomol. Soc. London* (A) 38:7–9

246. Turner, J. R. G. 1971. Experiments on the demography of tropical butterflies. II. Longevity and home-range behavior in *Heliconius erato. Biotropica* 3:21–31

247. Urquhart, F. A. 1960. *The Monarch Butterfly.* Toronto: Univ. Toronto Press. 361 pp.

248. Van Someren, V. G. L., Jackson, T. H. E. 1959. Some comments on protective resemblance amongst African Lepidoptera (Rhopalocera). *J. Lepid. Soc.* 13:121–50

249. Van Someren, V. G. L., Van Someren, R. A. L. 1926. The life-histories of certain African nymphalid butterflies of the genera *Charaxes, Palla,* and *Euxanthe. Trans. Entomol. Soc. London* 74:333–54

250. Varlet, R., Losser, D. 1964. Étude statistique de population de *Parnassius apollo* (L.). *Bull. Soc. Entomol. Mulhouse* 1964:103–14

251. Varley, G. C., Gradwell, G. R., Hassell, M. P. 1974. *Insect Population Ecology, An Analytical Approach.* Berkeley & Los Angeles: Univ. Calif. Press. 212 pp.

252. Vestergaard, D. A. 1970. De ruimtelijke diversiteit van de dagvlinder fauna (Lep., Rhopalocera) in het kustgebild van Voorne. *Entomol. Ber.* 30:236–40

253. Vielmetter, W. 1954. Die Temperaturregulation des Kaisermantels in der Sonnenstrahlung *Naturwissenschaften* 41:535–36

254. Wagner, W. H. Jr. 1973. An orchid attractant for monarch butterflies (Danaidae). *J. Lepid. Soc.* 27:192–96

255. Waldbauer, G. P., Sheldon, J. K. 1971. Phenological relationships of some aculeate Hymenoptera, their dipteran mimics, and insectivorous birds. *Evolution* 25:371–82

256. Warnecke, G. 1935. *Chrysophanus (Heodes) dispar* Haw. ein gefährdeter Tagfalter. *Entomol. Z.* 49:137–40

257. Warnecke, G., Harz, K. 1954. Zur Kenntnis der Populationsdynannik und des Migration-sverhaltens von *Vanessa atalanta* L. im paläarktischen Raum (Lep.). *Beitr. Entomol.* 14:155–58

258. Watt, W. B. 1968. Adaptive significance of pigment polymorphisms in *Colias* butterflies. I. Variation of melanin pigment in relation to thermoregulation. *Evolution* 22:437–58

259. Watt, W. B., Hoch, P. C., Mills, S. 1974. Nectar resource use by *Colias* butterflies: chemical and visual aspects. *Oecologia* 14:353–74

260. Weissman, K. G. 1972. The population biology of *Coenonympha tullia* on Jasper Ridge: Strategies of a bivoltine butterfly. PhD thesis. Stanford Univ., Stanford, Calif. 75 pp.

261. Wheeler, L. R. 1939. Deaths among butterflies. *Proc. Linn. Soc. London* 1939: 79–88

262. White, R. R. 1973. Community relationships of the butterfly *Euphydryas editha*. PhD thesis. Stanford Univ., Stanford, California. 90 pp.

263. White, R. R. 1974. Food plant defoliation and larval starvation of *Euphydryas editha*. *Oecologia* 14:307–15

264. White, R. R., Singer, M. C. 1974. Geographical distribution of hostplant choice in *Euphydryas editha* (Nymphalidae). *J. Lepid. Soc.* 28:103–7

265. Wicklund, C. 1973. Host plant suitability vs. the mechanism of host selection in *Papilio machaon* larvae. *Entomol. Exp. Appl.* 16:232–42

266. Wilbert, H. 1959. *Apanteles glomeratus* (L.). als Parasit von *Aporia crataegae* (L.). (Hymenoptera, Braconidae). *Beitr. Entomol.* 19:874–98

267. Wilbur, H. M., Tinkle, D. W., Collins, J. P. 1974. Environmental certainty, trophic level, and resource availability in life history evolution. *Am. Nat.* 108:805–17

268. Williams, C. B. 1930. *The Migration of Butterflies.* Edinburgh: Oliver & Boyd

269. Williams, C. B. 1964. *Patterns in the Balance of Nature.* New York: Academic. 324 pp.

270. Williams, C. B., Cockbill, G. F., Gibbs, M. E., Downes, J. A. 1942. Studies in the migration of Lepidoptera. *Trans. R. Entomol. Soc. London* 92:101–282

271. Wiltshire, E. P. 1946. Studies in the geography of Lepidoptera. III. Some Middle East immigrants, their phenology and ecology. *Trans. R. Entomol. Soc. London* 96:163–82

272. Wohlfahrt, T. A. 1968. Über das Zusammenivirken von Licht und Temperatur bei der Austrosung des Schlüpfens von *Iphiclides podalirius* (L.) (Lep. Papil.). *Verh. Zool. Bot. Ges. Wien* 31:434–39

273. Young, A. M. 1972. Community ecology of some tropical rainforest butterflies. *Am. Midl. Nat.* 87:146–57

274. Young, A. M., Thomason, J. H. 1974. The demography of a confined population of the butterfly *Morpho peleides* during a tropical dry season. *Stud. Neotrop. Fauna* 9:1–34

AUTHOR INDEX

A

Abbott, D. P., 224
Abbott, I., 224
Abel, E., 177, 226, 228, 233
Abramoff, P., 357
Abrosov, N. S., 322
Adams, M., 374
Adams, R. M., 139
Adler, H., 183
Adler, J., 172, 180
Adlerz, W. C., 157
Afzal, M., 154
Aginsky, B. W., 123
Ahmadjian, V., 299
Akin, D. E., 46
Akinla, O., 131
Akkermans-Kruyswijk, J., 43
Albrecht, H., 220
Alcock, J., 181
Alcorn, S. M., 154
Aldrich, J. W., 295
Aleem, A. A., 223, 225
Alexander, A. J., 380, 387
Alexander, C. L., 50
Alexander, R. D., 130, 175, 355
Alimen, H., 273
Allan, J. D., 295
Allen, G. R., 215, 216, 230, 232
Allen, J. F., 17
Allen, K. R., 323
Allison, M. J., 40, 42, 43, 45, 46
Alm, G., 89, 92, 93
Alpatov, U. V., 152
Al-Rabbat, M. F., 56
Amsel, H. G., 374
Anasiewicz, A., 153
Anderson, E., 160
Anderson, N. H., 295
Anderson, R. S., 87, 92, 94, 99, 101
Anderson, W. W., 159
Andrewartha, H. G., 7, 9
Angus, T. A., 367
Antonovics, J., 348, 349
Apfelbach, R., 17
Aptekar, H., 113
Aranki, A., 42
Arbogast, R. T., 300
Arcand, C., 42, 43
Archimowitsch, A., 154
Armitage, K. B., 178
Arms, K., 381
Armstrong, J. S., 314
Armstrong, J. T., 87, 89, 96

Armstrong, R., 204
Arnason, U., 241
Aron, W. I., 234
Arthur, A. P., 367
Aschner, M., 300
Asell, B., 332
Askew, R. R., 352, 354, 355
Atwood, C. E., 352, 353, 355, 356
Aubreville, A., 271
Auchmuty, J. F. G., 190
Aufsess, A. V., 148
Autrum, H., 108
Awadallah, A. M., 366, 385
Axelrod, H. R., 230
Ayala, F. J., 301, 356
Ayers, W. A., 57
Azab, A. K., 366, 385

B

Bachmann, K., 345, 347
Baile, C. A., 54
Bailey, M. E., 145
Bainbridge, R., 181
Baker, B. H., 254
Baker, D. G., 15, 25
Baker, F., 60
Baker, H. G., 139-43, 145, 146, 154, 156-58, 160, 381
Baker, I., 141, 145, 146, 156, 160, 381
Baker, R. J., 340, 347, 348
Baker, R. R., 183, 366, 367, 371-74
Bakus, G. J., 223-25, 233
Balch, C. C., 48
Balduf, W. V., 378, 381, 382, 387
Baldwin, R. L., 43, 47, 48, 55, 56
Balikci, A., 109, 129
Balsano, J. S., 357
Banks, C. J., 97
Banner, M. W., 347
Banse, K., 181
Baptista, L. F., 31-33
Barbehenn, K. R., 302
Bardach, J. E., 214, 223, 227, 233, 235
Barlow, G. W., 177, 215, 216, 227, 228, 231, 234
Barnard, E. A., 40
Barnes, M. M., 353
Barr, L. G., 356

Barrow, J. H. Jr., 297
Barth, F. G., 182
Barth, R. H., 175
Bartholomew, B., 224, 296
Bartholomew, G. A., 142-44
Bartlett, M. S., 194
Basch, P. F., 295
Bässler, U., 178
Basykin, A. D., 352
Bateman, A. J., 153, 154
Bates, M., 111
Bateson, P. P. G., 15, 23, 24
Battisini, R., 273
Bauchop, T., 45, 57
Baur, G. N., 70
Bawa, K. S., 158
Bayliss-Smith, T., 125, 133
Baynton, H. W., 76
Beadle, M., 16
Beardmore, J. A., 32
Beaver, R. Z., 88
Beebe, W., 370, 381, 382
Behrens, W. W. II, 314
Beirne, B. P., 375
Bendat, J. S., 191
Benignus, V. A., 191
Bennett, A. E., 220
Bennett, I., 224
Benson, W. W., 369-71, 385, 386
Bentley, D. R., 353
Ben-Tuvia, A., 233
Benzie, D., 41, 47
Berg, R. T., 54
Bergen, W. G., 57
Berger, L., 295
Berger, M., 143
Berndt, R., 19
Bernhard, F., 72
Bernhard-Riversat, F., 73
Berryman, A. A., 314, 315, 321
Bertram, B. C. R., 234
Berube, D. E., 147
Bethel, W. M., 297
Betts, A. D., 143, 153
Beucher, F., 271, 273
Beukema, J. J., 180
Beutler, R., 144-47, 150, 151
Beveridge, R. J., 54
Bhatia, I. S., 59
Bigelow, R. S., 341
Bigger, M., 200
Binford, L. R., 112
Bingham, C., 196
Bink, F. A., 366, 367

Birch, L. C., 7, 9
Birdsell, J. B., 113, 120, 121
Birdsong, R. S., 233
Birky, C. W. Jr., 96
Bischof, N., 178
Bishop, J. E., 73
Bishop, W. W., 263
Bjärvall, A., 19
Blackburn, T. H., 43, 45, 46
Blackman, R. B., 191, 195, 198
Blair, W. F., 345
Blake, G. H. Jr., 157
Blanchard, C. O., 70
Blasco, F., 76-78
Blaxter, J. H. S., 177
Bledsoe, L. J., 318, 319, 332
Blest, A. D., 230, 301
Blewett, M., 300
Blondheim, S. A., 348, 351
Bloom, S. G., 320
Bloomfield, J. A., 332
Boden, B. P., 218
Bohart, G. E., 139, 146, 152
Böhlke, J. E., 213, 222, 231, 232
Bohn, G. W., 148, 297, 301
Boice, R., 90
Boldyrev, M. I., 98, 99
Boling, R. H. Jr., 316, 318
Boller, E. F., 352
Bonhomme, A., 41
Bonnefille, R., 264
Booth, A. H., 275
Booth, R. S., 332
Borisov, A. I., 349
Bormann, F. H., 71, 73
Borowski, S., 367
Borrie, W. D., 123, 124
Botkin, D. B., 332
Boucque, C. V., 59
Bouma, C., 44
Bourne, D. W., 181
Bovbjerg, R. V., 291
Bovey, P., 353
Bowers, J. H., 340
Bowles, G. T., 126, 128
Box, G. P., 196, 200, 207
Boycott, B. B., 178
Bradner, N. R., 154
Bradshaw, A. D., 349
Braestrup, F. W., 19, 29
Bragg, A. N., 88, 92, 96, 97, 99, 102
Brahy, B. D., 217, 220
Brain, D., 153
Bray, J. R., 72, 77
Breder, C. M., 216, 220, 222, 226, 227
Breedlove, D. E., 224, 367, 380, 384

Brennan, R. D., 329
Brian, A. D., 149, 151, 153, 291
Brian, M. V., 290
Briggs, J. C., 211
Bright, T. J., 235
Brinkhurst, R. O., 92, 94, 303
Brinley, F. J., 220
Briscoe, C. B., 71
Britten, R. J., 347
Brock, M. L., 292
Brock, T. D., 292, 299
Brock, V. E., 217, 218, 227, 229
Brockmann, H. J., 215
Bronson, G., 17
Brough, B. E., 44
Brower, J. V. Z., 382, 386, 387
Brower, L. P., 88, 90, 92, 93, 236, 365, 369, 379, 382, 386, 387
Brown, H. D., 97
Brown, J. H., 250, 290
Brown, J. J., 381
Brown, J. M., 263
Brown, K. S., 374, 386
Brown, L. F., 333
Brown, R. A., 224
Brown, W. H., 73, 75-80
Brown, W. L. Jr., 340, 348
Brownlie, L. E., 40
Brues, C. T., 377
Bruggemann, P. F., 146
Brussard, P. F., 367, 368, 370, 371
Bryant, A. M., 47
Bryant, M. P., 42-44, 46, 58, 59
Bryson, R. A., 206, 259
Buchman, S. L., 157
Bucholtz, H. F., 57
Buck, J., 144
Bucklin, J. A., 45
Buckman, N. S., 216, 217, 220, 221
Budelmann, B. U., 178
Buechner, H. K., 59
Bulkley, R. V., 89, 91
Bunnell, F., 312, 314, 315, 318
Bunnell, F. L., 333
Bünning, E., 172
Burdick, D., 46
Buresch, I., 374
Burger, J., 179
Burger, M. C., 179
Burgess, P. F., 76
Burgess, W., 230
Burghardt, G. M., 16
Burke, K., 271
Burkey, L. A., 59
Burnham, C. P., 75-77
Burns, J. M., 366
BUSH, G. L., 339-64;

352-56
Butler, C. G., 139, 147, 151, 154
Butler, G. D. Jr., 154
Butzer, K. W., 263, 268, 269
Buzzard, C. N., 154

C

Cahen, L., 263
Cahn, P. H., 227
Cain, S. A., 68
Cairns, J., 303
Calder, W. A., 142, 144
Caldwell, D. R., 42-44
Calhoun, J. B., 108
Calow, P., 181
Cameron, G. N., 291
Campbell, C. A., 301
Campos, J. J., 179
Candlish, E., 43
Carcasson, R. H., 253, 374
Cardé, R. T., 384, 356
Cardon, S. Z., 191
Cardoso de Oliveira, R., 130
Carlson, B. A., 230, 232
Carnahan, B., 328
Carney, W. P., 297
Carpenter, F. L., 159
Carpenter, G. D. H., 386
Carpenter, S. J., 95, 97
Carroll, E. J., 56
Carroll, V., 125
Carr-Saunders, A. M., 111, 113
Carson, H. L., 339, 345-47
Carthy, J. D., 172
Case, T. J., 284-86, 288
Casey, T. M., 143, 144
Caspari, E., 356
Castenholz, R. W., 289, 292, 293, 296, 299, 301, 303
Castle, E. J., 48
Caswell, H., 313, 315, 323, 329
Catchpole, C. K., 18
Cates, R. G., 295, 297, 380
Catsky, J., 204
Cawthon, D. A., 97
Cazier, M. A., 290, 299
Cello, R. M., 42, 58
Chabora, A. J., 352
Chadwick, L. E., 145
Chagnon, N. A., 130
Chalupa, W., 43, 44
Chance, B., 190
Chanter, D. O., 385, 387
Chao, S., 227
Chaplin, C. C. G., 211, 213, 222
Chavaillon, J., 273
Chen, C. W., 332
Cheng, K. J., 42, 58

Cheng, T. C., 297, 300
Cheng, Y., 233
Chevalier, J. R., 89, 91-93, 97
Chew, F. S., 379
Chippendale, G. M., 381
Chivers, D. J., 81
Choat, J. H., 222, 229
Choate, J. R., 351
Christenson, P., 139, 144, 147
Chu, H., 44
Chua, K. E., 303
Church, D. C., 39
Churchill, D. M., 139, 144, 147
Clark, D. P., 90
Clark, E., 222, 223, 227
Clark, G. C., 381
Clark, J. D., 251, 252, 257, 270
Clark, J. L., 48
Clark, J. P., 325
Clark, L. R., 88
Clark, S., 301, 303
Clarke, B., 248, 349, 386
Clarke, R. T. J., 42, 58
Clarke, T. A., 220, 227
Clatworthy, J. N., 295
Clay, C. S., 191
Clayton, R. K., 175
Clench, H. K., 366, 376, 384
Clymer, A. B., 312
Cochran, W. T., 195
Cockbill, G. F., 375
Cody, M. L., 139, 154, 180, 281, 282
Coetzee, J. A., 256, 258, 268
Cohen, D., 211, 212
Cohen, J., 271
Cole, C. J., 351
Cole, S., 272
Coleman, G. S., 41, 42
Colin, P. L., 226, 233, 235
Colinvaux, P. A., 87
Collenette, C. L., 383, 387
Collett, T. S., 180
Collette, B. S., 231, 233, 234
Collins, J. P., 369
COLWELL, R. K., 281-310; 139, 159, 160, 282, 283, 285, 288, 290, 294, 303, 380
Connell, J., 7, 9, 12
Connell, J. H., 70, 173, 282, 284, 291, 296, 301, 302
Connor, D. J., 333
Conrad, G., 271, 273
Conrads, K., 31
Conrads, W., 31
Conway, G. R., 369
Cooch, F. G., 32
Cook, L. M., 349, 369

Cook, S. A., 290
Cook, S. F., 108, 111
Cooke, F., 32
Cooke, H. B. S., 271, 272
Cooley, J. W., 195
Coons, L. W., 341
Cooper, C. F., 333
Cooper, G. C., 97
Cooper, J. C., 183
Coppinger, R. P., 386
Corbet, P. S., 95, 96
Cornell, H., 302
Costerton, J. W., 42
Cott, H. B., 178, 228, 229
Cottyn, B. G., 59
Cours Darne, G., 76
Cowsert, R. L., 54
Craddock, E. M., 346, 351, 353
Craig, J. P., 345, 347
Crane, M. B., 154
Crisp, D. J., 174
Crisp, D. T., 92
Croft, B. A., 87, 97, 101
Crook, J. H., 130
Crosby, J. L., 349
Croze, H. J., 180
Cruden, R. W., 152
Crumpacker, D. W., 182
Culberson, W. L., 299
Cullen, J. M., 227
Culvenor, C. C. J., 382
Culver, D. C., 285, 295
Cummings, W. C., 217
Cummins, K. W., 316, 318
Curds, C. R., 325
Czerkawski, J. W., 59

D

Dahlberg, M. L., 303
Dahlman, R. C., 311, 333
Danilevski, A. S., 384
Darcy, G. H., 231, 233
Darnell, R. M., 357
da Silva, N. T., 68
Daumer, K., 148, 174
Davenport, D., 377
Davidson, E. H., 347
Davidson, R. S., 312
Davies, M. S., 281
Davies, R. W., 301
Davis, B. L., 340, 347, 348
Davis, D. M., 295
Davis, G. N., 148, 297, 301
Davis, J., 291
Davis, R. A., 145
Davis, W. P., 225, 231, 233, 234
Dawkins, M., 181
Dawson, E. Y., 223, 225
Dawson, P. S., 87, 97, 98
Dawson, W. R., 142, 143
Day, J. H., 218
Day, P. R., 353

Dayton, P. K., 99, 101, 301-3
Deacon, J. E., 346
DeBach, P., 282, 297
de Castro, L. C., 80
de Crespigny, C. C., 226
Degens, E. T., 262, 266
Dehority, B. A., 43, 54
Dekker, R. F. H., 54
De Lattin, G., 074
Delius, J. D., 177
Del Moral, R., 295, 297
Delpino, F., 140
Delsman, H. C., 220
Dement, W. A., 102
Demeyer, D. I., 56, 59
Demoll, R., 150
Dempster, J. P., 88, 99, 101, 366-69, 374, 376
Den Boer, M., 352
DENMAN, K. L., 189-210; 201-3
De Oliveira, A. E., 130
de Oliveira Castro, G. A., 68
de Ploey, J., 271
de Ruiter, L., 178, 301
Dethier, V. G., 152, 180, 295, 353, 366, 377, 380
Devlin, T. J., 43
De Wit, C. T., 329
DICKEMAN, M., 107-37
Dickie, L. M., 201
Dickinson, H., 348, 349
Dickson, C. G. C., 381
Dickson, K. L., 303
Dijkgraaf, S., 178
Dixon, A. F. G., 92
Dobzhansky, T., 339, 340, 341, 346, 356
Docters van Leeuwen, W. M., 156
Dodson, C. H., 139, 140, 155, 157
Dodson, G. J., 81
Doetsch, R. N., 44
Dogiel, V. A., 41
Dolinger, P. M., 224, 380
Domm, A. J., 234
Domm, S. B., 234
Dorn, H. F., 109
Dornstreich, M. D., 89
Doty, M. S., 224
Dougherty, R. W., 39, 40, 42, 45, 46, 58-60
Douglas, M., 129, 131
Doull, K. M., 152
Dowdeswell, W. H., 369
Dowding, P., 333
Downes, A. M., 47
Downes, J. A., 179, 375, 376, 381
Downey, J. C., 367, 378, 380, 387
Downie, C., 254
Dozier, B. J., 204
Drenowski, A. K., 374
Dressler, R. L., 139

Driedger, A., 47
Driver, P. M., 180
Drost, R., 18
Duffey, E., 366-68, 376
Dugdale, R. C., 313, 318, 329, 331
Dukepoo, F. C., 119
Duncan, C., 295
Duncan, O. D., 111
Dunham, W. E., 149, 150
Dunn, D. B., 378
Durbin, J., 198
Durotoye, A. B., 271
Durrant, A. J., 158
Dutton, J. A., 206
Duvigneaud, P., 271

E

Eadie, J. M., 42
Earle, S. A., 223
Eason, E. H., 87
Easter, St. S., 178
Eaton, J., 120
Ebert, E. E., 220
Ebert, H., 387
Eckert, J. E., 147, 150, 151, 156
Edgar, J. A., 382
Edgar, W. D., 88, 91, 92
Edge, B. L., 196
Edington, J. M., 295
Edwards, E. D., 384
Egan, M. E., 341
Ehrenfeld, J. G., 356
Ehrlich, A. H., 216, 227, 228, 382
EHRLICH, P. R., 211-47; 212, 216, 218-20, 224, 232, 234, 236, 295, 348, 367-71, 374, 375, 377, 380, 382, 384, 385, 393
Ehrman, L., 341, 356
Eibl-Eibesfeldt, I., 215, 226-28, 230-32
Eickwort, K. R., 88, 99
Eisenberg, J. F., 81, 82, 345
Eisner, E., 204
Eisner, T., 225
el-Din, M. Z., 47, 57
Eliot, J. M., 91, 93
El-Kifli, A. H., 89, 90
El-Lakwak, F., 89
Ellenberg, H., 294
Ellis, W. C., 47
El-Nahal, A. K. M., 366, 385
Elsden, S. R., 47, 57
el-Shazly, K., 47, 57
Elton, C., 281
Emerson, S., 154
Emery, A. R., 220, 225, 226, 228
Emery, R. S., 43
Emlen, J. M., 9, 139, 152, 174, 283

Emmanuel, B., 43, 45
Emmel, J. F., 375, 379, 387
Emmel, T. C., 369, 370, 374, 379, 385, 387
Endean, R., 224, 225
Endler, J. A., 348
Endo, A., 43, 60
Enochson, L. D., 199
Enright, J. P., 172
Epting, R. J., 144
Ercolini, A., 181
Erickson, J. M., 380
Eriksen, C. H., 295
Esaki, T., 374
Evans, G., 154
Evans, G. C., 70
Every, D. D., 44
Ewing, A. W., 345
Ewing, S., 317
Eyles, A. C., 295

F

Faegri, K., 140, 156
Fagerstrom, T., 332
Fahn, A., 146
Fahy, E., 89, 99
Farkas, S. R., 177
Farner, D. S., 142, 144
Faure, H., 271, 273
Feddern, H. A., 220, 221, 223
Feeny, P., 377, 380, 381
Feeny, P. P., 341, 353
Fellows, D. P., 294, 295
Ferguson, D. E., 181
Ferkovich, S. M., 353
Fieger, E. A., 145
Filion, C., 204
Filmer, R. S., 152, 159, 299, 302
Finkner, M. D., 154
Firth, R., 123, 124, 132
Fischer, Z., 89, 92
Fischer-Klein, K., 179
Fishelson, L., 215, 220, 222, 233-35
Fisher, L. J., 47
Fisher, R. A., 369
Fitch, W. L., 224, 380
Flaherty, B. R., 300
Flamm, W. G., 345
Flannery, M. A., 387
Flechsig, A. O., 227
Flint, R. F., 255, 268
Fock, G. J., 268, 269
Fogden, M. P. L., 82
Follmer, D., 295
Forbes, L., 174
Force, D. C., 354, 355
Ford, C., 114, 115, 117, 119, 127
Ford, E. B., 365, 368, 369, 371, 375-77, 384, 386
Ford, E. R. S., 369
Ford, H. A., 345

Ford, H. D., 368
Forman, B., 365
Forney, J. L., 89, 91, 95
Forrester, J. W., 314, 328
Forster, W., 374
Forsyth, G., 41
Fossey, D., 181
Fournier, L. A., 80
FOX, L. R., 87-106; 89-92, 95, 98, 99, 101
Fox, R., 127, 130
Fraenkel, G., 145, 172, 178, 224, 300
Frakes, R. V., 154
Fraleigh, P. C., 325, 330
Franck, D., 179
Frank, K., 369
Frank, P. W., 301
Frankie, G. W., 156, 157
Franklin-Klein, W., 43
Franzisket, L., 229
Fraser, C., 47
Fraser Brunner, A., 214
Free, J. B., 139, 147, 149, 151-53, 156, 158, 159
Freeman, H. A., 381
Freeman, T. N., 375
Freire-Maia, N., 117, 118
Freter, R., 42
FRETWELL, S. D., 1-13; 9
Fricke, H., 226, 228, 231, 232
Friedman, J., 295
Friedmann, H., 20, 33
Frisch, K. V., 182
Frisch, R. E., 122, 127
Fritts, H. C., 206
Frohawk, F. W., 387
Frost, W. E., 89, 93, 97, 355
Fryer, G., 266, 351
FUENTES, E. R., 281-310
Fukuda, H., 149, 151
Futuyma, D. J., 282, 283, 290, 380
Fye, R. E., 152

G

Gadgil, M., 286, 341
Gale, W. A., 268
Galle, O. R., 108
Gallopin, G. C., 331
Gallun, R. L., 353
Ganderman, N., 234
Garfinkel, D., 312, 317, 321
Garnaud, J., 220, 226
Garrels, R. M., 250
Gary, N. E., 147, 151
Gaston, L. K., 177
Gates, D. M., 74
Genoways, H. H., 351
Geyh, M. A., 272
Ghiselin, M. T., 339
Ghosh, A. K., 190
Gibbs, M. E., 375

Gibson, J. B., 352
Gilbert, J. J., 87, 96
Gilbert, L., 149
GILBERT, L. E., 365-97;
149, 150, 160, 215, 218-20,
232, 234, 236, 295, 298,
367-72, 374, 375, 377-79,
381-83, 385, 387
Gill, D. E., 295, 299, 302
Gillett, J. D., 00
Gilpin, M. E., 284-86, 288
Glansdorff, P., 190, 208
Glombitza, R., 197, 200
Glynn, P. W., 235, 236
Godfrey, M. D., 196
Gohar, H. A. F., 226
Goitein, R., 348, 350, 351
Goldman, B., 212, 233,
234
Goldman, C. R., 204
Goldrich, N., 152
Goldschmidt, R., 340, 347,
357
Goldstein, R. A., 81, 332,
333
Goodall, D. W., 312, 315,
316, 318, 322, 331-33
Goodman, L. J., 177, 178
Goodman, N. R., 199
Goodpasture, C., 378
Goodwin, D., 32
Goodwin, E. B., 32
Gordon, J. G., 57
Gorham, E., 72, 77
Görner, P., 179, 182
Gosline, W. A., 218, 233
Gosz, J. R., 71
Gottlieb, G., 33
Gottlieb, L. D., 349
Gottsberger, G., 157
Götz, K. G., 178
Gouws, L., 59
Gove, W. R., 108
Gradwell, G. R., 9, 366,
367, 386
Graham, K., 300
Grant, K. A., 140
Grant, P. R., 284, 291, 295,
341
Grant, V., 139, 140, 152,
153, 339-41, 347-49,
356
Grantzberg, G., 116
Grau, F. H., 40
Green, G., 224
Greenberg, B., 295
Greenhalgh, J. F. D., 54
Greenwood, P. H., 212, 263,
354, 355
Gregory, D. P., 152
Grenke, W. C., 75
Grieve, C. M., 54
Griffiths, A., 95, 96
Grigg, R. W., 227
Grobe, J., 233
Grodzinski, W., 87, 88, 96,
99

Gross, J. E., 314
Grove, A. T., 267, 272
Grovum, W. L., 55
Grubb, P. J., 68, 69, 75
Gudger, E. W., 226
Gunn, D. L., 172
Gutierrez, J., 42
Gwinner, E., 175, 183

II

Haaker, U., 181
Hadfield, W., 69
Haffer, J., 275
Hagen, H.-O., 177, 181,
182
Hailman, E., 179
Hailman, J. P., 19
Hainsworth, F. R., 139,
141-46, 148, 149, 151, 174
Hair, J. D., 286
Hall, D. J., 181
Hall, K. R. L., 181
Hall, W. P., 351
Hallander, H., 91
Halligan, J. P., 296
Halstead, B. W., 223, 225
Hambleton, J. J., 159
Hamilton, A. C., 256, 259
Hamilton, W. D., 98
Hamilton, W. J., III, 150,
181, 228, 230
Hampton, S. H., 96
Hanson, W. D., 348
Häntschel, W., 181
Harcourt, D. G., 366, 367
Harden-Jones, F. R., 179
Hardin, G., 132
Harger, J. R. E., 292
Harlan, H. V., 281
Harper, G., 269
Harper, J. L., 284, 285,
295, 296
Harper, J. W., 349
Harrington, B. A., 345,
347
Harris, B. J., 142, 143,
146, 156
Harris, N., 263
Harris, P., 352
Harrison, C. J. D., 20
Hart, J. S., 143
Hartline, A. C., 227
Hartline, P. H., 227
Harz, K., 369
Hasler, A. D., 17, 183
Hassell, M. P., 9, 366,
367, 369, 386
Hasselrot, T. B., 153
Hatch, F. T., 345
Hatchett, J. H., 353
Hatfield, E. E., 47
Hatheway, W. H., 75
Hauser, P. M., 111
Hawkes, O. A. M., 92, 97
Hawkins, R. P., 156
Hayne, D. W., 318-20,

322-24, 332
Hazlett, B., 217, 235
Heath, J., 368, 376
Heatwole, H., 178, 290, 295,
303
Heckenlively, D., 178
Hecky, R. E., 262, 266
Hedberg, O., 256
Hediger, H., 232
Heed, W. B., 294, 295,
353
Heer, D. M., 110
Heezen, B. C., 181
HEINRICH, B., 139-70; 139,
141-55, 157, 160, 161, 174,
298, 376
Heithaus, E. R., 141, 145,
146, 154-56, 160
Helfman, G., 321, 323
Helfrich, P., 220
Helgren, D. M., 269
Heller, H. C., 291, 295
Hembry, F. G., 48
Hemsley, J. A., 47
Henderickx, H. K., 56, 59
Henderson, C., 59
Hendry, L. B., 356
Henry, S. M., 299, 300
Henwood, K., 67
Herald, E. S., 226, 231
Heran, H., 144, 145, 179
Herne, D. H. C., 95
Herrnkind, W. F., 181
Heslop, I. R. P., 367, 383
Hespenheide, H. A., 9
Hess, B., 190
Hess, E. H., 16, 17, 19, 21-
23, 32
Hessler, R. R., 99, 101
Heusser, H., 302
Hiatt, R. W., 215, 223, 226,
229, 233
Hickman, J. C., 139, 150
Higuchi, M., 43, 59, 60
Hildebrand, S. F., 212
Hildén, O., 18, 19
Hill, G. R., 301, 303
Hill, M. O., 204
Hills, H. G., 139
Hinich, M. J., 191
Hinton, H. E., 387
Hinton, S., 226
Hird, F. J. R., 60
Hironaka, R., 42, 58
Hirst, E. L., 41
Hisada, M., 177
Hladik, A., 72, 81
Hladik, C. M., 81
Ho, F. K., 97
Hobbs, G. A., 152, 153,
290
Hobson, E. S., 217, 227,
231-34
Hobson, P. N., 42-44, 47,
57
Hoch, P. C., 146, 149, 152,
381, 382

Hocking, B., 139, 142, 145, 146
Hoffman, M. A., 295
Hoffman, R. J., 385
Hoffmann, M., 179
Hoffmann, R. S., 291
Hofmann, R., 40
Hofstetter, F. B., 29
Hogan, J. P., 54
Holdeman, L. V., 42
Holdridge, L. R., 75
Holdship, S., 264
Holliday, F. G. T., 177
Holling, C. S., 316, 317, 327, 386
Holloway, J. D., 374
Holm, S. N., 149
Holmes, F. O., 152
Holmes, I., 60
Holmes, J. C., 286-88, 297, 302
Holmes, R. T., 81
Holst, E. V., 177
Hoogenraad, N. J., 60
Hoogmoed, M. S., 352
Hoover, W. H., 46
Hopgood, M. F., 42, 47
Hopkins, B., 72
Horn, E., 179
Horn, E. G., 295
Horn, G., 15, 25
Horn, H. S., 7, 77
Horridge, G. A., 175
Hovanitz, W., 369, 374-76
Howard, B. H., 44
Howard, R. A., 76, 77
Howell, N., 121, 122
Hoyle, A. C., 271
Hubbell, S. P., 151, 152, 288
Huber, H., 146
Hubricht, L., 160
Huettel, M. D., 352, 353
Huey, R. B., 341
Huffaker, C. B., 88, 94, 297, 302
Huisingh, J., 44
Hullah, W. A., 43, 46
Humbert, H., 76
Humphries, D. A., 180
Humphries, D. W., 254
Hungate, D. P., 45, 59
HUNGATE, R. E., 39-66; 39, 41-45, 47, 50, 51, 55-59
Hungerford, D. A., 347
Hunt, J. H., 291
Hurd, P. D. Jr., 139, 143
Hurlbert, S. H., 91, 96, 97
Hurst, H. E., 263
Hutchinson, G. E., 6, 7, 108, 173, 181, 284, 288
Huttel, C., 73
Hyde, G. E., 367, 383
Hyman, L. H., 87

I

Ianotti, E. L., 46
Iberall, A. S., 189, 191, 208
Iles, T. D., 266, 351
IMMELMANN, K., 15-37; 15, 16, 21-25, 27, 34
Inger, R. F., 295
Innis, G. S., 312, 314, 319-23, 325, 329
Iperti, G., 97
Isaac, G. L., 263
Istock, C. A., 89, 97, 99, 101
Iversen, J., 262
Ivlev, V. S., 295

J

Jackson, J. L., 190
Jacobs, W., 178
Jaeger, R. G., 284, 285, 288
Jahn, Th., 177
Jain, S. K., 349
Jäkel, D., 272
Jakob, H., 299
JANDER, R., 177-88; 173-75, 177-80
Jander, U., 179
Janota-Bassalak, L., 60
Jantz, R. L., 119, 130
Janzen, D. G., 299, 302
Janzen, D. H., 76, 154, 155, 160, 299, 302, 355, 385
Jardine, N., 374
Jarvis, P. J., 204
Jayasuriya, G. C. N., 45
Jeffers, J. N. R., 319
Jeffree, E. P., 154
Jenkins, D. W., 95, 97
Jenkins, G. M., 191, 195, 196, 198-200, 207
Jensen, R. W., 295
Johannes, R. E., 235
Johansson, T. S. K., 154
John, B., 348
Johnes, D. D., 178
Johnson, C. G., 182, 369, 371-73
Johnson, G. B., 346
Johnson, L. K., 151, 152, 288, 290
Johnson, M. P., 375
Johnson, M. W., 218
Johnson, N. M., 73
Johnson, P. L., 73
Johnson, R. K., 217
Johnson, R. R., 54
Johnson, W. E., 339
Johnston, F. E., 119, 130
Johnston, R. F., 356
Jónasson, P. M., 92, 94
Jones, C. E., 157
Jones, F. M., 370
Jones, J. M., 341

Jones, K. W., 347
Jones, M. B., 300
Jones, R., 201
Jones, R. S., 215, 216, 218, 221, 233, 234
Jovanovic, M., 56
Joyner, A. E., 47
Judd, B. H., 347
Jumalon, J. N., 369
Junquiera, P. C., 130

K

Kaddou, I. K., 90, 97, 99
Kafkewitz, D., 46
Kähling, J., 174
Kaiser, H., 178
Kaisila, J., 374
Kalmus, H., 154
Kami, H. T., 233
Kammer, A., 141, 143, 144
Kanapi, C., 346
Kanayama, R. K., 223
Kanegasaki, S., 42, 58
Kano, T., 374
Karr, J. R., 81
Katz, I., 43
Kaufman, D. W., 345, 348
Kaufman, J. H., 180
Kaushik, N. K., 303
Kay, R. N. B., 41, 47
Kaye, S. V., 332
Keay, R. W. J., 271
Keeney, M., 43
Keenleyside, M. H. A., 215, 220
Keeton, W., 183
Keever, C., 295, 303
Keister, M., 144
Keiter, F., 130
Keith, A. D., 375
Keller, E. C., 370
Kendall, R. L., 256, 260-62, 270
Kennedy, J. J., 180
Kenney, E. B., 42
Kenny, R., 225
Kensinger, K. M., 119, 130
Kepner, R. E., 300
Kernaghan, R. P., 356
Kerner, A., 139, 140, 156
Kerster, H. W., 161
Kessler, S., 341
Kevan, P. G., 376
Key, G. S., 217
Khan, A. H., 154
Kiang, Y. T., 154
Kim, Y. J., 348, 351
King, C. E., 87, 98
King, J. R., 31, 32, 142, 144
Kipling, C., 89, 93, 97
Kira, T., 71
Kircher, H. W., 295, 353
Kiritani, K., 366
Kirkpatrick, T. W., 88, 91

Kistner, A., 59
Kleber, E., 146
Klinge, H., 72
Klopfer, P., 18, 19, 33
Klopfer, P. H., 172
Kluckhohn, C., 110, 124
Knerer, G., 352, 353, 355, 356
Knight, G. R., 341
Knight-Jones, D. W., 176
Knuth, P., 140, 155
Kobayashi, K., 124, 127
Kobayashi, S., 366
Kochansky, J., 356
Kochara, A. S., 59
Koening, H. E., 313, 315
Kokshaiski, N. V., 144
Kolata, G. B., 124, 127
Kolmogorov, A. N., 201
Koltermann, R., 148, 152
Konishi, M., 32
Koonce, J. F., 332
Koopman, K. R., 341
Koplin, J. R., 291
Koshland, D. E., 180
Kovrov, B. G., 322
Kowal, N. E., 315, 319, 333
Kowalski, W., 385
Kozhevnikov, B. T., 347
Kozlowski, T. T., 295
Kraai, A., 158
Kraus, R., 91
Krebs, C. J., 250
Krebs, J. R., 9, 180
Kremer, J. N., 332
Krieger, R. I., 380
Krowitz, A., 179
Kruckeberg, A. R., 291
Kruuk, H., 88, 94
Krzywicki, L., 114
Kugler, H., 148, 152, 161, 298
Kühn, A., 172
Kuile, A. T., 352
Kullenberg, B., 139
Kurtz, E. B. Jr., 154
Kusnezov, N. J., 377
Kutzbach, J. E., 259

L

Labine, P. A., 369
La Bounty, J. F., 346
Labuschagne, J. P. L., 59
Lachner, E. A., 226
Lack, D., 6, 9, 20, 108, 220
LaCroix, L. J., 43
Lakshaman, S., 43
Lamborn, W. A., 387
Land, M. F., 180
Lane, C., 381
Langenheim, J. H., 270
Langer, A., 179
Langer, W. L., 109
Langness, L. L., 130

Langston, R. L., 374, 384
Larkin, P. A., 88, 99, 101
Lasiewski, R. C., 142-44
Lassiter, R. R., 318-20, 322-24, 332
Laven, H., 356
Lawton, J. H., 91, 93
Lawton, R. M., 267
Lea, A., 170
Leck, C. F., 385, 387
Le Cren, E. D., 93, 94, 99, 101
Lederer, G., 381, 387
Lederhouse, R. C., 381
Lee, R. B., 121, 122
Lee, W. R., 151
Lees, A. D., 384
Lees, E., 376
LeFevre, E. A., 141
LEIGH, E. G. JR., 67-86; 68, 75
Lemee, G., 73
Lenneberg, E. H., 15, 16
Lenz, F., 368
Leppik, E. E., 161
Levandowsky, M., 293
Levin, A. A., 320
Levin, D. A., 147, 159, 161, 299, 301, 353, 354
Levin, M. D., 151, 152, 154
Levin, S., 235
Levin, S. A., 281-83
Levine, S., 15
Levins, R., 4, 7, 281, 283, 290, 293, 301, 303, 313, 318, 345, 349, 380
Levinton, J. S., 224, 235
Lewis, D. H., 356
Lewis, G. W., 15
Lewis, H., 157, 159, 339, 340, 348-50
Lewis, K. R., 348
Lewontin, R. C., 339, 341
Lex, Th., 148, 157
Leyhausen, P., 16
Liang, T., 75
Lie, K. J., 295
Liebherr, J., 356
Likens, G. E., 71, 73
Lim, M. T., 72
Limbaugh, C., 220, 231, 233
Lincicome, D. R., 303
Lindauer, M., 179, 182
Lindenbaum, S., 131
Line, L. J. S., 156
Linhart, Y. B., 141, 154, 160
Linsenmair, K. E., 179
Linsenmair-Ziegler, Ch., 179
Linsley, E. G., 152, 157, 290, 299
Linton, D. L., 269
Lipari, J. J., 46

Littleford, R. A., 17
Littlejohn, M. J., 341
Liu, P. C., 196
Liversidge, R., 20
Livingston, R. T., 234
LIVINGSTONE, D. A., 249-80; 254, 256, 261, 266, 267, 274
Lloyd, J. R., 68, 69, 75
Loftus-Hills, J. J., 341
Loh, W., 145
Löhrl, H., 17
Longhurst, W. M., 300
Longley, W. H., 212, 213, 228
Lorenz, K., 21, 24, 228, 230
Lorimer, F., 110, 120, 125
Lorkovic, Z., 365, 377, 384
Losey, G. S., 231-33
Loucks, O. L., 68, 75
Lovell, J. H., 148, 157, 161
Low, R. M., 214, 295
Lowe, C. H., 351
Lowe-McConnell, R. H., 345
Ludwig, W., 176, 348, 352
Lukschanderl, L., 30
Luther, H. A., 328

M

Macan, T. T., 88, 92
MacArthur, J. W., 7, 9
MacArthur, R. H., 3, 4, 6-12, 108, 173, 179, 250, 274, 281, 282, 285, 288, 293, 295, 317, 321, 341, 346, 366, 380
Macdonald, G. E., 15, 25
MacGregor, A., 45, 59
Macior, L. W., 139, 153, 159, 161, 298
Mackensen, O., 152
Mackenzie, F. T., 250
Mackinnon, J., 181
MacLellan, C. R., 88
MacMillen, R. E., 159
Macnab, R., 180
Macneill, C. D., 372, 382
Macrae, J. C., 55
MacSwain, J. W., 152, 157
Madden, W. D., 231
Madsen, B. L., 295
Maeda, B. R., 204
Magnus, D., 269
Maguire, B. Jr., 281, 283, 294
Magus, D. B., 226
Mah, R. A., 43, 56
Maisel, E., 231, 233
Maitland, P. S., 295
Mak, S., 233
Malcolm, L. A., 118
Maley, E., 7
Maley, J., 271, 272
Malicky, H., 366, 387

Malthus, T. R., 110, 120
Mamdani, M., 133
Manguin, E., 271
Mankin, J. B., 332, 333
Mann, S. O., 44
Manning, A., 148, 154
Mantel, N., 130
Manzi, J. J., 87, 88, 97
Marchant, R., 355, 356
Margalef, R., 319, 322
Mariscal, R. N., 215, 226
Marish, D., 180
Marler, P., 16, 31, 32
Marples, T. G., 72
Marsh, D. J., 189, 208
Marsh, N., 382, 385, 387
Marshall, G. A. K., 365, 386
Marshall, J. A., 217, 235
Marshall, L., 121, 122
Marshall, N. B., 345
Marston, J., 147, 151
Martin, A. R. H., 269
Martin, G. D., 314, 324, 329
Martini, M. L., 281
Marty, R. J., 59
Marwaha, S. R., 59
Mather, K., 154
Mathews, D. M., 73
Matrone, G., 44
Matthews, G. V. T., 183
Mattoni, H. T., 370
Matuszynska, G. M., 42, 60
Mauny, R. A., 272
Maxson, L. R., 345, 347
May, R. M., 7, 281, 301, 318, 322, 369
Mayberry, W. R., 56
Maybury-Lewis, D., 118, 130
Mayer, A. J., 120
Mayer, J., 54
Maynard-Smith, J., 352
Mayr, E., 22, 27, 28, 30, 339-41, 345, 346, 348, 351, 353, 354, 356
Mazanov, A., 333
Mazrimas, J. A., 345
McAllan, A. B., 42
McArthur, N., 133
McArthur, J. W., 122, 127
McBee, R. H., 300
McBride, B. C., 43
McCormack, J. C., 93, 94
McCormick, J. F., 290
McCosker, J. E., 217, 231, 232, 235
McDonald, I. W., 39, 40, 42, 44, 46
McDougall, I. J., 42, 43
McElroy, L. W., 54
McErlean, A. J., 214
McEvoy, P. B., 387
McFee, A. F., 347
McGregor, S. E., 154

McKay, F. E., 96
McKechnie, S. W., 215, 218-20, 234, 236
McKell, C. M., 295
McKenzie, J. A., 349
McLay, C. L., 295, 303
McLeod, J., 311, 314
McMurtry, J. A., 87, 97, 101
McNab, B. K., 142, 345
McNally, C. M., 32
McNaughton, I. H., 295
McNeill, J. J., 44
McPhail, J. D., 26
McPherson, J. M., 108
McQueen, D. J., 92
McWhirter, K., 380
Meadows, D. H., 314
Meadows, P. L., 314
Mecom, J. O., 92
Medina, E., 80
Medler, J. T., 152, 159
Medway, L., 18, 78, 82
Meggitt, M., 131
Meijer, W., 79, 80
Meinwald, J., 225
Mell, R., 374
Menahan, L. A., 43
Mendonça, A., 271
Menge, B. A., 291
Menge, J. L., 291
Menshutkin, V. V., 331, 332
Menzel, D. W., 223, 227
Menzel, R., 148
Mertz, D. B., 87, 88, 92, 93, 96-99
Mewaldt, L. R., 295
Meyer, D. L., 178
Michael, O., 370
Michaels, G. E., 46
Michelbacher, A. E., 367
Milborrow, B. V., 172
Miles, J., 295
Miller, P. C., 333
Miller, R. S., 332
Milligan, L. P., 43, 45
Millis, N. F., 60
Mills, J. N., 172
Mills, S., 381, 382
Mills, S. G., 146, 149, 152
Milne, L. J., 178
Milne, M., 178
Milner, C., 314, 331
Minato, H., 43, 60
Minderhoud, A., 154
Mirsky, P. J., 32
Mitchell, R., 296
Mittelstaedt, H., 177
Moawad, G. M., 89, 90
Mobley, C. D., 318
Mohr, H., 172
Moir, R. J., 54
Mommers, J., 139, 154, 158
Monteith, J. L., 204

Montgomery, G. G., 81, 82
Montgomery, M. J., 54
Mooney, H. A., 102
Moore, C. W. E., 294
Moore, W. E. C., 42
Moore, W. S., 96
Morales, R., 329
Moran, V. C., 175
Moreau, R. E., 18, 253, 274
Moriga, K., 149, 151
Morren, G. E. B., 89
Morris, R. D., 295
Morrison, M. E. S., 256, 258
Morse, D. H., 292, 295
Mortelmans, G., 271
Mosquin, T., 159
Moss, J. E., 367
Moss, R., 87
Mrosovsky, N., 181
Mulkern, G. B., 295
Mulla, M. S., 91, 96, 97
Mullen, R. R., 141
Mullenax, C. H., 40, 46
Muller, C. H., 295
Muller, D., 71
Müller, H., 140
Müller, K., 182
Müller, W., 355, 387
Mulligan, J. A., 16
Murdoch, W. W., 101, 139, 152, 301
Murray, J., 341, 348, 349, 352
Murray, M. G., 47
Muspratt, J., 88, 97
Muyshondt, A., 370, 385
Muyshondt, A. Jr., 370, 385
Myers, G. S., 212
Myrberg, A. A. Jr., 214, 215, 220, 235

N

Nader, C. J., 57
Nag, M., 115, 131
Nagy, J., 332
Nakamura, K., 42, 45, 48
Nakasuji, F., 366
Nasr, El-S. A., 89, 90
Nasr, H., 60
Neck, R. W., 379
Neel, J. V., 108, 109, 112, 117-19, 130
Neill, S. R. S. J., 227
Neill, W. E., 301
Nevo, E., 348, 350, 351
Newell, N. D., 219
Newman, W. A., 225
Nicholson, A. J., 9
Nicolai, J., 20, 34
Nicolis, G., 190
Nielsen, E. T., 369
Nielsen, J., 71
Niering, W., 70

Nikolic, J. A., 56
Nikolskii, G. V., 88, 97
Nilsson, E., 253, 265
Nixon, S. W., 332
Noirot-Timothee, C., 41
Norris, D. M., 353
Norris, M. J., 381, 383
Nott, K. H., 15
Nottebohm, F., 32
Nowak, F., 29
Nowakowski, J. T., 352
Nummi, W. O., 153, 290
Nuñez, J. A., 151
Nydal, R., 271
Nye, W. P., 146, 152

O

Odum, E. P., 108, 118,
  233
Odum, H. T., 71, 73, 77,
  81, 233, 312, 315, 316,
  327, 328
Ogawa, H., 71
Ogden, J. C., 212, 216, 217,
  220, 221, 224, 231, 233, 235
Ogino, K., 71
Oh, H. K., 300
Okuno, R., 215, 228
Olson, J. S., 312
O'Neill, R. V., 317-19, 323,
  332, 333
Oosting, H. J., 293
Ootomo, Y., 43, 60
Opler, P. A., 141, 145, 146,
  154-58, 160, 374, 384
Oppenheimer, J. R., 222
Orenski, S. W., 299
Orians, G. H., 380
Orlob, G. T., 332
Orpin, C. G., 44
Orskov, E. R., 41, 47
Osmaston, H., 253-56
Oster, G., 318
Otte, D., 175
Overton, W. S., 312, 332
Owen, D. F., 314, 366, 369,
  377, 383-87
Oxford, A. E., 41

P

Padden, R. C., 108
Paine, R. T., 80, 88, 89,
  92, 95, 224, 301, 303
Pajunen, V. I., 88, 92
Pakhurst, D. F., 68, 75
Pakrasi, K. B., 130
Pallister, J. W., 263
Palmer-Jones, T., 156
Pant, H. C., 57
Pantle, C., 145
Papageorgis, C., 382, 386
Paperna, I., 302
Pardi, L., 175
Parish, R. C., 59
Park, R. A., 332

Park, T., 87, 88, 96, 99
Park, W., 147
Parker, B. C., 301
Parker, F. D., 367
Parker, R. A., 332
Parkin, D. T., 345
Parks, D. R., 236, 374,
  393
Parsons, P. A., 349
Partridge, L., 19
Parzen, E., 191
Patrick, R., 303
Patriquin, D. G., 97
Patten, B. C., 312, 313,
  316, 318, 319, 332, 333
Patton, J. L., 348, 351
Paulian, R., 79
Paulik, G. J., 331
Pavlovsky, O., 341, 356
Payne, R. B., 20, 32, 34
Payne, W. J., 56
Paynter, M. J. B., 47
Pearce, G. R., 54
Pearl, R., 113
Pearre, S., 181
Pearson, O. P., 141, 143,
  148
Pedersen, M. W., 146
Peel, J. L., 42
Peitzmeier, J., 19
Penney, M. M., 295
Pennington, T. D., 68, 69,
  75
Percival, M. S., 145-47
Perry, T. O., 70
Peterman, R. M., 228, 230,
  331
Petersen, B., 383
Petersen, R. C., 316, 318
Phillips, G. D., 45, 55,
  59
Phillips, P. A., 353
Phillipson, A. T., 39
Phleger, C. F., 290
Pianka, E. R., 9, 10, 179,
  341, 345, 349
Pichi-Sermolli, R. E. G., 271
Pichler, H., 144
Pick, E. E., 179
Pienaar, L. V., 314, 315,
  331
Pienkowski, R. L., 97
Pierce, R. S., 73
Piersol, A. G., 191
Pietraszek, A., 42
Pimentel, D., 33, 348
Pires, J. M., 68
Pirie, R., 47
Pittman, K. A., 43
Plath, O. E., 149
Platt, R. B., 290
PLATT, T., 189-210; 201,
  202, 204
Platt, W. J., 301, 303
Pliske, T. E., 382
Polgar, S., 111
Pollard, E., 367, 376

Pollard, E. C., 176
Pontin, A. J., 291
Popper, D., 222, 234
Portin, P., 340
Portugal, A. V., 43
Post, J., 233
Potts, G. W., 227
Poulson, G. W., 15
Poulson, T. L., 88, 89
Poulton, E. B., 370, 381
Powell, T. M., 202, 205
Power, A., 231
Powers, D. A., 295
Press, J. W., 300
Pressick, M. L., 290, 303
Preston, E. M., 301
Preston, R. L., 48
Price, P. W., 303, 352,
  354, 355
Priesner, E., 139, 356
Prigogine, I., 190, 208
Pringle, J. W. S., 178
Prins, R. A., 43, 44, 55-57,
  59
Prior, R. L., 40
Pritchard, G., 93
Prochazka, G. J., 56
Proctor, M. C. F., 140,
  156
Prus, T., 87, 88, 96, 99
Putnam, W. L., 95
Putwain, P. D., 284, 295
Pye, E. K., 190

Q

Quadagno, D., 180
Quattrin, E. V., 329
Quiguer, J., 233

R

Rabinowitch, V., 17, 22
Rabinowitz, J. C., 45
Radakov, D. V., 227
Radford, P. J., 329
Raines, G. E., 320
Rainey, R. C., 183
Randall, H. A., 213, 220,
  221, 228-30, 232
Randall, J. E., 213-15, 218,
  220, 221, 223, 224, 226,
  228-32, 235
Randall, R. H., 233
Randers, J., 314
Rapoport, A., 139
Rappaport, R. A., 125, 129,
  131
Rary, J. M., 347
Rasa, O. A. E., 214, 215
Raven, P. H., 139, 142, 157,
  212, 224, 236, 295, 298,
  348, 375, 377
Ray, A., 57
Read, C. P., 287
Reddinguis, J., 313
Rees, P. J., 47

Reese, E. S., 214, 220
Reichhoff, J., 370
Reichl, J., 55-57
Reichle, D. E., 81, 332
Reid, C. S. W., 41, 58
Reid, G. W., 54
Reid, R. S., 47
Reighard, J., 222, 228
Reinboth, R., 221
Rensch, B., 174
Rerberg, M. S., 322
Rescigno, A., 7
Resh, J. A., 313, 315
Rettenmeyer, C. W., 301
Ribbands, C. R., 147, 156
Rice, E. L., 295
Richards, G. N., 54
Richards, K. W., 146, 147, 160
Richards, O. W., 366, 367
Richards, P. W., 67, 68, 70, 73, 76, 77
Richardson, A. E., 265
Richardson, I. W., 7
Richardson, J. L., 263, 265, 266
Richardson, R. H., 346, 352, 353
Richerson, P. J., 204
Richmond, R. C., 339, 352
Richter, G., 181
Ricklefs, R. E., 282
Ries, E., 300
Riffenburgh, R. H., 227
Roades, D. F., 380
Robblee, A. R., 54
Roberts, A. M., 176, 177
Roberts, D. W. A., 42
Robertson, A., 341, 352
Robertson, C., 159
Robertson, D. R., 221, 222
Robertson, F. W., 348
Robertson, J. R., 98
Robins, C. R., 231
Robins, E. J., 368
Robinson, I. M., 42, 43
Robinson, J. P., 44
Robinson, M. H., 178
Roccati, A., 253
Rock, J., 178
Rockwood, E., 346
Rodrigues, W. A., 72
Roede, M. J., 213, 221
Roelofs, W. L., 356
Roer, H., 369, 371, 372
Roessler, C., 233
Rognon, P., 271, 273
Root, R. B., 281-83
Rose, S. P. R., 15, 25
Rosen, D. E., 212, 220, 222
Rosenberg, N. J., 173
Rosenblatt, R. H., 222
Rosenzweig, M. L., 7, 73, 99, 295, 300
Ross, C., 73
Ross, G. G., 322

Ross, G. N., 369, 370, 387
Ross, Q. E., 313, 315
Roth, L. M., 225
Rothschild, M., 382, 385-87
Rothstein, S. I., 9
Roughgarden, J., 235, 281, 284, 298
Rufener, W. H., 59
Russell, B. C., 220, 230, 232, 234
Ryder, N. V., 111

S

Sack, R., 317, 321
Sagar, G. R., 295
Saila, S. B., 313, 314, 331
Sale, P. F., 215, 218, 220, 234
Salesky, N., 224
Salisbury, F. B., 73
Salzano, F. M., 117-19, 130
Sammarco, P. W., 224, 235
Sandeman, D. C., 178
Santos, G. A., 224
Sarich, V. M., 345, 347
Sasmal, B., 130
Saucier, J. F., 131
Scali, V., 380, 384
Scapini, F., 181
Scavia, D., 332
Schaal, B. A., 161, 299, 301
Schad, G. A., 286
Schaefer, K. P., 178
Schalke, H. J. W. G., 269, 275
Schaller, G. B., 88
Schapera, I., 121, 122
Scharloo, W., 352
Scheerer, E., 178
Scheifinger, C. C., 46
Scheltema, R. S., 220
Schlatterer, E. F., 295
Schlichter, D., 226
Schlissing, R. A., 154
Schmidly, D. J., 348, 351
Schneider, G., 178
Schnell, R., 67, 76
Schoener, A., 303
Schoener, T. W., 139, 174, 281
Scholes, R. B., 303
Scholze, E., 144
Schöne, H., 177, 178
Schontäg, A., 146
Schoonhoven, L. M., 380
Schott, D., 178
Schroeder, R. E., 226, 229, 234
Schroeter, G. L., 348, 351
Schull, W. J., 118
Schultz, J. C., 380
Schultz, L. H., 43
Schultz, L. P., 213

Schulz, E., 272
Schulz, J. P., 70
Schutz, F., 21, 28, 32
Scott, G. C., 59
Scott, J. A., 366, 367, 369, 372, 374, 376
Scott, P., 211
Scoullar, K. A., 333
Scudder, S. H., 383
Scudo, F. M., 323
Scullen, H. A., 146
Searle, R. B., 223, 224
Segel, L. A., 190
Seghers, B. H., 227
Seibt, U., 181
Seifarth, E., 182
Seiger, M. B., 32
Seitz, A., 386
Sekiguchi, K., 149, 151
Selander, R. K., 32, 339, 345, 346, 348, 351
Sellers, H. E., 70
Seneca, E. D., 290
Servant, M., 271
Servant, S., 271
Servant-Vildary, S., 271
Serventy, D. L., 18
Sestak, Z., 204
Seward, M. J. B., 174
Seybold, A., 157
Shallenberger, R. J., 231
Shankman, P., 89
Shapiro, A. M., 369, 384
Sharitz, R. R., 290
Sharp, M. A., 236, 367, 374, 393
Sharplin, C. D., 145, 146
Shaw, C. R., 348, 351
Shaw, E., 227
Shaw, F. R., 146
Sheldon, J. K., 385
Sheppard, D. H., 295
Sheppe, W., 291
Sherman, P. W., 385
Shewan, J. M., 303
Shields, O., 366, 381, 385
Shkolnik, A., 350, 351
Shontz, J. P., 293
Shorey, H. H., 177
Shorrocks, B., 207
Shorthouse, J. D., 376
Shreve, F., 77, 79
Shuel, R. W., 146
Shugart, H. H., 332, 333
Shull, E. M., 213
Shushkina, E. A., 331
Siccama, T. G., 71
Sieger, M. S. B., 370
Sijpesteijn, A. K., 59
Silberbauer, G. B., 121, 122
Silberglied, R. E., 365
Silver, P. H., 177
Simberloff, D. S., 303
Simeone, E., 81
Simesen, M., 56
Simpson, E. E., 345, 347

Simpson, T., 139
SINGER, M. C., 365-97; 215, 218-20, 234, 236, 295, 366-72, 374, 375, 377, 378, 382, 385
Singh, J. S., 314, 331
Skellam, J. G., 313, 314
Slansky, F. Jr., 369-74, 380, 384
Slatkin, M., 340
Slobodkin, L. B., 108, 139, 233, 317
Sluckin, W., 32
Small, N., 44
Smigel, V. W., 295
Smit, A. F. J., 270
Smith, C. C., 295
Smith, C. L., 217, 218, 222, 233, 234
Smith, D. A. S., 386
Smith, F. E., 322, 324
Smith, F. V., 15
Smith, G. J. C., 33, 348
Smith, J. N. M., 139, 152, 180
Smith, M. H., 340
Smith, N., 303
Smith, N. E., 48, 55
Smith, R. F., 367, 381
Smith, R. H., 42
Smith, S. H., 234
Smith, W., 45
Smith-Vaniz, W. F., 232
Smith-White, S., 356
Smyly, W. J. P., 89, 91
Smythe, N., 72, 73, 80, 82, 159
Snaydon, R. W., 281, 295
Snow, D. W., 159
Snyder, E. E., 45
Snyder, R. L., 347
Soans, J., 33, 348
Sokal, R. R., 339
Solbrig, O. T., 286, 341
Sollins, P., 332
Somchoudhury, A. K., 87, 89
Somero, G., 219
Sommerville, R. I., 286
Sondheimer, J. B., 81
Sotavalta, O., 144
Soulé, M., 219
Southwick, C. H., 88
Southwood, T. R. E., 179, 354, 369, 372
Spencer, K. A., 354
Spencer-Booth, Y., 147, 151, 158
Spillius, J., 123-25
Spires, J. Y., 217
Spites, J. Y., 235
Sprengel, C. K., 140, 148
Springer, S., 227
Springer, V. G., 214, 227, 232
Spurway, H., 357
Stacy, B. D., 48

Stang, G., 177
Starck, W. A., 226
Starck, W. A. II, 229-32, 234
Starr, C. R., 130
Stasko, A. B., 183
Stebbins, G. L., 139, 345, 355
Steele, J. H., 201, 312, 314, 322, 331
Stein, A., 178
Steinitz, H., 233
Stephen, W. P., 154
Stephens, J. S., 217
Stephenson, W., 224, 225
Stern, V. M., 381
Sterner, P. W., 295
Steven, D. M., 197, 200
Stevenson, R. A., 220, 226, 230
Stewart, R. E., 295
Stewart, R. H., 235
Stewart, W. D. P., 300
Stiles, F. G., 139, 141, 144-49, 151, 155, 160
Stimmel, J. F., 356
Stimson, J., 291
Stockley, R. E., 367, 383
Stockton, C. W., 206
Stoffers, P., 264
Stoll, L. M., 213
Stone, W., 346
Straatman, R., 378
Strasburg, D. W., 215, 223, 226, 229, 233
Straw, R. M., 139, 159
Strawinski, K., 88
Stresemann, E., 29
Strobel, M. G., 15, 25
Stuckenrath, R., 268, 269
Sturges, F. W., 81
Sudd, J. H., 174
Sugden, B., 41
Sugiyama, Y., 108
Summers, R., 42, 43, 47, 57
Sundby, R. A., 282
Sunquist, M. E., 81, 82
Sussman, R. W., 116
Sutherland, J. P., 303
Sutherland, T. M., 47
Suthers, R. A., 143
Suzuki, N., 177
Swank, W. T., 73
Swartzman, G. L., 314, 331
Swerdloff, S. N., 226
Swiecimskz, Z., 379
Swynnerton, C. F. M., 370, 386
Syed, S. A., 42
Synge, A. D., 152
Szent-Ivany, J. J. H., 383
Szmant, A. M., 227

T

Taeuber, I., 109, 125-28

Taieb, M., 273
Tait, D. E. N., 333
Takahashi, H., 45, 58
Takata, N., 368
Takeuchi, M., 76
Talbot, F. H., 220, 225, 231, 233-35
Talbot, G., 383, 387
Taljaard, T. L., 47
Talling, I. B., 264
Talling, J. F., 264
Tamasige, M., 177
Tamura, M., 31, 32
Tanner, J. T., 8
Tarakanov, B. V., 60
Tast, J., 29
Tavolga, W. N., 235
Taylor, C. D., 43
Taylor, K. M., 347
Teitelbaum, M. S., 130
Temple, P. H., 260
Terborgh, J. W., 75, 76
Teytaud, A. R., 226
Thaeler, C. S. Jr., 348, 351
Thibault, R. E., 90, 96
Thielcke, G., 27
Thies, S. A., 154
Thoday, J. M., 348, 352
Thomas, J. A., 366-68, 376
Thomas, S. P., 143
Thomason, J. H., 369
Thompson, B. W., 251
Thompson, G. B., 48
Thompson, R., 197
Thomson, J. M., 220
Thorington, R. W. Jr., 81, 82
Thorp, R. W., 157
Thorpe, W. H., 16, 18-20
Thresher, R. E., 214, 233
Tieszen, L., 333
Tietze, C., 120
Tinkle, D. W., 369
Tinnin, R. O., 285
Tisdale, E. W., 295
Toporkova, L. Ya., 295
Tosi, J. A. Jr., 75
Tracey, M. L., 356
Trei, J. E., 59
Trexler, R. C., 109, 130
Tribe, D. E., 57
Trickler, R. A. R., 3
Trites, R. W., 201
Trivers, R. L., 130
Trlica, M. J., 333
Tsang, N., 180
Tsao, W.-W., 180
Tschudin, E., 146
Tucker, A., 227
Tucker, V. A., 143, 144
Tukey, J. W., 191, 195-98
Tuleschkow, K., 374
Turing, A. M., 190
Turner, B. L., 301
Turner, B. V., 43

Turner, C. H., 220
Turner, J. A., 117
Turner, J. R. G., 369, 370, 374
Turpin, R. A., 154
Turvey, N. D., 73
Twine, P. H., 91, 98
Tyler, J. C., 217, 218, 233, 234

U

Uemura, T., 43, 59, 60
Ullyett, G. C., 94, 99, 100
Ulyatt, M. J., 55
Urquhart, F. A., 369, 371
Utter, J. M., 141

V

Vadas, R. L., 224
Vaillant, F., 290
Vallois, H. V., 129
Vance, R. R., 295
Van den Bosch, R., 297
Vandermeer, J. H., 281, 282, 284, 295, 301, 302
van der Pijl, L., 140, 155-57
Van Dyne, G. M., 312, 332, 333
van Golde, L. M. G., 43
van Gylswyk, N. O., 59
Van Hook, R. I. Jr., 81
van Nevel, C. J., 56, 59
Vansell, G. H., 147, 151, 156
van Soest, P. J., 43
Van Someren, R. A. L., 382
Van Someren, V. G. L., 382
van Steenis, C. G. G. J., 76
van Vugt, F., 59
van Zinderen Bakker, E. M., 256, 267, 270
Varley, G. C., 9, 366, 367, 386
Vayda, A. P., 125
Verivey, D. J., 226
Vestergaard, D. A., 387
Vielmetter, W., 144, 178, 376
Viktorov, G. A., 295
Vincent, P. M., 271
Vine, P. J., 216, 225, 234
Vinogradov, M. E., 331
Virostek, J. F., 153, 290
Visser, S. A., 311, 333
Vogel, B., 145
Vogel, S., 141, 143
Vollrath, F. W., 177
Volpe, E. P., 96
von Abdel-Salam, F., 89
Von Bier, K., 355
von Frisch, K., 145, 146, 151
von Herzen, R. P., 266
von Holst, D., 108

von Kirchner, O., 140

W

Wadington, C. H., 341
Wadsworth, F. H., 71
Wagner, H. O., 144
Wagner, R. P., 339
Wagner, W. H. Jr., 382
Wahrman, J., 348, 350, 351
Wainwright, A. A. P., 15
Waldbauer, G. P., 385
Walk, R. D., 179
Walker, D. J., 42, 47, 57
Walker, G. F., 119, 130
Walker, P. M. B., 345
Walker, T. J., 341
Wallace, B., 339, 341, 348
Waller, W. T., 303
Wallraff, H. G., 175
Walls, G. L., 178
Walsh, J. J., 313, 318, 329, 331
Walter, C. P., 179
Walter, H., 67-70, 73, 76, 79
Walters, C. J., 314, 316, 331, 332
Warakomska, Z., 153
Warburton, K., 182
Warnecke, G., 369
Warner, A. C. I., 48
Washbourn, C. K., 265
Washbourn-Kamau, C., 263
Waterman, T. H., 179
Watson, A., 87
Watson, G. S., 356
Watt, K. E. F., 150, 181, 314, 315, 317, 318, 321, 323, 324, 331, 333
Watt, W. B., 145, 146, 149, 152, 376-82, 385
Wavre, M., 303
Way, M. J., 87
Weaver, N., 142, 147, 150-52, 154
Webb, T., 259
Wecker, S. C., 18
Weinstein, E. D., 118, 119
Weir, W. C., 56
Weis-Fogh, T., 141-44
Weiss, P., 178
Weissman, K. G., 384
Weitzman, S. H., 212
Welch, P. D., 195
Werner, E. E., 181
Westoby, M., 81
Weston, R. H., 54
Wetzel, R. L., 328
Weyer, E. M., 108, 129
Weyl, N., 129
Wheeler, A. G. Jr., 356
Wheeler, M., 346
Whitaker, T. W., 152
White, B., 133
White, D. C., 44

White, E. G., 88, 94
White, J., 349
White, M. J. D., 339-41, 345-51, 357
White, R. R., 215, 218-20, 234, 236, 366-68, 370, 371, 375, 377
Whitehead, D. R., 158
Whiteman, A. J., 271
Whiting, J. W., 115, 131
Whitmore, T. C., 68-70, 75-77
Whittaker, R. H., 70, 71, 281-83, 341, 353
Whitten, B. H., 300
Wickler, W., 181, 220, 223, 228, 230, 232, 301
Wickstrom, E. E., 296
WIEGERT, R. G., 311-38; 72, 296, 311, 314, 316-18, 320, 322, 323, 325-28, 330, 332
Wiener, N., 191
Wiens, J. A., 17
Wilbert, H., 385
Wilbur, H. M., 301-3, 369
Wilcockson, W. H., 254
Wilde, W. H. A., 98, 99
Wilkes, J. O., 328
Wilkinson, C. F., 380
Wilkinson, P., 254
Willard, D. E., 130
Williams, C. B., 369, 371, 375
Williams, G. C., 227
Williams, I. H., 158
Williams, J. S., 182
Williams, M. M., 227
Williams, N. H., 139
Williams, R. B., 320, 332
Williams, R. D., 154
Williams, S. L., 348
Williams, V. F., 55
Williams, W. A., 329
Williamson, K.L., 387
Willis, E. O., 82
Willson, H. R., 91, 96, 97
Wilson, A. C., 345, 347
Wilson, D. S., 301
Wilson, E. O., 7, 10, 250, 274, 282, 285, 290, 303, 341, 346, 348
Wilson, F. G., 146, 148
Wiltschko, M., 175
Wiltshire, E. P., 385
Winkel, W., 19
Winn, H. E., 213, 216, 220, 221, 227
Witherell, P. C., 147, 151
Witkamp, M., 311, 333
Wohlfahrt, T. A., 376
Wohlwill, J. F., 108
Wolf, H. G., 178
Wolf, L. L., 139, 141-46, 148, 151, 174
Wolfe, A. B., 114
Wolfe, R. S., 43

Wolin, M. J., 46, 59
Wong, H. K., 266
Wong, Y. K., 70
Wood, G. W., 149
Wood, W. A., 43
Woodhill, A. R., 356
Woodwell, G. M., 72, 321
Woolf, C. M., 117, 119
Wright, D. E., 43
Wright, S., 98, 349
Wuenscher, J. E., 283
Wykes, G. R., 145
Wynne-Edwards, V. C., 9, 108, 111

Y

Yarwood, A., 15
Yasmineh, W. G., 345
Yates, F. E., 189, 208
Yates, H. S., 79, 80
Yeo, P. F., 140, 156
Yoda, K., 71
Yom-Tov, Y., 88, 91, 92, 97
Yonge, C. M., 211
Yoshiba, K., 108
Young, A. M., 93, 369, 387
Youngbluth, M. J., 231-33, 300

Youthed, G. J., 175
Yu, I., 45
Yunis, J. J., 345, 355
Yurkiewicz, W. J., 145

Z

Zelwer, M., 80
Zilch, A., 268
Zimmerman, L., 118
Zlobin, V. S., 200
Zouros, E., 339
Zuckerman, S., 108
Zumpe, D., 228
Zwölfer, H., 352

SUBJECT INDEX

A

Abortion, 109-11, 113, 114, 123
Abudefduf
  saxatilis, 219, 220
  troischelii, 219
  zonatus, 215
Acacia, 261, 299
Acalypha, 256
Acanthochromis, 219
  polyacanthus, 218, 220, 230
Acanthuridae, 215
Acanthurus, 212
  achilles, 216
  dussumieri, 215
  mata, 215
  pyroferus, 229
  sohal, 216
  triostegus, 215, 216
  xanthopterus, 215
Acraea, 384
  encedon, 369
Acrocephalus scirpaceus, 17
Actinote, 369
Aedes
  Australian, 356
  Aeschna cyanea, 178
Africa
  central
    plateaus of, 267
    climatic change in, 249-80
  tropical
    geography of, 250-53
Afrocrania, 258
Agelena, 182
Aglais urticae, 371, 372
Agromyzidae, 354

Agrotis ipsilon, 89, 90
Albinism, 119
Alchemilla, 256
Alchornea, 261
Alfalfa, 152
Algae
  blue-green, 200, 224, 289, 292, 299-301
Aliasing
  definition of, 196, 197
Amphibia
  ceratophyrid, 357
Amphiprion, 215, 226
  clarkii, 220
Anaerovibrio lipolytica, 44
Anagasta kiihniella, 94-95
Analysis
  Blackman-Tukey, 195-96
  ecosystem, 171
  Fourier, 192, 194-96, 199
  pollen, 256-60
  spectral, 189-210
  power, 191-96
Anampses
  meleagrides, 231
  twisti, 231
Anemonefish, 226
Anencephaly, 118
Angelfish, 222, 229, 231
Anguria, 160, 383
Anophthalmos-microphthalmos, 118
Anoptichthys jordani, 174
Anura, 177
Anurans
  cannibalism in, 88
Anthias squamipinnis, 222
Ant-lions, 175

Ants
  wood, 175
Apanteles glomeratus, 385
Aphid, 144
  wingless, 174
Apis, 182
  mellifera, 144, 298
Apogon, 212
  quadrisquamatus, 226
Apogonidae, 222
Apoidea, 152
Aporia crataegi, 385
Archilochus colubris, 150
Argynnis spp., 384
Aristolochia spp., 378
Artemia salina, 177
Artemisia, 258
  afra, 259
Arthropod, 88
Artibeus jamaicensis, 81
Asclepis syriaca, 160
Asimina, 68
Aspidontus, 233
  taeniatus, 232
Asplanchna sieboldi, 96, 97
Aster
  yellow, 293
Atresia ani, 118
Aulostomus, 231
Autocovariance, 192, 193

B

Baboon, 181
  hamadryas, 108
Backswimmer, 91
  freshwater, 90, 95
Bacteria

aerobic, 42
euryoxic, 42
rumen, 42-44
sulfur, 175
Bacteroides
amylophilis, 44
ruminicola, 44, 46
succinogenes, 43, 44, 46, 59
Badis badis, 177
Balistidae, 217, 222
Barnacles, 174
Bass
yellow, 91
Basslet, 222
Batrachoididae, 226
Bats
fruit
Jamaican, 81
group-foraging, 154
insectivorous, 174
pollinators, 143
Bees
Euglossine, 154
honey, 182
pollinators, 144-45
short-tongued, 148
Beetles
coccinellid, 88
Bigeye, 214
Biology
evolutionary, 357
marine, 211-47
population, 211-47
Biome
grassland, 332-33
tundra, 332
Birds
African indigo, 33
cannibalism in, 88
frugivorous, 81, 82
pollinators, 143-44
Blackbird, 27
Blackman-Tukey analysis
see Analysis, Blackman-
Tukey
Blattisocius tarsalis, 94
Blenniidae, 217, 223
Blenny, 217, 230
saber-toothed, 232, 234
Blowfly
sheep, 94
Bodianus, 230
Bolbometopon bicolor, 230
Boloria napaea, 375
Bombus, 149-52
fervidus, 148
sp., 144
spp., 290
terricola, 148
Borer
corn
European, 356
Boxbush, 77
Bug
milkweed, 90-91
Bumblebee, 144, 147, 148,
152-55, 161, 290, 298

queen, 142-43
Bunting
ortolan
European, 31
Bushmen
Kalahari
infanticide among, 121-
22
Butterflies, 155
cabbage, 183
ecology of, 365-97
Heliconius, 160
Butterfly, 144, 178
blue, 378
Monarch, 90, 371
Queen, 90
swallowtail, 379
Butterflyfish, 212, 214, 222,
231
Butyrivibrio, 44
fibrisolvens, 43

C

Cactus
saguaro, 154
senita, 295
Calloplesiops altivelis, 230
Calothrix, 289, 299, 302
Calypte anna, 143
Canis, 341
Cannibalism
demographic consequences
of, 92-96
distribution of, 87-88
factors affecting, 89-91
infant, 107-37
in mammals, 108
in natural populations, 87-
106
Canthigaster, 226
valenti, 230
Carangidae, 222
Cardinalfish, 217, 222, 226
Carnivora, 341
Carpinus, 68
Carpodacus mexicanus, 33
Caterpillar, 179
Caulerpa, 224
spp., 224
Celerio, 385
Cellumonas, 60
Celtis, 256, 261
Centipedes
cannibalism in, 87
Centriscidae, 226
Centropyge
flavissimus, 229
vrolikii, 229
Cephalopods
orientation in, 178
Chaetodon, 212
auriga, 228
ephippum, 230
vagibundus, 228
Chaetodontidae, 222, 228
Chaffinch, 27

Chalcidoidea, 355
Chamaedaphne calyculata,
160
Characters
learned
biological significance of,
26-27
Charaxinae, 382
Cheilinus, 226
Chelone glabra, 148
Chelydra serpentina, 16
Chen caerulescens, 32
Chiliothrichium diffusum,
302
Chimpanzees
cannibalism in, 108
Chironomus anthracinus, 94
Chlosyne lacinia, 379, 380
Chromis
margaritifer, 230
multilineata, 220
Chrysococcyx, 20
Chrysomyia albiceps, 94
Chrysopids, 88
Chubs, 222, 223
Ciguatera, 224
Circumcision, 116
Clamator jacobinus, 20
Clarkia, 157, 349
biloba, 159
Clingfish, 226
Clostridium
lochheadii, 43, 59
longisporum, 60
Clover
red, 152
Cockroach, 175
Coenonympha tullia, 384
Coevolution
plankton-planktivore, 225
plant-herbivore, 223-24
predator-prey, 225-31
Colias, 147, 365, 381, 385
Columbarium
definition of, 283
Community
algal-bacterial, 322
diversity, 6
multispecies
evolution of, 376
Competition, 8
Compositae, 148, 256, 267,
268
Conyza canadensis, 293
Copaifera mildbraedii, 270
Copepoda, 177
Copepods
cannibalism in, 87
Coral reef
fishes of, 211-47
Coris aygula, 230
Corvus corone, 91
Crab
fiddler, 181
Crataegus, 68
Cricket, 175, 353
field, 355

mole, 351
Cross-covariance function, 198
Cross-pollination, 140, 158
Crow, 91
Crustacea, 177
    orientation in, 178
Ctenochaetus strigosus, 218
Cuckoo
    European, 19
    glossy, 20, 33
    jacobine, 20
Cuculus canorus, 19
Culex pipiens, 356
Cupienus salei, 182
Cutworm, 90
    black, 89
Cyclops, 177
Cyclotella stelligera, 204
Cymodocea manatorum, 223
Cymopogon, 268
Cyperaceae, 256

D

Damselfish
    bicolor, 214, 230
    yellowtail, 230
Damselfly, 89
Danainae, 382
Danus
    gilippus, 379
        berenice, 90
    plexippus, 90, 371, 372, 379
Dascyllus
    aruanus, 215, 218, 219, 229
    melanurus, 218, 219
    reticularus, 220
Dasytricha, 42
Demography, 107-37
Desulfovibrio, 44
Diadema, 223
    antillarum, 224
Diapause
    as alternative to migration, 174
Diatom, 265
    nanno-planktonic, 204
Dictyota spp., 224
Dineutes
    assimilis, 99
    horni, 99
    nigrior, 99, 101
Diodontidae, 217, 222
Dipsacaceae, 368
Diptera, 88, 141, 145
    orientation in, 178
Divergence
    allochronic, 355
    parapatric, 348
Diversity
    community, 6
    species, 8, 9
DNA, 345
Dove

collared, 29
Dragonfly, 91, 177, 178
Drepanididae, 143, 159
Drill
    oyster, 97
Drosophila, 207, 341, 346, 353
    arizonensis, 294
    equinoxialis, 356
    funebris, 349
    melanogaster, 349, 352
    mojavensis, 294
    nigrospiraculata, 294
    pachea, 295
    paulistorum, 356

E

Ear-worm
    corn, 90, 91
Ecology, 1-13
    butterfly, 365-97
    community, 233-35, 376-87
    dogma of, 6-8
    evolutionary, 6
    experimental, 281-310
    frontiers of, 8-10
    historical, 206
    impact of Robert MacArthur on, 1-13
    orientation, 171-88
    population, 214-23
        insect, 365-76
    predictive, 10
    spectral analysis in, 189-210
    stochastic models in, 206-7
Ecosystem
    analysis, 171
    microbial, 39-66
Ecosystems
    algal-fly, 316
    aquatic, 332
    energy flow and nutrient cycling in, 6
    ontogeny of, 6
    perturbed, 204-6
    simulation models of, 311-38
    structure of, 8
    theory, 7
Ectocarpus indictus, 223
Eel
    moray, 217, 230
Elacatinus, 231
Emberiza hortulana, 31
Energetics
    pollinator, 139-70
Entodinium, 41
Environment
    definition of, 281
    manipulation of, 283-84
Ephestia kühniella, 20
Epidinium, 42
Epilobium, 154
    angustifolium, 160

Epiphytes, 75
Equation
    Lotka-Volterra, 323-24
    Michaelis-Menten, 324-25
Erebia
    cassioides, 384
    epipsodea, 371
    nivalis, 384
    tyndarus, 384
Ericaceae, 100, 207
Eriogonum, 378
Erpodbella octoculata, 91, 93
Escherichia coli, 172
Esox lucius, 89
Estrildidae, 20
Eubacterium cellulosolvens, 43, 59
Eucalyptus, 70
    mellidora, 294
    rossii, 294
Eucaryotes, 176
Eueides, 380
    aliphera, 369
    spp., 383
Euphydryas, 379
    aurinia, 368
    chalcedona, 383
    editha, 366-71, 374, 375, 377, 378, 380, 383-85
    bayensis, 375
Eupleura caudata, 97
Eupomacentrus, 212, 228
    leucostictus, 215
    partitus, 214, 230
    planifrons, 214
    spp., 214
Evolution
    chromosome, 341
    life history, 10

F

Fecundity
    restriction of, 111
Felis, 341
Fermentation
    continuous, 47-49
    rumen, 44-47
Fertility
    effective, 109
    extramarital, 119
    human, 115
    potential, 111
Fertilization
    cross, 357
Ficedula
    albicollis, 17
    hypoleuca, 19
Filefish, 217
    saddled, 230
Finch
    Bengalese, 25
    house, 33
    society, 25
    zebra, 17, 22, 25, 32
Finches

Darwin's, 345
viduine, 20
Fir
Douglas, 70
Fish
cannibalism in, 88
cave, 174
cichlid, 351
coral reef, 211-47
teleost, 357
teleostean, 212
Fishery
haddock, 331
Fixation
host
see Imprinting, host
Flounder, 228
Fly
fruit, 355
hawthorn, 353
Flycatcher
collared, 17
pied, 19
Forest
cloud, 77, 79
dipterocarp, 78, 80
Liriodendron, 81
midmontane, 80
montane
dwarfed, 77
moss, 79
Mucambo, 68
podocarp, 80
rain
Ipassa, 72
lowland, 67-74, 270-71
montane, 74-80
tropical, 67-86
Zairean, 253, 270-71
temperate, 67, 70
Formica, 175
Fourier analysis
see Analysis, Fourier
Frincrilla coelebs, 27
Frogs
hylid, 357
Fruitfly, 353
Fuchsia, 143

G

Gammarus lacustris, 297
Gasterosteus aculeatus, 26
Genetics
population, 6, 7
Geology
glacial
highland, 253-56
Geomorphology, 270
Gerridae, 94
Gerris najas, 94
Glaciation
in Africa, 253-56
Glaucopsyche lygdamus, 378
Goatfish, 220, 222
spotted, 216
Gobies, 217, 226

Gobiidae, 217
Gobiosoma, 231, 232
Goose
snow, 32
Gopher
pocket, 351
Gorilla
mountain, 181
Grammistidae, 226
Grasshopper
morabine, 350
Grouper, 222, 228, 231
Growth
population, 323
Grunt, 212, 216, 222, 225
Gulls
herring, 17, 18
ring-billed, 17
Gymnothorax meleagris, 230

H

Haemulon, 212
album, 217
flavolineatum, 216
plumieri, 216
Hagenia, 256, 258
Hairstreak
woodland, 376
Halichoeres, 212, 213
centriquadratus, 230
sp., 220
Hamlet, 222, 228, 230
barred, 216
yellowtail, 230
Haplochromis, 260, 351
Haplopappus divaricatus, 293
Harelip-cleft palate, 118
Hawthorn, 68
Heliconius, 149, 160, 371,
379-83, 387
charatonius, 178
erato, 370
ethilla, 367-70
Heliothis
armigera, 90, 91
punctigera, 90
Helleborus niger, 157
Hemlock
eastern, 70
Heniochus, sp., 229
Herring
common, 177
Heterogeneity
temporal, 323
Heterosis, 6
negative, 350
Hibernation
as alternative to migration,
174
Hieracium sp., 148, 155
Holly, 70
Holoptelea, 261
Homo sapiens, 108
Homoptera, 381
Honey
white clover, 147
Honeybees, 144, 147, 152-54,

298, 302
Honeycreeper, 143, 159
Honeyeaters, 144
Hopper
beach, 175
Horseweed, 293
Hospitalitermes sharpi, 174
Hummingbird, 144, 148, 149,
155
Anna, 143
ruby-throated, 150
Hybridization
interspecific, 356
Hymenolepsis dimunuta, 286,
287
Hymenoptera, 18, 141, 144,
154, 182
Hypochera, 33
Hypoplectrus, 213, 228
chlorurus, 230
puella, 216
spp., 230

I

Ilex, 70, 256
Impatiens biflora, 150, 160
Imprintability
individual differences in,
32
Imprinting
biological significance of,
26-33
characteristics of, 16, 21-
26
ecological, 16-20, 28-30
ecological consequences of,
28-33
ecological significance of,
1-37
food, 22, 25
habitat, 19
host, 19-20
locality, 19
motoric, 16
nonecological
ecological significance of,
30-32
sexual, 32
Infanticide
definition of, 109-10
in man, 107-37
twin, 116
Information
deterministic, 173
Insects
cannibalism in, 88
stick, 351
Ironwood, 68
Isloation
geographic, 357
Ithomiinae, 382

J

Jacks, 217, 222
Jellyfish, 175

Junco, 291
Juniperus, 256

K

Kalmia angustifolia, 160
Kyphosidae, 222, 223

L

Labridae, 220
Labroides
  bicolor, 232
  dimidiatus, 221, 222, 232, 233
  phthirophagus, 214, 231
Lachnospira multiparus, 44
Lagoda camilla, 375, 376
Lake
  Kivu, 266
  Mobutu Sese Seko, 262-63
  Rudolf, 263-65
  Tanganyika, 266
  Victoria, 260-62
Language
  simulation, 328-29
Langurs
  cannibalism in, 108
Larus argentatus, 18
Laspeyresia pomonella, 353
Lathyrus odorata, 298
Laurencia obtusa, 224
Leaf-miners
  coffee, 200
Leafworm
  cotton, 89
Leech, 91
  freshwater, 93
Lepidoptera, 20, 141, 152, 177, 385
Lepomis macrochirus, 181
Leptidea sinapis, 375, 376
Lestes nympha, 89
Lethe
  appalachia, 384
  eurydice, 384
Leucoptera
  caffeina, 200
  meyricki, 200
Lindera, 68
Lionfish, 226, 231
Liphyra brassolis, 381, 382
Liquidambar, 69
Liriodendron, 81
Lizardfish, 217
Lizards, 351
Locust, 177, 178
Locusta pardalina, 179
Lonchura striata, 25
Lophocereus schottii, 295
Loranthaceae, 156
Lotka-Volterra equations, 323-24
Lutjanidae, 222
Lutjanus, 216
Lycaena dispar, 367
Lycaenidae, 381, 382, 387

Lygaeus sp., 91
Lysandra
  bellargus, 375
  coridon, 375

M

Macaranga, 258
Mackerel
  Jack, 227
Maculinea arior, 381
Mahoma Lake
  glaciation, 254-55, 258, 259
Malaclemys centrata, 17
Mallard, 19, 32
Mammals
  cannibalism in, 88
  homeothermic
    metabolic rates of, 142
Margate, 217
Marpesia berania, 369
Medicago sativa, 152
Megarhinus, 95, 97
Megasphaera elsdenii, 43
Melitaea, 385
  harristii, 366
Melliphagidae, 144
Methanobacterium
  mobilis, 43
  ruminantium, 43, 46
Methanogenesis
  rumen, 59
Metrosideros collina, 146, 159
Michaelis-Menten equations, 324-25
Micromonospora, 60
Microspathodon chrysurus, 230
Microtus, 291
Midge
  gall-forming, 355
Mimicry, 228, 230, 386
Mimics
  Batesian, 301
Mite, 294
  cannibalism in, 87
Modeling
  stochastic, 317-18
Models
  deterministic, 317-18
  population, 207-8
  simulation, 311-38
  stochastic, 206-7, 317-18
Moniliformis dubius, 286, 287
Monkey, 81
  howling, 82
Montane, 74-80
Moraines
  Katabarua, 225
Moray, 222
Mosquito, 95, 97
Moth
  codling, 353
  sphinx, 144, 154, 155

Mouse
  deer, 18
  house, 89
  cannibalism in, 108
Mullidae, 222
Mulloidichthys martinicus, 216
Muraenidae, 217, 222
Mus musculus, 89
Muskrat
  cannibalism in, 108
Mussel, 292, 293
Mutualism, 387
  definition of, 298
Myrica, 256, 258, 259
Myrmeleon, 175
Myrsinaceae, 256
Myrtaceae, 267
Mytilis
  californianus, 292
  edulis, 292

N

Nandidae, 177
Navanax inermis, 95
Nectarinidae, 143, 156
Nemeritis canescens, 20
Neutroptera, 88
Niche
  coextensive, 293-95
  definition of, 282
  ecological, 281
  studies of
    experimental, 281-310
Nile River, 260-65
Nitzschia fonticola, 265
Nothofagus, 77
Notonecta hoffmanni, 90, 91, 95
Nymphalidae, 382
Nymphalinae, 382
Nyssa, 69

O

Oak, 69
Ochroma lagapus, 146
Oeneis chryxus, 375
Oenothera, 157
  organensis, 154
Ohia trus, 146
Olea, 256, 258, 267
Oncorhynchus spp., 17
Ontogeny, 27
  ecosystem, 6
  see also Succession
Orientation
  cropping, 181
  ecology, 171-88
  endokinetic, 173
  exokinetic, 173
  extrinsic, 173
  fitness, 173
  geographic, 175, 182-83
  habitat, 176
  home range, 176

intrinsic, 173
invertebrate, 172-73
object, 176, 179-82
positional, 176-79
search, 180
spatial, 171-88
strato, 176, 181
topographic, 176, 182
transverse gravity, 178
transverse light, 178
vector, 182
zonal, 176, 181-82
Ornithoptera
urvilleanus, 378, 379
victoriae, 378, 379
Orthoptera, 141
Oscillatoria, 293, 299, 301
terebriformis, 289, 296
Ostracion, 226
Ostrinia nubilalis, 356

P

Paleoecology, 206
Panthera, 341
Papilio, 213
indra, 379
machaon, 374, 375
rudkini, 379
Papilionidae, 144, 374, 382
Paraluteres prionurus, 230
Paramecium, 176, 299
Pararge aegeria, 376
Parasites
phytophagous, 352
zoophagous, 352
Parasitism, 295-97
Parasitoids, 352, 355
Pardosa lugubris, 91, 92
Parides, 369
Parrotfish, 212
princess, 221
redband, 221
striped, 221
yellowtail, 221
Parrots, 143
Partula, 349, 351
Parus
ater, 19
caeruleus, 19
Passerines
metabolic rates of, 142
Passiflora, 379
Patrilocality, 116
Pedicularis, 378
Penguins
Magellanic, 302
Perca
flavescens, 95
fluviatilis, 89, 91, 93
Perch, 89, 91
European, 93, 94
yellow, 95
Peromyscus, 291, 351
maniculatus bairdi, 18
Phaulacridium vittatum, 90
Philosophy

of science
definition of, 1
Philotes sonorensis, 370
Phlox glabberima, 147
Physiology
ecosystem, 6
Phytoplankton, 200-4, 325
Pieridae, 382
Pieris, 365
brassicae, 183, 371, 385
napi, 379, 383, 384
occidentalis, 379
protodice, 374
rapae, 183, 366-68, 372-
74, 381, 383, 384
spp., 371
Pigeons, 144, 177
Pike, 89, 94
Pine
longleaf, 315
Pipefish, 231
Plagiotremus, 230, 232, 234
rhynorhynchos, 234
tapeinosoma, 234
Planaria
cannibalism in, 87
Plantago erecta, 377
Plasmon
definition of, 356
Plassiflora, 380
Platanus, 69
Plebejus, 379
acmon, 378
icarioides, 367, 378
Plectognathi, 217
Plethodon
cinereus, 288
richmondi shenandoah, 288
Pleurocapsa, 289, 299, 301,
302
Ploceidae, 20
Podocarpus, 77, 256, 258,
259, 265
Poeciliopsis
lucida, 89, 96
monacha, 89, 96
Polecat
European, 17
Pollination
energetics of, 139-70
Pollinators
energetics of, 140-42
energy expenditure by, 142-
45
Polmadasyidae, 225
Polydactyly, 118
Polyethism, 30
see also Polymorphism,
behavioral
Polygonia c-album, 375
Polygoniaceae, 150
Polygonum cascadense, 150
Polygyny, 116
Polymorphism
behavioral, 30
Polymorphus paradoxus, 297
Polyploidy, 356, 357

Pomacanopercle, 229
Pomacanthidae, 223
Pomacanthus
arcuatus, 229
paru, 229
Pomacentridae, 215, 223
Pomacentrus, 212, 214, 228
flavicauda, 214, 218
jenkinsi, 214, 215
lividus, 216
wardi, 218, 219
Pomadasyidae, 212, 222
Pomo
infanticide among, 122-23
Population
dynamics, 8
natural, 200-1
regulation of, 6-8
Populations
biology of, 211-47
natural
cannibalism in, 87-106
panmictic, 218
species
dynamics of, 316
Porcupinefish, 217, 222
Porgies, 222
Potamocypris, 289, 296, 299,
301
Precis, 384
Predation, 295-97
Priacanthus, 231
phthirophagus, 214
Primates
nonhuman
cannibalism in, 108
infanticide in, 108
Procecidochares, 353
Promicrops lanceolatus, 231
Propionibacterium, 47
Protozoa
cannibalism in, 87
rumen, 41-42
Prunella vulgaris, 155, 160
Pseudomyrmex spp., 299
Pseudupeneus maculatus,
216, 220
Psithyrus, 149, 150
Pterois, 231
Puffer, 217, 222, 226
saddled, 230
Puffinus teniurostris, 18
Putorius putorius, 17
Pygeum, 258
Pyrrhosoma nymphula, 91

Q

Quercus, 69
petraea, 354
robae, 354

R

Races
ecogeographic, 340
Rana cascadae, 17

Rat
  mole, 350, 351
  Norway, 90
Rattus norvegicus, 90
Redwood, 71
Reef
  coral, 211-47
Reticulum
  rumen, 39-41
Rhagoletis, 233
  pomonella, 353
Rhinoseius, 294
Rhododendron angustifolia,
  160
Rhodospirillum rubrum, 175
Rhopalomyia californica, 175
Rhus toricodendron, 70
Roccus mississippiensis, 91
Rockfish, 227, 228
Rosa nitida, 157
Rotifer
  cannibalism in, 87
Rubus hispidus, 160
Rumen
  ecosystem, 39-66
Ruminococcus
  albus, 43, 46, 59
  flavefaciens, 46
Runula, 234

          S

Sahara
  climatic change in, 271-73
Salamander, 288, 297
Salmon
  North Pacific, 17
Salmonella typhimurium, 180
Sawfly, 355
Scaphiopus, 96
  holbrooki, 96
Scaridae, 230
Scarus, 212
  croicensis, 216, 217, 221
  taeniopterus, 221
  vetula, 234
Sceloporus, 351
Schistocerca americana, 178
Scorpaenidae, 226
Scorpionfish, 226, 227
Scyphozoa, 175
Seahorse, 226
Selasphorus flammula, 143
Selection
  archetypic, 8, 9
  ecotypic, 8, 9
  frequency-dependent, 6
Selenomonas, 46, 47
  ruminantium, 44
Serranidae, 213, 222
Shearwater
  short-tailed, 18
Shrews, 351
Shrimpfish, 226
Siganidae, 223
Simulation
  analog, 327-28

digital, 327-28
  models, 311-38
Skua
  arctic, 32
Sloth, 82
Snails
  cannibalism in, 87
  littoral, 181-82
Snapper, 216, 217, 222
Soapfish, 226
Sourgum, 69
Spalex, 350, 351
Sparidae, 222
Sparisoma, 212, 216
  rubripinne, 221
  viride, 221
Sparrow
  chingolo, 31
  chipping, 18
  golden-crowned, 291
  rufous-collared, 31
  white-crowned, 31, 32
Spartina, 332
Speciation
  allopatric, 340-48
  animal
    modes of, 339-64
  geographic, 340
  parapatric, 348-51
  stasipatric, 348
  subdivision, 341-45
  sympatric, 351-57
Species
  cultural, 34
  ecology of, 16
Sphecidae, 182
Spheniscus magellanicus,
  302
Sphingidae, 144
Spicebush, 68
Spider
  wandering, 91
  web, 182
Spina bifida manifesta, 118,
  119
Spizella passerina, 18
Spodoptera littoralis, 89
Squirrelfish, 217
Starfish
  crown-of-thorns, 226
Stephanomeria, 349
Stercorarius parasiticus, 32
Stickleback
  three-spined, 26-27
Stizostedion vitreum, 91,
  95
Stoebe, 268
Stonefish, 226
Stratification, 176
Streptococcus bovis, 44, 58
Stretopelia decaocto, 29
Strider
  water, 94
Strymonidia prumi, 376
Succession, 6
  community, 7
Succinimonas amylolytica, 44

Succissa pratensis, 368
Sunbird, 143
Sunfish, 181
Surgeonfish, 215
Sweetgum, 69
Sycamore, 69
Sylvia borin, 175
Syndactyly, 119
Synechoccus, 293, 299
  lividus, 209, 292
Syngnathidae, 231
Syrphids, 88

          T

Tadpole
  spadefoot, 96
Talitrus saltator, 175
Tapeworm, 286-88, 302
Tephritidae, 352
Termite
  nasute, 174
Terrapin
  diamondback, 17
Tetraodontidae, 217, 222
Tetraodontiformes, 217
Thalassia testudinum, 223
Thalassoma
  amblycephalus, 226
  bifasciatum, 213, 231
  lunare, 222
Thaumetopoea, 179
Themeda, 268
Thrush
  European mistle, 19
Tikopia
  infanticide among, 123-25
Tits
  blue, 19
  coal, 19
Toadfish, 226
Toxorhynchites, 95-96
Trachurus symmetricus,
  227
Trees
  wind-pollinated, 158
Tribolium, 96, 98
  castaneum, 93, 97
  confusum, 93, 97
Trichodesmium thiebaudii,
  200
Trifolium
  incarnatum, 157
  pratense, 147-49, 152,
    155
Triggerfish, 217, 222
Triturus viridescens, 297
Trochilidae, 144
Truckfish, 226
Trumpetfish, 217, 231
Trunkfish, 217
Trypanosoma diemyctyli,
  297
Tsuga, 70
Turdus
  merula, 27
  viscivorus, 19

Turtle
  snapping, 16
Typha, 268
Typhlodromus
  caudiglans, 95
  occidentalis, 97

U

Uca maracoani, 181
Uraeginthus cyanocephalus, 32
Urchin
  sea, 223, 224, 226
Urosalpinx cinerea, 97
Urticaceae, 256
Uvularia sessifolia, 149

V

Vanessa
  atalanta, 371

cardui, 371
Variability
  genetic, 8
  morphological, 8
Variation
  allozyme, 346
  morphological
    intraspecific, 6
Veillonella alkalescens, 43,
    44
Vespidae, 182
Vetch, 147
Vibrio succinogenes, 46
Viccia, sp., 147
Viellonella, 47
Virachola livia, 385

W

Walleye, 91, 95

Warbler
  garden, 175
  reed, 17
Waxbill, 20
  blueheaded, 32
Worm
  spiny-headed, 286
Wrasse, 214, 225, 230-
    32
  blueheaded, 213
  cleaner, 221

Z

Zanclus, 228
Zonation, 176
Zonotrichia
  capensis, 31
  leucophrys, 31

CUMULATIVE INDEXES

CONTRIBUTING AUTHORS VOLUMES 2-6

A

Alexander R. D., 5:325-83
Allen, M. B., 2:261-76
Ashlock, P. D., 5:81-99
Ayala, F. J., 5:115-38
Ayres, R. U., 2:1-22

B

Baker, H. G., 5:1-24
Bayly, I. A. E., 3:233-68
Berlin, B., 4:259-71
Bliss, L. C., 2:405-38;
4:359-99
Bush, G. L., 6:339-64

C

Campbell, C. A., 5:115-38
Carroll, C. R., 4:231-57
Chapman, A. R. O., 5:65-80
Cody, M. L., 4:189-211
Colwell, R. K., 6:281-310
Colwell, R. R., 4:273-300
Connell, J. H., 3:169-92
Courtin, G. M., 4:359-99
Cracraft, J., 5:215-61

D

Davis, G. E., 2:111-44
Denman, K. L., 6:189-210
Dickeman, M., 6:107-37

E

Ehrlich, P. R., 6:211-47
Eisenberg, J. F., 3:1-32
Eriksson, E., 2:67-84
Estabrook, G. F., 3:427-56

F

Faller, A. J., 2:201-36
Farris, J. S., 2:277-302
Flannery, K. V., 3:399-426
Fox, L. R., 6:87-106
Fretwell, S. D., 6:1-13
Fuentes, E. R., 6:281-310

G

Gallucci, V. F., 4:329-57

Georghiou, G. P., 3:133-68
Gilbert, L. E., 6:365-97
Gould, S. J., 3:457-98
Grant, P. R., 3:79-106

H

Hamilton, W. D., 3:193-232
Harper, J. L., 5:419-63
Heed, W. B., 3:269-88
Heinrich, B., 6:139-70
Hespenheide, H. A., 4:213-29
Holling, C. S., 4:1-23
Horn, H. S., 5:25-37
Hungate, R. E., 6:39-66

I

Immelmann, K., 6:15-37

J

Jackson, R. C., 2:327-68
Jander, R., 6:171-88
Janzen, D. H., 2:465-92;
4:231-57
Johnson, G. B., 4:93-116
Johnson, W. E., 4:75-91
Johnston, R. F., 3:457-98
Jordań, C. F., 3:33-50

K

King, J. A., 4:117-38
Kleiman, D. G., 3:1-32
Kline, J. R., 3:33-50
Kneese, A. V., 2:1-22

L

Leigh, E. G. Jr., 6:67-86
Lillegraven, J. A., 5:263-83
Livingstone, D. A., 6:249-80
Lugo, A. E., 5:39-64

M

Maguire, B. Jr., 2:439-64
McBee, R. H., 2:165-76
Meier, R. L., 3:289-314

Mertz, D. B., 3:51-78
Monsi, M., 4:301-27
Mooney, H. A., 3:315-46
Morse, D. H., 2:177-200
Müller, K., 5:309-23

N

Noy-Meir, I., 4:25-57; 5:195-214

O

Oikawa, T., 4:301-27
Otte, D., 5:385-417

P

Paine, R. T., 2:145-64
Pattie, D. L., 4:359-99
Peet, R. K., 5:285-307
Pianka, E. R., 4:53-74
Platt, T., 6:189-210

R

Rappaport, R. A., 2:23-44
Riewe, R. R., 4:359-99
Rohlf, F. J., 5:101-13
Rowlands, I. W., 4:139-63

S

Schoener, T. W., 2:369-404
Selander, R. K., 4:75-91
Simberloff, D. S., 5:161-82
Singer, M. C., 6:365-97
Snedaker, S. C., 5:39-64
Soulé, M., 4:165-87
Spieth, H. T., 3:269-88
Staley, T. E., 4:273-300
Stebbins, G. L., 2:237-60
Swartzman, G. L., 3:347-98

T

Taub, F. B., 5:139-60

U

Uchijima, Z., 4:301-27

419

V

van den Bosch, R., 2:45-66
Vandermeer, J. H., 3:107-
  32
Van Dyne, G. M., 3:347-98

Vayda, A. P., 5:183-93

W

Warren, C. E., 2:111-44
Weir, B. J., 4:139-63

White, J., 5:419-63
Whitfield, D. W. A., 4:359-99
Widden, P., 4:359-99
Wiegert, R. G., 6:311-38
Williams, W. T., 2:303-26
Witkamp, M., 2:85-110

CHAPTER TITLES VOLUMES 2-6

VOLUME 2 (1971)
  Economic and Ecological Effects of a Stationary
    Economy   R. U. Ayres, A. V. Kneese   1-22
  The Sacred in Human Evolution   R. A. Rappaport   23-44
  Biological Control of Insects   R. van den Bosch   45-66
  Compartment Models and Reservoir Theory   E. Eriksson   67-84
  Soils as Components of Ecosystems   M. Witkamp   85-110
  Laboratory Stream Research: Objectives,
    Possibilities, and Constraints   C. E. Warren, G. E. Davis   111-44
  The Measurement and Application of the Calorie
    to Ecological Problems   R. T. Paine   145-64
  Significance of Intestinal Microflora in Herbivory   R. H. McBee   165-76
  The Insectivorous Bird as an Adaptive Strategy   D. H. Morse   177-200
  Oceanic Turbulence and the Langmuir Circula-
    tions   A. J. Faller   201-36
  Adaptive Radiation of Reproductive Characteris-
    tics in Angiosperms, II: Seeds and Seedlings   G. L. Stebbins   237-60
  High-Latitude Phytoplankton   M. B. Allen   261-76
  The Hypothesis of Nonspecificity and Taxonomic
    Congruence   J. S. Farris   277-302
  Principles of Clustering   W. T. Williams   303-26
  The Karyotype in Systematics   R. C. Jackson   327-68
  Theory of Feeding Strategies   T. W. Schoener   369-404
  Arctic and Alpine Plant Life Cycles   L. C. Bliss   405-38
  Phytotelmata: Biota and Community Structure
    Determination in Plant-Held Waters   B. Maguire Jr.   439-64
  Seed Predation by Animals   D. H. Janzen   465-92

VOLUME 3 (1972)
  Olfactory Communication in Mammals   J. F. Eisenberg, D. G. Kleiman   1-32
  Mineral Cycling: Some Basic Concepts and
    Their Application in a Tropical Rain Forest   C. F. Jordan, J. R. Kline   33-50
  The Tribolium Model and the Mathematics of
    Population Growth   D. B. Mertz   51-78
  Interspecific Competition Among Rodents   P. R. Grant   79-106
  Niche Theory   J. H. Vandermeer   107-32
  The Evolution of Resistance to Pesticides   G. P. Georghiou   133-68
  Community Interactions on Marine Rocky
    Intertidal Shores   J. H. Connell   169-92
  Altruism and Related Phenomena, Mainly in
    Social Insects   W. D. Hamilton   193-232
  Salinity Tolerance and Osmotic Behavior of

Animals in Athalassic Saline and Marine
Hypersaline Waters   I. A. E. Bayly   233-68
Experimental Systematics and Ecology of
Drosophila   H. T. Spieth, W. B. Heed   269-88
Communications Stress   R. L. Meier   289-314
The Carbon Balance of Plants   H. A. Mooney   315-46
An Ecologically Based Simulation-Optimization
Approach to Natural Resource Planning   G. L. Swartzman, G. M. Van Dyne 347-98
The Cultural Evolution of Civilizations   K. V. Flannery   399-426
Cladistic Methodology: A Discussion of the
Theoretical Basis for the Induction of
Evolutionary History   G. F. Estabrook   427-56
Geographic Variation   S. J. Gould, R. F. Johnston   457-98

VOLUME 4 (1973)
Resilience and Stability of Ecological Systems   C. S. Holling   1-23
Desert Ecosystems: Environment and Producers I. Noy-Meir   25-51
The Structure of Lizard Communities   E. R. Pianka   53-74
Genetic Variation Among Vertebrate Species   R. K. Selander, W. E. Johnson   75-91
Enzyme Polymorphism and Biosystematics:
The Hypothesis of Selective Neutrality   G. B. Johnson   93-116
The Ecology of Aggressive Behavior   J. A. King   117-38
Reproductive Strategies of Mammals   B. J. Weir, I. W. Rowlands   139-63
The Epistasis Cycle: A Theory of Marginal
Populations   M. Soulé   165-87
Character Convergence   M. L. Cody   189-211
Ecological Inferences from Morphological Data   H. A. Hespenheide   213-29
Ecology of Foraging by Ants   C. R. Carroll, D. H. Janzen   231-57
Folk Systematics in Relation to Biological
Classification and Nomenclature   B. Berlin   259-71
Application of Molecular Genetics and Numerical
Taxonomy to the Classification of Bacteria   T. E. Staley, R. R. Colwell   273-300
Structure of Foliage Canopies and Photosynthesis M. Monsi, Z. Uchijima,
                                                      T. Oikawa   301-27
On the Principles of Thermodynamics in Ecology V. F. Gallucci   329-57
Arctic Tundra Ecosystems   L. C. Bliss, G. M. Courtin,
                                                      D. L. Pattie, R. R. Riewe,
                                                      D. W. A. Whitfield, P. Widden 359-99

VOLUME 5 (1974)
The Evolution of Weeds   H. G. Baker   1-24
The Ecology of Secondary Succession   H. S. Horn   25-37
The Ecology of Mangroves   A. E. Lugo, S. C. Snedaker   39-64
The Ecology of Macroscopic Marine Algae   A. R. O. Chapman   65-80
The Uses of Cladistics   P. D. Ashlock   81-99
Methods of Comparing Classifications   F. J. Rohlf   101-13
Frequency-Dependent Selection   F. J. Ayala, C. A. Campbell   115-38
Closed Ecological Systems   F. B. Taub   139-60
Equilibrium Theory of Island Biogeography and
Ecology   D. S. Simberloff   161-82
Warfare in Ecological Perspective   A. P. Vayda   183-93
Desert Ecosystems: Higher Trophic Levels   I. Noy-Meir   195-214
Continental Drift and Vertebrate Distribution   J. Cracraft   215-61
Biogeographical Considerations of the Marsupial-
Placental Dichotomy   J. A. Lillegraven   263-83
The Measurement of Species Diversity   R. K. Peet   285-307
Stream Drift as a Chronobiological Phenomenon
in Running Water Ecosystems   K. Muller   309-23
The Evolution of Social Behavior   R. D. Alexander   325-83
Effects and Functions in the Evolution of Signaling
Systems   D. Otte   385-417
The Demography of Plants   J. L. Harper, J. White   419-63

VOLUME 6 (1975)
The Impact of Robert MacArthur on Ecology   S. D. Fretwell   1-13
Ecological Significance of Imprinting and Early
Learning   K. Immelmann   15-37

The Rumen Microbial Ecosystem   R. E. Hungate   39-66
Structure and Climate in Tropical Rain Forest   E. G. Leigh Jr.   67-86
Cannibalism in Natural Populations   L. R. Fox   87-106
Demographic Consequences of Infanticide in Man M. Dickeman   107-37
Energetics of Pollination   B. Heinrich   139-70
Ecological Aspects of Spatial Orientation   R. Jander   171-88
Spectral Analysis in Ecology   T. Platt, K. L. Denman   189-210
The Population Biology of Coral Reef Fishes   P. R. Ehrlich   211-47
Late Quaternary Climatic Change in Africa   D. A. Livingstone   249-80
Experimental Studies of the Niche   R. K. Colwell, E. R. Fuentes   281-310
Simulation Models of Ecosystems   R. G. Wiegert   311-38
Modes of Animal Speciation   G. L. Bush   339-64
Butterfly Ecology   L. E. Gilbert, M. C. Singer   365-97